The Neuroendocrine Immune Network in Ageing

Neuroimmune Biology, Volume 4

# Neuroimmune Biology

*Series Editors*

I. Berczi, A. Szentivanyi

# The Neuroendocrine Immune Network in Ageing

*Volume Editors*

Rainer H. Straub
Eugenio Mocchegiani

*University Medical Center Regensburg, Regensburg, Germany*
*and*
*Immunology Center, INRCA, Ancona, Italy*

ELSEVIER
2004

AMSTERDAM – BOSTON – LONDON – NEW YORK – OXFORD – PARIS
SAN DIEGO – SAN FRANCISCO – SINGAPORE – SYDNEY – TOKYO

**ELSEVIER B. V.**             ELSEVIER Inc.           ELSEVIER Ltd              ELSEVIER Ltd
**Sara Burgerhartstraat 25**   525 B Street            The Boulevard            84 Theobalds Road
**P.O. Box 211, 1000 AE**      Suite 1900, San Diego   Langford Lane, Kidlington   London WC1X 8RR
**Amsterdam, The Netherlands** CA 92101-4495, USA      Oxford OX5 1GB, UK       UK

First edition 2004

Library of Congress Cataloging in Publication Data
A catalog record is available from the Library of Congress.

British Library Cataloguing in Publication Data
A catalogue record is available from the British Library.

ISBN: 0-444-51617-4
ISSN: 1567-7443 (Series)

♾ The paper used in this publication meets the requirements of ANSI/NISO Z39.48-1992 (Permanence of Paper).
Printed in The Netherlands.

# List of Contributors

**Marcus W. Agelink**
*Institute of Biological Psychiatry & Neuroscience, Ruhr University Bochum, Germany*

**Marta Balietti**
*Neurobiology of Aging Laboratory, INRCA Research Department, Ancona, Italy*

**A. Bartke**
*Department of Medicine, School of Medicine, Southern Illinois University, Springfield, Illinois, USA*

**Denise L. Bellinger**
*Department of Pathology and Human Anatomy, Loma Linda University School of Medicine, Loma Linda, CA, USA*

**Istvan Berczi**
*Department of Immunology, Faculty of Medicine, The University of Manitoba, Winnipeg, MB, Canada*

**Peter Berger**
*Institute for Biomedical Aging Research, Austrian Academy of Sciences, Innsbruck, Austria*

**Carlo Bertoni-Freddari**
*Neurobiology of Aging Laboratory, INRCA Research Department, Ancona, Italy*

**Stefan R. Bornstein**
*Department of Endocrinology, University Hospital Düsseldorf, Düsseldorf, Germany*

**Hella Bruunsgaard**
*Department of Infectious Diseases, Rigshospitalet, University of Copenhagen, Denmark*

**Stephen K. Butcher**
*MRC Centre for Immune Regulation, Birmingham University Medical School, Birmingham, UK*

**Giuseppina Candore**
*Dipartimento di Biopatologia e Metodologie Biomediche, University of Palermo, Palermo, Italy*

**Calogero Caruso**
*Dipartimento di Biopatologia e Metodologie Biomediche, University of Palermo, Palermo, Italy*

**Tiziana Casoli**
*Neurobiology of Aging Laboratory, INRCA Research Department, Ancona, Italy*

**Catia Cipriano**
*Immunology Center, Research Department INRCA, Ancona, Italy*

**Marco Colasanti**
*Department of Biology, University of Roma 3, Roma, Italy*

**Giuseppina Colonna-Romano**
*Dipartimento di Biopatologia e Metodologie Biomediche, University of Palermo, Palermo, Italy*

**Maurizio Cutolo**
*Division of Rheumatology, Department of Internal Medicine and Medical Specialties, University of Genova, Italy*

**Giuseppina Di Stefano**
*Neurobiology of Aging Laboratory, INRCA Research Department, Ancona, Italy*

**Klaus Dinkel**
*Department of Neuropharmacology, Leibniz Institute for Neurobiology, Magdeburg, Germany*

**F. Dominici**
*Instituto de Quimica y Fisicoquimica Biologicas, Facultad de Farmacia y Bioquimica, Buenos Aires, Argentina*

**Alessia Donnini**
*Laboratory of Immunology, INRCA Gerontology Research Department, Ancona, Italy*

**Magdy El-Salhy**
*Institute of Public Health and Clinical Medicine, University Hosptial, Umea, Sweden*

**Patrizia Fattoretti**
*Neurobiology of Aging Laboratory, INRCA Research Department, Ancona, Italy*

**Tamas Fulop**
*Centre de Recherche sur le Vieillissement, Université de Sherbrooke, Sherbrooke, Québec, Canada*

**Robertina Giacconi**
*Immunology Center, Research Department INRCA, Ancona, Italy*

**Belinda Giorgetti**
*Neurobiology of Aging Laboratory, INRCA Research Department, Ancona, Italy*

**Erhard Haus**
*Health Partners, St Paul Ramsey Medical Center, St. Paul, MN, USA*

**M. Heiman**
*Lilly Research Labs, Corporate Center, Indianapolis, Indiana, USA*

**Michel Hofman**
*Netherlands Institute for Brain Research, Amsterdam, The Netherlands*

**J.J. Kopchick**
*Edison Biotechnology Institute, Konneker Research Laboratories, Ohio University, Ohio, USA*

**Karen Krabbe**
*Department of Infectious Diseases, Rigshospitalet, University of Copenhagen, Denmark*

**Valeria Lamounier-Zepter**
*Department of Endocrinology, University Hospital Düsseldorf, Düsseldorf, Germany*

**Anis Larbi**
*Centre de Recherche sur le Vieillissement, Université de Sherbrooke, Sherbrooke, Québec, Canada*

**David Lascelles**
*MRC Centre for Immune Regulation, Birmingham University Medical School, Birmingham, UK*

**Domenico Lio**
*Dipartimento di Biopatologia e Metodologie Biomediche, University of Palermo, Palermo, Italy*

**Janet M. Lord**
*MRC Centre for Immune Regulation, Birmingham University Medical School, Birmingham, UK*

**Dianne Lorton**
*Hoover Arthritis Center, Sun Health Research Institute, Sun City, AZ, USA*

**Kelley S. Madden**
*Department of Psychiatry, Center for Psychoneuroimmunology Research, University Rochester School of Medicine and Dentistry, Rochester, NY, USA*

**Stephan Madersbacher**
*Department of Urology and Andrology, Donauspital, Vienna, Austria*

**Stefano Mariotti**
*Endocrinology Department of Medical Sciences, University of Cagliari, Cagliari, Italy*

**Sofia Mariotto**
*Department of Neuroscience and Vision, University of Verona, Verona, Italy*

**Caroline A.M. Mattheij**
*Netherlands Institute for Brain Research, Amsterdam, The Netherlands*

**Massimo Miscusi**
*Department of Neurosurgery, University of Roma, "La Sapienza", Roma, Italy*

**Eugenio Mocchegiani**
*Immunology Center, Research Department INRCA, Ancona, Italy*

**Elisa Muti**
*Immunology Center, Research Department INRCA, Ancona, Italy*

**Mario Muzzioli**
*Immunology Center, Research Department INRCA, Ancona, Italy*

**Graham Pawelec**
*University of Tübingen, Center for Medical Research, Tübingen, Germany*

**Tiziana Persichini**
*Department of Biology, University of Roma 3, Roma, Italy*

**Meinrad Peterlik**
*Department of Pathophysiology, University of Vienna Medical School, Vienna, Austria*

**Aldo Pinchera**
*Department of Endocrinology, University of Pisa, Pisa, Italy*

**Giovanni Pinna**
*Endocrinology Department of Medical Sciences, University of Cagliari, Cagliari, Italy*

**Eugen Plas**
*Department of Urology and Andrology, Krankenhaus Lainz, Vienna, Austria*

**Mauro Provinciali**
*Laboratory of Immunology, INRCA Gerontology Research Department, Ancona, Italy*

**Francesca Re**
*Laboratory of Immunology, INRCA Gerontology Research Department, Ancona, Italy*

**Rixt F. Riemersma**
*Netherlands Institute for Brain Research, Amsterdam, The Netherlands*

**Dirk Sanner**

Institute of Biological Psychiatry & Neuroscience, Ruhr University Bochum, Germany

**Robert M. Sapolsky**

Department of Biological Sciences, Stanford University, Stanford, CA, USA

**Jürgen Schölmerich**

Laboratory of Neuroendocrinoimmunology, Department of Internal Medicine I, University Hospital, Regensburg, Germany

**Rafael Solana**

Department of Immunology, Faculty of Medicine, Reina Sofia University Hospital, Cordoba, Spain

**Moreno Solazzi**

Neurobiology of Neurobiology of Aging Laboratory, INRCA Research Department, Ancona, Italy

**Eus J.W. van Someren**

Netherlands Institute for Brain Research, Amsterdam, The Netherlands

**Rainer H. Straub**

Laboratory of Neuroendocrinoimmunology, Department of Internal Medicine I, University Hospital, Regensburg, Germany

**Hisanori Suzuki**

Department of Neuroscience and Vision, University of Verona, Verona, Italy

**Dick F. Swaab**

Netherlands Institute for Brain Research, Amsterdam, The Netherlands

**Andor Szentivanyi**

Department of Internal Medicine, Faculty of Medicine, The University of South Florida, Tampa, FL, USA

**Yvan Touitou**

Department of Biochemistry and Molecular Biology, Faculty of Medicine Pitié-Salpêtrière, Paris Cedex 13, France

**D. Turyn**

Instituto de Quimica y Fisicoquimica Biologicas, Facultad de Farmacia y Bioquimica, Buenos Aires, Argentina

**Keqing Wang**

MRC Centre for Immune Regulation, Birmingham University Medical School, Birmingham, UK

**Dan Ziegler**

*The German Diabetic Research Institute, Heinrich-Heine University, Düsseldorf, Germany*

# Contents

## VI. The ageing process and chronic inflammatory diseases

## VII. Conclusions

# I.    INTRODUCTION

# Evolutionary Aspects for the Neuroendocrine Immune Network in Ageing

RAINER H. STRAUB

*Laboratory of Neuroendocrinoimmunology, Dept. of Internal Medicine I, University Hospital, 93042 Regensburg, Germany*

ABSTRACT

In this chapter we shortly mention two important evolutionary aspects for ageing of the neuroen-docrine-immune network:
1) In the literature, typical age-related alterations of single cells and cellular substructures are often described. Such alterations are extremely important for single cell longevity. In multi-cellular organisms, when two or more cells have different properties (for example in *Volvox Carteri*), ageing of one cell type can be accompanied by ageing of another cell type because secreted vital products of one cell type are not supplied to the other. This situation is even more complicated in organisms with many different organs which are connected by supersys-tems such as the nervous system, the immune system, and the endocrine system. Age-related changes of these interconnecting supersystems may now lead to global ageing of the entire body. During evolution, complex organisms appeared where age-related changes of one supersystem are accompanied by ageing of another supersystem or an organ system.
2) Advantageous mechanisms for healthy ageing and longevity are not evolutionary conserved because most aged beings, albeit healthy, can not reproduce themselves so that possible posi-tive genes are not retained in the offspring. Thus, natural selection has limited the opportunity to exert a direct influence over the ageing process. In addition, genes which may be beneficial in the early period of life including the reproductive phase may be deleterious in the elderly.
These facets demonstrate that it is necessary to understand the ageing process (of the neuroendo-crine-immune network) in order to consider seriously the evolutionary aspects of ageing.

1.   AGEING OF ONE SUPERSYSTEM AFFECTS AGEING OF ANOTHER SUPER-SYSTEM

During phylogenesis and ontogenesis, a single cell was the origin of complex organisms. From the perspective of a single cell, ageing will obviously depend on the cell's capacity to manage oxidative damage, non-enzymatic glycosylation, mitochondrial mutations, defects in cell cycle control, mitotic dysregulation, genome instability, telomere shortening, and other chromosomal pathologies. Thus, research often focuses on single cells with intracellular DNA-*fugal* and DNA-*pedal* signaling cascades. However, during evolution single cells started to form networks of specialized cell types such as, for example, in the *Volvox Carteri* with somatic cells and germ cells [1]. The somatic cells, similar to most cells of our body, are bound to die whereas germ cells create a new *Volvox* and, thus, survive. Under consideration of multicellular organisms, extracellular interactions by means of paracrine factors such as cytokines, hormones and others became important. An ageing cell may have influenced a cell in the neighborhood by means of secreted soluble factors. The distant cell may have sensed this information and may have reacted with another set of secreted mediators. In such a situation, ageing of one single cell in the tissue can lead to age-related processes in the entire tissue.

During evolution, the situation became more complex when specific tissue appeared leading to the development of organs. Organs moved away from each other to build relatively independent sections in a body. When distant organs developed, new supersystems became necessary in order to connect these distant regions. Important supersystems are vessels for blood or lymph (humoral pathway) and nerves (neuronal pathway, often called "hard-wiring"). With respect to the neuronal pathway, we distinguish between efferent and afferent pathways knowing that afferent sensory nerves can secrete neurotransmitters at nerve endings in the periphery (this is an efferent function). In inflamed tissue, for example, local cells produce mediators which can appear in the circulation or which can stimulate sensory nerves in order to announce the local problem to distant sites. The distant site, for example the brain, reacts by stimulating the hypoth-alamic-pituitary-adrenal (HPA) axis or the hypothalamus – autonomic nervous system axis [2–6]. In a very similar way, age-related changes of one organ, due to age-dependent changes of a group of single cells within the organ, are announced to distant sites in the body. Thus, multiple interactions of distant organs appear during ageing and may contribute to the overall ageing process in a non-linear dynamic way [7]. For example, ageing of the adrenal glands leads to a dramatically reduced secretion of dehydroepiandrosterone sulfate (DHEAS) [8,9]. This hormone is converted to downstream androgens and estrogens in peripheral cells [10], which is the most important source of sex hormones during the ageing process when gonadal glands have undergone gradual involution [11]. Sex hormones are important for bone homeostasis. Thus, ageing of the adrenal and gonadal glands with a loss of DHEAS and other sex hormones support osteoporosis in the elderly. This example demonstrates that ageing of one part of the endocrine supersystem is accompanied by ageing of another organ system.

## 2.     ADVANTAGEOUS MECHANISMS FOR AGEING AND LONGEVITY ARE NOT EVOLUTIONARILY CONSERVED

The reproductive phase of human subjects is the period during which women and men normally cohabit and get children. From the view point of *homo habilis*, who lived 2 million years ago, the reproductive period lasts normally from 12 to 25 years of age. Under consideration of this short reproductive phase, advantageous mechanisms for ageing and longevity were not evolu-tionarily conserved because possible advantageous genes for senescence were not retained in the offspring. On the other hand, many deleterious genes, which are responsible for diseases in the elderly, can accumulate over generations with little or no safety tests. In addition, genes which may be beneficial in the early period of life including the reproductive phase may be deleterious in the elderly. We can generalize that no reaction of the body is specifically conserved to support healthy ageing. Natural selection has limited the opportunity to exert a direct influence over the process of ageing [12].

## 3.    PREVIEW FOR THIS BOOK

In this book, we will distinguish three supersystems which integrate the different reactions in the body to maintain general homeostasis. Supersystems transmit signals from one area of the body to an other area of the body. The immune system, the nervous system, and the endocrine system are supersystems. These supersystems integrate many reactions which were evolutionarily conserved for transient non-life threatening inflammatory events (defense), energy supply, reproduction, and others.

In three different sections, examples for ageing of the three supersystems are given. In chapter V, ageing phenomena of one supersystem are linked to ageing processes of another supersystem. In chapter VI, ageing of the neuroendocrine immune network is linked to chronic inflammation and inflammatory conditions: It becomes more and more obvious that chronic inflammation leads to an accelerated ageing process. In the final chapter, we try to integrate all aspects to give possible recommendations for new anti-ageing strategies.

## REFERENCES

1.    Kirk MM, Stark K, Miller SM, Muller W, Taillon BE, Gruber H, Schmitt R, Kirk DL. regA, a Volvox gene that plays a central role in germ-soma differentiation, encodes a novel regulatory protein. Development 1999;126:639–647.
2.    Chrousos GP. The hypothalamic-pituitary-adrenal axis and immune-mediated inflammation. N Engl J Med 1995;332:1351–1362.
3.    Besedovsky HO Del Rey A. Immune-neuro-endocrine interactions. Endocr Rev 1996;17: 64–102.
4.    Straub RH, Westermann J, Schölmerich J, Falk W. Dialogue between CNS and immune system in lymphoid organs. Immunol Today 1998;19:409–413.
5.    Elenkov IJ, Wilder RL, Chrousos GP, Vizi ES. The sympathetic nervous system - an integrative interface between two supersystems: the brain and the immune system. Pharmacol Rev 2000;52:595–638.
6.    Madden KS. Catecholamines, sympathetic nerves, and immunity. In: Ader R, Felten DL, Cohen N, editors. Psychoneuroimmunology. San Diego: Academic Press, 2001;197–216.
7.    Straub RH, Cutolo M, Zietz B, Schölmerich J. The process of ageing changes the interplay of the immune, endocrine and nervous systems. Mech Ageing Dev 2001;122:1591–1611.
8.    Orentreich N, Brind JL, Rizer RL, Vogelman JH. Age changes and sex differences in serum dehydroepiandrosterone sulfate concentrations throughout adulthood. J Clin Endocrinol Metab 1984;59:551–555.
9.    Straub RH, Konecna L, Hrach S, Rothe G, Kreutz M, Schölmerich J, Falk W, Lang B. Serum dehydroepiandrosterone (DHEA) and DHEA sulfate are negatively correlated with serum interleukin-6 (IL-6), and DHEA inhibits IL-6 secretion from mononuclear cells in man in vitro: possible link between endocrinosenescence and immunosenescence. J Clin Endocrinol Metab 1998;83:2012–2017.
10.    Schmidt M, Kreutz M, Löffler G, Schölmerich J, Straub RH. Conversion of dehydroepiandrosterone to downstream steroid hormones in macrophages. J Endocrinol 2000;164:161–169.
11.    Greenspan FS. Basic and clinical endocrinology. East Norwalk: Appleton and Lange, 1991.

12. Kirkwood TB, Austad SN. Why do we age? Nature 2000;408:233–238.

# II.    AGEING OF THE IMMUNE SYSTEM

*The Neuroendocrine Immune Network in Ageing*
Edited by R.H. Straub and E. Mocchegiani

# Immunosenescence

RAFAEL SOLANA[1] and GRAHAM PAWELEC[2]

[1]*Department of Immunology, Faculty of Medicine, "Reina Sofía" University Hospital, Cordoba, Spain;* [2]*University of Tübingen, Center for Medical Research, Tübingen, Germany*

## ABSTRACT

The term immunosenescence refers to the age-associated decline of the immune system that may contribute to the increased incidence and severity of infectious diseases and possibly certain cancers in the elderly. It is clearly established that aged individuals have a decreased or negligible thymic output, which contributes to age-associated immunodeficiency, although lifetime´s accumulated exposure to infectious agents, autoantigens and cancer antigens may also induce T cell replicative senescence and clonal exhaustion. Furthermore different studies over the last years have associated changes in one or other immunological parameter with longevity, although it is still difficult to control for nutritional and psychological factors in the groups studied. Longitudinal studies have established that significantly lower numbers of CD4(+) and CD8(+)CD28(+) cells, and larger numbers of CD8(+)CD28(−) apoptosis-resistant cells, as well as inverted CD4/8 ratios, are characteristic of the elderly compared to the healthy middle-aged. Clustering these with certain other parameters has led to the emergence of the concept of an "immunological risk phenotype" (IRP) as a tool to predict incipient mortality in the very elderly. In this chapter we will review the main cellular and molecular aspects of the immune system that contribute to T cell immunosenescence and their possible influence in age-associated clinically relevant situations.

## 1.    INTRODUCTION

Average life expectancy throughout developed countries has rapidly increased during the last decades and geriatric infectious diseases have become an increasingly important issue [1,2]. It is clearly established that infections in the elderly are not only more frequent but also more severe. Cumulative evidence suggests that the reasons for increased susceptibility include immunosenescence. This term refers to the age-associated decline of the immune system (for a comprehensive review see [3]), particularly the T cell compartment, that may contribute to the increased incidence and severity of infectious diseases and possibly certain cancers in the elderly. Moreover, immunosenescence may be the cause of infection but infection can also be the cause of immunosenescence. Thus, in young individuals, chronic exposure to antigens, as occurs in chronic infections, HIV infection and also in autoimmune diseases and cancer, may lead to T cell clonal exhaustion reflecting the inability of T cell clones to divide indefinitely.

It is clearly established that aged individuals have a decreased or negligible thymic output, which contributes to age-associated immunodeficiency. However, recent studies have demonstrated that naive cells also undergo a process of postmitotic senescence, and that a lifetime´s accumulated exposure to infectious agents, autoantigens and cancer antigens may induce T cell replicative senescence and subsequent clonal exhaustion caused by antigen-stimulated proliferative stress, that may also contribute to age-associated immunodeficiency [4].

Here we will review the main cellular and molecular aspects of the immune system that contribute to T cell immunosenescence.

## 2. THE EFFECT OF AGEING ON THYMOPOIESIS

It has been known for a long time that although the thymus is extremely active during the fetal and perinatal periods, thymic function declines in an age-dependent manner, starting shortly after birth. This could be related to alterations in early stages of hematopoiesis, affecting the number and function of lymphoid-committed precursor cells in the bone marrow and/or to altered thymic differentiation.

Although basal hematopoiesis is maintained through age, cumulative evidences in humans indicates that ageing is associated with lower numbers of progenitor cells, decreased proliferative activity, and reduced capacity to produce colony-stimulating factors. In particular, ageing is associated with reduced numbers of lymphoid-committed hematopoietic progenitor cells [5]. Similarly, de Haan et al. [6] showed that ageing significantly alters hematopoiesis in mice, as ageing is associated with reduced proliferative activity and a compensatory increase in stem cell number. These changes are strain-dependent and relate both to the mean strain longevity and to the individual age of the mouse.

Furthermore, although hematopoietic stem cells express functionally-active telomerase [7] there is a decrease in the length of the telomeric repeats in hematopoietic stem cells from adult bone marrow compared to fetal liver-derived or umbilical cord-derived stem cells [8]. Telomere length of peripheral blood leukocytes, including naive T lymphocytes, also decreases with age in different species studied supporting the notion that telomerase may not be able to completely maintain telomere length in ageing stem cells.

One of the reasons that the numbers of peripheral naïve T cells decrease with age is thymic involution, a process in which the volume of the thymic area where active thymopoiesis occurs decreases and is replaced gradually by fat tissue with a subsequent marked reduction in output of mature naïve T lymphocytes. Nonetheless, work over the past decade has shown that the involuting thymus can retain a certain degree of function in adults even at an advanced age, and that this is different in different individuals. One study of thymic samples from donors from one week to 50 years old showed an early decrease of cellularity although the adult thymus still contained thymocytes with similar surface phenotypes to those seen in young donors [9]. In addition, Jamieson et al. [10] have shown that the TCR families that are present in immature and mature thymocytes of adults are as diverse as those that are present in fetal thymocytes, indicating that ageing does not significantly affect TCR rearrangement activity required to generate functional thymocytes. These data clearly indicate that the thymus can be active in adults for at least 60 years.

A new assay based on the detection and quantitation of the excised TCRdelta DNA locus from within the TCRalpha chain genes can be used to estimate the presence of newly-generated alpha beta(+) T cells. These TCR-rearrangement excision circles (TREC) are present at higher levels in naive T cells than in memory T cell populations, are not duplicated during division and diluted out within the total peripheral alpha/beta(+) T cell pool with advancing age, therefore permitting the assessment of thymic output [11] (on the asumption that T cells cannot be generated elsewhere than in the thymus). The analysis of TREC in CD4 and CD8 cells from donors of different ages revealed decreases in their numbers from birth to 80 years of age. Nevertheless, high levels of TRECs can also be found in the thymocytes of some elderly individuals, showing that old thymi can still generate functional T cells with actively rearranged TCR genes [12]. The number of TRECs correlated with the number of "naive" T cells [13] in children and adults. The use of a mathematical model to quantify age-dependent changes both in the number of recent thymic emigrants (RTE) generated per day and in TREC concentration during an 80-year lifespan demonstrate that RTE and peripheral T cell division have the same potential to affect TREC concentration at any age in healthy people, also showing that, during ageing, thymic involution primarily induces an age-dependent decline in TREC concentrations within both CD4(+) and CD8(+) T cell populations [14]. Another study has demonstrated a striking unequal content of TRECs in two different subsets of naive Th cells, indicating different peripheral proliferative histories. Whereas TRECs are highly enriched in peripheral naive CD45RA(+) Th cells coexpressing CD31, they can hardly be detected in peripheral naive CD45RA(+) Th cells lacking CD31 expression. These CD31(-) CD45RA(+) Th cells, distinguished from CD45RA(+) effector-memory cells on the basis of their expression of other surface molecules such as CD62L, represent the main compartment of the naive peripheral Th cell pool during ageing [15].

Recent studies in mice also showed that TREC levels decline in ageing mice with significant mouse strain-dependent differences [16]. The absolute number of thymocytes and the number of molecules of TREC per thymocyte dramatically decreased with age. However, while the TRECs in peripheral CD8 spleen cells fell with age they remained constant in the CD4 subset [17]. Furthermore, TREC analysis has permitted the assessment of the efficacy of IL-7 as an agent to reverse thymic atrophy. Although treatment of old mice with IL-7 did not produce changes in the total number of alpha beta(+) T cells in the peripheral T cell pool, it altered the component subsets. Mice treated with IL-7 showed increases in the number of TREC-positive T cells indicating improved thymic output, and splenic T cells from IL-7-treated mice performed significantly better in in vitro functional assays compared to controls [18–20]. These results further indicate the essential role of IL-7 produced by stromal cells in T-cell development and thymopoiesis, supporting the survival of early T-cell progenitors, and also controlling the differentiation of T cells in the thymus.

3.    SENESCENCE OF NAIVE T CELLS

After T cells have been generated in the thymus, they are released to the periphery to maintain T cell homeostasis. These so-called "naïve" cells remain in the periphery until stimulated by cognate antigen. However, naïve T cells require survival signals and slowly undergo replication. Whether this process may ever result in replicative senescence is unknown. Although many age-associated changes in T cell phenotype and function can be explained by the process of replicative senescence (see below), the possiblitity that "non-replicative" senescence also occurs in T lymphocytes from the aged cannot be discarded. Thus, the effect of ageing on naive CD4+ cells from mice transgenic for a TCR recognizing pigeon *cytochrome c* that have not been in contact with the antigen revealed that although these cells maintain a naive phenotype rather than acquiring a prototypical aged memory phenotype, they are decreased in number and have a decreased antigen responsiveness [21]. Thus, naive CD4 cells from aged mice expanded poorly, produced low levels of IL-2, and, although they generated highly polarized effector populations that secreted high levels of cytokines, when these aged effector populations were transferred to adoptive hosts, they generated poorly functioning memory cells [22]. These results suggest that intrinsic changes occur with ageing that do not depend on encounter with antigen but reflect senescence processes not associated with vigorous replication

## 4.  SIGNAL TRANSDUCTION IN AGED T LYMPHOCYTES

Because antigen processing and presentation capacity is essentially intact or even enhanced in most systems investigated in the elderly [23], problems arising in T cell activation may reside in the T cell's signal transduction apparatus. A high proportion of T cells from the elderly has been reported to exhibit multiple defects early in the signal transduction pathway, such as reduced enzymatic activity of tyrosine kinases Fyn and Lck, leading to a reduction in the generation of phospholipase Cgamma-1 and a decreased formation of second messengers such as IP3 and dia-cylglycerol; the alternative Ras-MAPK-ERK protein kinase activation pathway is also compro-mised during ageing (for review see [24]). Furthermore, defects in cytoskeletal reorganization triggered by initial contact between TCR and peptide-bearing APC have been shown [25]. This defect precedes, and presumably contributes to, defective activation of protein kinase-mediated signals in the activation of T cells from the aged. However, in interpreting experimental data in this respect, we must be careful to distinguish between the effects of age and the differences in signal transduction pathways applying in different subpopulations of T cells, the relative abun-dance of which may be different in samples from people of different ages, nutritional and health status.

## 5.  AGE-ASSOCIATED ALTERATIONS OF ALPHA/BETA T LYMPHOCYTES

The first of these considerations, namely age-associated alterations in relative abundance of T cell subsets, is well-known to apply in humans, where the composition of the T cell compart-ment changes during ageing. It is probably as a result not only of the decreased thymic output but also of repeated antigen exposure, clonal expansion, and accumulation of memory and effector T cells, as well as health status. Although there may be a slight reduction in overall numbers of mature peripheral CD3[+] T cells with age, several studies of very healthy elderly donors, includ-ing centenarians, show that the absolute number of lymphocytes is maintained within tightly defined limits irrespective of age [26]. Reports of decreased overall numbers of CD3 cells may have been due to examining populations with underlying disease, as reviewed in [27]. However, analysis of lymphocyte subsets from humans and mice shows that ageing is associated with an increase in the number of T cells with a phenotype of memory or effector T cells, which may compensate for the reduction seen in naive T cell numbers produced by the thymus.

Analysis of the effect of ageing on mature T cells clearly demonstrates that they are also sub-ject to ageing processes. Thus, T cells that undergo repeated proliferation cycles in response to intermittent clonal expansion for effective immune responses, or in response to chronic stimu-lation, are affected by "replicative senescence". This may also be consequent to extrathymic homeostatic proliferation of memory T cells required to replace naive T cells whose numbers decline with thymic atrophy, in order to maintain absolute lymphocyte numbers. Furthermore, long-lived, quiescent T cells are affected by "post-mitotic senescence". Thus both types of senescent T cells will be found associated to ageing.

The analysis of T cell subsets in elderly individuals clearly reveals an increase in the proportion T cells expressing the CD45RO marker of many memory cells, and a decrease in the proportion of T cells expressing the CD45RA marker. Furthermore there is a decrease in the number of CD62L+ T consistent with a decreased number of naive cells. The analysis of other markers such as CD95, mainly expressed on cells that have been stimulated by antigen, or the increased expression of CD60 on memory cells has helped to demonstrate the presence of an increased proportion of memory T cells in the elderly. In particular the analysis of naive-memory phenotypes within the CD8 subpopulation of T lymphocytes reveals a very marked shift to the memory phenotype. True naive CD8+ T cells as assessed by TREC analysis have recently been characterised as being CD45RO(–),CD103(+), CD11a(dim), CD95(dim), CD27(+) and CD62L+ [28]. According to these criteria, most CD8 cells from elderly individuals are memory cells as the vast majority of them strongly express CD95 [29].

The expression of CD28 on T lymphocytes also decreases with age. CD28 is perhaps the closest to a biomarker of ageing found for human lymphocytes. Both in vivo and in vitro, the proportion of CD28(+) cells decreases with age. Particularly the CD8 subset shows progressively decreasing CD28 expression with age [29,30]. CD28 is an important costimulatory molecule, so cells lacking CD28 may be refractory to optimal activation. However, the fraction of CD8(+)CD28(–) cells in centenarians is higher than in the 70–90 year-old population [31], despite commonly retained functional integrity of overall immunity, so the biological relevance of these findings remains questionable. Telomere lengths in the CD28-negative cells are lower than in the CD28+ cells from the same donors, clearly suggesting that they have had a longer proliferative history [32]. They may accumulate because of dysregulated apoptosis [33] ie. they are abnormally resistant to clonal deletion but cannot proliferate or function optimally. This type of proliferative senescence may therefore be responsible for the commonly observed accumulation of CD28-negative oligoclonal populations in elderly people [34]. This potential problem is not limited to CD8 cells, because CD4 T cells, almost all of which are CD28(+) in young adults, also show an increasing CD28-negative fraction in the elderly, albeit not so marked [31].

Furthermore, several markers originally described on NK cells including the MHC specific inhibitory or activating receptors [35] may also be expressed on CD8+ T cells, in particular in those cells that do not express CD28. The expression of these markers is increased in T cells from elderly donors probably as a consequence of the expansion of CD8(+)CD28(–) cytotoxic effector cells [36,37]. Whether they are relevant to the ageing process is not yet clear, but in other contexts of chronic antigenic stimulation such as HIV infection [38,39], rheumatoid arthritis [40,41] or antitumor immune responses [42] they may play very important roles either inhibiting or activating T cells.

The recent application of tetramer technology, where fluorescently-labelled soluble complexes of MHC molecules and specific peptide epitopes can be used to directly identify antigen-specific T cells ex vivo, has provided a critical insight into changes in the T cell repertoire during ageing. It has been known for many years that elderly people (and mice) show oligoclonal expansions of T cells, particularly CD8 cells [34]. However, until recently, the reason for this expansion and specificity of these cells was unknown. It now seems likely that a major driving force in humans for this accumulation of a limited diversity of CD8 cells is the persisting immunogenicity of latent Herpes viruses, especially CMV.

In monoclonal populations, mainly found in individuals over 65, the density of expression of CD28 is also decreased. Although originally described only in CD8 cells, the number of individuals with such nonmalignant clonal expansions in both CD4 and CD8 cells was found to be very similar and these expansions are stable over a two-year observation period [43]. CD8 clonal expansions in healthy elderly people contained mostly CD28(−) cells proliferating poorly in culture [44]. The CD28(−) cells contained larger amounts of perforin than the CD28(+), consistent with the hypothesis that the CD28(−) cells may represent terminally-differentiated effector CTL. As mentioned above, in humans a major factor driving the expansion of such clones may be cytomegalovirus [45,46]. However, other persistent viruses, eg. EBV, clearly also have a role [47].

## 5.1.    Age-associated alterations in the T cell repertoire

One consequence of the expansion/constriction of T lymphocyte subpopulations discussed above is the restriction in the T cell repertoire, as memory cells with their repertoire of previously seen antigens increases with age and may come to dominate the T cell pool. Although early reports both in humans and mice indicated that it was the CD8 cell repertoire rather than the CD4 which were primarily affected in this way, recent analysis using more sensitive methods of TCR analysis (CDR3 spectratyping) showed that oligoclonal expansions affect CD4 as well as CD8 cells to the same extent [48]. In aged humans, up to 30% of the entire population of CD8 T cells are of oligo- or mono-clonal cells expressing the same TCR-Vß markers [34]. These populations are included in the CD28(-) subset, have shorter telomeres, can express NK associated markers, including inhibitory receptors [49], and can include expansions of virus specific T cells [47,50]. The decreased expression of CD28 in these cells may lead to altered T cell activation because initial contact of a T cell with the antigen-presenting cell requires the interaction between CD28 on the T cell surface and CD80 on the APC.

## 5.2.    "Replicative senescence" of in vivo expanded T cells

T cells, like other somatic cells, have a limited proliferative lifespan and can only undergo a finite number of divisions. Thus, after a number of divisions T lymphocytes undergo replicative senescence characterized by the decreased proliferative capacity in response to antigenic re-stimulation.

Senescent T lymphocytes are characterized by their decreased capacity to proliferate, that is associated with shorter telomere length [51–53]. They can also be phenotypically recognized by their expression of the killer cell lectin-like receptor G1 (KLRG-1), a type II transmembrane glycoprotein with an extracellular domain homologous to C-type lectins and an amino-terminal cytoplasmic tail that contains an ITIM motif, which recruits the hematopoietic cell phosphatase SHP-1 to the complex, triggering an inhibitory signal. Unlike most other negative receptors of this type, KLRG-1 is expressed by senescent CD4 as well as CD8 T cells, which are present in increasing numbers in mice as they age. Studies in humans and mice show that KLRG-1+ T cells are unable to proliferate, even those few which are still CD28(+), and, at least in mice, these cells may not be rapidly lost from the peripheral pool and so may gradually accumulate with age [54,55]. Therefore, it was interesting to investigate the KLRG-1 phenotype of the CMV-specific cells accumulating in the elderly, referred to above. While 50–60% of CMV-specific CD8 cells in the young were KLRG-1(+), this figure increased to >90% in the old. Moreover, not only were the KLRG-1(+) cells in the elderly compromised in proliferation, unlike the KLRG-1(+) cells in the young, they were essentially unable to secrete IFN-γ after antigenic stimulation [56]. Therefore, KLRG-1(+) cells not only accumulate in the elderly but are functionally compromised. They may therefore fill the "immunological space" and reduce the T cell repertoire for response to new antigens.

## 6. AGE-ASSOCIATED ALTERATIONS OF GAMMA/DELTA T LYMPHOCYTES AND NKT CELLS

Not only mainstream T lymphocytes are altered with age. The number and function of other T cell subpopulations are also affected by ageing. Thus, the absolute number of $\gamma\delta$ T cells in peripheral blood of elderly subjects is reduced, although they show an enhanced expression of CD69 and an increased production of TNF-$\alpha$ [57,58]. It has been suggested that the basal activation of $\gamma\delta$ T cells is due to the "inflamed" environment and that these cells may act as a first line of defense to compensate for defects in specific immunity described above.

NKT cells, a novel lineage of T cells characterized by the expression of V$\alpha$24/J$\alpha$Q gene segments in humans and V$\alpha$14/J$\alpha$281+ in mice, the recognition of glycolipids associated with CD1d and their production of IFN-$\gamma$ and IL-4 [59], are also decreased in peripheral blood from healthy elderly individuals [60] and in murine liver [61], indicating that these cells do not escape the process of immunosenescence.

## 7. FUNCTIONAL CHANGES RESULTING FROM THE ABOVE AGE-ASSOCIATED ALTERATIONS AND THE EFFECT OF PYSCHOLOGICAL FACTORS

One of the most striking functional consequences of immunosenescence is the altered cytokine milieu, partially but not entirely as a result of alterations to T cell cytokine secretion. Age-associated alterations in T cell function may be related to changed molecular regulation of cytokine production due to differences in signal transduction pathways in old cells as outlined above. A major dysfunction in T cells from elderly donors is a selectively decreased ability to secrete T cell growth factors, first reported many years ago [62]. More recently, it has been proposed that there are age-associated changes in immune responsiveness characterised by a change from a predominant type 1 phenotype in adults to a type 2 phenotype in the elderly. However, data are discordant, for many reasons excellently reviewed recently by Gardner & Murasko [63]. One of the main reasons for these discrepancies lies with donor selection, illustrating a serious general problem in biogerontological studies of humans. Thus, early, and even some current studies employed apparently healthy elderly donors without rigorously excluding underlying illness, or assessing nutritional or psychological status. Even strict selection criteria, such as the SENIEUR protocol, may fail to take psychological and nutritional factors sufficiently into account. Age-associated decreased immunological function may be linked to psychological factors but very few studies have addressed this issue [64–66]. The effects of stress may also be important, eg. recent studies are beginning to suggest clinically relevant alterations, eg. in the decreased antibody responses to influenza vaccination of chronically stressed elderly carers of dementia patients [67] as well as lower IL 1 and IL 2 production in the elderly caregivers [68]. Several immune parameters may be more profoundly impaired in older patients with clinical depression, eg. proliferation to mitogens, lower NK activity, altered lymphocyte subsets and numbers [69]. A correlation between increased plasma levels of TNF-$\alpha$ and depression in the elderly has also been reported [70].

What might be the mechanism connecting the immune and neurological systems and their changes in ageing? This is still a new area of investigation, with many of the main players in this field represented by their contributions to this book. Here, we can only say that some data are beginning to emerge concerning possible mechanisms for such interactions, eg. on the regulatory effects of neuropeptides on cytokine secretion in young and elderly donors [71]. Although the significance of these findings is unclear, studies of this type may begin to help shed some light on the mechanisms responsible for neuroimmunological communication and age-associated alterations. That this could be a two-way communication is illustrated by the ability of IFN-γ and other cytokines to modulate the production of melatonin [72]. Such studies will surely be the way forward in this difficult area. An example of how such studies may have a clinical impact concerns Alzheimer's Disease and inflammation in ageing. Thus, local inflammatory processes can exert direct neurotoxicity, interfere with β-amyloid expression and metabolism and maintain chronic intracerebral acute phase protein secretion, in turn favouring formation of β-amyloid fibrils [73,74]

## 8. CONCLUSION: HOW IMPORTANT IS IMMUNOSENESCENCE CLINICALLY?

Many studies over the years have associated changes in one or other immunological parameter with longevity [75–78], but detailed longitudinal studies in humans have been rare. By studying the same donors over time, some variables confounding cross-sectional studies may be reduced, although it is obviously still difficult to control for nutritional and psychological factors. Unsurprisingly, there are very few longitudinal studies from which data are currently available and even fewer including immunological data. One such study series is the Swedish OCTO (over 80 years old) and current NONA (over 90) project being carried out in Jönköping, Sweden, in the very elderly [79–82]. Obviously, the very elderly are already selected as longer-than-average survivors, but nonetheless the data collected at regular intervals over the last decade can be very valuable, perhaps not only for the already-successfully aged. The OCTO and NONA studies have established that significantly lower numbers of CD4(+) and CD8(+)CD28(+) cells, and larger numbers of CD8(+)CD28(−) apoptosis-resistant cells, as well as inverted CD4/8 ratios, are characteristic of the elderly compared to the healthy middle-aged. Clustering these with certain other parameters has led to the emergence of the concept of an "immunological risk phenotype" (IRP) as a tool with power to predict incipient mortality in the very elderly. One of the current challenges is to refine the IRP and extend the concept to younger individuals to assess whether IRP parameters established in the aged are informative for younger people [83]. Our own investigations are suggesting that the accumulation of CMV-specific, CD28(−), KLRG-1(+) CD8 cells mentioned above, which are unable to secrete IFN-γ, and are probably apoptosis-resistant anergic cells, is an important parameter of the IRP. These cells may represent senescent CD8 clones which can neither divide nor delete – and hence may result in dysregulated responses. Deleting such cells, by analogy with a transgenic mouse model, may "rejuvenate" the immune system [84]

## ACKNOWLEDGEMENTS

The work of RS and GP was performed under the aegis of the EU 5th FP projects "Immunology and Ageing in Europe, ImAginE", contract no. QLK6-CT-1999-02031 and "T-Cells in Ageing, T-CIA", contract no. QLK6-CT-2002-02283. RS was also supported by the Spanish Ministries of Health (FIS98/1052, FIS01/0478) and Education (PM98/0168) and GP by the Deutsche Forschungsgemeinschaft (DFG Pa 361/7-1) and the VERUM Foundation.

## REFERENCES

1.  Gavazzi G, Krause KH, Ageing and infection. Lancet Infect Dis 2002;2:659–666.
2.  High KP, Infection in an ageing world. Lancet Infect Dis 2002;2:655.
3.  Pawelec G, Barnett Y, Forsey R, Frasca D, Globerson A, McLeod J, Caruso C, Franceschi C, Fulop T, Gupta S, et al, T cells and aging, January 2002 update. Front Biosci 2002;7: d1056–d1183.
4.  Pawelec G, Solana R, Immunoageing – the cause or effect of morbidity. Trends Immunol 2001;22:348–349.
5.  Keating A, The hematopoietic stem cell in elderly patients with leukemia. Leukemia 1996;10 Suppl 1:S30–S32.
6.  de Haan G, Nijhof W, Van Zant G, Mouse strain-dependent changes in frequency and proliferation of hematopoietic stem cells during aging: correlation between lifespan and cycling activity. Blood 1997;89:1543–1550.
7.  Engelhardt M, Kumar R, Albanell J, Pettengell R, Han W, Moore MA, Telomerase regulation, cell cycle, and telomere stability in primitive hematopoietic cells. Blood 1997;90:182–193.
8.  Vaziri H, Dragowska W, Allsopp RC, Thomas TE, Harley CB, Lansdorp PM, Evidence for a mitotic clock in human hematopoietic stem cells: loss of telomeric DNA with age. Proc Natl Acad Sci U S A 1994;91:9857–9860.
9.  Bertho JM, Demarquay C, Moulian N, Van Der Meeren A, Berrih-Aknin S, Gourmelon P, Phenotypic and immunohistological analyses of the human adult thymus: evidence for an active thymus during adult life. Cell Immunol 1997;179:30–40.
10. Jamieson BD, Douek DC, Killian S, Hultin LE, Scripture-Adams DD, Giorgi JV, Marelli D, Koup RA, Zack JA, Generation of functional thymocytes in the human adult. Immunity 1999;10:569–575.
11. Livak F, Schatz DG, T-cell receptor alpha locus V(D)J recombination by-products are abundant in thymocytes and mature T cells. Mol Cell Biol 1996;16:609–618.
12. Douek DC, McFarland RD, Keiser PH, Gage EA, Massey JM, Haynes BF, Polis MA, Haase AT, Feinberg MB, Sullivan JL, et al, Changes in thymic function with age and during the treatment of HIV infection. Nature 1998;396:690–695.
13. Steffens CM, Al-Harthi L, Shott S, Yogev R, Landay A, Evaluation of thymopoiesis using T cell receptor excision circles (TRECs): differential correlation between adult and pediatric TRECs and naive phenotypes. Clin Immunol 2000;97:95–101.
14. Ye P, Kirschner DE, Reevaluation of T cell receptor excision circles as a measure of human recent thymic emigrants. J Immunol 2002;168:4968–4979.
15. Kimmig S, Przybylski GK, Schmidt CA, Laurisch K, Mowes B, Radbruch A, Thiel A, Two subsets of naive T helper cells with distinct T cell receptor excision circle content in human adult peripheral blood. J Exp Med 2002;195:789–794.
16. Broers AE, Meijerink JP, van Dongen JJ, Posthumus SJ, Lowenberg B, Braakman E, Cornelissen JJ, Quantification of newly developed T cells in mice by real-time quantitative PCR of T-cell receptor rearrangement excision circles. Exp Hematol 2002;30:745–750.
17. Sempowski GD, Gooding ME, Liao HX, Le PT, Haynes BF, T cell receptor excision circle assessment of thymopoiesis in aging mice. Mol Immunol 2002;38:841–848.

18

18. Aspinall R, Andrew D, Pido-Lopez J, Age-associated changes in thymopoiesis. Springer Semin Immunopathol 2002;24:87–101.

19. Pido-Lopez J, Imami N, Andrew D, Aspinall R, Molecular quantitation of thymic output in mice and the effect of IL-7. Eur J Immunol 2002;32:2827–2836.

20. Aspinall R, Henson SM, Pido-Lopez J, My T's gone cold, I'm wondering why. Nat Immunol 2003;4:203–205.

21. Linton PJ, Haynes L, Klinman NR, Swain SL, Antigen-independent changes in naive CD4 T cells with aging. J Exp Med 1996;184:1891–1900.

22. Haynes L, Eaton SM, Swain SL, Effect of age on naive CD4 responses: impact on effector generation and memory development. Springer Semin Immunopathol 2002;24:53–60.

23. Castle SC, Uyemura K, Crawford W, Wong W, Makinodan T, Antigen presenting cell function is enhanced in healthy elderly. Mech Ageing Dev 1999;107:137–145.

24. Pawelec G, Hirokawa K, Fulop T, Altered T cell signalling in ageing. Mech Ageing Dev 2001;122:1613–1637.

25. Garcia GG, Miller RA, Age-dependent defects in TCR-triggered cytoskeletal rearrangement in CD4+ T cells. J Immunol 2002;169:5021–5027.

26. Hulstaert F, Hannet I, Deneys V, Munhyeshuli V, Reichert T, De Bruyere M, Strauss K, Age-related changes in human blood lymphocyte subpopulations. II. Varying kinetics of percentage and absolute count measurements. Clin Immunol Immunopathol 1994;70: 152–158.

27. Remarque E, Pawelec G, T cell immunosenescence and its clinical relevance in man. Rev Clin Gerontol 1998;8:5–11.

28. McFarland RD, Douek DC, Koup RA, Picker LJ, Identification of a human recent thymic emigrant phenotype. Proc Natl Acad Sci U S A 2000;97:4215–4220.

29. Fagnoni FF, Vescovini R, Passeri G, Bologna G, Pedrazzoni M, Lavagetto G, Casti A, Franceschi C, Passeri M, Sansoni P, Shortage of circulating naive CD8(+) T cells provides new insights on immunodeficiency in aging. Blood 2000;95:2860–2868.

30. Fagnoni FF, Vescovini R, Mazzola M, Bologna G, Nigro E, Lavagetto G, Franceschi C, Passeri M, Sansoni P, Expansion of cytotoxic CD8+ CD28– T cells in healthy ageing people, including centenarians. Immunology 1996;88:501–507.

31. Boucher N, Dufeu-Duchesne T, Vicaut E, Farge D, Effros RB, Schachter F, CD28 expression in T cell aging and human longevity. Exp Gerontol 1998;33:267–282.

32. Batliwalla FM, Rufer N, Lansdorp PM, Gregersen PK, Oligoclonal expansions in the CD8(+)CD28(-) T cells largely explain the shorter telomeres detected in this subset: analysis by flow FISH. Hum Immunol 2000;61:951–958.

33. Gupta S, Molecular steps of cell suicide: an insight into immune senescence. J Clin Immunol 2000;20:229–239.

34. Posnett DN, Sinha R, Kabak S, Russo C, Clonal populations of T cells in normal elderly humans: the T cell equivalent to benign monoclonal gammapathy. J Exp Med 1994;179: 609–618.

35. Borrego F, Kabat J, Kim DK, Lieto L, Maasho K, Pena J, Solana R, Coligan JE, Structure and function of major histocompatibility complex (MHC) class I specific receptors expressed on human natural killer (NK) cells. Mol Immunol 2002;38:637–660.

36. Borrego F, Alonso MC, Galiani MD, Carracedo J, Ramirez R, Ostos B, Pena J, Solana R, NK phenotypic markers and IL2 response in NK cells from elderly people. Exp Gerontol 1999;34:253–265.

37. Tarazona R, Delarosa O, Alonso C, Ostos B, Espejo J, Pena J, Solana R, Increased

expression of NK cell markers on T lymphocytes in aging and chronic activation of the immune system reflects the accumulation of effector/senescent T cells. Mech Ageing Dev 2000;121:77–88.

38. Galiani MD, Aguado E, Tarazona R, Romero P, Molina I, Santamaria M, Solana R, Pena J, Expression of killer inhibitory receptors on cytotoxic cells from HIV-1-infected individuals. Clin Exp Immunol 1999;115:472–476.

39. Tarazona R, Delarosa O, Casado JG, Torre-Cisneros J, Villanueva JL, Galiani MD, Pena J, Solana R, NK-associated receptors on CD8 T cells from treatment-naive HIV-infected individuals: defective expression of CD56. AIDS 2002;16:197–200.

40. Snyder MR, Muegge LO, Offord C, O'Fallon WM, Bajzer Z, Weyand CM, Goronzy JJ, Formation of the killer Ig-like receptor repertoire on CD4+CD28null T cells. J Immunol 2002;168:3839–3846.

41. Weyand CM, Fulbright JW, Goronzy JJ, Immunosenescence, autoimmunity, and rheumatoid arthritis. Exp Gerontol 2003;38:833–841.

42. Ikeda H, Lethe B, Lehmann F, van Baren N, Baurain JF, de Smet C, Chambost H, Vitale M, Moretta A, Boon T, et al, Characterization of an antigen that is recognized on a melanoma showing partial HLA loss by CTL expressing an NK inhibitory receptor. Immunity 1997;6: 199–208.

43. Schwab R, Szabo P, Manavalan JS, Weksler ME, Posnett DN, Pannetier C, Kourilsky P, Even J, Expanded CD4+ and CD8+ T cell clones in elderly humans. J Immunol 1997;158: 4493–4499.

44. Chamberlain WD, Falta MT, Kotzin BL, Functional subsets within clonally expanded CD8(+) memory T cells in elderly humans. Clin Immunol 2000;94:160–172.

45. Khan N, Shariff N, Cobbold M, Bruton R, Ainsworth JA, Sinclair AJ, Nayak L, Moss PA, Cytomegalovirus seropositivity drives the CD8 T cell repertoire toward greater clonality in healthy elderly individuals. J Immunol 2002 Aug 15 ;169 (4 ):1984–92 2002;169: 1984–1992.

46. Ouyang Q, Wagner WM, Wikby A, Walter S, Aubert G, Dodi AI, Travers P, Pawelec G, Large numbers of dysfunctional CD8+ T lymphocytes bearing receptors for a single dominant CMV epitope in the very old. J Clin Immunol 2003;23:247–257.

47. Ouyang Q, Wagner WM, Walter S, Muller CA, Wikby A, Aubert G, Klatt T, Stevanovic S, Dodi T, Pawelec G, An age-related increase in the number of CD8(+) T cells carrying receptors for an immunodominant Epstein-Barr virus (EBV) epitope is counteracted by a decreased frequency of their antigen-specific responsiveness. Mech Ageing Dev 2003;124: 477–485.

48. Mosley RL, Koker MM, Miller RA, Idiosyncratic alterations of TCR size distributions affecting both CD4 and CD8 T cell subsets in aging mice. Cell Immunol 1998;189: 10–18.

49. Tarazona R, Delarosa O, Alonso C, Ostos B, Espejo J, Pena J, Solana R, Increased expression of NK cell markers on T lymphocytes in aging and chronic activation of the immune system reflects the accumulation of effector/senescent T cells. Mech Ageing Dev 2000;121:77–88.

50. Ouyang Q, Wagner WM, Wikby A, Remarque E, Pawelec G, Compromised IFN-gamma production in the elderly leads to both acute and latent viral antigen stimulation: contribution to the immune risk phenotype? Eur Cytokine Netw 2002;13:387–3.

51. Effros RB, Boucher N, Porter V, Zhu X, Spaulding C, Walford RL, Kronenberg M, Cohen D, Schachter F, Decline in CD28+ T cells in centenarians and in long-term T cell cultures:

a possible cause for both in vivo and in vitro immunosenescence. Exp Gerontol 1994;29: 601–609.

52.  Effros RB, Insights on immunological aging derived from the T lymphocyte cellular senescence model. Exp Gerontol 1996;31:21–27.

53.  Effros RB, Pawelec G, Replicative senescence of T cells: does the Hayflick Limit lead to immune exhaustion? Immunol Today 1997;18:450–454.

54.  Voehringer D, Koschella M, Pircher H, Lack of proliferative capacity of human effector and memory T cells expressing killer cell lectinlike receptor G1 (KLRG1). Blood 2002;100: 3698–3702.

55.  Voehringer D, Blaser C, Brawand P, Raulet DH, Hanke T, Pircher H, Viral infections induce abundant numbers of senescent CD8 T cells. J Immunol 2001;167:4838–4843.

56.  Ouyang Q, Wagner WM, Voehringer D, Wikby A, Klatt T, Walter S, Muller CA, Pircher H, Pawelec G, Age-associated accumulation of CMV-specific CD8(+) T cells expressing the inhibitory killer cell lectin-like receptor G1 (KLRG1). Exp Gerontol 2003;38:911–920.

57.  Colonna-Romano G, Potestio M, Aquino A, Candore G, Lio D, Caruso C, Gamma/delta T lymphocytes are affected in the elderly. Exp Gerontol 2002;37:205–211.

58.  Romano GC, Potestio M, Scialabba G, Mazzola A, Candore G, Lio D, Caruso C, Early activation of gammadelta T lymphocytes in the elderly. Mech Ageing Dev 2000;121: 231–238.

59.  Godfrey DI, Hammond KJ, Poulton LD, Smyth MJ, Baxter AG, NKT cells: facts, functions and fallacies. Immunol Today 2000;21:573–583.

60.  Delarosa O, Tarazona R, Casado JG, Alonso C, Ostos B, Pena J, Solana R, Valpha24+ NKT cells are decreased in elderly humans. Exp Gerontol 2002;37:213–217.

61.  Tsukahara A, Seki S, Iiai T, Moroda T, Watanabe H, Suzuki S, Tada T, Hiraide H, Hatakeyama K, Abo T, Mouse liver T cells: their change with aging and in comparison with peripheral T cells. Hepatology 1997;26:301–309.

62.  Gillis S, Kozak R, Durante M, Weksler ME, Immunological studies of aging. Decreased production of and response to T cell growth factor by lymphocytes from aged humans. J Clin Invest 1981;67:937–942.

63.  Gardner EM, Murasko DM, Age-related changes in Type 1 and Type 2 cytokine production in humans. Biogerontology 1903;3:271–290.

64.  Guidi L, Bartoloni C, Frasca D, Antico L, Pili R, Cursi F, Tempesta E, Rumi C, Menini E, Carbonin P, Impairment of lymphocyte activities in depressed aged subjects. Mech Ageing Dev 1991;60:13–24.

65.  Guidi L, Tricerri A, Frasca D, Vangeli M, Errani AR, Bartoloni C, Psychoneuroimmunology and aging. Gerontology 1998;44:247–261.

66.  Goya RG, Hormones, genetic program and immunosenescence. Exp Clin Immunogenet 1992;9:188–194.

67.  Vedhara K, Cox NK, Wilcock GK, Perks P, Hunt M, Anderson S, Lightman SL, Shanks NM, Chronic stress in elderly carers of dementia patients and antibody response to influenza vaccination. Lancet 1999;353:627–631.

68.  Kiecolt-Glaser JK, Glaser R, Gravenstein S, Malarkey WB, Sheridan J, Chronic stress alters the immune response to influenza virus vaccine in older adults. Proc Natl Acad Sci U S A 1996;93:3043–3047.

69.  Herbert TB, Cohen S, Depression and immunity: a meta-analytic review. Psychol Bull 1993;113:472–486.

70.  Vetta F, Ronzoni S, Lupattelli MR, Fabbriconi B, Ficoneri C, Cicconetti P, Bruno A,

Russo F, Bollea MR, Tumor necrosis factor-alpha and mood disorders in the elderly. Arch Gerontol Geriatr 2001;33 Suppl 1:435–442.

71. Hernanz A, Tato E, De La Fuente M, de Miguel E, Arnalich F, Differential effects of gastrin-releasing peptide, neuropeptide Y, somatostatin and vasoactive intestinal peptide on interleukin-1 beta, interleukin-6 and tumor necrosis factor-alpha production by whole blood cells from healthy young and old subjects. J Neuroimmunol 1996;71:25–30.

72. Maestroni GJ, The photoperiod transducer melatonin and the immune-hematopoietic system. J Photochem Photobiol B 1998;43:186–192.

73. Akiyama H, Barger S, Barnum S, Bradt B, Bauer J, Cole GM, Cooper NR, Eikelenboom P, Emmerling M, Fiebich BL, et al, Inflammation and Alzheimer's disease. Neurobiol Aging 2000;21:383–421.

74. Franceschi C, Valensin S, Lescai F, Olivieri F, Licastro F, Grimaldi LM, Monti D, De Benedictis G, Bonafe M, Neuroinflammation and the genetics of Alzheimer's disease: the search for a pro-inflammatory phenotype. Aging (Milanob) 2001;13:163–170.

75. Shigemoto S, Kishimoto S, Yamamura Y, Change of cell-mediated cytotoxicity with aging. J Immunol 1975;115:307–309.

76. Roberts-Thomson IC, Whittingham S, Youngchaiyud U, Mackay IR, Ageing, immune response, and mortality. Lancet 1974;2:368–370.

77. Wayne SJ, Rhyne RL, Garry PJ, Goodwin JS, Cell-mediated immunity as a predictor of morbidity and mortality in subjects over 60. J Gerontol 1990;45:M45–M48

78. Bender BS, Nagel JE, Adler WH, Andres R, Absolute peripheral blood lymphocyte count and subsequent mortality of elderly men. The Baltimore Longitudinal Study of Aging. J Am Geriatr Soc 1986;34:649–654.

79. Wikby A, Johansson B, Ferguson F, Olsson J, Age-related changes in immune parameters in a very old population of Swedish people: a longitudinal study. Exp Gerontol 1994;29: 531–541.

80. Olsson J, Wikby A, Johansson B, Lofgren S, Nilsson BO, Ferguson FG, Age-related change in peripheral blood T-lymphocyte subpopulations and cytomegalovirus infection in the very old: the Swedish longitudinal OCTO immune study. Mech Ageing Dev 2000;121: 187–201.

81. Wikby A, Maxson P, Olsson J, Johansson B, Ferguson FG, Changes in CD8 and CD4 lymphocyte subsets, T cell proliferation responses and non-survival in the very old: the Swedish longitudinal OCTO-immune study. Mech Ageing Dev 1998;102:187–198.

82. Wikby A, Johansson B, Olsson J, Lofgren S, Nilsson BO, Ferguson F, Expansions of peripheral blood CD8 T-lymphocyte subpopulations and an association with cytomegalovirus seropositivity in the elderly: the Swedish NONA immune study. Exp Gerontol 2002;37:445–453.

83. Pawelec G, Ouyang Q, Colonna-Romano G, Candore G, Lio D, Caruso C, Is human immunosenescence clinically relevant? Looking for 'immunological risk phenotypes'. Trends Immunol 2002;23:330–332.

84. Zhou T, Edwards CK, Mountz JD, Prevention of age-related T cell apoptosis defect in CD2-fas-transgenic mice. J Exp Med 1995;182:129–137.

*The Neuroendocrine Immune Network in Ageing*
Edited by R.H. Straub and E. Mocchegiani

# Zinc-Binding Proteins (Metallothionein and α-2 Macroglobulin) as Potential Biological Markers of Immunosenescence

EUGENIO MOCCHEGIANI, ROBERTINA GIACCONI, ELISA MUTI, MARIO MUZZIOLI
and CATIA CIPRIANO

*Immunology Center, Section Nutrition, Immunity and Ageing, Research Department,
I.N.R.C.A., Ancona, Italy*

## ABSTRACT

Zinc is the most relevant trace element for the maintenance of numerous homeostatic mechanisms in the body. In particular, zinc plays a pivotal role in the function of the entire immune system for the entire life of an organism. The binding of zinc with some proteins, such as metallothioneins [MT] and α-2 macroglobulin [α-2M] is crucial immune efficiency during ageing and also in age-related diseases [cancer and infection]. These proteins may turn from a role of protection against cellular oxidative damage in young-adult age to a harmful role in ageing. Indeed, these proteins increase abnormally during ageing and according to their original biological function, they continuously sequester intracellular zinc. This provokes abnormally low zinc ion bio-availability for the immune system either at thymic or extrathymic level. The latter is prominent in ageing. As a result, an impaired immune function occurs in ageing. Physiological zinc supplementation restores the immune system, which leads to an increased rate of survival in old mice. Zinc supplementation does not interfere with the high levels of MT or α-2M. Therefore, zinc supplementation is of benefit in ageing and in age-related diseases [cancer and infection]. Moreover, zinc-binding proteins regain their original role of cellular protection against oxidative damage after zinc supplementation.

## 1.    INTRODUCTION

Zinc is a important trace element in the body. Among the most relevant findings, to be noted are the role of zinc as a catalytic component of more than 200 enzymes, its requirement as structural constituent of many proteins and, likely, its function of preventing free radical formation. Zinc is relevant to cell division and differentiation, as well as to programmed cell-death, for gene transcription, for biomembrane functioning, and obviously for many enzymatic activities. This has led to the consideration that zinc is an important element in assuring the correct functioning of various tissues, organs and systems. Indeed zinc is an essential nutritional component, which is required for normal development and for the maintenance of immune function in man and animals [1,2]. In particular zinc is essential for thymic function because it confers biological activity to a zinc-dependent thymic hormone called thymulin [3]. This hormone is required for

Table I    Structural action of zinc as "zinc fingers" for proteins or peptides DNA replication in various biological functions.

| Proteins or peptides | Biological function |
|---|---|
| SP1, A-20, ALR, MMPs, BP1O, BP1, Staf 50 | cell proliferation |
| TIMPs, alpha-2M | |
| GATA family, mp1, glycoprotein IIb/IIIa | cell differentiation |
| Nil-2-a, TGF-beta | cell growth arrest |
| Cyclin T1 | cell division |
| Protein Kinase-C (PKC), ZNF 162 | signal transmission |
| EGF, IL-1, IL-2, IL-4, IL-5, IL-6, IL-7, IL-12, IFN-α, TNF-α | cell growth |
| p53, p21, c-myc, c-fos,GFI-1, HIC-1, c-jun, bcl-2 | protooncogene for prevention or activation apoptosis |
| RANTES | chemokine production |
| NF-kB, AP-1, EGR-1, EGR-2, BTE | nuclear transcriptional |
| KS-1, WT-1, TF III A, Finb, TRAF-2, ZEB | factors activation |
| Melatonin, Growth Hormone, IGF-I | hormone nuclear receptor |
| Glucocorticoids, prostaglandins | superfamily codification |

the differentiation and maturation of T-cells [4]. Zinc is also relevant to the liver extrathymic T-cell pathway [2,5], a function that is prominent in old age [6]. With advancing age, zinc, thymic and extrathymic functions progressively decline [2]. Among the zinc-binding proteins, Metallothioneins [MT] [7] and α-2 Macroglobulin [α-2M] [8] are important for immune efficiency in ageing. These proteins have a cellular protective role against oxidative damage and they bind preferentially zinc with higher affinity [kd] [7] than does thymulin [see review 2]. It has been reported that MT is of protection in zinc deficiency [9] and donors of zinc for thymulin reactivation in thymic epithelial cells [TECs] [10]. However, both MT [11] and α-2M [12] are increased in ageing. Thus, a shift of zinc towards these zinc-binding proteins may occur with subsequent low zinc ion bioavailability for thymulin activation and, in general, for immune function. This phenomenon occurs in various conditions with persistent inflammation, such as ageing, cancer and infection [13]. Zinc-bound MT is not releasable for biological purposes [14]. Therefore, these proteins may turn from a role of protection in young-adults to harmful agents for the immune system in ageing and age-related diseases [13]. Here we report the role of zinc, MT and α-2M in immune function during ageing and in age-related diseases [e.g. cancer and infections]. In these conditions low zinc ion bio-availability and impaired immune function are regular events [15]. The possible efficacy of physiological zinc supplementation in ageing and in age-related diseases will also be reported and discussed.

## 2.    THE BIOLOGY OF ZINC

Zinc is one of the most important trace elements in the body, but its presence in nature does not exceed 0.02%. [16]. The major characteristics of zinc include a highly concentrated charge, a small radius [0.65 A], no variable valence [low risk of free radical production], ready passage from one symmetry in its surroundings to another without exchange, rapid exchange of ligands [on and off reactions], and binding mostly to 5- and N-donors in biological systems. These properties enable zinc to play three major biological roles, as catalysts, structural and regulatory ions [17]. With regard to the catalytic action, zinc is required for the biological function of more than 200 enzymes; in particular it is essential and directly involved in catalysis by the enzyme. The removal of the catalytic zinc results in an active apoenzyme that usually retains the native ter-

Table II    Binding affinity (kd) for zinc of some compounds relevant in the immune efficiency.

| Compounds | kd (mol/L) |
|---|---|
| Superoxide dismutase | $5.0 \times 10^{-4}$ |
| Nucleoside phoshorylase | $2.0 \times 10^{-5}$ |
| Alkaline-phosphatase | $1.0 \times 10^{-6}$ |
| Ecto-5 nucleotidase | $1.3 \times 10^{-6}$ |
| Endonucleases | $1.5 \times 10^{-6}$ |
| DNA, RNA polymerase | $2.0 \times 10^{-6}$ |
| Metalloproteinases (MMPs) | $1.0 \times 10^{-6}$ |
| Insulin-like growth factor I (IGF-I) | $1.7 \times 10^{-7}$ |
| Thymulin (ZnFTS) | $5.0 \pm 2 \times 10^{-7}$ |
| Protein Kinase- C | $1.2 \times 10^{-8}$ |
| P-450 enzyme system | $1.0 \times 10^{-9}$ |
| α-2 Macroglobulin | $1.0 \times 10^{-10}$ |
| Metallothioneins (I, II) | $1.4 \times 10^{-13}$ |

tiary structure. The catalytic zinc is usually bound by three protein ligands and a water molecule signifying an open co-ordination site, which is considered essential for the function of zinc in catalysis [17]. Indeed the function of the zinc is both in polarizing the substrate and in activating $H_2O$ molecule which then acts as nucleophile. Because of its flexible co-ordinator geometry, zinc ions act as template to bring together the substrate and the nucleophile [18].

As structural role, zinc is involved in DNA replication and reverse transcription, as well as in a number of eukariotic transcription factors where the potential binding domains are referred to as zinc fingers [19]. In this context, a plethora of data show zinc fingers involved in DNA replication of many proteins involved in cell proliferation, in cell differentiation, in cell growth and arrest, in cell division, in signal transmission, in growth factors production, in protooncogenes activation, in chemokines production, in codifying hormone nuclear receptor, in nuclear transcriptional factors [Table I], in mRNA stability and in maintaining the extracellular matrix [17,19,20]. In addition, zinc binds to enzymes, proteins and peptides with different binding affinity [kd] ranging from $10^{-2}$ to $10^{-14}$ mol/L. Some of them are relevant for immune efficiency [Table II] [2]. In particular, zinc binds thymulin [ZnFTS], a well-defined thymic hormone, which is responsible for the performances of cell-mediated immunity [4]. The absence of the zinc binding confers the inactivity of all these compounds [21]. These findings demonstrate the fundamental role played by structural zinc in the maintenance of many body homeostatic mechanisms. Moreover, they pinpoint the importance of structural zinc also in cancer and infection because many proteins, peptides, protooncogenes and transcriptional factors reported in Table I are also involved in tumor growth, apoptosis, and altered cytokine gene expression. [see review 2].

With regard to the regulatory role, zinc regulates both enzymatic activity and stability of the protein as an activator or as an inhibitor ion. In this context, gene over expressions of MMPs are under the control of the gene expression of some tissue inhibitors of matrix metalloproteinases [TIMPs] [22], of α-2M [23] and of 13-amyloid precursor protein [24], which are, in turn, codified by zinc-finger motifs [25]. Therefore, a good balance in the gene expression of the metalloproteinases [either as activator or as inhibitor] is necessary for an optimal function of many systems. Indeed, an unbalance between MMPs and TIMPs or α-2M is characteristic of cancer [26], inflammation [25] and ageing [27]. As free zinc ion bio-availability is essential for the efficiency of the immune system [1], we will focus on the role of two zinc-binding proteins [MT and α-2M] in immunosenescence. Despite of their cellular protective role against oxidative damage

[7], they abnormally increase in ageing [11,12], their zinc-binding affinity is higher than that of thymulin [Table II], and therefore, thymulin, immune functions and zinc ion bio-availability decrease in ageing [2]. These decrements often lead to the appearance of cancer and infection, which, in turn, lead to increased MT [28], enhanced α-2M [29], impaired immune functions and low zinc bio-availability [2,13].

## 3. THE BIOLOGY OF ZINC-BINDING PROTEINS

### 3.1. Metallothioneins

Metallothioneins [MT], a group of low-molecular-weight metal-binding proteins with 61 amino-acids, 20 of which are cysteines, play pivotal roles in metal-related cell homeostasis because of their high affinity for metals, in particular zinc and copper [7]. All 20 cysteines occur in reduced form and bind seven zinc atoms through mercaptide bonds that have the spectroscopy charac-teristics of metal thiolate clusters [30]. The zinc/cysteine interactions form mainly two different types of clusters: they are either bridging or form terminal cysteine thiolate. In the beta-domain cluster, three bridging and six terminal cysteine thiolates provide a coordinated environment that is identical for each of the three zinc atoms. In the alpha-domain clusters, there are two different zinc sites; two of them have one terminal ligand and three have bridging ligands, respectively, while the other two have two terminal and two bridging ligands [30]. MT exists in different iso-forms characterized by the length of aminoacid chain: isoform I, II, III and IV mapped on chro-mosome 16 in man and on chromosome 8 in mice with complex polymorphisms [31]. The more common isoforms are I and II while the isoform III, called GIF, is present only into the brain and the distribution of IV has not been clarified. MT binds zinc with high thermodynamic stability [kd=$1.4\times10^{-13}$ M) [7] (Table II). Following these biochemical and genetic characteristics, MT distributes intracellular zinc because zinc undergoes rapid inter- and intracluster exchange [32]. Moreover, MT acts as antioxidant against a wide spectrum of stressors, because the zinc-sulfur cluster is sensitive to changes in the cellular redox state and oxidizing sites in MT [e.g. reduced thiol groups] induce the transfer of zinc from its binding sites in MT to those of lower affinity in other proteins [30]. This transfer occurs, for example, in conferring biological activity to antioxi-dant metalloenzymes, such as superoxide dismutase [33]. Therefore, the redox properties of MT and their effect on zinc in the clusters are crucial for the biological functions of MT. Indeed, MT is peculiar to cellular proliferation and protects cells against cytotoxic effects of reactive oxygen species, ionizing radiations, electrophilic anti cancer drugs, mutagens and of heavy metals [7]. Consistent with this role, the gene expression of MT isoforms is regulated by the transcription factor MTF-I [34], which has six zinc-finger motifs [30],and has been identified in the evolution-ary tree from mollusks to humans [Table III]. MT is transcriptionally induced by a wide range of stressors, like metals, glucocorticoids, oxidative agents, strenuous exercise, cold exposure and irradiation: virtually by all agents that affect invertebrates and vertebrates [7].

These evolutionary findings suggest the importance of MT as a protective agent. Protection is peculiar to transient stress, as it may occur in young adult-age [35]. By contrast, the stress-like condition is not transient but permanent in ageing. MT is increased in the aged liver [11,14] and aged brain [37]. There is a constant presence of oxidative damage in ageing that gave rise to the free radical theory of ageing [36]. Moreover, high MT occurs in cancer, infection and dementia [28,38,39] and this is associated with unfavorable prognosis [40]. On the basis of these findings, it is clear that zinc cannot be transferred from MT to other proteins in ageing. Rather, zinc is con-

Table III     Evolutionary tree of Metallothionein isoforms.

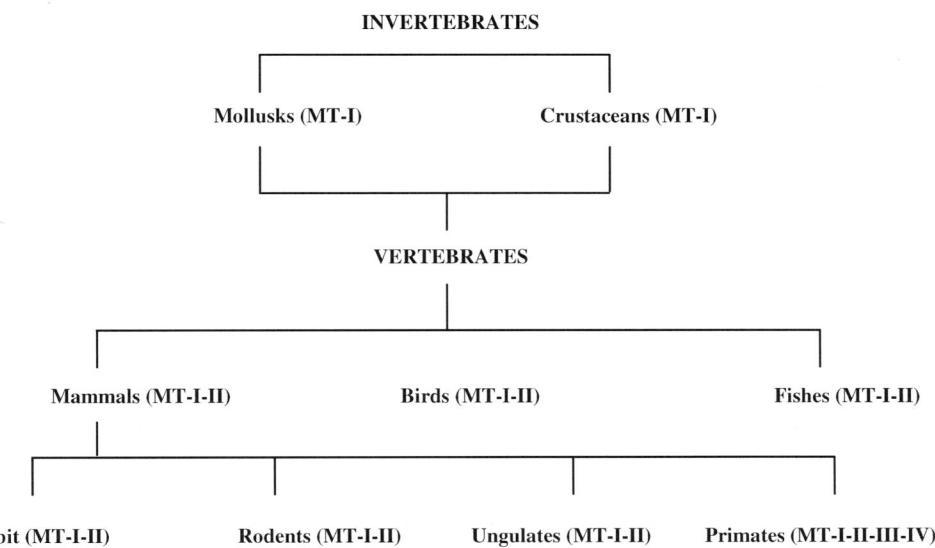

tinuously sequestered by MT in order to maintain the original role of protection and also because zinc ions must be spared as much as possible. However, this continuous sequestration provokes low zinc ion bio-availability. This alters the role of MT from protective in young-adult-age to harmful in old age. Indeed, it has been recently found that MT binds preferentially zinc, rather than copper, in ageing [14], and therefore, zinc bioavailability is low, but not that of copper [41]. Abnormal high MT and low zinc ion bio-availability are also present in constant inflammation and in syndromes of accelerated ageing [Down's syndrome] [11]. By contrast, low MT and satisfactory zinc ion bio-availability occur in successful ageing [11]. Therefore, the altered role of zinc-bound MT in ageing may be crucial for the immunocompetence because both zinc and immune function decrease in ageing, and this leads to the appearance of age-related diseases [2,13].

## 3.2.    α-2 macroglobulin

α-2 macroglobulin [α-2M] is an inhibitor of matrix metalloproteinases [MMPs] [23]. The mechanism of inhibition is unique. α-2M is cleaved by the proteinases in a specific region termed "bait region" and then undergoes conformational changes that activate thiol ester bounds, providing sites for covalent attachment of proteinases, which are entrapped [42]. This role of α-2 M occurs when other inhibitors of MMPs, such as tissue inhibitors of matrix metalloproteinases [TIMPs], are deficient. This suggests a central role of α-2M, which functions like a reserve proteinase inhibitor during emergencies [43]. This biological function of α-2 M is present in all animal kingdoms [42] and is required to remove proteolytic potential from the circulation when MMPs increase, forming α-2M-proteinases complexes [23]. The elimination occurs with the internalization of these complexes into the cells with subsequent trasfer of the proteinases into endosomes through specific α-2 M receptors, called LRP [44], which are present on the cell surface. A chaperone, called receptor-associated protein [RAP], binds, in turn, with high affinity to the heavy chain of LRP and induces a conformational change in LRP with a subsequent strong reduction of the affinity between α-2M-complex and LRP [45]. This modulates the binding of α-2M-complex to LRP. This causes a lack of the internalization of α-2M-proteinase complexes and the prevention of their transfer into cellular endosomes. Consequently these complexes will be present in the circulation, and may be harm many homeostatic mechanisms in the body [46]. Indeed, there are high concentrations of α-2M and MMPs in ageing and in age-related diseases, such as infection, cancer, and dementia [47,48]. This fact is crucial during ageing and in age-related diseases for two reasons. First, because intracellular RAP [RAP lp] is involved in the modulation of the telomere length [49], which in turn is relevant to the maintenance of chromosome integrity and stability [50]. Therefore, the length of telomeres are reduced in ageing, cancer and infections that leads to a chromosome mutations with subsequent cellular alteration or cell-death [51]. In these conditions, shorter telomeres are indirectly correlated with increased RAP lp [52], suggesting that RAP and its potential binding to α-2M/LRP receptor play a crucial role in modulating the internalization of α-2M-protease complexes into cells during ageing. Moreover, the recent discovery of 22 polymorphisms of the human LRP/AP1 gene coding for RAP in chromosome 4p [53] is a further support for the relevance of RAP to ageing, and to age-related diseases. They are of special importance in centenarians for successful ageing. Second, α-2M binds zinc with a higher affinity than do MMPs [25] [Table II]. As α-2M protects against increments of MMPs, the continuous sequestration of zinc by α-2M might induce low zinc bio-availability in the long run with subsequent alterations of zinc-dependent body homeostatic systems, including the immune system. Thus, possible different roles of α-2M might exist: turning from protective in young-adults to harmful in ageing. Indeed, α-2 M increases in ageing [12] coupled with alterations at cellular and systemic level [48]. Thus, while on one hand, these findings pin-point a continuous shifting of zinc towards α-2M in order to eliminate potential proteolytic substances, on the other hand they suggest, like for MT, that abnormal increments of α-2M are harmful leading to the appearance of age-related diseases. Like reported for MT, abnormal increments of α-2M in cancer and infections [48] are indeed an index of unfavorable prognosis [29]. In healthy young-adult individuals, α-2M plasma levels are of $178.8 \pm 8.5$ mg/ml [48]. Moderate increments are observed in nonagenarian/centenarian subjects [$201 \pm 7.5$ mg/ml] in comparison to old individuals [$260 \pm 9.3$ mg/ml] and old infected patients [$305 \pm 7.4$ mg/ml] [E. Mocchegiani, unpublished results]. Therefore, the role played by zinc-bound MT and α-2M is crucial for the immune efficiency in ageing taking also into account that low zinc bioavailability is a risk factor for the appearance of infectious diseases in ageing [54].

## 4.    ZINC, IMMUNE SYSTEM AND AGEING

A lot of data in experimental animals and in humans indicate that zinc is required for the maintenance of immunocompetence [1,2,13]. Zinc deficient diets cause major immune abnormalities that include decreased chemotaxis by neutrophils and monocytes, the specific impairment of cell-mediated immune functions, including thymic function, natural killer [NK] cell activity, and cytokine production. In particular, IL-2, IL-12, IFN-α and IFN-γ decrease, whereas pro-inflammatory cytokines, such as TNF-α, IL-1, IL-6, increase [see review 2]. Zinc has more influence on the cytokines production by Th1 cells than on those produced by Th2 lymphocytes [1]. That zinc has a beneficial effect on IFN-α production by Th1 cells is supported by the recent discovery in virus transfected cells, indicating that a protein, Staf-50, that contains two zinc

Table IV    "In vitro" effect of inactive FTS (FTS), of inactive FTS + Zinc$^{++}$ and zinc-bound active FTS (ZnFTS) on NK cell activity and PHA response in spleen cells from old mice.

| Spleen cells of mice | NK activity | PHA response |
|---|---|---|
| Young control | 73.2.0 ±14.1 | 28.2 ± 5.2 |
| Old control | 25.2 ±7.5* | 9.5 ± 1.3* |
| Old + inactive FTS ($10^{-6}$ M) | 24.5 ± 6.8* | 8.7 ± 1.5* |
| Old + inactive FTS ($10^{-6}$ M) + Zn$^{++}$($10^{-6}$ M) | 71.2 ± 13.7 | 23.3 ± 4.3 |
| Old + active ZnFTS ($10^{-6}$ M) | 71.4 ± 13.5 | 22.8 ± 3.8 |

*p< 0.01 when compared to young controls (paired Student's t test)

Preincubation time 4h. Data of PHA are expressed in [cpm/cult] × $10^3$. Data of NK assay are expressed in Lytic Unit [L.U. 20/$10^7$] against tumoral target YAC-1. The dose used of $10^{-6}$M of ZnFTS is the minor in order to obtain significant thymulin biological activity in thymulin assay [58]. The choice of zinc$^{++}$ $10^{-6}$M because zinc is present in thymulin molecule in equimolar ratio [3] (data redrawn from ref. [58,62] with permission).

finger motifs, is involved in a new family of IFN-$\alpha$ production [55]. This observation suggests that the Thl/Th2 balance is biased towards Th2 cytokine production during zinc deficiency [1]. Moreover, zinc is also involved in the recognition of superantigens by MHC class II molecules [56]. These findings support the importance of zinc ion bioavailability for the immune function even if the effect of zinc upon the immune system is related to competition phenomena with other metals, in particular copper. However, progressive depletion as well as excessive intake of copper and zinc result in impaired immunocompetence [2]. Zinc, but not copper, decreases in ageing [41]. This fact does not support the hypothesis of competition between zinc and copper for the maintenance of immune function during zinc deficiency. Rather, it is indicated that the lack of zinc ions is involved in immunosenescence [2]. Indeed, the copper pool is positive in young and old age [57]. By contrast, the zinc pool measured through the crude zinc balance [the difference between zinc intake and zinc excretion], is negative in old mice [58] and old humans [57] as compared to positive values observed in young-adult age. This negativity is coupled with reduced thymic functions, decreased NK cell activity, and imbalance in the Thl/Th2 ratio [59], increments in memory T-cells [CD45RO] and lack in naive T-cells [CD45RA] [60]. As a consequence, a lack in the homing capacity for new T-cell differentiation and maturation occurs allowing for a constant impairment of immune function in a large number of old people, in comparison to exceptional individuals like centenarians [61]. Among the immune abnormalities in ageing, the reduced thymic endocrine activity is of interest because it affects T-cell differentiation, NK cell activity and cytokines production [4]. Indeed, one of the best-known thymic hormones, called thymulin [ZnFTS], requires zinc for its biological activity. Zinc not bound to FTS has an inhibitory action on the active form ZnFTS. The inactive form of FTS is increased during ageing, whereas the active ZnFTS decreases. The "in vitro" addition of zinc ions [200nM] to plasma samples from old donors containing FTS restores circulating plasma levels of ZnFTS to values present in young-adult age (Table IV) [58, 62]. Therefore, the lack of normal ZnFTS levels because of low zinc ion bioavailability in ageing provokes impairments in peripheral immune functions (NK cell activity and mitogenic response to phytohaemoagglutinin) [58].

These findings demonstrate, on one hand, that NK cell activity and PHA response are under the control of ZnFTS, on the other hand, they pinpoint the relevance of zinc ion bioavailability for peripheral immune functions, which are peculiar for host defense against tumors and infections [63]. It has been also demonstrated in aged individuals and cancer patients that zinc affects the synthesis and release of IL-2 [48,64], which is relevant for NK cell activity [63]. Thymulin stimulates IL-2 receptor expression [CD25] by NK cells [65]. Zinc has been proposed to affect the immune system by many other mechanisms. Direct mechanisms through the activation of DNA- and RNA-polymerases, nucleoside phosphorylase, ecto-5 nucleotidase and protein kinase-C [PKC] have been documented. Indirect mechanisms involving transcriptional factors [NF-kB and AP-1] and hormones [melatonin, growth hormone, IGF-I and thyroid hormones] have also been reported [2]. In all cases, zinc ion bioavailability and immune efficiency are closely linked. Zinc ion bioavailability is also crucial for the liver-extrathymic-T-cell pathway, which is prominent in ageing, cancer and infections [6]. Young partially hepatectomized mice were used to study this problem because they resemble old mice with atrophic thymuses [66]. In such animals the liver extrathymic T-cell pathway was impaired coupled with abnormally high levels of zinc-bound MT and low zinc ion bioavailability [5]. Therefore, zinc-MT homeostasis [see below], is fundamental for immunocompetence [thymic and extrathymic] during ageing.

## 5. METALLOTHIONEIN, THE IMMUNE SYSTEM AND AGEING

MT induces cytokine secretion from macrophages, such as IL-1, IL-6, IFN-$\alpha$, TNF-$\alpha$, which, in turn, provoke new synthesis of MT in the liver [67]. At the same time zinc status is also altered [68]. These findings indicate an interaction amongst zinc, MT and the immune system. Moreover, IL-1 is involved in MTmRNA production in thymic epithelial cells [TECs] from young man [10] via PKC, which is zinc-dependent [10] and participates in the process of metal-induced MT gene expression [69]. Moreover, MT releases zinc for thymulin reactivation in TECs [10]. The reaction of MT null mice exposed to endotoxins [LPS] further supports the impotance of MT for immune function. In particular, MT null mice display decreased GM-CSF production as well as impaired GM-CSF receptor function [70] coupled with enhanced thymus apoptosis [71]. Conversely, young mice exposed to endotoxins [LPS] display increased MTmRNA, enhanced production of cytokines [IL-1, IL-6] and chemokines [RANTES, MIP classes, TECK and SDF-1] and a prompt immune response [72]. The interplay between MT and immune system attains relevance in ageing and in age-related diseases [cancer and infection], which are characterized by high levels of MT and pro-inflammatory cytokines, impaired immune efficiency and low zinc ion bioavailability [14]. MT protects against stresses [7] and has a balancing effect during zinc deficiency and zinc toxicity, serves as a type of zinc reservoir and is capable of sequestering excess of zinc, respectively [9]. However, the high MT and zinc content in the liver and atrophic thymus of old mice, suggests that MT plays a different role in immunosenescence: it is turning from protective to dangerous. [2]. In Table V zinc and MT levels are shown in the liver and in the thymus of very old mice, which supports the above assumption. An increment of MTmRNA gene expression and impaired NK cell activity also occur in the liver of old mice. By contrast, low MTmRNA and enhanced liver NK cell activity are observed in very old mice [11]. These findings suggest an efficient liver extrathymic T-cell pathway in very old age because of low zinc-bound MT levels. For the same reason, the thymus may be still functioning despite it is atrophic. Moreover, the increments of MT in lymphocytes from old people and from subjects with Down's syndrome and the decrements of MT in lymphocytes from centenarians [11] support further the different role of MT in ageing. Indeed, findings of Kelly et at. [9] were obtained in young-adult age which may mimic transient abnormal conditions. By contrast, zinc deficiency and impaired immune functions are persistent in ageing [2]. Therefore, MT may constantly recall zinc in order to exert their protective role. This provokes low zinc bioavailability and to impaired immune response at the thymic and extrathymic levels [2]. This assumption is supported by the

Table V    Tissue zinc and metallothionein (MT) concentrations in the thymus and liver from young, old and very old mice.

| | Young | Old | Very Old |
|---|---|---|---|
| Zinc in the thymus (μg/gr) | 62.3 ± 11.2 | 107.4 ± 27.5* | 77.4 ± 8.7 |
| Zinc in the liver (μg/gr) | 110.6 ± 23.4 | 125.6 ± 21.4 | 115.7 ± 15.5 |
| MT in the thymus (μg/gr w.w.) | 3.5 ± 0.4 | 8.6 ± 0.7* | 4.3 ± 0.5 |
| MT in the liver (μg/gr w.w.) | 5.3 ± 1.0 | 12.5 ± 4.7* | 5.8 ± 0.7 |

*$p < 0.01$ as compared to young mice [paired Student's t test]

Liver and thymus MT concentrations were determined with silver saturation method.
Content zinc tissue was determined with Atomic Absorption Spectropothometry [AAS].
Since AAS tests zinc-bound and zinc-unbound, the major increments of zinc in the thymus and liver from old mice is due to high zinc-bound MT. (Data redrawn from Ref.[11] with permission).

following findings: i] the addition of zinc to plasma samples from old donors *in vitro* restores the bioactivity of thymulin [58]; ii] in transgenic mice that over-express MT the thymus was atrophic as if they underwent to constant stress [14]; iii] endogenous zinc is required for T-cell growth in adults and in old individuals [73] and for TEC restoration number in old thymus [58] and for the restoration of NK cell activity, via active thymulin, in old mice [62]. It is consistent with these findings that MT is not a donor of zinc in ageing but rather, it sequesters of zinc. MT abnormally increases in the thymus and liver of old mice with a possible harmful role. This phenomenon has been proposed to be responsible for thymic involution in ageing [2]. On the other hand, increased MT levels induce down-regulation of many other biological functions related to zinc, such as metabolism, gene expression and signal transduction [7]. MT has a higher binding affinity for zinc than does thymulin (Table II). The atomic absorption spectrophotometer does not distinguish between bound and unbound zinc, which explains the high thymus and liver zinc content in old mice (Table V)] coupled with low zinc bioavailability. This is one of the crucial causes of immunodefficiency in ageing. Moreover, in the liver of old mice MT binds preferentially zinc when compared to copper [14]. It is clear that MT turns from protective to harmful by aggravating zinc deficiency and not of copper, in ageing [1, 41]. MT does not only sequester zinc continuously in ageing but the subsequent release of zinc by MT is also inhibited [14]. This problem is still not resolved at present in full detail. It is known that a correct balance between iNOS and cNOS induces zinc release by MT [74]. In ageing, iNOS increases with respect to cNOS, which induces an imbalance of NO synthases with very limited zinc release by MT [13]. Preliminary *in vitro* experiments with old lymphocytes showed that the addition of zinc [$10^{-6}$M] corrects the imbalance of NO synthases [E. Mocchegiani, unpublished results]. This may occur because NO synthases is regulated by zinc finger motifs [75]. Therefore, zinc-bound MT may be considered a biological and genetic marker of immunosenescence in age-related diseases because it shows abnormally increased levels in cancer and in infections [38].

## 6.    α-2 MACROGLOBULIN, THE IMMUNE SYSTEM AND AGEING

In recent decades various biological functions have been described for the plasma protein α-2M. The earliest function to be discovered was its proteinase-binding capacity with an inhibitory action on proteinases themselves [23]. One decade later, its ability to modulate T-cell responses was described, giving a role to α-2M within the immune system. α-2M is synthesized by macrophages and monocytes [76]. In particular α-2M-proteinase complexes, but not α-2M by itself, decrease T-cell responses to mitogens, antigens and alloantigens [77]. It has been proposed that macrophages down-regulate MHC class-II antigens via α-2M [78], however this was not confirmed by others [77]. It has been shown subsequently that the target of α-2M-proteinase complexes was T-cell growth factors [i.e. cytokines]. Indeed, IL-2 was degraded and inactivated by α-2M-bound proteinases, which was suggested to cause decreased T-cell response [79]. Degradation doesn't occur with other cytokines, in particular for pro-inflammatory cytokines [IL-1, IL-6 and TNF-α], even though these cytokines are bound to α-2M [79]. α-2M-cytokine complexes have been suggested to have a clearance functions of cytokines, protection from the immediate toxic effect of cytokines, the protection of cytokines from extracellular proteolysis and from the loss of cytokines through the kidney and the targeting of cytokines to α-2M-receptor-bearing cells [79]. α-2M inhibits the formation of MMPs. Indeed, MMPs are temporarily produced in response to exogenous signals by pro-inflammatory cytokines [80]. The binding of α-2M with pro-inflammatory cytokines occurs in order to induce latency of cytokines themselves minor MMPs production [81]. In addition, the possible over-expression of these cytokines induced by MMPs may be also avoided by the α-2M-cytokine complex through α-2MR/LRP [82]. These findings clearly suggest and further confirm the role of protection of α-2M against exogenous stimuli, other than the formation of α-2M-protease complexes, through the formation of also α-2M-cytokines complex. Pro-inflammatory cytokines [IL-1, IL-6 and TNF-α] increase in inflammation [83], and IL-6 regulates α-2M gene expression [84]. Therefore, the interrelationship between α-2M and immune system is evident giving further support to the role of protection of α-2M against toxic effects by abnormal production of pro-inflammatory cytokines during inflammation. This role is out of doubt in transient inflammation, as it may occur in young-adult age, but it may be questioned in chronic inflammation [ageing, cancer and infection], with a particular attention to the relationships with the immune functions, for the following reasons. **i]** α-2M requires zinc into its molecules in order to be active and zinc, as reported above, is important for the efficiency of the entire immune system [1,2]. **ii]** zinc is also required for the binding of α-2M with cytokines [85]. **iii]** α-2M plasma levels are strongly increased in ageing [12], in old infected patients [see above] and in cancer [48] coupled with damaged immune functions and low zinc bioavailability (Table VI). Intriguing findings in human centenarians have shown that the relatively low plasma levels of α-2M [see above] are coupled with satisfactory thymic endocrine activity, zinc ion bio-availability, NK cell activity and diminished IL-6 together with its sub-unit receptor gp130 [11]. As α-2M is involved in IL-2 degradation and in conferring latency to pro-inflammatory cytokines. These findings support the protective role of α-2M against toxic effects by an over-expression of pro-inflammatory cytokines. This protection is harmful for immune function in ageing and in age-related diseases [cancer and infections], whish is due to a continuous sequestration of zinc. An abnormal increase of α-2M is a diagnostic index for unfavorable prognosis in cancer [29], and zinc supplementation in ageing, cancer and infectious disease restores thymic and peripheral immune efficiency, including the altered cytokines production [38]. Therefore, α-2M may have dual functions for immune function: it turns from protective in young adult-age to harmful during ageing [59]. Th1 cells produce anti-inflammatory cytokines [IL-2], and Th2 cells produce pro-inflammatory cytokines [IL-6] [86], and α-2M impairs T-cells response via IL-2 degradation [79], α-2M may suppress Th1 cells. This leads to an imbalance of Thl/Th2 cells with increased pro-inflammatory cytokines production in ageing [87]. Moreover, α-2M is an acute phase protein and is expressed in the liver [25], which in turn is a prominent site of extrathymic T-cells in ageing, cancer and infection [6]. Elevated levels of α-2M may be used, like MT, as a marker of immunosenescence

Table VI      Zinc, α-2 macroglobulin, active thymulin (ZnFTS), NK cell activity, IL-2 and IL-6 in patients affected by cervical carcinoma.

|  | Healthy controls | Cervical carcinoma |
|---|---|---|
| Active thymulin (ZnFTS)(log $_2$) | 3.0 ± 0.3 | 1.5 ± 0.5* |
| IL-2 (pg/ml) | 315.1 ± 87.2 | 106.4 ± 65.3* |
| IL-6 (pg/ml) | 2.5 ± 0.6 | 21.4 ± 7,7* |
| NK cell activity (L.U.×$10^6$ cells) | 9.3 ± 2.2 | 3.0 ± 2.1** |
| α-2 Macroglobulin (mg/dl) | 178.8 ± 8.5 | 280.0 ± 22.8* |
| Plasma zinc (µg/dl) | 92 .5 ± 6.8 | 72.4 ± 5.3* |

*p< 0.01 and **p< 0.05 when compared to healthy controls [paired Student's t test]

Age of cancer patients and healthy controls 52 ± 7 yrs (Data redrawn from Ref. [48] with permission).

and of age-related diseases.

## 7.    ZINC SUPPLEMENTATION

The important role of zinc for immunocompetence and the presence of a negative crude zinc balance in old mice and humans [57,58], stimulated experiments on dietary zinc supplementation in old mice and in elderly patients in order to improve their immunocompetence. In rodents zinc supplementation for life prevented the development of age-related cell-mediated immune deficiency [88]. Such treatment induced thymus re-growth and functionality [58,89] and restored peripheral NK cell activity [58]. Physiological levels of zinc prevented thymocytes apoptosis in old mice [90], by acting on endonuclease enzymes [91]. In human prostate carcinoma, zinc [at the physiological dose of 100 µg/dl] inhibited tumor cell growth by arresting the cell cycle in the G2/M phase and by enhancing apoptosis through increased p21 expression [92] or of c-myc [93], and MT and α-2M increase in tumor cells [28,48]. Therefore, a larger bio-availability of zinc ions through zinc supplementation might, on one hand, regenerate the protective role to MT and α-2M, on the other hand, it might cause apoptosis in cancer cells. Therefore, zinc is a crucial element for maintaining immune function during ageing and for the possible prevention of age-related diseases. Indeed, old zinc-treated mice display an increased rate of survival and significant decreases in the incidence of cancer and infection [2].

For elderly patients, zinc supplementation yielded contradictory data regarding its beneficial effect. This may be due to different doses of zinc used and to variation in the length of the treatment [94–96]. Indeed, physiological doses of zinc for a long periods or high doses of zinc for short periods may induce a zinc accumulation in various organs and tissues with subsequent toxic effect of zinc on the immune function and on cytokine production [1]. In our experience, zinc treatment at the dose of 15mg Zn$^{++}$/day [RDA] for one month in Down's syndrome subjects, in elderly and in old infected patients restores thymic functions, peripheral immune efficiency [54, 97] and DNA-repair [98]. At the clinical level, significant reductions of infectious episodes and infection relapses occur in Down's syndrome [94] in the elderly and in old infected patients [57]. All the supplementation experiments and clinical trials demonstrate the pivotal role of zinc for the maintenance of immune function in ageing and in the alleviation of age-related diseases.

However, one may ask the question, whether or not zinc supplementation may further increase zinc-binding proteins with major harmful effects. Recent experiments in old zinc treated mice have shown no further increments of liver MTmRNA, suggesting that the high levels of MT in ageing may be completely saturated by pre-existing zinc ions with no further harmful effect [14]. Old hepatectomized mice show no modifications of the already high MT levels during liver regeneration when compared to young hepatectomized mice [5]. Moreover, physiological zinc doesn't induce any further increments of MTmRNA induced by IL-1 in young thymus [10], and IL-1 increases in ageing [2]. With regard to α-2M, no similar data are available. However, α-2M is unaffected in lymphocytes of cancer patients after treatment with zinc *in vitro* [48]. Thus, the possible harmful effect of MT and α-2M on the immune function due to the continuous seques-tration of intracellular zinc might be avoided by physiological zinc supplementation in ageing animals and humans and thus to avoid age-related diseases.

## 8.    CONCLUSIONS AND PREDICTIONS FOR THE FUTURE

Zinc is an essential trace element for the maintenance of the entire immune system at thymic and extrathymic levels. T-cells partially compensate for thymic failure during ageing. The role of normal zinc levels are fundamental for the maintenance of immunocompetence for the entire life of an organism. Zinc is also required for many other biological functions, that include many enzymes and proteins, in particular for MT and α-2M, which protect the cells against oxida-tive damage and thereby prevent cell destruction or cell death. MT exerts its role by binding preferentially zinc in thiolate groups that form clusters, whereas α-2M binds MMPs and thus forms α-2M-proteinase complexes. This is internalized by cells through α-2M/LRP receptors and eliminated in endosomes. These proteins protect against a transient exposition to reactive oxygen species. In order to exert this task, MT and α-2M takes up intracellular zinc. A stress-like condition is prevalent in that increases the risk of age-related diseases. During ageing MT and α-2M abnormally increase and sequester constantly zinc ions. This leads to low zinc ions bioavailability and immune deficiency [thymic and extrathymic] during ageing. Therefore, MT and α-2M is protection in young-adult animals/humans but it becomes harmful during ageing. This phenomenon is in agreement with the evolutionary theories of ageing, which suggest that a trade off may occur between the early beneficial effects and late negative outcomes of vital func-tions due to genetic and molecular mechanisms [99]. The inability of MT in zinc release during ageing due to an unbalance between iNOS and cNOS strengthens this purpose. Physiological zinc supplementation in ageing corrects the defect, improves immune functions and increases the rate of survival in old mice. The benefits of supplementing zinc can be also extended to age-related diseases [cancer and infections] because exogenous zinc does not affect the already high MT levels because of their complete saturation by pre-existing zinc ions.

As high MT levels are present in old atrophic thymus and high MT and α-2M gene expres-sions are found in aged liver, future and intriguing directions will be in the study of their role in age-related thymus involution as well as in extrathymic T-cell pathway being, this latter, promi-nent in ageing and age-related diseases. It has been recently shown that DNA polymorphisms in the α-2M genes are associated with Alzheimer dementia [100]. DNA polymorphisms in the MT gene are actually under study in our lab. As final results, genetic markers of immunosenescence may thus be given up-stream of the immune efficiency because, at this regard, a bulk of data in markers of immunosenescence exclusively exists at down-stream, resulting as such misleading. Studies are in progress in our lab.

## ACKNOWLEDGMENTS

This paper was supported by INRCA and Italian Health Ministry (RF 216/02 to E.M).

## REFERENCES

1. Wellinghausen N, Kirchner H, Rink L. The immunobiology of zinc. Immunol Today 1997;18:519–521.
2. Mocchegiani E, Muzzioli M, Cipriano C, Giacconi R. Zinc, T-cell pathways, ageing: role of metallothioneins. Mech Ageing Dev 1998;106:183–204.
3. Dardenne M, Pleau JM, Nabarra B, Lefrancier P, Derrien M, Choay J, Bach JF. Contribution of zinc and other metals to the biological activity of the serum thymic factor. Proc Natl Acad Sci U S A 1982;79:5370–5373.
4. Hadden JW. Thymic endocrinology. Ann N Y Acad Sci 1998;840:352–8.
5. Mocchegiani E, Verbanac D, Santarelli L, Tibaldi A, Muzzioli M, Radosevic-Stasic B, Milin C. Zinc and metallothioneins on cellular immune effectiveness during liver regeneration in young and old mice. Life Sci 1997;61:1125–1145.
6. Abo T, Kawamura T, Watanabe H. Physiological responses of extrathymic T cells in the liver. Immunol Rev 2000;174:135–149.
7. Kagi JH, Schaffer A. Biochemistry of metallothionein. Biochemistry 1988 Nov 15;27(23): 8509–15.
8. Borth W. Alpha 2-macroglobulin, a multifunctional binding protein with targeting characteristics. FASEB J 1992;6:3345–3353.
9. Kelly EJ, Quaife CJ, Froelick GJ, Palmiter RD. Metallothionein I and II protect against zinc deficiency and zinc toxicity in mice. J Nutr 1996;126:1782–1790.
10. Coto JA, Hadden EM, Sauro M, Zorn N, Hadden JW. Interleukin 1 regulates secretion of zinc-thymulin by human thymic epithelial cells and its action on T-lymphocyte proliferation and nuclear protein kinase C. Proc Natl Acad Sci U S A 1992 ;89:7752–7756.
11. Mocchegiani E, Giacconi R, Cipriano C, Muzzioli M, Gasparini N, Moresi R, Stecconi R, Suzuki H, Cavalieri E, Mariani E. MtmRNA gene expression, via IL-6 and glucocorticoids, as potential genetic marker of immunosenescence: lessons from very old mice and humans. Exp Gerontol 2002;37:349–357.
12. Rink L, Seyfarth M. Characteristics of immunologic test values in the elderly. Z Gerontol Geriatr 1997;30:220–225.
13. Mocchegiani E, Muzzioli M, Giacconi R. Zinc and immunoresistance to infection in aging: new biological tools. Trends Pharmacol Sci 2000;21:205–208.
14. Mocchegiani E, Giacconi R, Cipriano C, Gasparini N, Orlando F, Stecconi R, Muzzioli M, Isani G, Carpene. Metallothioneins (I+II) and thyroid-thymus axis efficiency in old mice: role of corticosterone and zinc supply. Mech Ageing Dev 2002;123:675–694.
15. Shankar AH, Prasad AS. Zinc and immune function: the biological basis of altered resistance to infection. Am J Clin Nutr 1998;68:447S–463S.
16. Mills CF (ed). Zinc in human biology. London : Springer Verlag; 1989;1–600.
17. Vallee BL, Falchuk KH. The biochemical basis of zinc physiology. Physiol Rev 1993;73: 79–118.
18. Lowther WT, Matthews BW. Metalloaminopeptidases: common functional themes in disparate structural surroundings. Chem Rev 2002;102:4581–4608.

19. Coleman JE. Zinc proteins: enzymes, storage proteins, transcription factors, and replication proteins. Annu Rev Biochem 1992;61:897–946.

20. Taylor GA, Blackshear PJ. Zinc inhibits turnover of labile mRNAs in intact cells. J Cell Physiol 1995;162:378–387.

21. Hopper NL (ed). Zinc and metalloproteases in health and disease. Taylor and Francis, London, 1996, pp. 1–750.

22. Kolkenbrock H, Orgel D, Hecker-Kia A, Noack W, Ulbrich N. The complex between a tissue inhibitor of metalloproteinases (TIMP-2) and 72-kDa progelatinase is a metalloproteinase inhibitor. Eur J Biochem 1991;198:775–781.

23. Sottrup-Jensen L. Alpha-macroglobulins: structure, shape, and mechanism of proteinase complex formation. J Biol Chem 1989;264:11539–11542.

24. Miyazaki K, Funahashi K, Numata Y, Koshikawa N, Akaogi K, Kikkawa Y, Yasumitsu H, Umeda M. Purification and characterization of a two-chain form of tissue inhibitor of metalloproteinases (TIMP) type 2 and a low molecular weight TIMP-like protein. J Biol Chem 1993 ;268:14387–14393.

25. Nagase H, Woessner JF Jr. Matrix metalloproteinases. J Biol Chem 1999 Jul 30;274(31): 21491–4.

26. Stetler-Stevenson WG, Liotta LA, Kleiner DE Jr. Extracellular matrix 6: role of matrix metalloproteinases in tumor invasion and metastasis. FASEB J 1993 Dec;7(15):1434–41.

27. Ashcroft GS, Herrick SE, Tarnuzzer RW, Horan MA, Schultz GS, Ferguson MW. Human ageing impairs injury-induced in vivo expression of tissue inhibitor of matrix metalloproteinases (TIMP)-1 and -2 proteins and mRNA. J Pathol 1997;183:169–176.

28. Ebadi M, Swanson S. The status of zinc, copper, and metallothionein in cancer patients. Prog Clin Biol Res 1988;259:161–175.

29. Matoska J, Wahlstrom T, Vaheri A, Bizik J, Grofova M. Tumor-associated alpha-2-macroglobulin in human melanomas. Int J Cancer 1988;41:359–363.

30. Maret W, Vallee BL. Thiolate ligands in metallothionein confer redox activity on zinc clusters. Proc Natl Acad Sci U S A 1998;95:3478–3482.

31. West AK, Stallings R, Hildebrand CE, Chiu R, Karin M, Richards RI. Human metallothionein genes: structure of the functional locus at 16q13. Genomics 1990;8: 513–518.

32. Otvos JD, Engeseth HR, Nettesheim DG, Hilt CR. Interprotein metal exchange reactions of metallothionein. Experientia Suppl 1987;52:171–178.

33. Lazo JS, Kuo SM, Woo ES, Pitt BR. The protein thiol metallothionein as an antioxidant and protectant against antineoplastic drugs. Chem Biol Interact 1998;111–112:255–262.

34. Palmiter RD. Regulation of metallothionein genes by heavy metals appears to be mediated by a zinc-sensitive inhibitor that interacts with a constitutively active transcription factor, MTF-1. Proc Natl Acad Sci U S A 1994;91:1219–1223.

35. Cai L, Satoh M, Tohyama C, Cherian MG. Metallothionein in radiation exposure: its induction and protective role. Toxicology 1999;132:85–98.

36. Ashok BT, Ali R. The aging paradox: free radical theory of aging. Exp Gerontol 1999;34: 293–303.

37. Mocchegiani E, Giacconi R, Cipriano C, Muzzioli M, Fattoretti P, Bertoni-Freddari C, Isani G, Zambenedetti P, Zatta P. Zinc-bound metallothioneins as potential biological markers of ageing. Brain Res Bull 2001;55:147–153.

38. Mocchegiani E, Muzzioli M, Giacconi R, Cipriano C, Gasparini N, Franceschi C, Gaetti R, Cavalieri E, Suzuki H. Metallothioneins/PARP-1/IL-6 interplay on natural killer cell

activity in elderly: parallelism with nonagenarians and old infected humans. Effect of zinc supply. Mech Ageing Dev 2003;124:459–468.

39. Zambenedetti P, Giordano R, Zatta P. Metallothioneins are highly expressed in astrocytes and microcapillaries in Alzheimer's disease. J Chem Neuroanat 1998;15:21–26.

40. Abdel-Mageed A, Agrawal KC. Antisense down-regulation of metallothionein induces growth arrest and apoptosis in human breast carcinoma cells. Cancer Gene Ther 1997;4: 199–207.

41. Sandstead HH. Requirements and toxicity of essential trace elements, illustrated by zinc and copper. Am J Clin Nutr 1995;61(3 Suppl):621S–624S.

42. Starkey PM, Barrett AJ. Evolution of alpha 2-macroglobulin. The demonstration in a variety of vertebrate species of a protein resembling human alpha 2-macroglobulin. Biochem J 1982;205:91–95.

43. Chu CT, Howard GC, Misra UK, Pizzo SV. Alpha 2-macroglobulin: a sensor for proteolysis. Ann N Y Acad Sci 1994;737:291–307.

44. Strickland DK, Ashcom JD, Williams S, Burgess WH, Migliorini M, Argraves WS. Sequence identity between the alpha 2-macroglobulin receptor and low density lipoprotein receptor-related protein suggests that this molecule is a multifunctional receptor. J Biol Chem 1990;265:17401–17404.

45. Williams SE, Ashcom JD, Argraves WS, Strickland DK. A novel mechanism for controlling the activity of alpha 2-macroglobulin receptor/low density lipoprotein receptor-related protein. Multiple regulatory sites for 39-kDa receptor-associated protein. J Biol Chem 1992;267:9035–9040.

46. Gliemann J, Nykjaer A, Petersen CM, Jorgensen KE, Nielsen M, Andreasen PA, Christensen EI, Lookene A, Olivecrona G, Moestrup SK. The multiligand alpha 2-macroglobulin receptor/low density lipoprotein receptor-related protein (alpha 2MR/LRP). Binding and endocytosis of fluid phase and membrane-associated ligands. Ann N Y Acad Sci 1994;737:20–38.

47. Gottschall PE, Deb S. Regulation of matrix metalloproteinase expressions in astrocytes, microglia and neurons. Neuroimmunomodulation 1996;3:69–75.

48. Mocchegiani E, Ciavattini A, Santarelli L, Tibaldi A, Muzzioli M, Bonazzi P, Giacconi R, Fabris N, Garzetti GG. Role of zinc and alpha2 macroglobulin on thymic endocrine activity and on peripheral immune efficiency (natural killer activity and interleukin 2) in cervical carcinoma. Br J Cancer 1999;79:244–250.

49. Krauskopf A, Blackburn EH. Control of telomere growth by interactions of RAP1 with the most distal telomeric repeats. Nature 1996;383:354–357.

50. Shore D. Telomeric chromatin: replicating and wrapping up chromosome ends. Curr Opin Genet Dev 2001;11:189–198.

51. Kveiborg M, Kassem M, Langdahl B, Eriksen EF, Clark BF, Rattan SI. Telomere shortening during aging of human osteoblasts in vitro and leukocytes in vivo: lack of excessive telomere loss in osteoporotic patients. Mech Ageing Dev 1999;106:261–271.

52. Marcand S, Gilson E, Shore D. A protein-counting mechanism for telomere length regulation in yeast. Science 1997;275:986–990.

53. Van Leuven F, Thiry E, Stas L, Nelissen B. Analysis of the human LRPAP1 gene coding for the lipoprotein receptor-associated protein: identification of 22 polymorphisms and one mutation. Genomics 1998;52:145–151.

54. Mocchegiani E, Muzzioli M, Gaetti R, Veccia S, Viticchi C, Scalise G. Contribution of zinc to reduce CD4+ risk factor for 'severe' infection relapse in aging: parallelism with

HIV. Int J Immunopharmacol 1999;21:271–281.

55. Tissot C, Mechti N. Molecular cloning of a new interferon-induced factor that represses human immunodeficiency virus type 1 long terminal repeat expression. J Biol Chem 1995;270:14891–14898.

56. Papageorgiou AC, Collins CM, Gutman DM, Kline JB, O'Brien SM, Tranter HS, Acharya KR. Structural basis for the recognition of superantigen streptococcal pyrogenic exotoxin A (SpeA1) by MHC class II molecules and T-cell receptors. EMBO J 1999;18:9–21.

57. Turnlund JR, Durkin N, Costa F, Margen S. Stable isotope studies of zinc absorption and retention in young and elderly men. J Nutr 1986;116:1239–1247.

58. Mocchegiani E, Santarelli L, Muzzioli M, Fabris N. Reversibility of the thymic involution and of age-related peripheral immune dysfunctions by zinc supplementation in old mice. Int J Immunopharmacol 1995;17:703–718.

59. Cakman I, Rohwer J, Schutz RM, Kirchner H, Rink L. Dysregulation between TH1 and TH2 T cell subpopulations in the elderly. Mech Ageing Dev 1996;87:197–209.

60. Franceschi C, Bonafe M, Valensin S. Human immunosenescence: the prevailing of innate immunity, the failing of clonotypic immunity, and the filling of immunological space. Vaccine 2000 ;18:1717–1720.

61. Franceschi C, Bonafe M. Centenarians as a model for healthy aging. Biochem Soc Trans 2003;31:457–461.

62. Muzzioli M, Mocchegiani E, Bressani N, Bevilacqua P, Fabris N. In vitro restoration by thymulin of NK activity of cells from old mice. Int J Immunopharmacol 1992;14:57–61.

63. Trinchieri G. Natural killer cells wear different hats: effector cells of innate resistance and regulatory cells of adaptive immunity and of hematopoiesis. Semin Immunol 1995;7: 83–88.

64. Tanaka Y, Shiozawa S, Morimoto I, Fujita T. Role of zinc in interleukin 2 (IL-2)-mediated T-cell activation. Scand J Immunol 1990;31:547–552.

65. Saha AR, Hadden EM, Hadden JW. Zinc induces thymulin secretion from human thymic epithelial cells in vitro and augments splenocyte and thymocyte responses in vivo. Int J Immunopharmacol 1995;17:729–733.

66. Ohteki T, Okuyama R, Seki S, Abo T, Sugiura K, Kusumi A, Ohmori T, Watanabe H, Kumagai K. Age-dependent increase of extrathymic T cells in the liver and their appearance in the periphery of older mice. J Immunol 1992 Sep 1;149:1562–1570.

67. Cui L, Takagi Y, Wasa M, Iiboshi Y, Inoue M, Khan J, Sando K, Nezu R, Okada A. Zinc deficiency enhances interleukin-1alpha-induced metallothionein-1 expression in rats. J Nutr 1998; 128:1092–1098.

68. Bui LM, Dressendorfer RH, Keen CL, Summary JJ, Dubick MA. Zinc status and interleukin-1 beta-induced alterations in mineral metabolism in rats. Proc Soc Exp Biol Med 1994;206:438–444.

69. Yu CW, Chen JH, Lin LY. Metal-induced metallothionein gene expression can be inactivated by protein kinase C inhibitor. FEBS Lett 1997;420:69–73.

70. Penkowa M, Giralt M, Moos T, Thomsen PS, Hernandez J, Hidalgo J. Impaired inflammatory response to glial cell death in genetically metallothionein-I- and -II-deficient mice. Exp Neurol 1999;156:149–164.

71. Deng DX, Cai L, Chakrabarti S, Cherian MG. Increased radiation-induced apoptosis in mouse thymus in the absence of metallothionein. Toxicology 1999;134:39–49.

72. Johnston CJ, Finkelstein JN, Gelein R, Oberdorster G. Pulmonary cytokine and chemokine mRNA levels after inhalation of lipopolysaccharide in C57BL/6 mice. Toxicol Sci

1998;46:300–307.

73. Miller GG, Strittmatter WJ. Identification of human T cells that require zinc for growth. Scand J Immunol 1992;36:269–277.

74. Zangger K, Oz G, Haslinger E, Kunert O, Armitage IM. Nitric oxide selectively releases metals from the amino-terminal domain of metallothioneins: potential role at inflammatory sites. FASEB J 2001;15:1303–1305.

75. Kroncke KD. Zinc finger proteins as molecular targets for nitric oxide-mediated gene regulation. Antioxid Redox Signal 2001;3:565–575.

76. Lysiak JJ, Hussaini IM, Gonias SL. alpha 2-Macroglobulin synthesis by the human monocytic cell line THP-1 is differentiation state-dependent. J Cell Biochem 1997;67: 492–497.

77. Heumann D, Vischer TL. Immunomodulation by alpha 2-macroglobulin and alpha 2-macroglobulin-proteinase complexes: the effect on the human T lymphocyte response. Eur J Immunol 1988;18:755–760.

78. Hoffman MR, Pizzo SV, Weinberg JB. Modulation of mouse peritoneal macrophage Ia and human peritoneal macrophage HLA-DR expression by alpha 2-macroglobulin "fast" forms. J Immunol 1987;139:1885–1890.

79. Borth W. Alpha 2-macroglobulin. A multifunctional binding and targeting protein with possible roles in immunity and autoimmunity. Ann N Y Acad Sci 1994;737:267–272.

80. Kahari VM, Saarialho-Kere U. Matrix metalloproteinases and their inhibitors in tumour growth and invasion. Ann Med 1999;31:34–45.

81. James K. Interactions between cytokines and alpha 2-macroglobulin. Immunol Today 1990;11:163–166.

82. Gonias SL, LaMarre J, Crookston KP, Webb DJ, Wolf BB, Lopes MB, Moses HL, Hayes MA. Alpha 2-macroglobulin and the alpha 2-macroglobulin receptor/LRP. A growth regulatory axis. Ann N Y Acad Sci 1994;737:273–290.

83. Dinarello CA. Reduction of inflammation by decreasing production of interleukin-1 or by specific receptor antagonism. Int J Tissue React 1992;14:65–75.

84. Horn F, Wegenka UM, Lutticken C, Yuan J, Roeb E, Boers W, Buschmann J, Heinrich PC. Regulation of alpha 2-macroglobulin gene expression by interleukin-6. Ann N Y Acad Sci 1994;737:308–323.

85. Gettins PG, Crews BC. Binding of epidermal growth factor to human alpha 2-macroglobulin. Significance for cytokine alpha 2-macroglobulin interactions. Ann N Y Acad Sci 1994;737:383–398.

86. Diehl S, Rincon M. The two faces of IL-6 on Th1/Th2 differentiation. Mol Immunol 2002;39:531–536.

87. Frasca D, Pucci S, Goso C, Barattini P, Barile S, Pioli C, Doria G. Regulation of cytokine production in aging: use of recombinant cytokines to upregulate mitogen-stimulated spleen cells. Mech Ageing Dev 1997;93:157–169.

88. Iwata T, Incefy GS, Tanaka T, Fernandes G, Menendez-Botet CJ, Pih K, Good RA. Circulating thymic hormone levels in zinc deficiency. Cell Immunol 1979;47:100–105.

89. Dardenne M, Boukaiba N, Gagnerault MC, Homo-Delarche F, Chappuis P, Lemonnier D, Savino W. Restoration of the thymus in aging mice by in vivo zinc supplementation. Clin Immunol Immunopathol 1993;66:127–135.

90. Provinciali M, Di Stefano G, Stronati S. Flow cytometric analysis of CD3/TCR complex, zinc, and glucocorticoid-mediated regulation of apoptosis and cell cycle distribution in thymocytes from old mice. Cytometry 1998;32:1–8.

91. Barbieri D, Troiano L, Grassilli E, Agnesini C, Cristofalo EA, Monti D, Capri M, Cossarizza A, Franceschi C. Inhibition of apoptosis by zinc: a reappraisal. Biochem Biophys Res Commun 1992;187:1256–1261.

92. Liang JY, Liu YY, Zou J, Franklin RB, Costello LC, Feng P. Inhibitory effect of zinc on human prostatic carcinoma cell growth. Prostate 1999;40:200–207.

93. Xu J, Xu Y, Nguyen Q, Novikoff PM, Czaja MJ. Induction of hepatoma cell apoptosis by c-myc requires zinc and occurs in the absence of DNA fragmentation. Am J Physiol 1996;270:G60–70.

94. Prasad AS, Fitzgerald JT, Hess JW, Kaplan J, Pelen F, Dardenne M. Zinc deficiency in elderly patients. Nutrition 1993;9:218–224.

95. Boukaiba N, Flament C, Acher S, Chappuis P, Piau A, Fusselier M, Dardenne M, Lemonnier D. A physiological amount of zinc supplementation: effects on nutritional, lipid, and thymic status in an elderly population. Am J Clin Nutr 1993;57:566–572.

96. Bogden JD, Oleske JM, Lavenhar MA, Munves EM, Kemp FW, Bruening KS, Holding KJ, Denny TN, Guarino MA, Holland BK. Effects of one year of supplementation with zinc and other micronutrients on cellular immunity in the elderly. J Am Coll Nutr 1990;9: 214–225.

97. Fabris N, Mocchegiani E, Albertini A. Psycoendocrine-immune interactions in Down's syndrome: role of zinc. In: Castell S and Wisnieski KE (eds.) Growth hormone treatment in Down's syndrome. New York: John Wiley and Sons, 1994; 203–218.

98. Chiricolo M, Musa AR, Monti D, Zannotti M, Franceschi C. Enhanced DNA repair in lymphocytes of Down syndrome patients: the influence of zinc nutritional supplementation. Mutat Res 1993;295:105–111.

99. Franceschi C, Valensin S, Fagnoni F, Barbi C, Bonafe M. Biomarkers of immunosenescence within an evolutionary perspective: the challenge of heterogeneity and the role of antigenic load. Exp Gerontol 1999;34:911–921.

100. Kovacs DM. alpha2-macroglobulin in late-onset Alzheimer's disease. Exp Gerontol 2000;35:473–479.

*The Neuroendocrine Immune Network in Ageing*
Edited by R.H. Straub and E. Mocchegiani

# Neutrophil Ageing and Immunosenescence

STEPHEN K BUTCHER, KEQING WANG, DAVID LASCELLES and JANET M LORD

*MRC Centre for Immune Regulation, Birmingham University Medical School, Birmingham B15 2TT, UK*

## ABSTRACT

In the last century life expectancy in the developed world has almost doubled, due in the main to a reduction in life-threatening infections as a result of improved hygiene and healthcare. However, in the ageing population of the 21st century infection is still a major cause of morbidity and mortality as normal human ageing is associated with a loss of immune function. In this chapter we focus upon the effect of ageing on innate immunity, more specifically on the neutrophil, which plays a key role in the defense against rapidly dividing bacteria, a major cause of infection in the elderly. The majority of data in the literature have identified a differential effect of ageing on neutrophil function, with a marked decline in neutrophil phagocytic function while superoxide generation and chemotaxis are unaffected. We also discuss the possible causes of this decline including age associated loss of receptors for antibody (CD16) and changes in cell signalling pathways such as protein kinase C. All of these studies of immunosenescence have been carried out on healthy elderly volunteers, to identify age-related rather than pathology related changes. It is now crucial to determine if this reduction in neutrophil function actually influences immunity. To this end we discuss the evidence for an effect of ageing on the body's response to trauma in the elderly. Mild trauma such as hip-fracture is well known to lead to increased infection rates in the elderly. We describe preliminary data showing an altered response to trauma in the elderly which leads to further suppression of neutrophil function. We suggest that this is mediated by an elevated cortisol to DHEAS ratio in the elderly.

## 1.    THE AGEING PROCESS

Ageing is usually defined as the progressive loss of function accompanied by decreasing fertility and increasing mortality with advancing age [1]. Such a trait, which impairs survival and fertility, is clearly bad for the individual. However, ageing by the above definition is not a universal phenomenon, as some species, for example the American lobster, shows no age associated increase in mortality or decline in fertility. Such animals are not immortal since they are still under constant threat of disease, predation and accidents. In the case of humans, the potential maximum life span has remained unchanged for the past millennium at about 125 years. What has changed is average life expectancy, which has risen from 49 years in 1900 to 76 years in 1997. This increase has been caused in the main by elimination of infectious diseases in youth, better hygiene and the discovery of antibiotics and vaccines.

Although many infectious diseases that were a severe cause of mortality in the last century have now been eradicated, infection is still a significant cause of increased morbidity and mortality in the elderly population. As stated in earlier chapters of this issue (see Pawelec and Solana) the ageing process includes a decline in immune function, immunosenescence, helping to explain one aspect of increasing frailty in the elderly. Understanding the molecular basis of this phenomenon should allow the development of therapeutic modalities to reduce mortality and improve quality of life in the ageing population of the 21st century. This chapter will focus upon the effect of ageing on a central element of innate immunity, the neutrophil, which plays a key role in the defense against rapidly dividing bacteria, yeast and fungi. As bacteria are a major cause of infection in the elderly it is clear that ageing has a significant impact upon neutrophil function and some of the possible causes will be discussed.

A widely accepted theory of how we age is the disposable soma theory, based on the physiological theory of ageing, which predicts that the mechanisms responsible for ageing are those types of damage for which maintenance and repair mechanisms are metabolically costly [2]. Several causative molecular mechanisms for such damage have been proposed, these include the accumulation of aberrant proteins, defective mitochondria, free radical damage and replicative senescence. Few of these mechanisms are likely to apply to neutrophils, with the exception of free radical damage. Whatever the mechanism of ageing, age-related changes in the function of individual tissues will have a differential impact on the organism as a whole, dictating whether or not the organism ages well. As stated above, one system which undergoes age-related changes that can have a significant impact upon life expectancy is the immune system. The latter is not surprising bearing in mind the sentinel role of the immune system in combating life threatening infections and the detection and killing of tumour cells. Indeed the rate of biological ageing in humans and many other species can be assessed by the rate of immune functional decline, with well preserved immune functions being associated with relatively successful ageing [3]. Moreover, as has been discussed in this issue, specific components of the immune system are affected by oxidative status and telomere shortening, indicating that the immune system is responsive to two of the key mechanisms of ageing.

## 2.    AGEING AND INFECTIOUS DISEASE

The increased incidence of infectious diseases in the elderly is well documented and has a significant impact upon morbidity and mortality in the over 65 years age group. In a recent study of 108 healthy elderly people, 48 (44%) developed infections during the 12 month study period and 11 of those patients died of their infections [4]. Thus not only are the elderly more susceptible to infection but they are also less able to resolve infections once they take hold. Increased susceptibility to infection in the elderly has been reported for both viral and bacterial species, suggesting that both innate and adaptive immune responses are compromised in ageing. For example, age-related changes in incidence have been observed with respect to the influenza virus, with an estimated 75% of the 10,000–40,000 deaths attributed to influenza annually in the United States occurring in persons over 65 years of age [5]. This also correlated with a decline in response to influenza virus vaccine, a 50% decrease in antibody response was observed in the over 65 years age group [6]. However pneumonia is the leading cause of mortality from infectious disease in the United States and bacteria, namely *Streptococcus pneumoniae* prevail as the most common infecting pathogens leading to this condition.

Numerous studies have examined the relationship between the incidence of pneumonia and

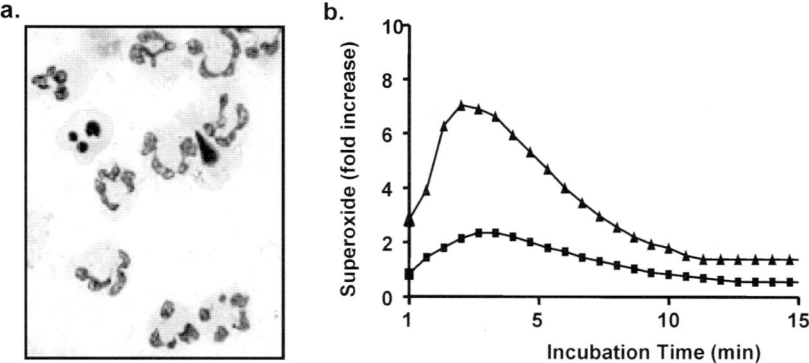

Figure 1. Neutrophil function declines as neutrophils age.

Neutrophils are short-lived cells that survive in the circulation for approximately 36 hours before dying by apoptosis. They are polymorphonuclear leucocytes with a distinctive tri-lobed nucleus (a). When freshly isolated they respond to the bacterial peptide fMLP with the generation of superoxide (b), which is significantly reduced when they are aged in culture for 20h.

age. In 1981 in the US it was estimated that there was a six-fold increase in the incidence of pneumonia in the over 65 years age group compared with younger members of society [7]. Also a study in the US demonstrated a 10-fold increase in morbidity associated with pneumonia in the elderly (over 65 years) population [8]. It can thus be seen that bacterial infections make a significant contribution to morbidity and mortality from infectious diseases in the elderly in the Western World. Furthermore, a given infectious disease may be associated with a more diverse range of pathogens among older patients than young patients. For example, in younger adults urinary tract infection is primarily an infection of young women and *E.coli* and *Staphylococcus saprophyticus* are the commonest uropathogens. However, with ageing, the incidence of urinary tract infection increases in both men and women, with uropathogens including not only *E.coli* but also other gram-negative bacteria such as *Proteus* and *Enterobacter* species [9]. These observations are suggestive of a general failing of the innate immune response to bacterial pathogens, which will include neutrophils, but also other innate mechanisms regulating their function including monocyte derived cytokines and the complement system. In order to understand the potential impact of age on neutrophil function it is important to consider the basic biology of neutrophils, which is significantly different from lymphocytes and elements of the adaptive immune system already described elsewhere in this issue (see Pawelec and Solana).

## 3. NEUTROPHILS AND NEUTROPHIL AGEING

Neutrophils are the most abundant leukocyte in peripheral blood and are produced in vast numbers in the bone marrow, with $1-2 \times 10^{11}$ cells released per day. They are polymorphonuclear with a characteristic tri-lobed, horseshoe shaped nucleus (Figure 1) and a cytoplasm densely packed with granules containing lytic enzymes. They are fully differentiated cells with no replicative capacity, as they lose a significant portion of their DNA during the final stages of maturation in the bone marrow. They also have relatively few mitochondria, which are thought to be vestigial with regard to ATP production and may instead be required only for apoptosis. Neutrophils are

44

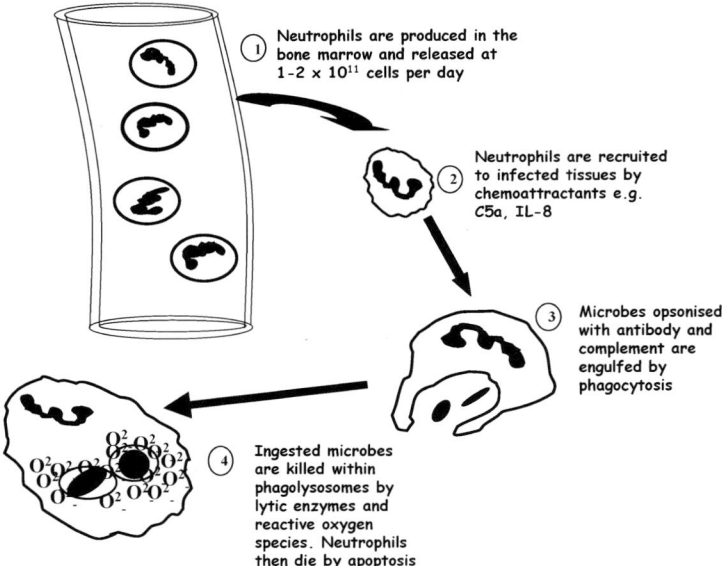

Figure 2. The key stages of neutrophil activation and the effect of age.

Neutrophils are released from the bone marrow in vast numbers each day and circulate until they are recruited to a site of infection. Neutrophil numbers are not reduced with age (1). During infection chemoattractants are produced which induce extravasation of the neutrophil from the circulation and chemotaxis towards the site of infection. There is no effect of each on adhesion molecules involved in extravasation, or in chemotaxis itself (2). Opsonised microbes are recognized and engulfed by phagocytosis. This step is reduced in the elderly and may in part be due to decreased expression of the opsonin receptor CD16 (3). Phagocytosis of microbes triggers bactericidal mechanisms, including supeoxide generation and this is reduced in the elderly (4).

very short-lived cells and will survive for approximately 36 hours in the circulation before dying by apoptosis, allowing them to be recognized and engulfed by macrophages [10]. However, if they are recruited to a site of infection their life-span is extended by pro-inflammatory cytokines [11], complement and bacterial components such as LPS [12] and this is important for maximizing their bactericidal function.

In order to kill invading bacteria neutrophils extravasate from the blood in response to chemotactic signals, including cytokines (TNF-$\alpha$, IL-8), complement (C5a) and bacterial products (fMLP). Neutrophils recognise bacteria by two distinct mechanisms: firstly via bacterial cell surface components recognized by receptors such as Toll-like receptors (TLR), for example LPS is recognized by TLR4; secondly via opsonin receptors that recognize complement C3b/C3bi or the Fc regions of antibody (IgG) coating the bacterial target. Ingestion of bacteria by phagocytosis then triggers internal bactericidal mechanisms, including release of granule bactericidal enzymes and generation of toxic reactive species, including superoxide and hydrogen peroxide (Figure 2). Activation of these mechanisms also induces apoptosis of the neutrophil allowing their removal by macrophages. As neutrophils age in the circulation they become functionally senescent, therefore an important consequence of infection is increased production and release of neutrophils from the bone marrow. The latter is stimulated by pro-inflammatory cytokines, notably GM-CSF, which is produced in the early stages of infection.

Much is now known of the mechanisms underlying functional senescence in ageing neutrophils. The ageing neutrophil shows altered expression of a variety of cell surface molecules, including those involved in adhesion and extravasation and in triggering the effector functions of neutrophils [13]. Molecules involved in chemotaxis, such as the receptor for the chemoattractant C5a are reduced in aged neutrophils [14] and this combined with a reduction in adhesion molecules such as CD31 and ICAM-3 [14], leads to an impaired ability to extravasate from the circulation and in to the infected tissue in aged neutrophils. Even if aged neutrophils do reach the site of infection they have a significantly reduced expression of phagocytic receptors including the receptors for antibody CD16 (FcγRIII) and CD32 (FcγRII) and the CR1 complement receptor CD35 [14]. These changes significantly reduce the ability of the neutrophil to phagocytose opsonised pathogens. In addition their ability to activate intracellular bactericidal mechanisms such as superoxide generation is minimal (Figure 1). It is therefore important that effete neutrophils die by apoptosis and are removed efficiently by macrophages, allowing replenishment from the marrow with young, functionally effective cells. The events which trigger apoptosis of effete neutrophils remain poorly understood. Neutrophil apoptosis itself is caspase dependent [15] and involves loss of mitochondrial membrane potential. The current literature would suggest that a loss of anti-apoptotic Bcl-2 family proteins, notably Mcl-1, with maintenance of pro-apoptotic proteins such as Bax, is the key driver for perturbation of mitochondrial stability and apoptosis [16]. As neutrophils possess relatively few mitochondria, this is still an area of hot debate.

## 4. AGEING AND NEUTROPHIL FUNCTION

From the preceding description of neutrophil biology and function it should be clear that defects in any of the aspects of neutrophil production and activation would seriously undermine the immune response to bacteria. We will now review the current understanding of the effect of age on neutrophil function and will also speculate on some of the potential causes of loss of neutrophil function with age. The latter will include consideration of the potential role of the neuroendocrine immune axis.

### 4.1. Neutrophil production

Immunosenescence is defined as the state of dysregulated immune function that contributes to the increased susceptibility of the elderly to infection and possibly to autoimmune disease and cancer. Despite the fact that neutrophil function does decline with age and will be a significant factor in immunosenescence and the well documented increase in bacterial infections in the elderly [9], there is relatively little known of the molecular basis of this loss of function [17]. As stated above, neutrophils are short-lived, fully differentiated leukocytes with no proliferative capacity that gradually senesce before dying by apoptosis. The rate of release of neutrophils from the bone marrow is therefore a major influence on circulating cell numbers and overall functional efficiency. Vulnerability to infection in the elderly could result from an age related decline either in neutrophil supply and/or function. Several studies have shown that neutrophil numbers in the blood and neutrophil precursors in the marrow [18] are not lowered in healthy elderly subjects, although the proliferative response of neutrophil precursor cells to G-CSF was reduced[18]. As responses to GM-CSF and IL-3 were unaltered with age, the reduced response to G-CSF is unlikely to affect the ability of the elderly to maintain normal neutrophil numbers. During acute infection we have also shown that the elderly can produce a normal neutrophilia

[19], providing further evidence that basic granulopoiesis is maintained in the elderly. However, during periods of chronic infection, neutropenia can arise in the elderly [20], and this could in part be caused by the reduced responsiveness to G-CSF. Therefore, reduced neutrophil supply is unlikely to be a major component of innate immunosenescence and is only likely to influence neutrophil responses during established chronic infection.

4.2.    Neutrophil extravasation and chemotaxis

*In vitro* studies of chemotaxis have found no consistent evidence of compromised migratory responses of neutrophils from healthy elderly subjects, with the majority of studies showing that chemotaxis was either unaltered [7] or only slightly reduced [21]. Measurements of adhesion molecule expression on neutrophils also suggest the lack of any negative effect of ageing on neutrophil recruitment processes. The level of CD15, which binds to E-selectin on vascular endothelium and mediates neutrophil rolling, was slightly increased with donor age [22], CD11a was found to be unaltered [23] and CD11b has been reported as either unaltered or slightly increased in the elderly [22,23]. Taken together, these data suggest that reduced extravasation and recruitment of neutrophils are not major factors contributing to increased risk of infection in the elderly.

4.3.    Neutrophil phagocytosis of pathogens

The data concerning phagocytic ability of neutrophils from the elderly are reasonably consistent. Studies measuring phagocytosis of opsonized bacteria or yeast and opsonized zymosan as the neutrophil target have all shown a significant reduction in phagocytic ability in the elderly [19,22,24]. Reduced phagocytic ability appears to comprise both reduced number of cells with phagocytic capacity and reduced uptake of targets on a per cell basis. The reduced response of neutrophils to *Staphylococcus aureus* [24] is of particular clinical importance because of the increased susceptibility to this pathogen in elderly subjects. The few studies that have not found decreased phagocytosis have in the main used targets either not opsonized with complement and antibody [25] or coated only with antibody [26]. Thus adequate recruitment to sites of infection must be off set by reduced phagocytic capacity and thus play a part in the poor resolution of infections in the elderly.

The molecular basis of reduced phagocytosis in neutrophils from elderly subjects is not fully understood. However, Emanuelli *et al* showed that neutrophil phagocytosis of unopsonized bacterial targets occurred at the same low level in young and old subjects [25]. Their data suggest that the receptors for innate recognition of bacterial components (CD14 and TLRs) are not affected by ageing and moreover, confirm that they do not contribute significantly to phagocytosis in the absence of complement and antibody. However, it is possible that some of these receptors are affected by ageing but that there is compensation by other receptors of similar function. In particular a recent study of TLR expression in macrophages in mice has shown a significant down regulation of TLRs and TLR function with age [27]. Toll-like receptors (TLR) are pattern recognition receptors that recognize conserved molecular patterns on microbes, for example TLR 1, 2 and 6 recognise components of the cell wall of gram-positive bacteria (peptidoglycan) and yeast and TLR 4 recognise the gram-negative component LPS. Macrophages from old mice were seen to have reduced expression of all nine TLRs and demonstrated reduced secretion of IL-6 and TNF-α after TLR ligand stimulation [27]. As neutrophils also use TLR for pathogen recognition it is possible that their expression is also affected by age, though this remains to be established.

The reduced response of neutrophils to opsonized targets [19,24,25] suggests that the relevant receptors may be present at reduced levels or have compromised signaling function, particularly as serum IgG and complement levels are within the normal range or slightly increased in the elderly [7,28]. Levels of complement receptors CD11b and CD11c, as stated above, are essentially unaltered in the elderly and there are no published data concerning expression of CR1/CD35. However, phagocytosis of yeast by neutrophils is reduced in the elderly [19,25] and as this process is mediated entirely via complement receptors, we would predict that complement receptor function would be affected by age. Thus signaling via complement receptors may be compromised in neutrophils from elderly subjects, though no data exist at the moment to support this proposal. There is also a paucity of data regarding Fcγ receptor expression or related signaling pathways in the elderly. Our recent work has examined the expression of Fcγ receptors on the neutrophils of healthy young and elderly donors [19]. This study identified a marked decrease in expression of the constitutive Fcγ receptor CD16 and this loss of cell surface expression correlated with reduced phagocytosis of *E.coli*.

4.4.    Neutrophil intracellular bactericidal processes

Neutrophil microbicidal activity has been examined by several groups and although data have often been conflicting, a majority support a decline in cytotoxicity towards bacteria and yeast with age [21,26]. What is not clear from these basic assays is whether the observed reduction in microbicidal function is additional to the decline in phagocytic capacity, or occurs secondary to this effect. Analysis of individual microbicidal processes addresses this question. Several authors have reported normal superoxide production by neutrophils of elderly subjects in response to fMLP [21,22], whilst others have shown reduced responses in mice and humans [29,30]. In contrast to the data regarding fMLP, studies concerning superoxide generation in response to particulate stimuli and Fc receptor ligation consistently report a reduced response in the elderly. In a publication by Wenisch *et al*, superoxide generation was decreased in response to *S. aureus*, but not to *Escherichia coli* [24], an observation with particular clinical relevance bearing in mind the reduced ability of the elderly to resolve infection to Gram-positive bacteria. It is possible that neutrophil superoxide responses to *E.coli* involve binding of LPS to CD14 [31] and TLR4 and may be unaffected by age, whilst responses to Gram-positive bacteria such as *S. aureus*, which are dependent upon complement and Fc receptor mediated superoxide response, are significantly reduced in the elderly [26]. Such data again identify Fc mediated responses as a significant factor in age related neutrophil functional decline.

4.4.    Neutrophil priming

Neutrophil responses to microbes and soluble microbial products (fMLP) at the site of infection are optimized by the agents that prime the neutrophil in the circulation and during extravasation, e.g. TNF-α and GM-CSF both increase superoxide generation in response to fMLP. Seres and colleagues have shown that neutrophils from elderly volunteers were not primed as efficiently by GM-CSF as those from young subjects [31]. The importance of these data extend beyond the priming of the neutrophil, as priming by cytokines such as GM-CSF is also able to extend the short life of the neutrophil at sites of infection by inhibiting apoptosis [11], improving its bactericidal efficiency. The ability of GM-CSF, G-CSF and LPS to delay neutrophil apoptosis was found to be significantly reduced in the elderly [32,33]. From these data it can be predicted that, compared with their younger counterparts, neutrophils from elderly donors will respond less well to infectious stimuli upon recruitment to the site if infection and will also die more rapidly, blunting the impact of neutrophilia induced in response to infection.

4.6.    Mechanisms underlying reduced neutrophil phagocytosis

The underlying cause of reduced neutrophil CD16 expression has not been established. It is possible that CD16, which is a GPI-linked cell surface protein, is shed in to the circulation or that neutrophils are released from the bone marrow with reduced CD16. Our preliminary investigations show that serum CD16 was not raised in the elderly (Butcher, unpublished observations) and furthermore that during post-infection neutrophilia, in which the blood contains a higher proportion of neutrophils freshly released from the marrow, there was no change in CD16 expression on peripheral blood neutrophils in the elderly [19]. The latter suggests that reduced CD16 expression arises mainly in the bone marrow. We have also compared bone marrow and circulating neutrophils from elderly individuals undergoing surgery after hip fracture and have observed no difference in expression thus indicating that this decline may occur in the bone marrow. Caution is still required in interpreting these data as both physical trauma, such as hip fracture, and infection result in increased levels of GM-CSF in the marrow and circulation and this could induce shedding of CD16 [34].

There is an alternative explanation for the data regarding reduced neutrophil CD16 expression in the elderly. As stated above, loss of CD16 expression has been demonstrated to be a marker of the ageing of neutrophils in the circulation and their approach to senescence and apoptosis [34]. It is therefore possible that reduced neutrophil phagocytosis could result from the accumulation of older, functionally senescent neutrophils in the circulation in the elderly. Furthermore this could be mediated by raised serum levels of cytokines that can extend the life-span of neutrophils, e.g. TNF-$\alpha$, GM-CSF. Unfortunately, there are currently no good biochemical markers of neutrophil age and therefore this hypothesis cannot be tested directly. However, Fulop and co-workers have shown previously that neutrophils from elderly humans entered apoptosis slightly quicker than those from young subjects [32], though the difference did not reach statistical significance. Moreover, whilst raised serum GM-CSF is not seen in elderly individuals, higher levels of circulating TNF-$\alpha$ are consistently reported [35]. Thus inappropriately raised levels of TNF-$\alpha$ in the serum may be sufficient to extend neutrophil survival in the circulation, without extending functional life-span.

It is unlikely that reduced CD16 expression accounts for the total decline in phagocytic capacity observed in the neutrophils of the elderly, as both the number of phagocytes and the uptake of bacteria per cell are affected. CD16 itself does not signal directly and instead functions as an accessory molecule to the high affinity Fc$\gamma$ receptor CD32, amplifying the signal provided by ligation of CD32 [36]. We would therefore propose that reduced CD16 could account for the reduced efficiency of phagocytosis (lower numbers of bacteria engulfed per cell), but would not explain the low number of phagocytic cells in the elderly. It is therefore important to examine the signaling pathways associated with phagocytic receptors, for example actin polymerization and calcium flux in response to whole bacteria in future studies.

Alvarez and co-workers made a detailed study of fMLP induction of superoxide response in rat neutrophils, which has provided data that may inform upon many aspects of reduced neutrophil responses in the elderly. They showed that the defect in activation of NADPH oxidase occurred proximal to the fMLP receptor, in particular membrane lipid content was altered with a decreased cholesterol to lipid ratio and a 50% decrease in phosphoinositides [30]. The former is significant, as it would affect membrane fluidity and lipid raft formation, which are both important for cell signaling. Lipid rafts are discrete membrane microdomains rich in cholesterol and sphingolipid and allow the assembly of signaling complexes within the plane of the cell membrane. Some cell membrane proteins, including GPI-anchored proteins such as CD16, are constitutively associated with lipid rafts and rely upon them for their association with other components of their signaling pathway. Thus altered membrane lipid composition in neutrophils in the elderly [30] could also affect signaling through phagocytic receptors. It is potentially significant that perturbation of lipid raft function and membrane lipid composition has been reported for both neutrophils and T cells [30,37] with age and may represent a common effect of the ageing process on immune cell membranes that impacts upon cell signaling and cell function. The reduced levels of phosphoinositides reported by Alvarez and co-workers could also impact upon signaling through protein kinase C, a signaling pathway involved in neutrophil superoxide generation, phagocytosis and apoptosis.

From the preceding review of the decline in neutrophil function with age it is clear that the mechanisms involved at the cellular level are various, including membrane lipid composition changes and alterations in cell surface effector receptor expression. Whilst some of the underlying causes of these changes are partially understood and may arise in the bone marrow, it is also possible that the environment of the ageing body also influences neutrophil function. It is important to integrate information on cellular alterations identified *in vitro* with global changes in the body of the elderly in which the cells function and interact *in* vivo. In this final section we will propose that changes in the hormonal environment, specifically that influenced by the hypothalamo-pituitary-adrenal axis, could contribute to immunosenescence in the neutrophil. Moreover, that manipulation of adrenocortical hormones could offer a therapeutic strategy in elderly patients with both trauma and age-associated immune suppression.

## 5.    AGEING AND ADRENAL STEROID PRODUCTION

As this topic is dealt with in detail elsewhere in this volume (see Bornstein), only a brief outline will be made here. Ageing is associated with a gradual but dramatic change in the serum ratio of adrenocortical steroids, namely the glucocorticoids cortisone and cortisol and the anti-glucocorticoid dehydroepiandrosterone (DHEA). The latter is present in serum together with its intraconversion product DHEA sulphate (DHEAS). In contrast to cortisol secretion in the elderly, which remains unchanged, DHEA secretion is reduced with age[38]. The serum levels of DHEA/DHEAS are very low in the first years of life and then progressively rise until reaching maximum levels in the third decade. After the age of thirty the levels of DHEA/DHEAS then decline at a rate of 1–2% per year, thus by the 8th and 9th decades only 20–30% of the maximal level remains. Therefore in later life there is a relative glucocorticoid excess in the serum, which could have profound effects on immune function bearing in mind the well documented immunosuppressive effects of glucocorticoids.

Cortisone is an adrenal glucocorticoid and is one of the stress hormones produced at elevated levels during physical and emotional trauma. Glucocorticoids are potent immunosuppressive agents and it has been suggested that the cortisone levels reached in response to traumatic injury are significantly immunosuppressive [39]. Most of the data regarding the effects of glucocorticoids on the immune system have examined the effects on the adaptive immune response. For example, dexamethasone has been shown to inhibit T-cell proliferation in response to antigen and thus block the clonal expansion necessary to amplify the primary response. This may relate to its ability to suppress IL-2 production and enhance IL-4 production in antigen specific cloned T cell lines [40]. Glucocorticoids are also potent inducers of T cell apoptosis, which would again blunt the adaptive immune response. Only very limited data are available on the effects of glucocorticoids on neutrophil function. Working with bovine neutrophils superoxide production has been examined utilizing a luminol dependent chemiluminescent assay. Cortisone, cortisol and dexamethasone all gave a significant reduction in superoxide production in response to PMA, although the levels of glucocorticoids were far in excess of physiological levels [41]. However, these findings have been reproduced in human neutrophils *in-vitro* with a decrease in superoxide production in response to fMLP after culture with cortisol at physiological levels [42]. Neutrophils therefore appear to be a target of glucocorticoid actions, though it is unclear whether they affect immune function in healthy elderly individuals. There are no reports of an effect of glucocorticoids on phagocytosis and this is the most consistent element of neutrophil function reported to decline with age.

Dehydroepiandrosterone (DHEA) is another steroid hormone produced by the adrenal gland. Various effects of DHEA have been observed including positive effects on the CNS, which have been most well characterized. A large-scale epidemiological study, found a correlation between people taking DHEA supplements and a feeling of well being. Other effects of DHEA recorded include reduction of body weight, inhibition of mammalian glucose-6 phosphate, anti tumour effects, and an anti diabetic effect through a reduced blood glucose mechanism [43]. With regard to immune function, DHEA is considered to have anti-glucocorticoid effects and would therefore be expected to be generally immune enhancing. Studies supporting the immune enhancing effects of DHEA have mainly involved mice and the beneficial effects in man are less clear. For example mice treated with a number of pathogens including herpes viruses, Coxsackie virus B4, *Enterococcus faecalius* and *Pseudomonas paravum*, have been protected against infection after treatment with DHEA [44]. These data suggest that DHEA supplementation may be beneficial, especially in those groups of elderly who are at increased risk of infection, for example following physical trauma such as hip fracture [45]. Indirect evidence for DHEA as an immune enhancing steroid in humans comes from studies of patients with hypopituitarism. The majority of these patients have had their pituitary removed due to malignancy and are supplemented for the major adrenocortical hormones, with the exception of DHEA. These patients provide a model for the effect of low serum DHEA, in the presence of normal levels of glucocorticoid, on immune function in the absence of other age-related effects. A recent retrospective study of 1014 patients in the UK with hypopituitarism, found that there was an excess mortality with respiratory disease being a major cause of excess mortality [46]. Whilst the cause of increased respiratory infections was not known the data could be explained by compromised immune function in these patients. These data do at least suggest that the presence of reduced serum DHEA could affect in immune function, distinct from other factors that are secondary to the broader decline in adaptive and innate responses in the elderly.

5.1. DHEA supplementation and immune function

DHEA supplementation studies in humans are sparse and limited in their design and need to be expanded to confirm the positive results found in mice. They have also tended to focus upon adaptive immune responses and there are consequently no published data concerning DHEA supplementation and neutrophil function or susceptibility to infection in humans. Our preliminary *in vitro* data have shown that DHEA is able to improve superoxide generation in response to fMLP and is also able to delay neutrophil apoptosis (Figure 3). These data do not allow us

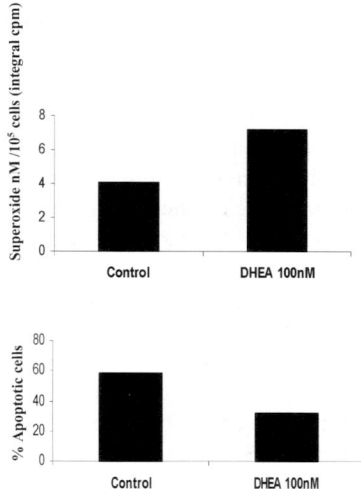

Figure 3. Effect of DHEA on neutrophil function.
Neutrophils were isolated and incubated with 100 nM DHEA for 1h prior to stimulation with fMLP and measurement of superoxide generation (a). Neutrophils were also incubated overnight with DHEA and apoptosis determined by analysis

to rule out an action of immediate DHEA metabolites such as androstenediol, which have been shown by others to have immune enhancing functions in mice [47]. Interestingly we have also found recently that the effects of DHEA on neutrophil apoptosis were dependent upon activation of protein kinase C [Wang unpublished observations] and the ability of DHEA to improve macrophage function has also been linked to modulation of PKC [48]. However, DHEA could not improve phagocytosis in neutrophils (data not shown) and so may not be able to restore neutrophil function in healthy elderly subjects.

## 5.2.  Trauma responses in the elderly

As immune function and resistance to infection are decreased in the elderly under specific situations, such as trauma [49], DHEA supplementation may actually be most beneficial at times of stress in the elderly, when the glucocorticoid to DHEA ratio is further exaggerated. As already discussed, trauma is associated with increased serum levels of immune-suppressive glucocorticoids, such as cortisol and cortisone. In the elderly this cannot be balanced by increased secretion of DHEA and one could predict that this would increase susceptibility to infection. The incidence, outcome and risk factors associated with infection in trauma patients have been examined extensively. It is well documented that after trauma patients become prone to infection. One study has demonstrated that hospitalized trauma patients are at a high risk for infectious morbidity, with 37% of patients developing at least one infection [50]. In one extensive study of 10,308 patients hospitalized with multiple trauma, the most common sites of infection were the lower respiratory tract (32%), urinary tract (17%) and surgical wound (12%) [51]. This study demonstrated the pattern of infecting organisms most frequently isolated were *S. aureus* (25%), *E.coli* (13%), *P. aeruginosa* (10%), and *Enterobacter* species (10%). *S. aureus* has consistently been found to be the most frequently encountered pathogen in the trauma patient population [50]. These studies have focused upon severe trauma, but it is also clear that mild, acute trauma also significantly increases morbidity and mortality in the elderly. Hip fractures are an acute trauma occurring with highest incidence in the elderly, with the rate of hip fracture doubling every 5 year increment. Infection has been established as a major cause of extended hospital stays [45], readmission to hospital and of excess mortality [52] in elderly hip fracture patients. Of particular interest is the finding that bacterial and fungal infections predominate in hip fracture patients [52], again suggesting that neutrophil function is suppressed in these patients.

Our research has examined the incidence of infection after fractured neck of femur [49]. The findings demonstrate that the combination of old age and trauma make individuals particularly prone to bacterial and fungal infection with almost 50% of patients presenting with bacterial or fungal infection within 6 weeks of trauma. These individuals showed a very low superoxide production in response to the bacterial peptide fMLP after trauma, combined with an age related low phagocytic capacity. Serum levels of cortisol were increased in these patients, increasing the cortisol:DHEA ratio more than 10 fold. As our *in vitro* studies showed that DHEA can increase superoxide generation (Figure 3), this may be a suitable prophylactic intervention therapy in elderly hip-fracture patients. Interestingly, DHEA supplementation has been shown to normalize the elevated levels of serum cortisone seen following injury and its anti-glucocorticoid actions may therefore include suppression of cortisone secretion [39]. These data add further support for a post-trauma utility for DHEA supplementation.

## 6.    CONCLUDING REMARKS

There is now little doubt that human ageing is accompanied by a gradual dysregulation of immune function which is likely to contribute to the increased incidence of infectious disease, chronic inflammatory disease and possibly also cancer, in the elderly. Ageing affects both the adaptive and innate immune systems. In the innate system there is a significant reduction in the bactericidal responses of neutrophils to both yeast and bacteria. Superimposed upon these changes is a gradual increase with age in the glucocorticoid (cortisol) to anti-glucocorticoid (DHEA) level, producing an immune suppressive environment. This detrimental steroid milieu is further increased at times of stress, increasing the risk of infection in the elderly. Correcting this imbalance with DHEA supplementation may provide a cheap, safe and beneficial therapy to reduce trauma associated infections and mortality in the elderly.

## ACKNOWLEDGEMENTS

Work in our laboratory is funded by grants from the Biotechnology and Biological Sciences Research Council (SKB), the PPP Foundation (KW) and the Dr Hadwen Trust for Humanity in Research (DL).

## REFERENCES

1.    Kirkwood TB, Austad SN. Why do we age? Nature 408, 233–238. 2000.
2.    Jennings BJ, Ozanne SE, Hales CN. Nutrition, oxidative damage, telomere shortening, and cellular senescence: individual or connected agents of aging? Mol Genet Metab 2000; **71**: 32–42.

3.  Bruunsgaard H, Pedersen AN, Schroll M, Skinhoj P, and Pedersen BK. Decreased natural killer activity is associated with atherosclerosis in elderly humans. Exp Gerontol 37, 127–136. 2001.

4.  Ogata K, An E, Shioi Y et al. Association between natural killer cell activity and infection in immunologically normal elderly people. Clin Exp Immunol 2001; 124:392–7.

5.  Bentley DW. Immunizations in older adults. Infectious Diseases in Clinical Practice 1996; 5:490–7.

6.  Remarque EJ. Influnza vaccination in elderly people. Exp Gerontol 1986;34: 445–452.

7.  MacGregor RR, Shalit M. Neutrophil function in healthy elderly subjects. J Gerontol 1990; 45:M55–M60.

8.  Marston BJ, Plouffe JF, File TM, Jr. et al. Incidence of community-acquired pneumonia requiring hospitalization. Results of a population-based active surveillance Study in Ohio. The Community-Based Pneumonia Incidence Study Group. Arch Intern Med 1997; 157: 1709–18.

9.  Yoshikawa TT. Perspective: aging and infectious diseases: past, present and future. J Infect Dis 1997;176: 1053–1057.

10. Savill JS, Wyllie AH, Henson JE, Walport MJ, Henson PM, Haslett C. Macrophage phagocytosis of aging neutrophils in inflammation. Programmed cell death in the neutrophil leads to its recognition by macrophages. J Clin Invest 1989; 83:865–75.

11. Brach MA, deVos S, Gruss H-J, and Herrmann F. Prolongation of survival of human polymorphonuclear neutrophils by granulocyte-macropahge colony stimulating factor is caused by inhibition of programmed cell death. Blood 1992;80: 2920–2924.

12. Lee A, Whyte MKB and Haslett C. Inhibition of apoptosis and prolongation of neutrophil functional longevity by inflammatory mediators. J Leuk Biol 1993;54: 283–288.

13. Haslett C, Savill JS, Whyte MK, Stern M, Dransfield I, Meagher LC. Granulocyte apoptosis and the control of inflammation. Philos Trans R Soc Lond B Biol Sci 1994; 345: 327–33.

14. Hart SP, Ross JA, Ross K, Haslett C, Dransfield I. Molecular characterization of the surface of apoptotic neutrophils: implications for functional downregulation and recognition by phagocytes. Cell Death Differ 2000; **7**:493–503.

15. Pongracz J, Webb P, Wang KQ, Deacon E, Lunn OJ, Lord JM. Spontaneous neutrophil apoptosis involves caspase 3-mediated activation of protein kinase C-delta. J Biol Chem 1999; 274:37329–34.

16. Moulding DA, Quayle JA, Hart CA, Edwards SW. Mcl-1 expression in human neutrophils: regulation by cytokines and correlation with cell survival. Blood 1998;92(7):2495–502.

17. Schroder AK, Rink L. Neutrophil immunity of the elderly. Mech Aging Dev 2003;124: 1–7.

18. Chatta GS, Andrews RG, Rodger E, Schrag M, Hammond WP, Dale DC. Hematopoietic progenitors and aging: alterations in granulocytic precursors and responsiveness to recombinant human G-CSF, GM-CSF, and IL-3. J Gerontol 1993;M207–M212.

19. Butcher S, Chahal H, Lord JM. The effect of age on neutrophil function: loss of CD16 associated with reduced phagocytic capacity. Mechanisms of ageing and development 2001; 122/14:1521–1532.

20. Finkelstein M, Petkun W and Friedman M. Pneumococcal bacteremia in the elderly. J Am Geriatr Soc 1983; 31,:19–24.

21. Corberand J, Ngyen F, Laharrague P et al. Polymorphonuclear functions and aging in humans. J Am Geriatr Soc 1981; 29:391–7.

22. Esparza B, Sanchez H, Ruiz M, Barranquero M, Sabino E, Merino F. Neutrophil function in elderly persons assessed by flow cytometry. Immunol Invest 1996; 25:185–90.

23. Rao KMK. Age-related decline in ligand-induced actin polymerisation in human leukocytes and platelets. J Gerontol 1986;41: 561–566.

24. Wenisch C, Patruta S, Daxbock F, Krause R, Horl W. Effect of age on human neutrophil function. J Leukoc Biol 2000; 67:40–5.

25. Emanuelli G, Lanzio M, Anfossi T, Romano S, Anfossi G, Calcamuggi G. Influence of age on polymorphonuclear leukocytes in vitro: phagocytic activity in healthy human subjects. Gerontology 1986; 32:308–16.

26. Fulop T, Foris G, Worum I, and Leovey A. Age-dependent alterations of Fc gamma receptor-mediated effector functions of human polymorphonuclear leucocytes. Clin Exp Immunol 1985;61:425–432.

27. Renshaw M, Rockwell J, Engleman C, Gewirtz A, Katz J, Sambhara S. Cutting edge: impaired Toll-like receptor expression and function in aging. J of Immunol 2002;169: 4697–4701.

28. De Greef GE, Van Tol MJ, Kallenberg CG et al. Influence of ageing on antibody formation in vivo after immunisation with the primary T-cell dependent antigen Helix pomatia haemocyanin. Mech Ageing Dev 1992; 66:15–28.

29. Braga PC, Sala MT, Dal Sasso M, Mancini L, Sandrini MC, Annoni G. Influence of age on oxidative bursts (chemiluminescence) of polymorphonuclear neutrophil leukocytes. Gerontology 1998; 44:192–7.

30. Alvarez E, Ruiz-Gutierrez V, Sobrino F, Santa-Maria C. Age-related changes in membrane lipid composition, fluidity and respiratory burst in rat peritoneal neutrophils. Clin Exp Immunol 2001;124(1):95–102.

31. Seres I, Csongor J, Mohacsi A, Leovey A, Fulop T. Age-dependent alterations of human recombinant GM-CSF effects on human granulocytes. Mech Ageing Dev 1993; **71**:143–54.

32. Fulop T, Fouquet C, Allaire P, Perrin N, Lacombe G, Stankova J, Rola-Plesczynski M, Gagne D, Wagner JR, Khalil A, and Dupuis G. Changes in apoptosis of human polymorphonuclear granulocytes with aging. Mech Ageing Dev 1997;96: 15–34.

33. Tortorella C, Piazzolla G, Spaccavento F, Pece S, Jirillo E, Antonaci S. Spontaneous and Fas-induced apoptotic cell death in aged neutrophils. J Clin Immunol 1998;18:321–9.

34. Dransfield I, Buckle AM, Savill JS, McDowall A, Haslett C, Hogg N. Neutrophil apoptosis is associated with a reduction in CD16 (Fc gamma RIII) expression. J Immunol 1994; 153: 1254–63.

35. Pedersen BK, Bruunsgaard H, Ostrowski K et al. Cytokines in aging and exercise. Int J Sport Med 2000; 21:S4–S9.

36. Chuang FYS, Sassaroli M, Unkeless JC. Convergence of Fc gamma receptor IIA and Fc gamma receptor IIIB signaling pathways in human neutrophils. J Immun 2000;164: 350–60.

37. Fulop T, Douziech N, Larbi A, Dupuis G. The role of lipid rafts in T lymphocyte signal transduction with aging. Cell Signaling, Transcription, and Translation As Therapeutic Targets 2002; 973:302–4.

38. Ferrari E, Casarotti D, Muzzoni B et al. Age-related changes of the adrenal secretory pattern: possible role in pathological brain aging. Brain Res Brain Res Rev 2001; 37: 294–300.

39. Catania RA, Angele MK, Ayala A, Cioffi WG, Bland KI, Chaudry IH.

Dehydroepiandrosterone restores immune function following trauma- haemorrhage by a direct effect on T lymphocytes. Cytokine 1999; 11:443–50.

40. Daynes RA, Araneo BA. Contrasting effects of glucocorticoids on the capacity of T cells to produce growth factors interleukin 2 and interleukin 4. Eur J Immunol 1989;19: 2319–2325.

41. Hoeben D, Burvenich C and Massart-Leen AM. Glucocorticosteroids and in vitro efefcts on chemiluminescence of isolated bovine blood granulocytes. Eur J Pharmacol 1998;354: 197–203.

42. Bekesi G, Kakucs R, Varbiro S, Racz K, Sprintz D, Feher J, Szekacs B. In vitro effects of different steroid hormones on superoxide anion production of human neutrophil granulocytes. Steroids 2000;65: 889–894.

43. Flynn MA, Weaver-Osterholtz D, Sharpe-Timms KL, Allen S, Krause G. Dehydroepiandrosterone replacement in aging humans. J Clin Endocrinol Metab 1999; 84:1527–33.

44. Loria RM, Padgett DA, Huynh PN. Regulation of the immune response by dehydroepiandrosterone and its metabolites. J Endocrinol 1996; 150 Suppl:S209–S220.

45. Khasraghi FA, Lee EJ, Christmas C, Wenz JF. The economic impact of medical complications in geriatric patients with hip fracture. Orthopedics 2003;26: 49–53.

46. Tomlinson JW, Holden N, Hills RK et al. Association between premature mortality and hypopituitarism. West Midlands Prospective Hypopituitary Study Group. Lancet 2001; 357:425–31.

47. Whitnall MH, Elliott TB, Harding RA et al. Androstenediol stimulates myelopoiesis and enhances resistance to infection in gamma-irradiated mice. Int J Immunopharmacol 2000; 22:1–14.

48. Corsini E, Lucchi L, Meroni M et al. In vivo dehydroepiandrosterone restores age-associated defects in the protein kinase C signal transduction pathway and related functional responses. J Immun 2002; 168:1753–8.

49. Butcher SK, Killampalli V, Chahal H, Kaya AE, Lord JM. Effect of age on susceptibility to post-traumatic infection in the elderly. Biochem Soc Trans 2003; 31:449–51.

50. Papia G, McLellan BA, El Helou P, Louie M, Rachlis A, Szalai JP, Simor AE. Infection in hospitalized trauma patients: incidence, risk factors and complications. J Trauma 1999;47: 923–927.

51. Caplan ES, Hoyt NJ. Identification and treatment of infections in multiply traumatised patients. Amer J Med 1985;79: 68–76.

52. Myers AH, Robinson EG, Vannatta ML, Michelson JD, Collins K, Baker SP. Hip fractures among the elderly – factors associated with in hospital mortality. Amer J Epidem 1991;134: 1128–1137.

# Apoptosis and Ageing

ANIS LARBI and TAMAS FULOP

*Centre de Recherche sur le Vieillissement, Institut Universitaire de Gériatrie, Université de Sherbrooke, 1036 rue Belvedere sud, Sherbrooke, Québec, Canada*

## ABSTRACT

With ageing we assist to an alteration of the immune response. Recently it became evident that apoptosis can modulate the functions of cells participating in the immune response. Thus it can be supposed that an alteration in apoptosis of T cells and polymorphonuclear neutrophils might explain some of the alterations found with ageing. It was shown that the activation induced cell death of T cells is altered with ageing, however the exact effect is not definitively determined for each T cell subpopulation, mainly for the CD8+CD28– T cells. It was also shown that PMN could not be rescued from apoptosis by pro-inflammatory agents such as GM-CSF. Moreover, an alteration in the signal transduction of pro-apoptotic and anti-apoptotic pathways was found in T cells and PMN of elderly subjects. The executioner phase of the apoptotic pathway involving the Bcl-2 family members and the caspases was also found to contribute to the altered apoptosis of immune cells with ageing. The contribution of each of these components to the alteration with ageing in T cells and PMN will be reviewed.

## 1. INTRODUCTION

Ageing is a very complex phenomenon which is genetically and environmentally determined, but the exact part of each is largely unknown [1]. Cells die either by apoptosis or necrosis [2]. Both necrosis (where the cell membrane is ruptured and the released cell content causes a massive inflammatory response) and apoptosis (where the cell content remains well contained in apoptotic bodies and inflammation does not occur) are well characterized at the cellular and molecular levels. Apoptosis is an active, gene directed cell suicide in response to various external as well as internal stimuli [3]. Apoptosis is a normal, physiological process which plays a crucial role during embriogenesis, tissue homeostasis and remodelling, or in shaping and maintaining the repertoire of immune cells [4]. Insufficient apoptosis can contribute to the pathogenesis of cancer, autoimmune disorders and viral infections. On the contrary, excessive apoptosis results in inadequate cell loss and consequent degenerative diseases such as Alzheimer disease or immune senescence [5,6]. Thus, apoptosis could play an important role, depending on the stage of development either in maintaining functions (activation induced cell death) or decreasing functions (immunosenescence). These phenomena are compatible with the decreased life expectancy observed in p53$^{+/m}$ mice [7] in accordance with the antagonistic pleiotropy [8]. Moreover cellular senescence, tumor growth and longevity are strictly interconnected and deeply related to apoptosis [9,10] .

It is well established that the immune response is declining with ageing. This decline affects all part of the immune system, however it seems that the decline in cellular immune response mediated by T lymphocytes is the basic one [11]. The immunological alterations with ageing include T cell proliferative response to mitogens and antigens (clonal expansion), altered cytokine expression and T cell subpopulations with an increase in the memory phenotype [12]. This accumulation of memory phenotype suggests that most of the peripheral T cells are already proliferated and manifest a decreased telomere length [13]. Moreover, the co-receptor CD28 is playing an important role in T cell full activation and its loss is leading to anergy and increased susceptibility to apoptotic stimuli, resulting in clonal deletion. An age-dependent increase in CD8+CD28− T cells, having a high cytotoxic capacity, both in percentage and absolute number, has been found in elderly people as well as in the oldest old [14]. These cells also show a short-ening of telomere length. The exact cause of these alterations is just starting to be elucidated. Several causes may be evoked such as the decline of thymus functions, the alteration of signal-ling through various receptors, the shift in T cell subpopulations which could be directly related to an alteration in T cell apoptosis.

Experimental data suggest that there also exists decreased receptor mediated functions in the innate immune system including monocytes and polymorphonuclear granulocytes (PMN) [15]. The phagocytic as well as the killing activities are altered. Recently, it was also shown that the polymorphonuclear granulocytes apoptosis delaying activity of various growth factors such as GM-CSF is altered with ageing [16,17].

Apoptosis is executed by caspases. There exists initiator and effector caspases. Caspases are activated via an extrinsic (death receptor signalling) pathway or an intrinsic (death receptor independent/mitochondrial) pathway [18]. There is overwhelming evidence suggesting that regulation of apoptosis is altered in ageing. This has led to the plausible explanation of several age-related changes, such as the decrease in immune functions. Scanty and controversial data exist concerning immune cells [19]. Nevertheless, profound change in apoptosis of T cells and PMN exist with ageing and we will review in details what are the possible determinants of these changes.

## 2.    T CELL APOPTOSIS AND AGEING

Apoptosis in T cells is well characterized, and recent studies indicate that with advancing age, defects in T cell apoptosis may correlate with increased autoimmune disorders and susceptibility to infections in older individuals [20]. Apoptosis is a fundamental part of normal T-lymphocyte maturation and selection. Physiologically any T cells that bind to self antigens or make non-functional receptors undergo apoptosis [21]. The age-dependent diminished synthesis of growth and survival factors, transmembrane signalling defects, default in the expression of particular genes implicated in the control of cell proliferation and the inability to cope with oxidative stress are susceptible to initiate the apoptotic process in ageing organisms [21]. In several studies ageing was found to increase CD8+ T cell apoptosis by overstimulation through the TCR [22]. Furthermore, with ageing the percentage of apoptotic cells collected at basal status is increased as the susceptibility to mitogenic stimulation (PHA, ant-CD3mAb) and to AICD [23,24]. Fur-thermore, interleukin-2 (IL-2) is also less efficient with ageing to rescue T cells from apoptosis. Lymphocyte apoptosis is mediated either by surface receptors such as Fas, TNF or by the mito-chondrial pathways.

## 2.1.    Death receptor pathway

### 2.1.1.  TNF-α, apoptosis and ageing

Tumor necrosis factor-α (TNF-α) is a pleiotropic pro-inflammatory cytokine which secretion is increased with ageing [25]. Gupta et al [26–28] showed that TNF-α induced more apoptosis in CD4+ and CD8+ cells with ageing in comparison to young subjects. This was also true for CD45 RA+ (naïve) and CD45RO+ cells (memory) [29]. This similar susceptibility to apoptosis manifested by naïve and memory T cells would suggest that the increased apoptosis observed with ageing is not the consequence of the age-related accumulation of memory type cells. In parallel, the expression of TNF-RI is increased, while that of TNF-RII is decreased. There exists two distinct TNF receptors, namely TNF-RI and TNF-RII [30] by which TNF-α exerts its biological activity. TNFRs belong to the family of nerve growth factor receptors and are type I transmembrane receptors. These receptors differ in their structure and signalling pathways as well as in their biological effects [31,32]. TNF-RI contains DD whereas TNF-RII does not. This composition determines that TNF-RI signals cell death and survival, whereas TNF-RII mediates cell survival.

#### 2.1.1.1.  Signalling through TNF-RI

Ligation TNF-RI will recruit an adapter molecule TNF-R associated death domain (TRADD) as the DD domain does not possess any intrinsic enzymatic activity [33,34]. Furthermore, an other adapter molecule is recruited the Fas-associated death domain (FADD). To this complex the procaspase 8 is recruited and this completes the formation of a death inducing signal complex (DISC). During the formation of this complex the procaspase 8 is activated, detached and serves as an enzyme for the down-stream effector caspases-3, -6 and -7 [35]. The activation of these effector caspases finally results in the morphological and biochemical features of apoptosis. Considering the importance of the apoptotic process, the whole signalling process is under tight control. Several inhibitory molecules might intervene at different steps of the activation. The apoptosis mediated via FADD is regulated by Flice-inhibitory protein (FLIP) which appears to inhibit the initiator caspases. On the other hand, it was also shown that FLICE activate survival signals by inducing the activation of NFkB and Erk [36]. Altogether, the activation of TNF-RI mainly mediates signals leading to cell apoptosis.

The expression of TNFRI was increased in T cells from elderly subjects compared with young subjects [26–28]. This leads to increased signalling resulting in increased apoptosis of T cells. Consequently, the TRADD, FADD and Bax expression was also found increased in T cells with ageing resulting in increased and early activation of both initiator (caspase-8) and effector (caspase-3) caspases. This increased apoptosis was similar in memory (CD45RO+) and naïve (CD45RA+) T cell subsets. The literature is rather contradictory concerning the individual susceptibility of T cell subsets to apoptosis [20,24,37]. This leads to conflicting results depending on the method of measurement, on the stimulating agent used and finally on the activation status of the cells. Altogether these results suggest that the increased apoptosis in T cells with ageing is not due to the shift in CD45RO+ T cells, but could nevertheless contribute to the decrease of CD45RA+ T cells.

With ageing it has been shown that a very specific T cell subsets is increasing with a CD8+CD28– phenotype [38]. CD28 is an important receptor that affects both T cell activation and susceptibility to apoptosis. McLoad et al. [39] have shown that costimulation via CD28 in superantigen blast cultures leads to resistance to CD95 ligation which may be due to increased levels of intracellular regulator Bcl-XL [40]. These subsets represent suppressor T cells. These cells are generated the most probably after repeated stimulation with allogeneic or xenogeneic antigen presenting cells. However, other experimental evidences seem to indicate that these cells do not derive from naïve cells but are directly generated in the bone morrow [41]. Gupta et al [26–29] observed that these are effector memory T cells and are more susceptible to TNF-α induced apoptosis than CD8+CD28+ T cells. Others also found that CD28 low cells are more susceptible to apoptosis than CD28+ T cells [39,42] However, this is equally true for CD8+CD28– T cells from young and elderly subjects. If this is the case what might explain the increase of the CD8+CD28– T cell accumulation with ageing? It appears that even this subpopulation can be divided between central and effector memory T cells and that ageing affects them differentially. In contrast, Effros's laboratory indicate that in parallel to the increase of CD28– T cells in elderly individuals these senescent T cells are quite resistant to apoptosis [43]. The exact nature of these contradictory results are unknown. An other phenomenon might be considered. It was stated that these CD8+CD28– T cells are in a state of replicative senescence with a certain functional capacity, but unable to proliferate or die by apoptosis. In fact replicative senescence itself may lead to a block in apoptotic pathway [43]. This may be attributed to an increased Bcl-2 expression. Finally, the contribution of the altered receptor signalling might be also evoked. Actually, there are no sufficient data to explain by the apoptotic phenomenon why these special subsets of T cells is accumulating with ageing.

### 2.1.1.2. Signalling through TNF-RII

As this receptor is lacking of cytoplasmic DD it is able to bind TNF-R associated factor-2 (TRAF2) which is able to recruit the cellular inhibitor of apoptosis proteins cIAP-1 and cIAP-2 [31,34,44]. This leads to the activation of the MAPK pathway and finally to the NFkB translocation to the nucleus. NFkB has been shown to inhibit apoptosis [45]. NFkB is regulating several immune response gene such as for cytokines, chemokines, cell surface receptors and adhesion molecules expression. In resting state the NFkB is kept in the cytoplasm in an inactive form by an inhibitory protein, IkB and when exposed to its inducers such as TNF-α, IkB is phosphorylated. After this phosphorylation IkB is degraded by the 26S proteasome. The free NFkB translocates to the nucleus where its activate the transcription of antiapoptotic genes, such as c-*iap1*, c-*iap2*, *xiap*, *gadd45β*, and *Bcl-XL*. This suppression of apoptosis by NFkB is important for biological functions of TNFα. Finally, the studies concerning the NFkB in T cells are showing a decreased phosphorylation of IkB and activation of NFkB in TNF induced T lymphocytes from aged humans [46]. Ponappan et al [47] have suggested that this decreased induction of NFkB in T cells with ageing could be due to decreased proteasome mediated degradation of Ik-B. All these data suggest that in T cells from aged subjects there is a decreased TNF-α mediated NFkB activity correlated with the decreased expression of TNFRII.

It was suggested that the TNF-RII may also play an important role in the regulation of apoptosis induced by TNF-RI. However, the exact mechanism of this interaction is still unknown. It is of note that the effect of these receptors is changing dependent of the cell studied. In T cells this receptor does not induce apoptosis, but modulate the TNF-RI apoptosis inducing effect through the modulation of RIP [48]. With ageing the TRAF-2 expression is decreased explaining partly the increased susceptibility to apoptosis induction of T cells via TNF-α equally for CD4+ and CD8+ T cells.

In summary, T cells from aged subjects show increased expression of TNFRI and decreased expression of TNFRII. The increased susceptibility of T cells to apoptosis with ageing under TNF-α stimulation might be due to the increased signalling through TNFRI while the increase of the antiapoptotic NFkB through TNFRII is decreased [49]. This is further supported by the work of Gupta et al [28] that the overexpression of TNFRII by transfection in T cells from aged subjects (to achieve levels comparable to young subjects) decreased the TNF-α induced apoptosis to a level comparable to the levels that was observed in young subjects. Modulation of the TNFRI or II could be a therapeutic target in the modulation of altered immune response with ageing.

## 2.1.2. The Fas/FasL receptors and ageing

The other very important apoptotic pathway in the immune system is triggered by CD95 (Fas/ApoI) [50]. Fas controls several immune processes, including selection of T cell repertoire, deletion of self reactive T cells and cytotoxicity against target cells or tissues. CD95 is a member of the TNF superfamily expressed on various tissues, whereas the expression of its ligand (CD95L) is more restricted to a few cell types including T cells, macrophages. CD95L is not present in resting T cells, but is highly expressed upon T cell activation [51]. Activated T cells may undergo apoptosis using the CD95/CD95L system. This system mediating the activation induced cell death (AICD) is known to play an important role in the maintenance of peripheral lymphocytes homeostasis. Thus, apoptosis occurring through CD95 is vital in the regulation of T cell numbers following antigenic stimulation.

CD95 ligation activates the adapter protein FADD. FADD then directly activates caspase-8 which acts directly on caspase-3 [26,52]. However caspase-8 cleavage of Bid results in translocation of Bid to the mitochondria with subsequent cytochrome c release. Thus, Fas induced apoptosis is either mitochondria independent or mitochondria dependent. Moreover, regulation of CD95 apoptotic signals occurs through intracellular proteins of the bcl-2 family from which Bax and Bad are pro-apoptotic and Bcl-2, Bcl-XL are anti-apoptotic.

It was shown that defective Fas-mediated apoptosis is one of the major mechanisms associated with the accumulation of senescent T cells in very old mice [53]. A decreased AICD of CD8 T cells was found in old mice compared to CD4+ T cells which may explain the expansion of CD8+ T cells with ageing. Thus CD8+ T cells of mice are more resistant to apoptosis compared to CD4+ T cells. This decreased apoptosis in CD8+ T cells is primarily associated with defect in the upregulation of FasL after activation and a decrease in Fas apoptosis signalling [53].

In contrast, it was found that lymphocytes from elderly subjects overexpressed the apoptosis molecule Fas (CD95), at the protein as well as at the mRNA levels, which induces apoptosis in the presence of FasL during AICD [37]. Both the density of CD95 expressed on lymphocytes as well as the number of CD95+ T cells is progressively increased during the course of ageing and that it reaches the highest levels in donors of the sixth through the seventh decades. Identical findings were reported under mitogenic stimulations of T cells from elderly subjects. It then decreases and in the oldest donors values are the same as those observed in adult subjects. The decrease is mainly due to the loss of CD4+ cells. Not only the expression of CD95 is increased with ageing, but this is linked to increased CD95-induced apoptosis as well as to constitutive activity of caspase-3 and 8 with ageing [54–58]. What about the expression of CD95L? There are contradictory results, because of Gupta [27] showed an increased expression in lymphocytes with ageing, while Pinti et al. [59] showed a decreased expression being more marked in centenarians. Thus, T cells of elderly subjects seem to be more susceptible to apoptosis than their younger counterparts, however the exact role of the Fas/FasL system is still unknown.

The discrepancy between experimental animals and humans concerning the Fas/Fas-L system may reflect differences between the *in vitro* and *in vivo* conditions, the nature of apoptosis stimuli, the stage of the development of cells or the sub-population of the cells under study.

There is no consensus on the modifications of the number of CD3+, CD4+ and CD8+ T cells in the course of development and of ageing and its relationship with survival. Moreover, the CD4+ and CD8+ T cells subpopulation demonstrate differential susceptibility to apoptosis under stimulation even at a high expression of CD95 in all cells. The CD4+ cells are more susceptible to apoptosis, while CD8+ cells are more resistant. Other studies have demonstrated similar susceptibility to apoptosis of these two cell populations. In contrast, it seems to exist a consensus that with ageing there is an increase in CD45R0+ (memory/preactivated) compared to CD45R0- (naïve) T lymphocytes and both of these populations demonstrate an increase in the expression of CD95 with ageing with a concomitant proneness to undergo apoptosis. These alterations could explain the decreased secondary immune response with ageing during vaccination and the overproduction of auto-antibodies.

### 2.1.3. Receptors and lipid rafts

When triggering CD95 by various stimuli such as CD95L, there occurs a clustering of this receptor [59]. Receptor clustering resembles a kind of polarizing phenomenon requiring asymmetric organization of the plasma membrane [60]. This clustering is essential for signalling as it promotes the recruitment of lipid rafts containing essential receptors and signalling proteins. Lipid rafts are small liquide ordered membrane microdomains composed of cholesterol, sphingolipids and GPI-anchored proteins [61,62]. In case of TNFR and Fas this clustering (trimeric receptor) leads to the formation of a death inducing signalling complex (DISC) [63]. Recent experimental data suggest that lipid rafts are required for efficient Fas signalling in peripheral lymphocytes by enhancing Fas clustering and signalling when Fas cross linking may be suboptimal [64]. This is further suggested by the experimental disruption of lipid rafts by methyl-β-cyclodextrine which extracts the cholesterol from the exofacial part of the plasma membrane. Recently, it was further shown that Fas and the components of the DISC were recruited to lipid rafts following Fas ligation and that rafts structures were required for the efficient propagation of apoptosis signals by Fas in human peripheral T cells [65]. It was also found that Fas is constitutively present in lipid rafts with low level of caspase 8. With ageing there is alteration in the composition as well as in the coalescence of lipid rafts in peripheral T cells [66]. Many signalling molecules are present in less amount and their recruitment is also decreased. This leads to a signalling defect for several receptors, but mainly for TcR/CD3 complex [67]. The cholesterol content is also increasing with ageing in lipid rafts. There is no actual data on the role of altered lipid rafts composition in peripheral T cells concerning the constitutive presence of Fas or caspase 8 molecules as well as on the recruitment of the DISC complex with ageing. However, an increase in Fas receptor expression and caspase 8 activation was found in ageing with an increased susceptibility to apoptosis. Thus, we can hypothesize that the lipid rafts alteration observed in the membrane of T cells may play a crucial role in the altered apoptosis proneness observed with ageing.

Furthermore, the capacity of a cell to polarize is directly linked to an interaction between the membrane and the cytoskeleton. There is no doubt that apoptosis has dramatic effects on the cytoskeleton. Through cleavage of intermediate filament proteins, apoptosis alters the integrity of the cytoskeleton and thereby affects overall cellular structure. On the other hand disruption of the cytoskeleton by itself can induce or accelerate apoptosis [59]. Disruption of the cytoskeleton causes apoptosis via activation of CD95. The cells are able to signal from inside to induce apoptosis. There is a lack of evaluation of the role of cytoskeleton in the T cell apoptosis changes with ageing.

Altogether as for TNFR, the Fas system mediated apoptosis is also increased with ageing. The data seem to suggest that different subpopulations of T cells are differentially affected in their susceptibility to apoptosis. Age-related changes in the signalling of these death receptors might greatly contribute as well as modulate these apoptosis alterations. Nevertheless, the differential apoptosis susceptibility favours the tendency to lymphopenia, the accumulation of memory type T cells and the decrease in naïve T cells setting the stage for the altered immune response with ageing.

## 2.2. The mitochondrial pathway

Mitochondria has been implicated in apoptosis in many ways either by accelerating or contributing to many forms of apoptosis [3]. In some form of apoptosis, mitochondria undergo major functional and structural changes that regulate and accelerate cell death. The mitochondrial inner transmembrane potential ($\Delta\psi_m$) collapses prior to classical morphological signs of apoptosis. Following specific mitochondrial proteins are released (e.g. cytochrome c and apoptosis inducing factor) for the activation of caspases [68]. Cytochrome c release from mitochondria is one of the most intensively studied pathway of apoptosis. Cytochrome c associates with Apaf-1, caspase 9 and adenosine trisphosphate to form a complex called an apoptosome. This complex formation would activate caspase 9 which in turn leads to the activation of the executioner caspases, such as caspase 3, 6 and 7. Caspase inhibitors control the activation of this pathway such as XIAP, C-IAP1 and c-IAP2 [69]. Another key protein released from mitochondria is the apoptosis inducing factor (AIF) [70]. Various apoptotic signals such as free radicals, staurosporin, c-Myc, etoposide and ceramide can lead to mitochondrial release of this proapoptotic protein. This protein translocates directly to the nucleus to cause chromatin alteration. Thus, this pathway is mitochondrial dependent but caspase-independent. The Bcl-2 family of proteins is able to regulate some forms of apoptosis by controlling the release of cytochrome c. Some proteins of this family such as Bcl-2, Bcl-xL are antiapoptotic whereas others such as bax, Bad, Bak are pro-apoptotic.

### 2.2.1. Role of free radicals in inducing apoptosis

The overproduction of free radicals can induce cell death [71,72]. Ageing, as stated in the free radical theory of ageing is characterized by an increased production of free radicals in several tissues or a decreased antioxidant defence leading to a chronic oxidative stress [73]. This increased production of free radicals may result in the induction of apoptosis. One important source of free radicals is mitochondria [3]. In mitochondria the complex I is the major site for the production of free radicals. This complex is encoded by mitochondrial DNA which is prone in ageing to mutations explaining the alterations of this complex I with ageing [74]. In T cells the mitochondrial pathway of apoptosis seems to correlate the death receptors induced cell death. T cells of elderly might be more susceptible to free radicals as with ageing the main antioxidant defence represented by the reduced glutathione (GSH) is decreased, while the oxidized glutathione (GSSG) increased [75]. This correlated with an increased DNA damage found in these T cells. However, what is the exact role of the altered free radical induced mitochondrial apoptotic pathway in the altered T cell apoptosis observed with ageing is still a matter of debate.

3.    APOPTOSIS OF POLYMORPHONUCLEAR NEUTROPHILS (PMN) AND AGEING

Beside the alterations in T cell functions with ageing, it became more and more evident that the functions of the cells of the innate immune response are also influenced by ageing [17,76–79]. It was shown by several groups that the effector functions of PMNs, such as intracellular killing, free radical production are decreased with ageing [76]. In FMLP stimulated neutrophils, age-related decreases have been observed for chemotaxis, superoxide production and lytic enzyme production [80]. There is a controversy whether the phagocytic activity is altered, but this seems to depend upon the agent to be phagocytosed [17,76]. These results clearly indicate that with ageing there is an alteration in PMN functions. The question has been raised why these short living cells are altered in their function.

It was shown that these PMN functions are dependent on receptors and that the receptor signalling is also altered with ageing [15]. The most studied receptors are the Fcγ, C3b, FMLP and GM-CSF. It was shown that the number of receptors as well as the composition of lipid rafts do not change with ageing [81]. However, the various signalling pathways such as the calcium, the MAPK and the JAK/STAT pathways were altered [82,83]. Beside these alterations the survival of neutrophils under stimulation could also play a very important role in the altered functioning. Thus one possible alteration could be the decrease of apoptosis rescue of neutrophils with ageing.

3.1.    Apoptosis of PMNs

Mature neutrophils undergo spontaneous apoptosis very rapidly when maintained *in vitro*, this apoptotic sensitivity regulates both their production and survival [84]. Although neutrophils appear to be committed to apoptotic death *in vitro* and *in vivo*, the life span and functional activity of mature PMN can be extended *in vitro* by incubation with pro-inflammatory cytokines, including LPS, IL-2, GM-CSF [85]. This rescue is essential for efficient effector functions of PMN. On the other hand for controlling this prolonged survival and to avoid the establishment of a chronic inflammatory process PMN should die. This means that the apoptotic process should be tightly regulated externally as internally. Thus, PMN survival represents an important factor for neutrophil efficacy.

Bcl-2, Bcl-X (Bcl-XL and Bcl-XS), Bad and Bax are members of the Bcl-2 family of proteins and play important roles in regulating cell survival and apoptosis. Bcl-2 and Bax are homologous proteins that have opposing effects on apoptosis [86]. Bcl-2 serves to prolong cell survival, but as PMN do not contain Bcl-2, however, other homologous proteins were identified such as A1 and Mcl-1 (novel hematopoietic specific homologues of Bcl-2) [87]. The ratio of Bax to Mcl-1 is important for death or survival of PMN [87]. GM-CSF was suggested to increase Mcl-1 expression. These proteins act as ion channels and adapter proteins regulating the release of mitochondrial proteins into the cytosol in order to activate the caspases that are the terminal effectors of apoptosis.

The caspase superfamilies have emerged as key players in the regulation of the apoptotic pathway [18]. Although caspase 3 is the most widely studied caspase it is becoming evident that there is a molecular ordering of caspases in the apoptotic program [88]. Activation of the upstream initiator caspase such as caspase 8 leads to the subsequent induction of effector phase of apoptosis via processing of the executioner caspase such as caspase 3, 6 or 7 [89].

We [16,23] and others [17] have recently demonstrated that the PMN apoptosis is altered with ageing. PMNs of elderly subjects were unable to delay the survival after proinflammatory cytokines stimulation compared with that of young subjects. Thus, this decreased rescue of PMN of elderly subjects from apoptosis could have important functional consequences.

### 3.1.1. GM-CSF and PMN apoptosis

GM-CSF is able to delay mature PMN apoptosis [85]. Prolongation of the functional life span of the PMN by GM-CSF is now thought to represent a further important mechanism whereby priming may enhance the PMN mediated response to injury or infection. PMNs express a low number of high affinity GM-CSF receptor. We have shown that the number of GM-CSF receptor ($\alpha$-subunit) measured by FACscan did not change with ageing. Several intracellular signalling pathways are known to be activated once the GM-CSF receptor is ligand activated. These include activation of the JAK/STAT pathway, multiple MAPK signalling pathways and phosphatidylinositol 3-kinase (PI3K) pathway. Some of these pathways have been implicated in the PMN apoptosis rescuing effect of GM-CSF [90].

We studied the GM-CSF elicited signal transduction pathways in relation to its apoptosis delaying activity and tried to elucidate whether they change with ageing. We have demonstrated that the Jak2-STAT5 signal transduction pathway is altered with ageing and this alteration contributes to the decreased rescue of PMN from apoptosis by GM-CSF with ageing. We also found that by modulating the Jak2 phosphorylation by an inhibitory agent AG490 we could blunt the apoptosis delaying effect of GM-CSF in PMN of young subjects, while this inhibitor had no effect on PMN rescue from apoptosis by GM-CSF of elderly. These experiments underline the importance of the JAK/STAT pathway in the rescue and the changes in ageing. The MAPK ERK1 /2 have been also shown to be important in the rescue of PMN by GM-CSF. Similarly to JAK/STAT pathway alterations in PMN of elderly, the MAPK pathway was also found to be altered. These signal transduction changes in PMNs of elderly under GM-CSF stimulation might explain the inefficacy of GM-CSF to rescue them from apoptosis [16].

In the later phase of the apoptotic pathway the bcl-2 family members, Mcl-1 and Bax, play an important role. We found that the expression of Bax was identical in young and elderly PMN undergoing spontaneous apoptosis. After GM-CSF stimulation for 18 hours Bax expression was significantly decreased by the GM-CSF and was almost identical to that found in freshly isolated PMN of young subjects. In contrast, in PMN of elderly subjects, GM-CSF was unable to modulate the expression of Bax. The MAPK ERK inhibitor PD98059 and the MAPK p38 inhibitor SB203580 had no effect on Bax expression in any age group. Mcl-1 was significantly increased in PMN of young subjects following 18 hours of GM-CSF stimulation compared to spontaneous apoptosis. In contrast, in the PMN of elderly subjects, there was a similar decrease after 18 hours of culture either during stimulation or spontaneous apoptosis. The ratio of Bax/Mcl-1 was 0.2 in the PMN of young after 18 hours of treatment with GM-CSF, while it was 0.84 in PMN of the elderly [91].

Our results suggest that the modulation by GM-CSF of the expression of the bcl-2 family members, Mcl-1 and Bax, as well as their relative ratio, is very important for the rescue of PMN from apoptosis. In PMN of young subjects, Mcl-1 predominates after 18 hours of GM-CSF treatment relative to Bax content. In contrast, in the PMN of the elderly, there is almost no difference between Mcl-1 and Bax content under GM-CSF stimulation. The relationship between these alterations in the expression of the bcl-2 family members with ageing and the alteration of the JAK2/STAT signaling pathway is not yet fully understood. Nevertheless, as JAK2 is related to the expression of Bcl-2 there might also be a link between JAK2 and Mcl-1.

Furthermore, we have studied the caspase-1 and caspase-3 expression as well as the caspase 3 activity in PMN of young and elderly subjects. GM-CSF was able to stabilize the pro-caspase-3 expression and decreases the activated form of caspases in PMN of young subjects compared to that observed during spontaneous apoptosis. This effect of GM-CSF on caspase-3 was much less pronounced in PMN of elderly subjects. Identical results were obtained when the caspase-3 activity was measured. Altogether GM-CSF could decrease the caspase-3 activity in PMN of young subjects while this was not the case in elderly [23].

All these results suggest that the inefficient rescue of PMN of elderly subjects from apoptosis is due either to altered signal transduction pathways ot to a dysregulation of the effector phase. Furthermore, all these alterations leading to the decreased rescue of PMN from apoptosis might contribute to the increased incidence of infection with ageing.

## 3.2.    Free radical production of PMNs

One of the most accepted theories of ageing is the free radical theory of ageing, suggested for the first time by Harman in 1956 [92]. Since then a huge amount of experimental evidence has been put together for supporting this theory. FMLP receptor-stimulation by FMLP induced the production of superoxide anion as well as hydrogen peroxide by phagocytic cells (monocytes and polymorphonuclear neutrophils) [82,93]. This free radical production participates in the destruction of invading organisms [80,94] and as such in host defence. Superoxide anion may also combine with NO, where the production is also induced by FMLP, to form peroxinitrite (NOO-), a very reactive free radical. Moreover, these free radicals may also contribute to the peroxidation of lipoproteins in blood vessel walls. It was demonstrated some time ago that the production of free radicals by PMN of elderly subjects was decreased under FMLP stimulation, while the number of FMLP receptors did not change, however being higher in the resting state compared to young subjects. Moreover, with ageing, superoxide anion production is also significantly decreased after 24 and 48 hours of PMN culture in the presence of GM-CSF. The increased free radical production in resting state by PMN of elderly subjects did not seem to influence their spontaneous apoptosis. However, could this increased free radical production at basal state contribute to the decreased rescue of PMN by GM-CSF with ageing is not fully elucidated. We have shown that antioxidants, such as Vitamin E or C or N-acetyl-cysteine or dehydroepiandrosterone were all able to rescue the PMN of elderly subjects from apoptosis [23].

## 4.    CONCLUSIONS AND FUTURE PERSPECTIVES

The decline of the immune response with ageing is revealed by a decrease in delayed-type hypersensitivity responses, diminished responses to vaccination, increased susceptibility to infections and cancers. The cause of these defects in age-related immune responses is unknown. A number of possibilities have been suggested, namely an alteration of the intracellular biochemical cascade pathways that transduce signals from surface receptors such as the TCR, cell surface co-stimulatory molecules, Fas or cytokine receptors (e.g. TNFR). It is well-documented that there is an alteration in several signalling pathways with ageing induced by these receptors. These alterations either in receptor expressions or in signalling lead to an alteration of apoptosis of T cells or PMNs with ageing. It seems that in elderly persons the T cells are more susceptible to apoptosis. However controversial results exist concerning the various T cell subpopulations. In PMNs there seems to be a consensus considering that PMN can not be rescued from apoptosis by proinflammatory stimuli. Recent findings conceptualising the interaction between ligands and receptors on T cells or neutrophils under the form of specialized plasma membrane domains called lipid rafts help to better understand the mechanism of signal transduction. The sparse data related to ageing in this context indicate an alteration in the composition and the signaling molecule content of lipid rafts and enforce the importance of cholesterol in the signal transduction of TCR, Fas, GM-CSF through lipid rafts.

Taken together one or more fundamental processes that are responsible for ageing may alter the regulation of immune cell apoptosis. However, whether the apoptotic response is increased or decreased by ageing may depend on both the inducing signal and the T cell subpopulation. Upon senescence some cell types might become resistant to certain apoptotic signals. This resistance to apoptosis may explain why some senescent cells can accumulate in tissues with age. Tissues may differ how much cells loss or disruption can be tolerated before tissue function declines. Nevertheless, the alteration in the susceptibility to apoptosis of immune cells will have serious functional consequences explaining at least in part the altered immune response with ageing. Further research is needed to elucidate the more precise role of apoptosis in ageing and in diseases associated with ageing. Finally, the separation of T cells into various well-defined sub-populations is needed to understand whether these changes are age-related only, or related to the increase of memory cells or other subsets during life, due to encounters of the immune system with pathogens. The understanding of these basic mechanisms will lead to the design of efficient interventions in the changing immune response of elderly subjects.

## ACKNOWLEDGEMENTS

This work was supported by a grant-in-aid from the National Science and Engineering Research Council of Canada (CRSNG No. 249549), the Research Center of Ageing of Sherbrooke, the Canadian Institute for Health Research (CIHR No 63149) and the ImAginE consortium.

## REFERENCES

1.  Troen BR.The biology of aging The Mount Sinai J Med 2003;70 :3–22.
2.  Soti C, Sreedhart AS, Csermely P. Apoptosis, necrosis and cellular senescenced : chaprone occupancy as a potential switch. Aging Cell 2003;2 :39–45.
3.  Pollack M, Leeuwenburgh C. Apoptosis and aging : role of the mitochondria. J gerontol 2001;56A :B475-B482.
4.  Cohen JJ, Duke RC. Apoptosis and programmed cell death in immunity Ann Rev Immunol 1992;10 :267–293.
5.  Thomson CB. Apoptosis in the pathogenesis and treatment of disease Science 1995;267 : 1456–1462.
6.  Gupta S. Molecular and biochemical pathways of apoptosis in lymphocytes and aged humans. Vaccine 2000;18 :1596–1601.
7.  Tyner SD, Venkatachalam S, Choi J, Jones S, Ghebranious N, Ingelmann H, Lu X, Soron G, Cooper B, Brayton C, Park SH, Thompson T, Karsenty G, Bradly A, Donehower LA. P53 mutant mice that display early ageing-associated phenotype. Nature 2002;415 : 45–53.
8.  Kirkwood TB. Human senescence Bioassays 1996;18 :1009–1016.
9.  Monti D, Triano E, Grassilli E, Agnesini1C, Tropea F, Barbieri D, Capri M, Salvioli S,

Ronchetti I, Bellomo G, Cossarizza A, Franceschi C. Cell proliferation and cell death in immunosenescence. Ann N Y Acad Sci 1992;663 :250–261.

10. Warner HR. Recent progress in understanding the relationships among ageing, replicative senescence, cell turnover and cancer. In vivo 2002;16 :393–396.

11. Miller RA. The ageing immune system: Primer and prospectus. Science 1996;273 : 70–74.

12. Grubeck-lobenstein B, Wick G. The ageing of the immune system. Ann Rev Immunol 2003; :248–284.

13. Weng NP, Palmer LD, Levine BL, Lane HC, June CH, Hodes RJ. Tales of tails: regulation of telomere length and telomerase activity during lymphocyte development, differentiation, activation, and ageing. Immunol Rev 1997;160:43–54.

14. Fagnoni FF, Vescivini R, Mazzola M, Bologna G, Nigro E, Lavagetto G, Franchesci C, Passeri M, Sansoni P. Expansion of cytotoxic CD8(+)CD28(-) T cells in healthy aging people including centenarians. Immunology 1996;88 :501–507.

15. Fulop T. Jr. Signal transduction changes in granulocytes and lymphocytes with aging. Immunol Lett 1994;40: 259–269.

16. Fülöp T. Jr, Fouquet C, Allaire P, Perrin N, Lacombe G, Stankova J, Rola-Pleszczinsky M, Wagner JR, Khalil A, Dupuis G. Changes in apoptosis of polymorphonuclear granulocytes with aging. Mech Age Dev 1997; 96: 15–34.

17. Lord JM, Butcher S, Killampali V, Lascelles D, Salmon M. Neutrophil ageing and immunosenescence. Mech Age Dev 2001;122:1521–1535.

18. Thornberry N. The caspase family of cystein proteases. Br Med Bull 1996;53 :478–490.

19. Pawelec G, Barnett Y, Solana R, Remarque E. T cell immunosenescence. Frontiers in Biosciences 1998;3:59–99.

20. Ginaldi L, De Martinis M, D'Ostilio A, Mariani L, Loreto MF, Corsi MP, Quaglino D. Cell proliferation and apoptosis in the immune system in the elderly. Immunol Res 2000;21 : 31–38.

21. Krammer PH. CD95's deadly mission in the immune system. Nature 2000;407 :789–795.

22. Potestio M, Caruso C, Gervasi F, Scialabbaa G, D'Annaa C, Di Lorenzoc G, Balistrerid CR, Candored G, Colonna Romanoa G. Apoptosis and ageing. Mech Age Dev 1998;102 : 221–237.

23. Fulop T. Jr, Desgeorges S, Goulet AC, Fallery C, Arcand M, Linteau A, Bergevin M, Douziech N. Apoptosis in immune cells with aging. In Current Concepts of Experimental gerontology. Eds. C. Bertoni-Freddari and H. Niedermuller, Vienna Aging Series Vol. 6. P.261, 2001;6:261–276.

24. McLeod JD. Apoptotic capability in ageing T cells. Mech Age Dev 2000;121 :151–159.

25. Stosic-Grujicic SS, Lukic ML. The production of TNF, IL-1 and IL-6 in cutaneous tissues during maturation and aging. Adv Exp Med Biol 1995;371:411–414.

26. Gupta S. Molecular and biochemical pathways of apoptosis in lymphocytes from aged humans. Vaccine 2000;18:1596–1601.

27. Gupta S. Tumor necrosis factor-alpha-induced apoptosis in T cells from aged humans: a role of TNFR-I and downstream signaling molecules. Exp Gerontol 2002; 37: 293–299.

28. Gupta S, Chiplunkar S, Kim C, Yel L, Gollapudi S. Effect of age on molecular signalling of TNF-alpha-induced apoptosis in human lymphocytes. Mech Age Dev 2003;124:503–509.

29. Gupta S. Molecular steps of cell suicide: an insight into immune senescence. J Clin Immunol 2000;20 :229–239.

30. Screaton G, Xu X.-N. T cell life and death signaling via TNF-receptor family members. Cur. Opin Immunol 2000;12:316–3222.

31. Yuan J. Transducing signals of life and death. Curr Opin Cell Biol 1997; 9: 247–251.

32. Rath PC, Aggarwal BB. TNF-induced signaling in apoptosis. J Clin Immunol 1999;19 : 350–364.

33. Hsu H, Shu HB, Pan MG, Goeddel DV. TRADD-TRAF2 and TRADD-FADD interactions define two distinct TNF receptor 1 signal transduction pathways. Cell 1996;84:299–308.

34. Darnay BG, Aggarwal BB. Early events in TNF signaling: a story of associations and dissociations. J Leukocyte Biol 1997;61:559–566.

35. Nicholson DW. Caspase structure, proteolytic substrates, and function during apoptotic cell death. Cell Death Differ 1999;6:1028–1042.

36. Kataoka T, Budd RC, Holler N, Thome M, Martinon F, Irmler M, Burns K, Hahne M, Kennedy N, Kovacsovics M, Tschopp J. The caspase-8 inhibitor FLIP promotes activation of NF-B and Erk signaling pathways. Curr Biol 2000;10:640–648.

37. Potestio M, Pawelec G, Di Lorenzoc G, Candora G, D'Annaa C, Gervasi F, Lio D, Tranchida G. Caruso C, Colonna Romana G. Age-related changes in the expression of CD95 (APO1/FAS) on blood lymphocytes. Exp Gerontol 1999;34:659–673.

38. Effros RB. Loss of CD28 expression on T lymphocytes: a marker of replicative senescence. Dev Comp Immunol 1997;21:471–478.

39. McLeod JD, Walker LSK, Patel YI., Boulougouris G, Sansom DM. Activation of human T cells with superantigen (SEB) and CD28 confers resistance to apoptosis via CD95. J Immunol 1998;160 :2072–2079.

40. Boise LH, Thompson CB. Bcl-XL can inhibit apoptosis in cells that have undergone Fas-induced protease activation. Proc Natl Acad Sci 1997; 94;3759-

41. Timm JA, Thoman ML. Maturation of CD4+ lymphocytes in the aged microenvironment results in a memory-enriched population. J Immunol 1999;162:711–717.

42. McLeod JD. Apoptotic capability in ageing T cells.Mech Age Dev 2000; 121:151–159.

43. Spaulding C, Guo W, Effros RB. Resistance to apoptosis in human CD8+ T cells that reach replicative senescence after multiple rounds of antigen-specific proliferation. Exp Gerontol 1999; 34:633–644.

44. Deveraux QL, Stennicke HR, Salvesen GS, Reed JC. Endogenous inhibitors of caspases. J. Clin Immunol 1999;19 :388–398.

45. DeSmaele E, Zazzeroni F, Papa S, Nguyen DU, Jin R, Cong R, Franzoso G. Induction ogadd45 beta by NFkB doen-regulates proapoptotic JNK signaling. Nature 2001;414: 308–313.

46. Pahlavani M, Harris MD. The age-related changes in DNA binding activity of AP-1, NF-B, and Oct-1 transcription factors in lymphocytes from rats. Age 1996;19:45–54.

47. Ponnappan U, Zhong M, Trebilcock GU. Decreased proteosome-mediated degradation in T cells from the elderly: a role in immune senescence. Cell Immunol 1999;192:167–174.

48. Weiss T, Grell M, Siekienski K, Muhlenbeck F, Durkop H, Pfizenmaier K, Scheurich P, Wajant H. TNFR80-dependent enhancement of TNFR60-induced cell death is mediated by TNFR-associated factor and is specific for TNFR60. J. Immunol. 1998;161 :3136–3142.

49. Aggarwal S, Gollapudi S, Gupta S. Increased TNF--induced apoptosis in lymphocytes from aged humans: changes in TNF- receptor expression and activation of caspases. J Immunol 1999; 162 :2154–2161.

50. AshkenaziA, Dixit VM. Death receptors : signaling and modulation. Science 1998; 281: 1305–1308.

51. Suda T, Takahashi T, Goldstein P, Nagata S. Molecular cloning and expression of the Fas ligand, a novel memberof the tumour necrosis factor family. Cell 1994;75:1169–1175.

52. Scaffidi C, Fulda S, Sirinivasan A, Friesen C, LiF, Tomaselli KJ, Debatin KM, Krammer PH, Peter ME. Two CD95 (Apo-1/Fas) signalling pathway. EMBO J 1998;17:1675–1687.

53. Hsu HC, Shi J, Yang P, Xu X, Dodd C, Matsuki Y, Zhang HG, Mountz JD. Activated CD8+ T cells from aged mice exhibit decreased activation-induced cell death. Mech Age Dev 2001;122:1663–1684.

54. Phelouzat MA, Laforge T, Arbogast A, Quadri RA, Boutet S, Proust JJ. Susceptibility to apoptosis of T lymphocytes from elderly humans is associated with increased in vivo expression of functional Fas receptors. Mech Ageing Dev 1997;96:35–46.

55. Aggarwal S, Gupta S.Increased apoptosis of T cell subsets in ageing humans: altered expression of Fas (CD95), Fas ligand, Bcl-2, and Bax. J Immunol 1998;160:1627–1637.

56. Aggarwal S, Gupta, S.Increased activity of caspase-3 and caspase-8 during Fas-mediated apoptosis in lymphocytes from ageing humans. Clin Exp Immunol 1999;117:285–290.

57. Herndon FJ, Hsu HC, Mountz JD.Increased apoptosis of CD45RO-T cells with aging. Mech Ageing Dev 1997; 94:123–134.

58. Lechner H, Amort M, Steger MM, Maczek C, Grubeck-Loebenstein B. Regulation of CD95 (Apo-1) expression and the induction of apoptosis in human T cells: changes in old age. Int Arch Allergy Immunol 1996;110:238–243.

59. Kulms D, Dussmann H, Poppelmann B, Stander S, Schwartz A, Schwartz T. Apoptosis induced disruption of the actin cystoskeleton is mediated via activation of CD95 (Fas/APO-1) Cell Death Differ 2002;9:598–608.

60. Parlato S, Giammarioli AM, Lozupone F, Matrrese P, Lucani F, Falchi M, Malomi W, Fais S. CD95 (APO-&/Fas° likage to the actine cytoskeleton through ezrin in human T lymphocytes: anovel regulatory mechanims of the CD95 apoptosis pathway. EMBO J 2000;19:5123–5134.

61. Simons K, Ikonen E. Functional rafts in cell membranes. Nature 1997;387:569–574.

62. Wang TY, Leventis R, Silvius JR. Fluorescence-based evaluation of the partitioning of lipids and lipidated peptides into liquid-ordered lipid microdomains: a model for molecular partitioning into "lipid rafts". Biophys J 2000;79:919–925.

63. Kischkel FC, Hellbardt S, Behrmann I, Germer M, Pawlita M, Krammer PH, Peter ME. Cytotoxicity-dependent Apo_1 (Fas/CD95-associated proteins form a death inducing signalling complex (DISC) with the receptor. EMBO J 1995;14:5579–5588.

64. Grassme H, Jekle A, Riehle A, Schwartz H, Berger J, Sandhoff K, Kolesnick R, Gulbins E. CD95 signaling via ceramide rich lipid rafts. J Biol Chem 2001;276:20589–20596.

65. Scheel-Toellner D, Wang K, Singh R, Majeed S, Raza K, Curnow SJ, Salmon M, Lord JM. The death-inducing signalling complex is recruited to lipid rafts in Fas induced apoptosis. Biochem Biophys Res Commun 2002;297:876–879.

66. Larbi A, Douziech N, Dupuis G, Khalil A, Pelletier H, Guerard KP, Fülöp T. Jr. Age-associated alterations in the recruitment of signal transduction proteins to lipid rafts in human T lymphocytes. J Leuk Biol 2004;75;373–381.

67. Fülöp, T. Jr. Douziech N, Larbi A, Dupuis G. The role of lipid rafts in T lymphocyte signal transduction with aging. Ann. NY Acad. Sci. 2002;973:302–304.

68. Green DR, Reed JC. Mitochondria and apoptosis. Science 1998;281:1309–1312.

69. Hay BA. Understanding IAP function and regulation: a view from drosophila. Cell Death Differ 2000;7:1045–1056.

70. Daugas E, Susin SA, Zamzami M. Kroemer G Mitochondrio-nuclear translocation of AIF in apoptosis and necrosis. FASEB J 2000;14:729–739.

71. Beckman KB, Ames RN. Mitochondrial aging: open questions. Ann N Y Acad Sci 1998;854:118–127.

72. Beckman KB, Ames RN. Endogeneous oxidative damage of mtDNA. Mutat Res 1999;424: 51–58.

73. Beckman KB, Ames RN. The free radical theory of aging matures. Physiol Rev 1998;854: 547581.

74. Barja G, Cadenas S, Rojas C, Lopez-Torres M, Perez Campo R. A decrease of free radical production near critical targets as a cause of maximum longevity in animals. Comp Biochem Physiol Biochem Mol Biol 1994;108501–512.

75. Lenton K, Thériault H, Fulop T. Jr., Payette H, Wagner R. Glutathione and and ascorbate are negatively correlated with oxidative DNA damage in human lymphocytes. Carcinogenesis 1999;20: 607–613.

76. Fulop T. Jr, Foris G, Worum I., Leovey A. Age-dependent alterations of Fc receptor mediated effector functions of human polymorphonuclear leukocytes. Clin. Exp. Immunol. 1985;61: 425–432.

77. Vlahos CJ, Matter WF. Signal transduction in neutrophil activation. Phosphatidylinositol 3-kinase is stimulated without tyrosine phosphorylation. FEBS Lett. 1992;309: 242–248.

78. Wenisch C, Patruta S, Daxbock F, Krause R, Horl W. Effect of age on human neutrophil function. J. Leuk. Biol. 2000;67: 40–45.

79. Tortorella C, Piazzolla G, Spaccavento F, Jirillo E, Antonaci S. Age-related effects of oxidative metabolism and cyclic AMP signaling on neutrophil apoptosis. Mech. Ageing Dev. 1999;110:195–205.

80. Varga Zs, Kovacs E, Paragh G, Jacob MP, Robert L, Fulop T. Jr. Effects of Kappa elastin and FMLP on polymorphonuclear leukocytes of healthy middle-aged and elderly. K-elastin induced changes in intracellular free calcium. Clin. Biochem. 1988;21:127–135.

81. Fulop T. Jr. Barabas, G, Varga, Zs, Csongor, J, Seres, I, Szucs, S, Szikszay, E, Penyige, A. Alteration of transmembrane signalling with aging. Identification of G proteins in lymphocytes and granulocytes. Cellular Signaling. 1993;5: 593–604.

82. Fulop T. Jr. Foris G, Worum I, Paragh G, Leovey A. Age-related variations of some PMNL functions. Mech. Age. Dev. 1985;29: 1–11.

83. Fulop T. Jr, Douziech N., Jacob M.P., Hauck M., Wallach J. and Robert L. Age-related alterations in the signal transduction pathways of the elastin-laminin receptor. Pathol. Biol. 2001;49: 339–348, 2001.

84. Akgul C, Moulding DA, Edwards SW. Molecular control of neutrophil apoptosis. FEBS Lett.. 2001;487:318–322.

85. Whyte M, Renshaw S, Lawson R, Bingle C. Apoptosis and the regulation of neutrophil lifespan. Biochem. Soc. Transactions 1999;27: 802–807.

86. Chang LC, Wang JP. Examination of the signal transduction pathways leading to activation of extracellular signal-regulated kinase by formyl-methionyl-leucyl-phenylalanine in rat neutrophils. FEBS Lett. 1999;454:165–168.

87. Chang LC, Wang JP. Activation of p38 mitogen-activated protein kinase by formyl-methionyl-leucyl-phenylalanine in rat neutrophils. Eur. J. Pharmacol. 2000;390:61–66.

88. Porter AG, Janicke RU. Emerging role of caspase-3 in apoptosis. Cell Death Differ 1999;6: 99–104.

89. Watson RWG, O'Neill A, Branningen A, Coffey R, Marchall JC, Brady HR, Fitzpatrick

JM. Regulation of Fas antibody induced neutrophil apoptosis is both caspase and mitochondrial dependent. FEBS Lett. 1999;453:67–71.

90. De Groot RP, Coffer PJ, Koenderman L. Regulation of proliferation, differentiation and survival by IL3/IL-5/GM-CSF receptor family. Cell Signal 1998;10:619–628.

91. Fulop T. Jr, Larbi A, Linteau A, Desgeorges S, Douziech N. The Role of Mcl-1 and Bax Expression Alteration in the Decreased Rescue of Human Neutrophils from Apoptosis by GM-CSF with Aging. Annals of the New York Academy of Sciences. 2002;973: 305–308.

92. Harman D. Aging: A theory based on free radical and radiation chemistry. J. Gerontol. 1956;11: 298–300.

93. Fulop T. Jr, Varga Z, Jacob M.P, Robert L. Effects of Lithium on superoxyde production and intracellular free calcium mobilization in elastin peptide (kappa-elastin) and FMLP stimulated human polymorphonuclear leukocytes. Effect of aging. Life Sci. 1997;60: PL325–332.

94. Varga Zs, Jacob MP, Robert L, Fulop T. Jr. 1997. Studies on effector functions of soluble elastin receptors of phagocytic cells at various ages. Exp. Gerontol. 1997;32 : 653–662.

*The Neuroendocrine Immune Network in Ageing*
Edited by R.H. Straub and E. Mocchegiani

# MHC-Unrestricted Cytotoxicity in Ageing

MAURO PROVINCIALI, ALESSIA DONNINI and FRANCESCA RE

*Laboratory of Tumor Immunology, INRCA Gerontol. Res. Dept., Via Birarelli 8, 60121
Ancona, Italy*

## ABSTRACT

The studies of age-associated immune remodeling have been concentrated for many years on
the adaptive response, particularly considering the relevance of post-pubertal thymus involution
that may determine the level of specific immunity. For the past few years, however, the interest
has been increasing in innate immunity, based on new knowledge of its integration with specific immune effector mechanisms and its key role in initiating the clonal response of adaptive
immunity.

This paper will examine the numerical and functional alterations of MHC-unrestricted immunity during ageing, particularly focusing on cytotoxic cells, and the role that alterations of these
cells may have in the age-related impairment of adaptive immunity.

## 1.    PREMISES

Experimental and clinical data have demonstrated that ageing is associated with a dysregulation
of the immune system, this is called immunosenescence. The age-related immune alteration has
been associated with the increase of infections, tumors, and autoimmune diseases observed in
the elderly population. For many years, studies on immunity during ageing have concentrated on
the adaptive response and its hallmarks, particularly considering the relevance that post-pubertal
thymus involution may determine at the level of specific immunity. Age-related changes in the
adaptive immune system have been well documented and include reduction in clonal expansion
and function of Ag-specific T and B cells and diminished and/or altered cytokine patterns [14].

The phylogenetically ancient defence mechanism, also known as the innate immune system,
has been considered for long time as a separate entity from the adaptive immune response and
has been regarded to be of less importance in the hierarchy of immune functions. For the past
few years, however, interest in innate immunity has grown enormously, based on new knowledge of its integration with specific immune effectors and its key role in stimulating the subsequent, clonal response of adaptive immunity. MHC-unrestricted cytotoxic cells, besides to their
direct killing of target cells, seem to play a pivotal role in providing signals that are required for
directing the adaptive immune response. On this ground, age-related alterations of the cellular
components of innate immunity might be involved in the impairment of the adaptive immunity
observed in the elderly.

In this article we will discuss the numerical and functional alterations of MHC-unrestricted immunity during ageing, particularly focusing on cytotoxic cells, and the role that alterations of these cells may have in the age-related impairment of adaptive immunity.

## 2.    MACROPHAGES

Macrophages are important cellular constituents of innate immunity that influence the priming environment in addition to their phagocytic and tumoricidal functions. They are able to kill bacteria, viruses, parasites and tumor cells either directly or through the release of mediators, such as IL-1, TNF-$\alpha$, IFN-$\gamma$, which, in turn, can activate other immune cells. The antigen presenting capacity of macrophages is linked to their ability to process antigens and to present digested peptides to T lymphocytes. The function of monocytes and macrophages in ageing has been less well studied in comparison with that of other leukocyte populations. Early studies conducted in the 1970–1980 in ageing mice showed normal or enhanced macrophage function [57]. Normal antimicrobial levels in activated mouse peritoneal macrophages, but delayed onset of activation, were found in Toxoplasma gondii-infected old mice [8]. The study of functional and physiological properties of monocytes from young and old subjects revealed that neither chemotaxis, nor phagocytosis, or the adhesion capabilities of peripheral blood monocytes were significantly altered in the elderly group [9]. Monocyte phagocytosis was found increased with ageing in two studies [10,11]. Furthermore, macrophages from old and young adults appeared to produce similar levels of cytokines [12,13]. Recent studies however, have suggested that macrophage number and function may indeed be altered with ageing. A significant expansion of CD14[dim]/CD16[bright] circulating monocytes, which are considered to have phenotypic evidence for activation, has been reported to occur in elderly persons [14]. In the same study, a significant increase of the constitutive production of cytokines including, interleukin-1 (IL-1)-$\beta$, IL-1 receptor antagonist, and IL-6, by nonstimulated monocytes from elderly was found. Another study reported a higher production of IL-6 and IL-1 receptor antagonist (IL-1Ra) in the elderly than in young control subjects, whereas the levels of IL-1$\beta$ and TNF-$\alpha$ where not correlated with age [15]. The intracellular levels of TNF-$\alpha$, IL-6, and IL-1$\beta$, when evaluated through flow cytometry, were increased in monocytes from elderly subjects [16]. However, in a recent study conducted in young and old subjects screened through the Senieur protocol, no age-related differences were noted in the absolute amounts of IL-1$\beta$ and IL-6 serum levels after normalizing for circulating monocytes [17]. A defect in macrophage function in old humans was indirectly suggested showing that replacement of "old" macrophages with other sources for activation, such as interleukin-2, or phorbol 12-myristate 13-acetate, enhanced T cell responses [18]. Monocytes from old donors, when compared with monocytes from young subjects, displayed decreased cytotoxicity against tumor cells after LPS activation, the impairment being associated with decrease in IL-1 secretion and production of reactive oxygen intermediates such as $NO_2$ and $H_2O_2$ [19]. These contradictory results may be explained by the differences in experimental conditions and by the health status of the population studied.

Alterations of macrophage number and function have been described also in aged rats and mice. Wound healing, a process regulated by macrophages, was shown to take twice as long in old mice than in young mice [20]. An age-associated impairment in TNF-$\alpha$ production was found in rat macrophages [21]. Similarly, macrophages from aged mice, in vitro activated with IFN-$\gamma$ and LPS, exhibited reduced antitumor activity and impaired capacity to produce TNF, IL-1 and nitric oxide, which are critical monokines and effector molecules able to directly inhibit tumor growth [22]. In another study, macrophages from old mice were found to be less responsive to the activation signal of LPS and IFN γ for tumoricidal activity and for the production of effector molecules, such as oncostatin M, TNF, and nitric oxide [23]. Differently, macrophages from aged rats were found to produce more TNF-α and prostaglandin I than those from young animals after activation with LPS [24]. A lower expression of the product of MHC class II gene was found on the cell surface of macrophages from aged mice after incubation with IFN-γ [25]. Finally, recent findings have been reported on the expression of toll-like receptors (TLR) on macrophages in ageing mice [26]. Both splenic and activated peritoneal macrophages from aged mice have been found to express significantly lower levels of TLRs. These cells also secreted lower levels of IL-6 and TNF-α when stimulated with known ligands for TLR when compared with those from young mice. Decreased expression and function of various TLRs may increase the susceptibility of the elderly to infections and tumors.

## 3.    POLYMORPHONUCLEAR LEUKOCYTES

The polymorphonuclear leukocytes (PMN) combat the various microorganisms that could invade the host and forms the first line of defence against rapidly dividing bacteria and fungi. However, elderly subjects have been found more susceptible to infection with these microbes and this fact has been ascribed to a decline in immune status. Bacterial killing by PMN is a complex process. The attachment to the cell surface of a bacterium fully opsonised by antibody and complement induces an increase in the concentration of intracellular free $Ca^{2+}$, which triggers a series of metabolic events and initiates the respiratory burst. The oxidative respiratory burst is characterized by a dramatic increase in oxygen consumption and an hexose monophosphate shunt activity resulting in the generation of the bactericidal products: superoxide anion, singlet oxygen, hydrogen peroxide, and hydroxyl radical. Vulnerability to infections in the elderly could result from an age-related decline either in neutrophil supply and/or functional efficiency.

Several studies showed that neutrophil number in the blood and neutrophil precursors in the marrow are not lowered in the healthy elderly, although the proliferative response of neutrophil precursors cells to G-CSF was reduced, whereas response to GM-CSF and IL-3 was not affected by age [27,28]. Another paper showed that age is not associated with a change in neutrophil number or activity in the absence of bacterial infection. Infection in both young and elderly produced a significant increase in neutrophil number and chemiluminescence activity [29]. In vitro studies of leukocyte chemotaxis have found migratory responses of neutrophils from healthy elderly subjects to be either unaltered [30,31] or only slightly reduced [32,33]. The data from the literature have reported differences concerning the effects of age on phagocytosis of PMNs. Although Corberand et al. [32] did not describe any age-related defect in endocytosis, other studies reported a dramatic decrease in the phagocytic activity of PMNs from aged individuals [34,35]. Bongrand et al. [36] showed a selective decrease of Fc-receptor-mediated phagocytosis by granulocytes in ageing people. In agreement with this finding Butcher and co-workers [37] proposed that reduced neutrophil CD16 expression and phagocytosis contribute to human immunosenescence.

Several groups have examined neutrophil microbicidal activity and although data are often conflicting [38,39] a majority of them support a decline in cytotoxicity towards bacteria and yeast with age [32,4043]. The decreased bactericidal activity and the reduced phagocytic ability of neutrophils in the elderly could be explained by an increase of intracellular calcium concentrations in resting neutrophils and/or a reduced hexose uptake [44]. Unchanged PMN bactericide activity in ageing was described in other papers [31,45,46].

## 4. NATURAL KILLER CELLS

Natural Killer (NK) cells are large granular lymphocytes (LGL) that represent a critical first line of defence against infectious diseases and tumors. NK cells are non-adherent, they kill by non-phagocytic mechanisms, and their cytotoxicity is not restricted by the major histocompatibility complex (MHC) system [47]. It is currently believed that NK cells arise from bone marrow, undergo further processing in lymph nodes and spleen, and finally are released into the peripheral blood. The thymus is not required for their processing, as demonstrated by the evidence that NK cells develop normally in nude (athymic) mice and in children with congenital thymic aplasia. Several abnormalities have been described with advancing age both in animals and humans at the level of NK cell number and activity.

In mice, a peak value of basal NK activity is observed at about 58 weeks of age while nearly undetectable levels are present in 25 month-old animals [48–50]. A similar finding is observed in rats, which have a peak NK cell activity at 5 to 20 weeks of age, with a decline at 36 weeks of age [51,52]. With regard to the age-related pattern of cytokine-induced NK cell activity, and in particular of IFN and IL-2 stimulation of NK cells, it has been pointed out that in old mice the sensitivity of spleen cells to IL-2 is generally maintained, though to a reduced level when compared to maturity, while the responsiveness to IFN is strongly reduced [50,53]. In agreement with this finding aged C57Bl/6 and Balb/c mice had a significantly reduced IFN-$\alpha$/$\beta$-stimulated NK cytotoxicity compared to adult mice [54]. Such defects do not seem to depend on the increased presence, in old mice, of cells with suppressor activity or to the requirement of other factors, which might lack in *the in vitro* system, since also *in vivo* administration of IFN is not effective in old mice [50]. As possible explanation for the decreased ability of NK cells of aged mice to respond to IFN-$\alpha$/$\beta$, either an altered number of IFN-$\alpha$/$\beta$ receptors or an increased percentage of NK cells undergoing apoptosis have been proposed [54]. Roder [55] suggested that the impaired NK activity of old mice is not due to a decrease of NK cell activity or of the number of binding sites, but rather is due to a decrease in the size of the splenic NK cell pool. In another paper, the decline of NK activity during ageing was attributed to a loss of competence to lyse targets rather than to a major decline in the actual number of NK cells [56]. The mechanism for the aged-related decline in NK activity in mice has been further explained by an increase in adherent cell suppressor function [57]. A declined NK cell response to poly(I:C) and LPS in aged animals was correlated with the decreased expression of the IFN-$\gamma$ found in these animals [58]. Independently by the mechanism, the defect of NK activity in aged mice does not represent an irreversible process, since it may be recovered by hormonal and nutritional treatment [59]. Among hormonal factors relevant for NK function, it has been observed that thymic peptides, in particular thymulin (Zn FTS), when administered *in vitro*, are able to restore the crippled NK cytotoxicity of spleen cells from old mice [60]. Similarly, the in vivo administration of *thyroid hormones*, but not of the pineal hormone melatonin, was able to recover the reduced NK activity in aged mice [61,62]. Among nutritional factors, either zinc or a lipid mixture able to increase membrane fluidity, called "active lipids", were found able to recover the impaired NK function in aged animals [63]. Other studies indicated that diet and the microbial flora can modulate NK cell activity of mice in ageing [64].Whether or not endocrine and nutritional factors have an additive effect or act through the same intracellular mechanism remains to be established. The first possibility seems more likely since the action of TSH and thyroid hormones is specifically directed towards lymphokine-boosted NK activity, while active lipids are able to prevent age-associated impairment of basal NK cytotoxicity [65]. Apart from IFN and IL-2, the effect of cytokines on the development of cytotoxic cells has been scarcely investigated during ageing.

In a study conducted with IL-12, this cytokine was able to boost both endogenous and IL-2-induced NK cell activity in young and old mice. The levels of cytotoxicity were lower in old than in young animals although the relative increase of IL-12 plus IL-2 versus IL-2 alone was greater for old mice [66]. These data confirm and extend previous findings obtained in the man and show that IL-12 is able to enhance NK cytotoxicity to the same degree in both young and elderly subjects, whereas the induction of IL-2-activated cytotoxic cells was decreased in the elderly compared to young individuals [67].

Less work has been done on the effect of ageing on the generation of lymphokine-activated killer (LAK) cells by IL-2 in mice. In several papers no age-related difference was found in the cytolytic response of murine spleen cells after stimulation with IL-2 [65,68–70]. However, the use of IFN-$\gamma$ with rIL-2 in the culture augmented LAK activity in bone marrow cells of young mice, but, inhibition was observed with bone marrow cells of old mice [71]. In contrast, Saxena et al. [53] and Bubenik et al. [70] described a decline in IL-2-induced LAK activity in aged mice.

The changes that may occur in human NK cell activity with advancing age are not yet well defined. The data reported in the literature differ according to the selection criteria of the elderly population studied and of the use of total lymphocyte populations or of purified NK cells.

Various studies have demonstrated the age-related numerical alterations of NK cells. Both the absolute and the relative numbers of circulating NK cells significantly increased in healthy elderly people in comparison with young adults. The increased percentage of NK cells in the elderly has been mainly related to the higher number of the CD56$^{dim}$ population, which represents the mature NK cell subset. In contrast, the number of CD56$^{bright}$ cells did not seem to be significantly altered in old subjects, thus suggesting the existence of a phenotypic shift in the maturity status of NK cells during ageing [72]. In another study, there was a significant decrease in the percentage of CD56$^{bright}$ cells amongst CD56$^{+}$ cells in the elderly with a relative sparing of the CD56$^{dim}$ subset. Moreover, the CD56$^{bright}$/CD56$^{dim}$ ratio representing NK cell maturity status, declined with age [73].

Either no change or decrement, or an increment of NK activity have been reported in PBMC from unselected elderly subjects in various studies [74]. However, NK cell activity was not found to be significantly different in elderly and young subjects when selection was done according to protocols specific for immunogerontological studies, such as the Senieur protocol, [75]. When centenarians were analyzed as a separate group, they were found to have higher NK and redirected killer activity than middle-aged donors [76]. A decrease of the cytolytic activity per cell was found using NK cell populations from elderly subjects purified through sorting or after cloning of NK cell precursors by limiting dilution [77]. The decreased cytotoxicity per NK cell seems to be time-dependent and probably related to a slowing down of the lytic phase since it was observed in short-term but not in long-term assays [78]. The impairment of NK cell activity was a post-binding event and was probably due to a defect of the activating signal through the cell membrane or a decreased activity of the lytic machinery. A possible role for perforin expression in the age-related impairment of NK activity was suggested. Both the proportion of perforin containing cells and the mean levels of perforin per cell were reduced in a population of NK cells from apparently healthy elderly people [79]. As reported for other immune parameters, the distribution of perforins in cytolytic NK cells, the number of cells producing perforins, and the ability to utilize perforins in the generation of cytolytic activity against tumor target cells, were found similar in NK cells from young and elderly subjects selected according to the Senieur protocol [80]. Other data from the literature have suggested an impairment of the metabolic pattern of activation as responsible for the modification of NK lytic activity in the elderly. In particular, a pronounced age-related decrease in the ability to generate total inositol monophosphates and, specially, inositol trisphosphates by NK cells following K562 stimulation has been reported. Phosphatidylinositol bisphoaphate hydrolysis was attenuated and delayed, while phosphoinositide turnover was preserved following Fc triggering [81].

Limited and contradictory findings are available on the response of NK cells from elderly humans to the boosting action of IFN or IL-2. Decreased [82] or unchanged [83] NK responses to IL-2 and to IFN boosting have been found in peripheral blood lymphocytes from old humans [84]. Stimulation of NK cytotoxicity against the NK-sensitive K562 cell line with IL-2 or IFN-$\alpha$ from "healthy" elderly donors yielded similar results to cells of young donors [67,78]. In one of these papers, similarly to IL-2 and IFN, the use of IL-12 was found to enhance NK cytotoxicity to the same degree in cells of both young and elderly subjects over a wide range of doses and incubation times [67]. A progressive decline in both IFN and IL-2 inducible NK activities appeared with increasing age in humans selected by means of the Senieur protocol [85]. The defect was more evident at the level of IFN rather than of IL-2 responsiveness of NK cells. This is similar to that observed in mice and suggests, that the IFN inducibility of NK cells is more affected by age than that of IL-2 inducibility. Differently, other data showed that the response of NK cells to IL-2 was impaired when proliferation, expression of CD69, and $Ca^{2+}$ mobilization where considered [72]. In agreement with these findings NK cells from elderly people showed a decreased proliferative response to IL-2 and impaired expression of the CD69 activation antigen [86].

In addition to their endogenous activity and to the short-term activation mediated by cytokines, NK cells have been extensively studied as precursors of "lymphokine-activated killer" (LAK) cells after long-term activation with IL-2. The LAK activity is mediated by broadly reactive cytotoxic cells that are not major histocompatibility complex restricted and are capable of lysing fresh and cultured tumors, including both NK-sensitive and NK-resistant targets. In humans most of the LAK activity is associated primarily with cells of the CD56+ and CD16+ phenotype, the precursor of which is similar to NK cells. The effect of ageing on the *in vitro* induction of LAK cell activity has not been extensively studied. Two initial studies indicated that LAK activity induced by short-term activation with IL-2 declines with age [70,87]. In another study, conducted as the previous in healthy subjects non-selected according to specific protocols for immunogerontology but only on the bases of exclusion of major diseases or assumption of potential immunosuppressive agents, a significant decrease of LAK activity was found after induction with IL-2, IL-12, or IFN-$\alpha$ in elderly donors in comparison with young subjects [67]. When the kinetics of development of LAK cells was evaluated in young and old healthy subjects screened in accordance with the Senieur protocol, no age-related difference was found in terms of proliferative capacity, cytotoxicity against Daudi tumor cells, and expression of p55 or p75 IL-2 receptors [88]. Differently, the proportion of CD56+ and CD16+ cells reached higher levels in old than in young donors. This suggests that an increased number of cytotoxic cells are required in old subjects to obtain the same levels of LAK cell activity what is present in young age [88]. The continuous culture or the short pulse of peripheral blood mononuclear cells with IL-2 was able to develop significant levels of LAK cell cytotoxicity also in elderly cancer patients [89].

In addition to their direct cytotoxic effects, NK cells have been shown to represent one of the first lines of defence during the early stages of immune activation because of their inducible secretory function. NK cells synthesize many cytokines and chemokines that exert autocrine and paracrine regulation of natural and adaptive immune functions. IL-2 induced NK cells from healthy elderly donors secreted only ¼ of IFN-$\gamma$ released by the cells from young subjects [90]. However, this ageing-related early phase secretory deficit could be overcome by chronic stimulation with IL-2 [73]. NK cells from nonagenarian healthy subjects were found to express the receptors for the chemokines MIP-1α, Rantes, and IL-8, to retain the ability to synthesize these chemotactic cytokines, and to up-regulate their production in response to stimulation by IL-12 or IL-2 cytokines, though the production remained lower in comparison with young subjects [91,92]. Similarly, NK cells from elderly subjects spontaneously produced detectable amounts of IL-8 and increased the production of this cytokine after activation with IL-2, but at lower levels than in young age [91]. Furthermore, early IFN-γ and chemokine production in response to IL-2 or IL-12 were also decreased. The TNF-α production by NK cells was not altered during ageing [93] whereas the IFN-γ production was found significantly augmented in centenarians [94].

## 5.    γδ T CELLS

T lymphocytes bearing the γδ T cell receptor (TCR) represent a minor population of human peripheral lymphocytes (1–10%), the majority of them expressing the CD3$^+$CD4$^-$CD8$^-$ phenotype [95–98]. The ability of γδ T cells to respond to nonprocessed and nonpeptidic phosphoantigens in a major histocompatibility complex (MHC)-unrestricted manner is an important feature distinguishing them from γδ T cells [99–103]. In human peripheral blood two main populations of γδ T cells have been identified based on the TCR composition. The predominant subset expresses the Vδ2 chain associated with Vγ9 and represent 70% of the circulating γδ T cells in adults, while a minor subset (approximately 30%) expresses a Vδ1 chain linked to a chain different from Vγ9. At birth the Vδ1 population predominates, while in adults there is a shift towards Vδ2 T lymphocytes, probably due to a selective response to environmental stimuli such as commonly encountered bacteria [104]. Although little is known about the physiologic significance of γδ T cells, their marked reactivity toward mycobacterial and parasitic antigens as well as tumor cells suggests that γδ T cells play a role in the anti-infectious and anti-tumoral immune surveillance [98]. Moreover it has been shown that in healthy donors, γδ T cell stimulated with nonpeptidic phosphoantigens such as isopentenylpyrophosphate (IPP), produced high levels of cytokines and mainly of IFN-γ and TNF-α [105]. Because of their cytokine production, γδ T cells have been proposed to be involved in coordinating the interplay between innate and adaptive immunity and in particular to guide the establishment of acquired immunity contributing to the definition of αβ T cell responses toward T helper cell type 1 (Th1) or Th2 phenotype [106,107].

The data from the literature on numerical or functional changes of γδ T cells during ageing are scarce and fragmentary. It has been reported that the complexity of the gamma delta T cell repertoire decreases with age as a consequence of the expansion of a few T cell clones [108]. The analysis of γδ T cell number and function in ageing has been examined by two groups with similar results. Colonna-Romano et al. [109] reported that the percentage of blood γδ T cells in lymphocytes from old subjects is decreased when compared with the young. Interestingly, these cells were found more activated in the elderly than in young subjects, as determined by the increased expression of the early activation marker CD69, on γδ T lymphocytes from old subjects. Argentati et al. [110] confirmed and extended these preliminary results not only in elderly people but also in centenarians. They demonstrated an age-dependent alteration of γδ T lymphocytes, with a lower frequency of circulating γδ T cells, an altered pattern of cytokine production, and an impaired in vitro expansion of these cells. The decrease of the γδ T cell number was due to an age-dependent reduction of Vδ2 T cells whereas the absolute number of Vδ1 T cells was unaffected by age. As a consequence, the Vδ2/ Vδ1 ratio was inverted in old subjects and centenarians. A higher percentage of γδ T cells producing TNF-α was found in old donors and centenarians whereas no age-related difference was observed in the IFN-γ production. After a 10-day in vitro expansion, a two-fold lower expansion index of γδ T cells, and particularly of Vδ2 but not of Vδ1 subset, was found in old people and centenarians in comparison with young subjects demonstrating the existence of a proliferative defect in γδ T lymphocytes from aged subjects. Differently, the cytotoxicity of sorted γδ T cells was preserved in old people and centenarians thus suggesting that, as previously observed for αβ T cells, the lytic function of γδ T cells is not impaired during ageing [111].

These results on numerical and functional changes of γδ T cells in ageing were successively confirmed by Colonna-Romano et al. [112] who suggested that the high level of basal activation of γδ T cells, evidenced by increased CD69 expression, was due to the "inflamed" environment of the elderly host.

6.    MHC-UNRESTRICTED IMMUNITY: IMPACT ON THE SENESCENCE OF ADAP-
      TIVE IMMUNE RESPONSES

In the past few years, interest in MHC-unrestricted immunity has grown enormously because of its role in the development of adaptive immune responses. In fact, besides to act in the early phase of immune defence, innate immunity in mammals appears to play a key role in stimulating and orienting the subsequent, clonal response of adaptive immunity [106,113,114].

The MHC-unrestricted immune cells control the initiation of the adaptive immune response through the regulation of costimulatory molecules on antigen-presenting cells (APCs) and instruct the adaptive immune system to develop a particular effector response by releasing chemokines and cytokines. It is well known that a signal received through the T cell antigen receptor is not sufficient on its own for the activation of naive lymphocytes since a second so-called costimulatory signal is required to induce effective immunogenicity. This costimulatory activity seems to be required to signal the presence of a pathogen to antigen-specific receptors on T lymphocytes. Among the factors able to regulate the expression of costimulatory activity on APCs are sets of receptors of the nonclonal innate recognition system called pattern-recog-nition receptors (PRRs). Among these, Toll-like receptors (TLRs) are receptors that recognize conserved molecular patterns, which are shared by large groups of microbial components and are perfectly able to discriminate self from non-self pathogen-associated structures and then to signal the presence of a pathogen to APCs. TLRs expression has been observed in several types of immune cells, mainly on dendritic cells, monocytes, and B cells. It seems then clear that APCs maturation and activation, which is in part mediated through the TLRs family signalling, represents a critical link between innate and adaptive immunity. Furthermore, APCs maturation is characterized by the production of proinflammatory cytokines (IL-12 and TNF-α), up-regula-tion of costimulatory molecules (CD40, CD80, and CD86), and altered expression of chemokine receptors (CCR2, CCR5, and CCR7), all of these being necessary for the activation and driv-ing of adaptive responses. Although in some study it has been shown that APCs function is not affected by ageing [115–121], in several other studies a reduced function and an alteration of the antigen presentation capacity of APCs has been demonstrated during ageing [122–127]. This impairment may be related to the decreased expression and function of TLRs which has been very recently demonstrated on APCs from aged mice [26]. Furthermore, the migratory capacity of DCs has also been found affected by the ageing process [115]. The microenvironment existing in the "old" host may certainly influence the differentiation and performance of either APCs or other immune cells. It has been observed that IL-10, a key cytokine that can suppress cell mediated immunity and maturation of DC subsets, is elevated in the healthy elderly. However, the production of IL-12 required for the initiation of T cell immune responses, declines in frail elderly along with APC antigen presenting function. These findings suggested that shifts in IL-10 and IL-12 may not only directly influence immune response but may also alter the balance and maturation of APC subsets. Although the absolute number and the expression of MHC I and II, CD80, and CD86 both on immature and mature APCs do not seem to be significantly different in young and old mice [128], dendritic cells in germinal centers of aged mice were found to lack expression of important costimulatory ligands such as CD86 [129], which would encourage the induction of anergy in the antigen-specific T cells with which they interacted. Furthermore, in APCs from old mice a lower expression of the mRNA for the migratory CCR7 chemokine receptor was found and a lower lymphocyte cytotoxicity and a reduced number of CD8$^+$ T cells producing IFN-$\gamma$ were induced in comparison with APCs from young animals [128]. The fact that CCR7 was greatly increased in mature APCs up to the levels found in young animals and that in vivo migration of APCs to regional lymph nodes was higher in old than in young mice, suggested that an increased migratory capacity of old APCs may be required to balance their reduced antigen presentation to cytotoxic lymphocytes [128]. Besides to APCs, other cellular components of innate immunity are involved in the modulation of the clonal immune response. Natural killer cells and $\gamma\delta$ T cells, once activated, may produce either effector cytokines, like IL-12 or IFN-$\gamma$, or proinflammatory cytokines, like IL-6. These cytokines are essential for the activation of the adaptive immune response and its direction into a particular effector type. The age-related alteration of these lymphocyte populations, as described above, stress the relevant role that they may have in senescence of T cell-mediated specific responses in the elderly and, in particular, in the dysregulation of the Th1/Th2 system, with a predominant production of Th2 cytokines, which has been reported in ageing [130–132]. Furthermore, the strict integration between the innate and the adaptive immune systems emphasizes the pivotal role that age-related alterations of the components of the innate immune system may have in the reduced development and effectiveness of clonal immune responses against infectious and tumor diseases in ageing [133–135].

## REFERENCES

1.  Thoman ML and Weigle WO. The cellular and subcellular bases of immunosenescence. In : Advances in Immunology. Dixon FJ Ed. Academic Press, San Diego, California, 1989, vol.46, pp. 221–261.
2.  Miller RA. The ageing immune system: primer and prospectus. Science 1996;273:70–74.
3.  Pawelec G and Solana R. Immunosenescence. Immunol. Today 1997 ; 18 :514–516.
4.  Burns EA and Goodwin JS. Immunological changes of aging. In: Comprehensive geriatric oncology. Balducci L, Lyman GH, Ershler W Eds. Harwood academic publishers, Amsterdam, The Netherlands. 1998; pp. 213–222.
5.  Cantrell W, Elko EE. Effect of age on phagocytosis of carbon in the rat. Exp Parasitol 1973;34:337–343.
6.  Jaroslow BN, Larrick JW. Clearance of foreign red cells from the blood of ageing mice. Mech Ageing Dev 1973;2:23–32.
7.  Bar-Eli M, Gallily R. Age-dependent macrophage functions in New Zealand black mice.

Cell Immunol 1979;45:309–317.

8.  Gardner ID, Remington JS. Ageing and the immune response. II. Lymphocyte responsiveness and macrophage activation in Toxoplasma gondii-infected mice. J Immunol 1978;120:944–949.

9.  Gardner ID, Lim ST, Lawton JWM. Monocyte function in ageing humans. Mech Ageing Dev 1981;16:233–239.

10. Wustrow TP, Denny TN, Fernandes G, Good RA. Changes in macrophages and their functions with ageing in C57BL/6J, AKR/J, and SJL/J mice. Cell Immunol 1982;69: 227–234.

11. Lovic M, North RJ. Effect of ageing on antimicrobial immunity: old mice display a normal capacity for generating protective T cells and immunologic memory in response to infection with Listeria monocytogenes. J Immunol 1985;135:3479–3484.

12. Delfraissey JF, Galanaud P, Dormont J, Wallon C. Age-related impairment of the in vitro antibody response in the human. Clin Exp. Immunol. 1980;39:208–214.

13. Delfraissy JF, Galanaud P, Wallon C, Balavoine JF, Dormont J. Abolished in vitro antibody response in elderly: exclusive involvement of prostaglandin-induced T suppressor cells. Clin Immunol Immunopathol. 1982;24:377–385.

14. Sadeghi HM, Schnelle JF, Thoma JK, Nishanian P, Fahey JL. Phenotypic and functional characteristics of circulating monocytes of elderly persons. Exp Gerontol 1999;34:959–970.

15. Roubenoff R, Harris TB, Abad LW, Wilson PW, Dallal GE, Dinarello CA. Monocyte cytokine production in an elderly population: effect of age and inflammation. J Gerontol A Biol Sci Med Sci 1998;53:M20–26.

16. O'Mahony L, Holland J, Jackson J, Feighery C, Hennessy TP, Mealy K. Quantitative intracellular cytokine measurement: age-related changes in proinflammatory cytokine production. Clin Exp Immunol 1998;113:213–219.

17. Ahluwalia N, Mastro AM, Ball R, Miles MP, Rajendra R, Handte G. Cytokine production by stimulated mononuclear cells did not change with ageing in apparently healthy, well-nourished women. Mech Ageing Dev 2001;122:1269–1279.

18. Beckman I, Dimopoulos K, Xu XN, Bradley J, Henschke P, Ahern M. T cell activation in the elderly: evidence for specific deficiencies in T cell/accessory cell interactions. Mech Ageing Dev 1990;51:265–276.

19. McLachlan JA, Serkin CD, Morrey KM, Bakouche O. Antitumoral properties of aged human monocytes. J Immunol 1995;154:832–843.

20. Danon D, Kowatch MA, Roth GS. Promotion of wound repair in old mice by local injection of macrophages. Proc Natl Acad Sci U S A 1989;86:2018–2020.

21. Corsini E, Battaini F, Lucchi L, Marinovich M, Racchi M, Govoni S, Galli CL. Reduction of transendothelial migration of mononuclear cells in interferon-beta1b-treated multiple sclerosis patients. Ann Neurol 1999;46:435.

22. Wallace PK, Eisenstein TK, Meissler JJ, Morahan PS. Decreases in macrophage-mediated antitumor activity with ageing. Mech Ageing Dev 1995;77:169–184.

23. Khare V, Sodhi A, Singh SM. Effect of ageing on the tumoricidal functions of murine peritoneal macrophages. Nat Immun 1996–97;15:285–294.

24. Tang Y, Di Pietro L. Feng Y, Wang X. Increased TNF-alpha and PGI(2), but not NO release from macrophages in 18-months-old rats. Mech Ageing Dev 2000;114:79–88.

25. Herrero C, Marques L, Lloberas J, Celada A. IFN-gamma-dependent transcription of MHC class II IA is impaired in macrophages from aged mice. J Clin Invest 2001;107:

485–493.

26. Renshaw M, Rockwell J, Engleman C, Gewirtz A, Katz J, Sambhara S. Cutting edge: impaired Toll-like receptor expression and function in ageing. J Immunol 2002;169: 4697–4701.

27. Chatta GS, Andrews RG, Rodger E, Schrag M, Hammond WP, Dale DC. Hematopoietic progenitors and ageing: alterations in granulocytic precursors and responsiveness to recombinant human G-CSF, GM-CSF, and IL-3. J Gerontol. 1993;48:M207–212.

28. Born J, Uthgenannt D, Dodt C, Nunninghoff D, Ringvolt E, Wagner T, Fehm HL. Cytokine production and lymphocyte subpopulations in aged humans. An assessment during nocturnal sleep. Mech Ageing Dev 1995;84:113–126.

29. Angelis P, Scharf S, Christophidis N. Effects of age on neutrophil function and its relevance to bacterial infections in the elderly. J Clin Lab Immunol 1997;49:33–40.

30. MacGregor RR, Shalit M. Neutrophil function in healthy elderly subjects. J Gerontol 1990;45:M55–60.

31. Phair JP, Kauffman CA, Bjornson A, Gallagher J, Adams L, Hess EV. Host defenses in the aged: evaluation of components of the inflammatory and immune responses. J Infect Dis 1978;138:67–73.

32. Corberand J, Ngyen F, Laharrague P, Fontanilles AM, Gleyzes B, Gyrard E, Senegas C. Polymorphonuclear functions and ageing in humans. J Am Geriatr Soc 1981;29:391–397.

33. McLaughlin B, O'Malley K, Cotter TG. Age-related differences in granulocyte chemotaxis and degranulation. Clin Sci (Lond) 1986;70:59–62.

34. Charpentier B, Fournier C, Fries D, Mathieu D, Noury J, Bach JF. Immunological studies in human ageing. I. In vitro functions of T cells and polymorphs. J Clin Lab Immunol 1981;5:87–93.

35. Antonaci S, Jirillo E, Ventura MT, Garofalo AR, Bonomo L. Non-specific immunity in ageing: deficiency of monocyte and polymorphonuclear cell-mediated functions. Mech Ageing Dev 1984;24:367–375.

36. Bongrand P, Bartolin R, Bouvenot G, Arnaud C, Delboy C, Depieds R. Effect of age on different receptors and functions of phagocytic cells. J Clin Lab Immunol 1984;15: 45–50.

37. Butcher SK, Chahal H, Nayak L, Sinclair A, Henriquez NV, Sapey E, O'Mahony D, Lord JM. Senescence in innate immune responses: reduced neutrophil phagocytic capacity and CD16 expression in elderly humans. J Leukoc Biol 2001;70:881–886.

38. Corberand JX, Laharrague PF, Fillola G. Neutrophils are normal in normal aged humans. J Leukoc Biol 1986;40:333–334.

39. Lipschitz DA, Udupa KB, Milton KY, Thompson CO. Effect of age on hematopoiesis in man. Blood 1984;63:502–509.

40. Esperaza B, Sanchez M, Ruiz M, Barranquero M, Sabino E, Merino F. Neutrophil function in elderly persons assessed by flow cytometry. Immunol Invest 1996;25:185.

41. Fulop T Jr, Foris G, Worum I, Leovey A. Age-dependent alterations of Fc gamma receptor-mediated effector functions of human polymorphonuclear leucocytes. Clin Exp Immunol 1985;61:425–432.

42. Fulop T Jr, Komaromi I, Foris G, Worum I, Leovey A. Age-dependent variations of intralysosomal enzyme release from human PMN leukocytes under various stimuli. Immunobiology 1986;171:302–310.

43. Lipschitz DA, Udupa KB, Boxer LA. The role of calcium in the age-related decline of neutrophil function. Blood 1988;71:659–665.

44. Wenisch C, Werkgartner T, Sailer H, Patruta S, Krause R, Daxboeck F, Parschalk B. Effect of preoperative prophylaxis with filgrastim in cancer neck dissection. Eur J Clin Invest 2000;30:460–466.

45. Palmblad J, Haak A. Ageing does not change blood granulocyte bactericidal capacity and levels of complement factors 3 and 4. Gerontology 1978;24:381–385.

46. Johnson DD, Renshaw HW, Warner DH, Browder EJ, Williams JD. Characteristics of the phagocytically induced respiratory burst in leukocytes from young adult and aged beagle dogs. Gerontology 1984;30:167–177.

47. Koren HS, Herberman RB. Natural killing-present and future. J Natl Cancer Inst 1983;70: 785–786.

48. Weindruch R, Devens BH, Raff HV, Walford RL. Influence of dietary restriction and ageing on natural killer cell activity in mice. J Immunol 1983;130:993–996.

49. Albright JW, Albright JF. Age-associated impairment of murine natural killer activity. Proc. Natl Acad Sci 1983;20:6371–6375.

50. Provinciali M, Muzzioli M, Fabris N. Timing of appearance and disappearance of IFN and IL-2 induced natural immunity during ontogenetic development and ageing. Exp Gerontol 1989;24:227–236.

51. Nunn ME, Djeu JL, Glaser M, Lavrin DH, Herberman RB. Natural cytotoxic reactivity of rat lymphocytes against syngeneic gross virus-induced lymphoma. J Natl Cancer Inst 1976;56:393–399.

52. Shellam GR, Hogg N. Gross-virus-induced in the rat. IV. Cytotoxic cells in normal rats. Int J Cancer 1977;19:212–224.

53. Saxena RK, Saxena QB, Adler WH. Interleukin-2-induced activation of natural killer activity in spleen cells from old and young mice. Immunology 1984;51:719–726.

54. Plett PA, Gardner EM, Murasko DM. Age-related changes in interferon-alpha/beta receptor expression, binding, and induction of apoptosis in natural killer cells from C57BL/6 mice. Mech Ageing Dev 2000;118:129–144.

55. Roder JC. Target-effector interaction in the natural killer (NK) cell system. VI. The influence of age and genotype on NK binding characteristics. Immunology 1980;41: 483–489.

56. Albright JW, Albright JF. Age-associated decline in natural killer (NK) activity reflects primarily a defect in function of NK cells. Mech Ageing Dev 1985;31:295–306.

57. Irimajiri N, Bloom ET, Makinodan T. Suppression of murine natural killer cell activity by adherent cells from ageing mice. Mech Ageing Dev 1985;31:155–162.

58. Hsueh CM, Chen SF, Ghanta VK, Hiramoto RN. Involvement of cytokine gene expression in the age-dependent decline of NK cell response. Cell Immunol 1996;173:221–229.

59. Fabris N and Provinciali M. Hormones. In: Natural immunity. (Nelson D Ed.) New York: Academic Press, 1989;306–347.

60. Muzzioli M, Mocchegiani E, Bressani N, Bevilacqua P, Fabris N. In vitro restoration by thymulin of NK activity of cells from old mice. Int J Immunopharmacol 1992;14:57–61.

61. Provinciali M, Muzzioli M, Di Stefano G, Fabris N. Recovery of spleen cell Natural Killer activity by thyroid hormone treatment in old mice. Nat Immun Cell Growth Regul 1991;10:22–236

62. Provinciali M, Di Stefano G, Bulian D, Stronati S, Fabris N Long-term melatonin supplementation does not recover the impairment of Natural Killer cell activity and lymphocyte proliferation in aging mice. Life Sciences 1997;67:857–864.

63. Provinciali M, Fabris, Pieri C Improvement of Natural Killer cell activity by in vitro active

lipids (AL 721) administration in old mice. Mech Ageing Develop 1990;52:245–254.

64. Bartizal KF, Salkowski C, Pleasants JR, Balish E. The effect of microbial flora, diet, and age on the tumoricidal activity of natural killer cells. J Leukoc Biol 1984;36:739–750.

65. Provinciali M, Pieri C, Fabris N. Reversibility of age associated NK defect by endocrinological and/or nutritional intervention. Ann Inst Super Sanita 1991;27:61–66.

66. Argentati K, Bartozzi B, Bernardini G, Di Stasio G, Provinciali M. Induction of Natural killer cell activity and perforin and granzyme B gene expression following continuous culture or short pulse with interleukin-12 in young and old mice. Eur Cytokine Netw 2000;11:59–65.

67. Kutza J, Murasko DM. Age-associated decline in IL-2 and IL-12 induction of LAK cell activity of human PBMC samples. Mech Ageing Dev 1996;90:209–222.

68. Kawakami K, Bloom ET. Lymphokine-activated killer cells and ageing in mice: significance for defining the precursor cell. Mech Ageing Dev 1987;41:229–240.

69. Bloom ET, Kubota LF. Effect of IL-2 in vitro on CTL generation in spleen cells of young and old mice: asialo GM1+ cells are required for the apparent restoration of the CTL response. Cell Immunol 1989;119:73–84.

70. Bubenik J, Cinader B, Indrova M, Koh SW, Chou CT. Lymphokine-activated killer (LAK) cells: I. Age-dependent decline of LAK cell-mediated cytotoxicity. Immunol Lett 1987;16:113–119.

71. Kawakami K, Bloom ET. Lymphokine-activated killer cells derived from murine bone marrow: age-associated difference in precursor cell populations demonstrated by response to interferon. Cell Immunol 1988;116:163–171.

72. Borrego F, Alonso MC, Galiani MD, Carracedo J, Ramirez R, Ostos B, Pena J, Solana R. NK phenotypic markers and IL2 response in NK cells from elderly people. Exp Gerontol 1999;34:253–265.

73. Krishnaraj R. Senescence and cytokines modulate the NK cell expression. Mech Ageing Dev 1997;96:89–101.

74. McNerlan SE, Rea IM, Alexander HD, Morris TC. Changes in natural killer cells, the CD57CD8 subset, and related cytokines in healthy ageing. J Clin Immunol 1998;18:31–38.

75. Ligthart GJ, Schuit HR, Hijmans W. Natural killer cell function is not diminished in the healthy aged and is proportional to the number of NK cells in the peripheral blood. Immunology 1989;68:396–402.

76. Sansoni P, Cossarizza A, Brianti V, Fagnoni F, Snelli G, Monti D, Marcato A, Passeri G, Ortolani C, Forti E, et al. Lymphocyte subsets and natural killer cell activity in healthy old people and centenarians. Blood 1993;82:2767–2773.

77. Facchini A, Mariani E, Mariani AR, Papa S, Vitale M, Manzoli FA. Increased number of circulating Leu 11+ (CD 16) large granular lymphocytes and decreased NK activity during human ageing. Clin Exp Immunol 1987;68:340–347.

78. Bruunsgaard H, Pedersen AN, Schroll M, Skinhoj P, Pedersen BK. Decreased natural killer cell activity is associated with atherosclerosis in elderly humans. Exp Gerontol 2001;37:127–136.

79. Rukavina D, Laskarin G, Rubesa G, Strbo N, Bedenicki I, Manestar D, Glavas M, Christmas SE, Podack ER. Age-related decline of perforin expression in human cytotoxic T lymphocytes and natural killer cells. Blood 1998;92:2410–2420.

80. Mariani E, Ravaglia G, Meneghetti A, Tarozzi A, Forti P, Maioli F, Boschi F, Facchini A. Natural immunity and bone and muscle remodelling hormones in the elderly. Mech

Ageing Dev 1998;102:279–292.

81. Mariani E, Mariani AR, Meneghetti A, Tarozzi A, Cocco L, Facchini A. Age-dependent decreases of NK cell phosphoinositide turnover during spontaneous but not Fc-mediated cytolytic activity. Int Immunol 1998;10:981–989.

82. Rabinowich H, Goses Y, Reshef T, Klajman A. Interleukin-2 production and activity in aged humans. Mech Ageing Dev 1985;32:213–226.

83. Bender BS, Chrest FJ, Adler WH. Phenotypic expression of natural killer cell associated membrane antigens and cytolytic function of peripheral blood cells from different aged humans. J Clin Lab Immunol 1986;21:31–36.

84. Sato T, Fuse A, Kuwata T. Enhancement by IFN of natural cytotoxic ativity of lymphocytes from human cord blood and peripheral blood of aged persons. Cell Immunol 1979; 45: 458–463.

85. Provinciali M, Di Stefano G, Bernardi A, Fabris N. Role of pituitary-thyroid axis on basal and lymphokine-induced NK cell activity in ageing. Int J Neurosci 1990;51:273–274.

86. Solana R, Alonso MC, Pena J. Natural killer cells in healthy ageing. Exp Gerontol 1999;34:435–443.

87. Bykovskaya SN, Abronina IF, Kupriyanova TA, Bubenik J. Down-regulation of LAK cell-mediated cytotoxicity: cancer and ageing. Biomed Pharmacother 1990;44:333–338.

88. Provinciali M, Di Stefano G, Fabris N. Evaluation of LAK cell development in young and old healthy humans. Natural Immunity 1995;14:134–144.

89. Provinciali M, Di Stefano G, Stronati S, Fabris N. Generation of human lymphokine-activated killer cells following an IL-2 pulse in elderly cancer patients. Cytokine 1998;10: 132–139.

90. Krishnaraj R, Bhooma T. Cytokine sensitivity of human NK cells during immunosenescence. 2. IL2-induced interferon gamma secretion. Immunol Lett 1996;50:59–63.

91. Mariani E, Pulsatelli L, Meneghetti A, Dolzani P, Mazzetti I, Neri S, Ravaglia G, Forti P, Facchini A. Different IL-8 production by T and NK lymphocytes in elderly subjects. Mech Ageing Dev 2001;122:1383–1395.

92. Mariani E, Meneghetti A, Neri S, Ravaglia G, Forti P, Cattini L, Facchini A. Chemokine production by natural killer cells from nonagenarians. Eur J Immunol 2002;32:1524–1529.

93. Solana R, Mariani E. NK and NK/T cells in human senescence. Vaccine 2000;18:1613–1620.

94. Miyaji C, Watanabe H, Toma H, Akisaka M, Tomiyama K, Sato Y, Abo T. Functional alteration of granulocytes, NK cells, and natural killer T cells in centenarians. Hum Immunol 2000;61:908–916.

95. Lanier LL, Ruitenberg J, Bolhuis RL, Borst J, Phillips JH, Testi R. Structural and serological heterogeneity of gamma/delta T cell antigen receptor expression in thymus and peripheral blood. Eur J Immunol 1988;18:1985–1992.

96. Groh V, Porcelli S, Fabbi M, Lanier LL, Picker LJ, Anderson T, Warnke RA, Bhan AK, Strominger JL, Brenner MB. Human lymphocytes bearing T cell receptor gamma/delta are phenotypically diverse and evenly distributed throughout the lymphoid system. J Exp Med 1989;169:1277–1294.

97. Parker CM, Groh V, Band H, Porcelli SA, Morita C, Fabbi M, Glass D, Strominger JL, Brenner MB. Evidence for extrathymic changes in the T cell receptor gamma/delta repertoire. J Exp Med 1990;171:1597–1612.

98. Poccia F, Gougeon ML, Bonneville M, Lopez-Botet M, Moretta A, Battistini L, Wallace

M, Colizzi V, Malkovsky M. Innate T-cell immunity to nonpeptidic antigens. Immunol Today 1998;19:253–256.

99. Brenner MB, McLean J, Scheft H, Riberdy J, Ang SL, Seidman JG, Devlin P, Krangel MS. Two forms of the T-cell receptor gamma protein found on peripheral blood cytotoxic T lymphocytes. Nature 1987;325:689–694.

100. Constant P, Davodeau F, Peyrat MA, Poquet Y, Puzo G, Bonneville M, Fournie JJ. Stimulation of human gamma delta T cells by nonpeptidic mycobacterial ligands. Science 1994;264:267–270.

101. Tanaka Y, Morita CT, Tanaka Y, Nieves E, Brenner MB, Bloom BR. Natural and synthetic non-peptide antigens recognized by human gamma delta T cells. Nature 1995;375:155–158.

102. Burk MR, Mori L, De Libero G. Human V gamma 9-V delta 2 cells are stimulated in a cross-reactive fashion by a variety of phosphorylated metabolites. Eur J Immunol 1995;25: 2052–2058.

103. Bukowski JF, Morita CT, Tanaka Y, Bloom BR, Brenner MB, Band H. V gamma 2V delta 2 TCR-dependent recognition of non-peptide antigens and Daudi cells analyzed by TCR gene transfer. J Immunol 1995;154:998–1006.

104. Cipriani B, Borsellino G, Poccia F, Placido R, Tramonti D, Bach S, Battistini L, Brosnan CF. Activation of C-C beta-chemokines in human peripheral blood gammadelta T cells by isopentenyl pyrophosphate and regulation by cytokines. Blood 2000;95:39–47.

105. Poccia F, Cipriani B, Vendetti S, Colizzi V, Poquet Y, Battistini L, Lopez-Botet M, Fournie JJ, Gougeon ML. CD94/NKG2 inhibitory receptor complex modulates both anti-viral and anti-tumoral responses of polyclonal phosphoantigen-reactive V gamma 9V delta 2 T lymphocytes. J Immunol 1997;159:6009–6017.

106. Fearon DT, Locksley RM. The instructive role of innate immunity in the acquired immune response. Science 1996;272:50–53.

107. Mak TW, Ferrick DA. The gammadelta T-cell bridge: linking innate and acquired immunity. Nat Med 1998;4:764–765.

108. Giachino C, Granziero L, Modena V, Maiocco V, Lomater C, Fantini F, Lanzavecchia A, Migone N. Clonal expansions of V delta 1+ and V delta 2+ cells increase with age and limit the repertoire of human gamma delta T cells. Eur J Immunol 1994;24:1914–1918.

109. Colonna-Romano G, Potestio M, Scialabba G, Mazzola A, Candore G, Lio D, Caruso C. Early activation of gammadelta T lymphocytes in the elderly. Mech Ageing Dev 2000;121: 231–238

110. Argentati K, Re F, Donnini A, Tucci MG, Franceschi C, Bartozzi B, Bernardini G, Provinciali M. Numerical and functional alterations of circulating gammadelta T lymphocytes in aged people and centenarians. J Leukoc Biol 2002;72:65–71.

111. Fagnoni FF, Vescovini R, Mazzola M, Bologna G, Nigro E, Lavagetto G, Franceschi C, Passeri M, Sansoni P. Expansion of cytotoxic CD8+ CD28- T cells in healthy ageing people, including centenarians. Immunology 1996;88:501–507.

112. Colonna-Romano G, Potestio M, Aquino A, Candore G, Lio D, Caruso C. Gamma/delta T lymphocytes are affected in the elderly. Exp Gerontol 2002;37:205–211.

113. Medzhitov R, Janeway CA Jr. Innate immunity. The virtues of a nonclonal system of recognition. Cell 1997;91:295–298.

114. Medzhitov R, Janeway CA Jr. Innate immunity: impact on the adaptive immune response. Curr Op Immunol 1997;9:4–9.

115. Steger MM, Maczek C, Grubeck-Loebenstein B. Morphologically and functionally intact

dendritic cells can be derived from the peripheral blood of aged individuals. Clin Exp Immunol 1996;105:544–550.

116. Saurwein-Teissl M, Schonitzer D, Grubeck-Loebenstein B. Dendritic cell responsiveness to stimulation with influenza vaccine is unimpaired in old age. Exp Gerontol 1998;33: 625–631.

117. Perkins EH, Massucci JM, Glover PL. Antigen presentation by peritoneal macrophages from young adult and old mice. Cell Immunol 1982;70:1–10.

118. Schwab R, Hausman PB, Rinnooy-Kan E, Weksler ME. Immunological studies of ageing. X. Impaired T lymphocytes and normal monocyte response from elderly humans to the mitogenic antibodies OKT3 and Leu 4. Immunology 1985;55:677–684.

119. Fagiolo U, Cossarizza A, Scala E, Fanales-Belasio E, Ortolani C, Cozzi E, Monti D, Franceschi C, Paganelli R. Increased cytokine production in mononuclear cells of healthy elderly people. Eur J Immunol 1993;23:2375–2378.

120. Steger MM, Maczek C, Grubeck-Loebenstein B. Peripheral blood dendritic cells reinduce proliferation in in vitro aged T cell populations. Mech Ageing Dev 1997;93:125–130.

121. Lung TL, Saurwein-Teissl M, Parson W, Schonitzer D, Grubeck-Loebenstein B. Unimpaired dendritic cells can be derived from monocytes in old age and can mobilize residual function in senescent T cells. Vaccine 2000;18:1606–1612.

122. Sprecher E, Becker Y, Kraal G, Hall E, Harrison D, Shultz LD. Effect of ageing on epidermal dendritic cell populations in C57BL/6J mice. J Invest Dermatol 1990;94: 247–253.

123. Belsito DV, Epstein SP, Schultz JM, Baer RL, Thorbecke GJ. Enhancement by various cytokines or 2-beta-mercaptoethanol of Ia antigen expression on Langerhans cells in skin from normal aged and young mice. Effect of cyclosporine A. J Immunol 1989;143: 1530–1536.

124. Choi KL, Sauder DN. Epidermal Langerhans cell density and contact sensitivity in young and aged BALB/c mice. Mech Ageing Dev 1987;39:69–79.

125. Haruna H, Inaba M, Inaba K, Taketani S, Sugiura K, Fukuba Y, Dai H, Toki J, Tokunaga R, Ikehara S. Abnormalities of B cells and dendritic cells in SAMP1 mice. Eur J Immunol 1995;25:1319–1325.

126. Villadsen JH, Langkjer ST, Ebbesen P, Bjerring P. Syngrafting skin among mice of similar and different ages increases the number of Langerhans cells and decreases responsiveness to 1,4-dinitrofluorobenzene. Compr Gerontol [A] 1987;1:78–79.

127. Effros RB, Walford RL. The effect of age on the antigen-presenting mechanism in limiting diluition precursor cell frequency analysis. Cell Immunol 1984;88:531–539.

128. Donnini A, Argentati K, Mancini R, Smorlesi A, Bartozzi B, Bernardini G, Provinciali M. Phenotype, antigen-presenting capacity, and migration of antigen-presenting cells in young and old age. Exp Gerontol 2002;37:1097–1112.

129. Miller C, Kelsoe G, Han S. Lack of B7–2 expression in the germinal centers of aged mice. Ageing: Immunol Infect Dis 1994;5:249–257.

130. Shearer GM. Th1/Th2 changes in aging. Mech Ageing Dev 1997;94:1–5.

131. Rink, L, Cakman I, Kirchner H. Altered cytokine production in the elderly. Mech Ageing Dev 1998;102:199–209.

132. Karanfilov C, Liu B, Fox CC, Lakshmanan RR, Whisler RL. Age-related defects in Th1 and Th2 cytokine production by humanT cells can be dissociated from altered frequencies of CD45RA$^+$ and CD45RO$^+$ T cell subsets. Mech Ageing Dev 1999;109:97–112.

133. Provinciali M, Argentati K, Tibaldi A. Efficacy of cancer gene therapy in aging:

adenocarcinoma cells enginnered to release IL-2 are rejected but do not induce tumor specific immune memory in old mice. Gene Therapy 2000;7:624–632.

134. Provinciali M, Smorlesi A, Donnini A, Bartozzi B, Amici A. Low effectiveness of DNA vaccination against HER-2/neu in ageing Vaccine. 2003;21:843–848.

135. Provinciali M, Di Stefano G, Colombo M, Della Croce F, Gandolfi MC, Daghetta L, Anichini M, Della Bitta R, Fabris N. Adjuvant effect of low-dose interleukin-2 on antibody response to influenza virus vaccination in healthy elderly subjects. Mech Ageing Dev 1994;77:75–82.

*The Neuroendocrine Immune Network in Ageing*
Edited by R.H. Straub and E. Mocchegiani

# Major Histocompatibility Complex Polymorphisms and Ageing

GIUSEPPINA CANDORE, CALOGERO CARUSO, GIUSEPPINA COLONNA-ROMANO
and DOMENICO LIO

*Gruppo di Studio sull'Immunosenescenza, Dipartimento di Biopatologia e Metodologie, Bio-
mediche, Università di Palermo, Palermo, Italy*

ASTRACT

Longevity seems to be directly correlated with optimal functioning of the immune system,
suggesting that some genetic determinants of longevity might reside in those polymorphisms
for the immune system genes that regulate immune responses. Accordingly, mouse lifespan is
influenced by MHC (major histocompatibility complex) genotype. The HLA (the human MHC)
region encompasses over 4 Mb of DNA on the chromosome band 6p21.3 and its extensive characterisation has recently culminated in the determination of the nucleotide sequence of the entire
region, confirming the presence of ~220 genes. The MHC is traditionally divided into the class I,
class II and class III regions. Most HLA genes involved in the immune response fall into classes I
and II, which encode highly polymorphic heterodimeric glycoproteins involved in graft rejection
and antigen presentation to T cells. However, many genes of the class III region are also involved
in the immune and inflammatory responses. .MHC polymorphisms have been the focus of a
vast number of ageing association studies. One of the limitations in the most reported studies is
the number of aged individuals available of both genders, and the lack of complete background
histories of health or ethnicity for study participants. However, also taking into account well
planned and designed studies, discordant results have been obtained. The discordant results of
MHC/ageing association studies might be due to distinct linkage in different cohorts or to other
interacting, genetic or environmental factors or to the heterogeneity of ageing. Nevertheless,
bearing all the reported studies in mind, there is no convincing evidence of a strong, direct association between longevity and any MHC alleles.

1.    INTRODUCTION: THE MAJOR HISTOCOMPATIBILITY COMPLEX

Longevity seems to be directly correlated with optimal functioning of the immune system, suggesting that some genetic determinants of longevity might reside in those polymorphisms for
the immune system genes that regulate immune responses. Accordingly, mouse lifespan is influenced by MHC (major histocompatibility complex) genotype [1]. The HLA (the human MHC)
region encompasses over 4 Mb of DNA (~0.1% of the genome) on the chromosome band 6p21.3
and its extensive characterisation has recently culminated in the determination of the nucleotide
sequence of the entire region, confirming the presence of ~220 genes. The MHC is traditionally
divided into the class I, class II and class III regions (Figure 1a). Most HLA genes involved in

the immune response fall into classes I and II, which encode highly polymorphic heterodimeric glycoproteins involved in graft rejection and antigen presentation to T cells. The class III region, which has been redefined as the ~730 kb of DNA extending from NOTCH4 to BAT1, is now known to contain at least 62 genes. As putative functions are ascribed to the products of these genes, it is becoming increasingly apparent that many of these are involved in the immune and inflammatory responses [2,3].

An intriguing feature of the MHC is the occurrence of particular combinations of alleles, at loci across this 4 Mb region, more frequently than would be expected based on the frequencies of individual alleles. This non-random association of alleles gives rise to highly conserved haplotypes that appear to be derived from a common remote ancestor, so called *ancestral haplotypes* (AH). This term underlines the fact that conserved, population-specific haplotypes are continuous sequences derived with little, if any, change from an ancestor of all those now carrying all or part of the haplotype. These AH determine a non-random assortment of alleles at neighbouring loci, referred to as *"linkage disequilibrium"* [4–6].

The class I genes code for the α polypeptide chain of the class I molecule; the β chain of the class I molecule is encoded by a gene on chromosome 15, the $\beta_2$-microglobulin gene. There are some 20 class I genes in the HLA region; three of these, HLA-A, B, and C, the so-called classic, or class Ia genes, are the main actors in the immunologic theatre. Besides, there are nonclassical HLA class Ib molecules, which are characterized by a limited polymorphism and a restricted expression pattern (Figure 1b). The class II genes code for the α and β polypeptide chains of the class II molecules (Figure 1b). Class Ia and class II molecules are highly polymorphic heterodimeric glycoproteins, which bind antigenic peptides intracellularly and emerge on the cell surface where they present processed peptide fragments to the T lymphocyte receptor. Thus they can restrict and regulate T cell responses against specific antigens. These are the bases for the antigen-specific control of the immune response. The response depends on the ability of histocompatibility molecules to bind some peptides and not others. Thus, survival and longevity might be associated with a positive or negative selection of alleles that respectively confer resistance or susceptibility to infectious disease(s). Besides, genetic studies have shown that persons who have certain HLA alleles have a higher risk of specific immune-mediated diseases than persons without these alleles. The associations vary in strength, and in all the diseases studied, several other genes in addition to those of the HLA region are likely to be involved [2,6,7].

HFE, the most telomeric HLA class Ib gene (Figure 1b), codes for a class I α chain, which seemingly no longer participates in immunity, because has lost its ability to bind peptides due to a definitive closure of the antigen binding cleft that prevents peptide binding and presentation. The HFE protein, expressed in crypt enterocytes of the duodenum, regulates the iron uptake by intestinal cells because it has acquired the ability to form complex with the receptor for iron-binding transferrin. Thus, it indirectly regulates immune responses, because iron availability plays a role in specific and non-specific immune responses [2,8].

The C282Y mutation (a cysteine to tyrosine mutation at amino acid 282) in this gene has been identified as the main genetic basis of *hereditary haemochromatosis* (HH). It destroys its ability to make up a heterodimer with β2-microglobulin. The defective protein fails to associate to the transferrin receptor and the complex cannot be transported to the surface of the duodenal crypt cells. As a consequence, in homozygous people two to three times the normal amount of iron is absorbed from food by the intestine. Another mutation, H63D (a histidine to aspartate at amino acid 63) appear to be associated with milder forms of HH. In fact, a small number of patients with HH were identified as homozygous for H63D or as compound heterozygous H63D/C282Y. It has been suggested that an estimated 60 to 70 generations ago, the C282Y mutation occurred

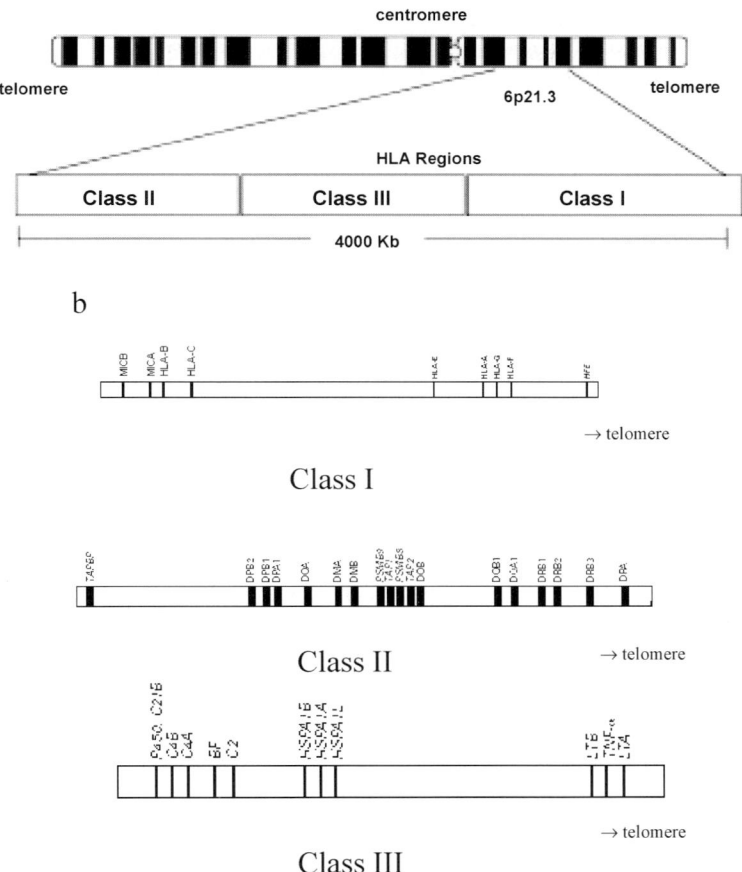

Figure 1a,b. The HLA region encompasses over 4 Mb of DNA (~0.1% of the genome) on the chromosome band 6p21.3 and its extensive characterisation has recently culminated in the determination of the nucleotide sequence of the entire region, confirming the presence of ~220 genes. The MHC is traditionally divided into the class I, class II and class III regions. There are some 20 class I genes in the HLA region; three of these, HLA-A, B, and C, the so-called classic, or class Ia genes, are the main actors in the immunologic theatre. Besides, there are nonclassical HLA class Ib molecules characterized by a limited polymorphism and a restricted expression pattern. HFE is the most telomeric HLA class Ib gene. The class II genes code for the and polypeptide chains of the class II molecules. The designation of their loci on chromosome 6 consists of three letters: the first (D) indicates the class, the second (M, O, P, Q, or R) the family, and the third (A or B) the chain ( or , respectively). The class III region, which has been redefined as the ~730 kb of DNA extending from NOTCH4 to BAT1, is now known to contain at least 62 genes.

in the HFE gene of a Celtic individual who is the ancestor of the more than 5% of Caucasoids now carrying the allele. It has been claimed that the great expansion of Celtic people could be in part explained by the widespread presence of this HFE gene mutation. This gave to heterozygous carriers selective advantages on the basis of improved survival during infancy, childhood, and pregnancy, by leading to increased iron absorption and accumulation of larger body iron stores because ancient diet consisted mainly in iron-poor grains and cereals, whereas meat was highly uncommon [2,9,10]. This evolutionary significance of these mutations suggests a their possible role in survival and longevity.

The pleiotropic pro-inflammatory cytokine tumour necrosis factor (TNF)-α maps within the HLA class III (Figure 1b). The TNF cluster genes encode three inflammation related proteins, TNF-α, TNF-β and lymphotoxin-β. All three are important mediators of the immune response with multiple biologic activities. Several polymorphic areas are documented within the TNF gene cluster. Notably, the −308A single nucleotide polymorphism (SNP) located in the promoter region of the TNF-α gene is reported to have an increased frequency in autoimmune and inflammatory diseases and is associated with stronger transcriptional activation than the -308G SNP. However, other functionally relevant polymorphisms have been described [5,6,11,12]. TNF genes are thought to determine the strength, effectiveness and duration of local and systemic inflammatory reactions, as well as repair and recovery from infectious and toxic agents and, as such, they are candidates to be involved in age-related disease processes and in ageing itself [1,13].

Three components of the complement system, which are the principal effector mechanism of humoral immunity and are important in the clearance of immune complexes, opsonization and cell lysis, are encoded in the class III region (Figure 1b). C2 and C4 participate in the classical pathway, which may be activated by the binding of C1 to antigen-antibody complexes or following the interaction of mannan binding protein with carbohydrate ligands. Factor B (Bf) is a component of the alternative pathway, which is activated by a diverse set of substances including components of yeast and bacterial cell walls. Most human chromosomes carry two C4 genes, C4A and C4B, which form part of a duplicated segment of DNA spanning ~75 kb. The C4A and C4B genes encode proteins that differ by only four amino acids, but nevertheless have profoundly different covalent binding activities. Both isotypes of C4 are highly polymorphic [3,14,15]. Any polymorphism resulting in the qualitative or quantitative changes in complement function might cause a differential sensitivity to infections, so affecting survival and longevity.

The HSP70-1 and -2 genes (HSPA1A and HSPA1B) that encode the major heat-inducible 70 kDa heat shock protein (HSP70) and the HSP70-HOM gene, (HSPA1L), that encodes a non-heat-inducible protein that is 90% identical to HSP70s, also map within the HLA class III (Figure 1b). HSPs are a group of conserved proteins that are induced in a variety of cellular stresses, including infections. All HSP70s can bind unfolded proteins and peptides and act as chaperones in the synthesis, folding, assembly, translocation and degradation of proteins during normal cellular processes and following stress. Members of the 70 kDa HSP family are consistently associated with defences against conditions involving *oxidative stress* such as ischemia, inflammation and the ageing process. HSP70s have been implicated in many diseases. For example, increased expression of HSP70s has been observed in autoimmune diseases such as scleroderma and lupus, although this may be a consequence of disease, rather than a contributory factor. The induction of HSPs, particularly HSP 70, declines in response to heat shock with ageing, therefore rendering the organism more vulnerable to stress damage [3,16,17]. Thus, any polymorphism that results in the alteration of qualitative or quantitative HSP 70 expression could cause a differential sensitivity to stresses and affect survival and longevity.

## 2.    HLA IA & II CLASS ALLELES IN AGEING

The studies performed on the association between longevity and HLA are generally difficult to interpret, owing to major methodological problems. However, as reported by Caruso et al. [7,18], some of them, well designed and performed, suggest an HLA effect on longevity. In studies performed in Caucasoids, an increase in HLA-DR11 (that is a HLA-DR5 split) in Dutch women over 85 years was observed [19]. The same laboratory performed a further study and by using a *'birth-place-restricted comparison'* in which the origin of all the subjects was ascertained, the authors were able to confirm that ageing in women was positively associated with HLA-DR5 [20]. Two French studies confirmed the relevance of HLA-DR11 to longevity in aged populations [21,2]). This increase is consistent with the protective effects of this allele in viral diseases, as HLA-DR5, or its subtype HLA-DR11, frequencies have been shown to be decreased in some viral diseases [7,18]. Finally, an association between longevity and the AH 8.1 (or part of this haplotype, i.e. HLA-B8,DR3) apparently emerges. In fact, an excess of this AH in the oldest old men has been reported in French and in North-Ireland populations [23,24]. This association appears to be gender-specific. In fact, a Greek study showed a significant decrease of 8.1 AH in aged women [25]. So, immune dysfunctions of the 8.1 AH should contribute to early morbidity and mortality in elderly women, more susceptible to autoimmune diseases than men, and to longevity in elderly men [6,7,18]. These associations are gender-related, but it is not unexpected on the basis of available data on the genetics of longevity, showing that the association of longevity with particular alleles may be found only in one gender [26]. So these studies seem to suggest that HLA-DR11 in women and HLA-B8, DR3 in men may be considered markers of successful ageing. However, in a longitudinal study, in which a total of 919 subjects aged 85 years and older, were HLA-typed and followed up for at least 5 years, no HLA-association with mortality was found [27]. Positive studies need confirmation in different Caucasoid populations with different ethnic background and/or replication in a new cohort of oldest old from the same population. However, in Sardinian centenarians we did not observe the associations demonstrated in the other well planned and designed studies above discussed [28]. Moreover, a study set out to specifically confirm the previously reported increase in the frequency of the 8.1 AH in aged men of the northern Irish population [24], did not reveal any statistically significant haplotype frequency differences between the aged cohort of individuals in comparison to the younger controls. However, a striking decrease was observed when the aged women (13.1%) were compared to the control women (17.8%) [29].

On the whole these findings clearly show that HLA/longevity associations are population-specific, being heavily affected by the population-specific genetic and environmental history (selection by different kinds of infectious diseases). So, HLA class Ia and II genes should be considered survival genes not longevity genes.

## 3.    HLA IB ALLELES IN AGEING

Concerning the class Ib gene HFE, we have recently reported that C282Y mutation may confer a selective advantage in term of longevity to Sicilian women. Interestingly in these subjects an increase of H63D polymorphism was also observed but the difference was not significant [30]. We have not been able to confirm in Sardinian centenarians the increase of C282Y mutation observed in Sicilian oldest old women. This result is not surprising because the HLA AH 7.1, which usually carries this mutation, is virtually absent in the Sardinian population [9]. However, we observed an increase of the other mutation H63D. This trend was not significant, but the cumulative frequency of H63D mutations in centenarian and very old women from Sardinia and Sicily was 22% vs. 11%, i.e. 30/136 vs. 23/210 (P = 0.008) [31].

It is seemingly puzzling that HFE mutations have been suggested to be involved in unsuccessful ageing too. In fact, these mutations have been suggested to be involved in Alzheimer's disease and coronary heart disease, although the data are seemingly contradictory [32,33]. However, it is not surprising that HFE mutations are associated both with longevity and diseases affecting life span. In fact, the available data from genetics of longevity show that longevity may be associated with alleles with increased risk to a variety of diseases in the younger phases of life (antagonistic pleiotropy) [18]. This theory is also a possible explanation for the apparent discrepancy with a recent study performed in Denmark. In a C282Y mutation high-carrier frequency population, as in Denmark, this mutation shows an age-related reduction in the frequency of heterozygotes for C282Y, which suggests that carrier status is associated with shorter life expectancy. The observed reduction in C282Y mutation carrier frequency persists until age 95 years; however, the frequency in the centenarian group was the same as that in the youngest group for both men and women [34]. So, in that study heterozygosity for C282Y has a higher mortality rate in the younger groups but becomes beneficial in the oldest old. On the other hand, in two other European studies, the observed frequency of C282Y homozygosity in oldest old male English and in oldest old male and female Dutch was not significant lower than that predicted [35,36]. All together these reports seem to strongly suggest that HFE effects on ageing depend on interaction between genetic and environmental factors that in different age, gender and population may result in a successful or unsuccessful ageing.

4.    HLA III CLASS ALLELES IN AGEING

The TNF-308A/G polymorphism has been assessed in 3 studies in European nonagenarian and centenarian subjects to assess association with ageing. In a study in 252 Finnish nonagenarians, Wang et al. [37] noted no difference in the frequency of the TNF-308A/G SNP in comparison to 400 blood donors of mixed age groups. In a study by Lio et al. [38], the TNF-$\alpha$ -308 A/G SNP was unchanged in frequency between 72 male and 102 female centenarians and 227 controls, though the AA homozygotes represented only 1%, which is somewhat lower than in other Caucasian groups. A recent study by Ross et al [29], comparing the same TNF-$\alpha$ -308A/G SNP frequency in Irish nonagenarians with a younger control group, also noted no significant differences in AA homozygotes (5% controls v 8% in aged subjects) with a similar 24% A allele frequency in controls and elderly subjects. Interestingly, women irrespective of age showed a trend for higher representation of A allele compared to men. In the same study [29], no change in the TNF-$\beta$ +252A/G SNP between Irish nonagenarians and a control group was reported. Overall these three studies, with between them 350 nonagenarians and 172 centenarians, appear to demonstrate no major shift in the genotype frequency of TNF-$\alpha$ -308 SNP as a function of ageing. These data suggest that this particular polymorphism of TNF genes does not give an advantage for survival in the last decade of life and is compatible with the hypothesis that this polymorphism constitutes a shield evolutionary conserved towards pathogen agents whose threatening power lasts for the entire life-span.

The incidence of allotypes of the genes of the fourth component (C4) and factor B of the complement system was compared in 252 persons less than 45 years of age ("young" group) with 482 people between 61 and 90 years of age ("old" group). One hundred people older than

90 years of age (nonagenarians) were also investigated. A striking difference was found between the "young" and "old" groups in the incidence (16.1% and 5.4%, respectively) of a silent gene of the C4B allele (C4B*Q0). This difference was even more marked among "young" and "old" men (17.6% vs. 3.4%). The incidence of the C4B*Q0 allele in women dropped to the level of the men only in the nonagenarian group. For the authors, the most probable explanation for this finding was that people carrying the C4B*Q0 allele die from as yet unidentified disease(s) in their middle age. Therefore, male (and to a lesser extent female) carriers of this allele may have a considerably shorter life expectancy than individuals without a silent gene in the C4B locus [39]. In a further study, the complement system and the distribution of class III alleles C4, BF were analysed in healthy aged people (77 centenarians and 89 elderly subjects). The authors also studied the alleles of C3, a complement component genetically unrelated to HLA, the immunochemical levels of C4 and C3 and serum functional hemolytic activity for classical (CH50) and alternative (AP50) complement pathways. The levels of C3 and C4 and the CH50 and AP50 were found to be within the normal range. The frequencies of C3, BF, and C4A alleles were similar in the cohorts that have been studied. For C4B null allele (C4BQ0) a trend toward an increase in the older cohort was observed, although the differences were not significant after statistical correction [40]. These data do not support the results obtained in the previous study, but suggest instead the importance of a well preserved complement system to attain survival and longevity.

The frequency of the functional polymorphism, T2437C transversion (Met→Thr), in the HSP 70-Hom gene was investigated within a healthy aged Irish population using oligonucle-otide probes. The 2437T SNP was observed to increase in the elderly, although not attaining statistical significance. The TT genotype was observed to be significantly increased within the Irish aged population (p=0.03), while conversely the TC genotype was significantly decreased in the aged subjects (p=0.01). These findings would support the theory that the change from a Met (non-polar and hydrophobic) residue to a Thr (polar and neutral) residue may disrupt the peptide-binding specificity of HSP 70-Hom and have an effect on its functional efficiency. One postulates that the highly significant p-value obtained for the TC genotype may infer that the presence of both the T and the C allele (heterozygosity) resulting in the generation of two differ-ent HSP 70-Hom protein species may negatively influence survival and longevity [41].

Besides, a recent article has presented data on HSP70-1. In eukaryotes, HSP70 gene expres-sion is regulated essentially through the activity of heat shock transcription factors, which inter-act with heat shock elements (HSEs) present in HSP promoters, resulting in a modulation of the transcription rates. An A/C polymorphism lies in the HSE region of the HSP70-1 gene at 110nt from the transcription-starting site. This polymorphism was the marker chosen for the study. A total of 591 southern Italian subjects were enrolled in the study (263 men and 328 women; age range 18–109 years). A significant age-related decrease of the frequency of allele (A)-110 was observed in women. The probability ratio of 0.403 computed by considering female centenar-ians as cases and young women (18–49 years old) as controls showed that the (A)-110 allele is unfavourable to longevity in women [42].

HSPs are crucial for maintenance of cell homeostasis and survival both during and after vari-ous stresses. These two studies suggest that polymorphisms in heat shock protein genes may affect the chance of survival of the individual.

## 5.    CONCLUDING REMARKS

As discussed above, a vast number of investigators studied MHC polymorphisms in relation to ageing. The limitations of most reported studies are the small number of aged individuals available of both genders, and the lack of complete background histories of health or ethnicity for the participants. However, well planned and designed studies were also performed, as discussed by Caruso et al. (7,18), yet discordant results have been obtained. The discordant results of MHC/ageing association studies might be due to distinct linkage in different cohorts or to other interacting, genetic or environmental factors or to the heterogeneity of ageing.. Nevertheless, bearing all the reported studies in mind, there is no convincing evidence of a strong, direct association between longevity and any MHC alleles. However, further examination of polymorphism of other class III genes such as receptor for advanced glycosilation end-products or Notch 4 might well provide interesting and informative data [3,42].

## ACKNOWLEDGEMENTS

The "Gruppo di Studio sull'immunosenescenza" coordinated by Professor C. Caruso is funded by grants from MIUR, Rome (ex 40%, to CC and DL; ex 60% to CC, DL, GC and GCR), and from Ministery of Health Projects "Immunological parameters age-related" and "Genetic and non genetic determinants of healthy ageing".

## REFERENCES

1.    Pawelec G, Barnett Y, Forsey R, Frasca D, Globerson A, McLeod J, Caruso C, Franceschi C, Fülöp T, Gupta S, Mariani E, Mocchegiani E, Solana R. T cells and aging, 2002 update. Front Biosci 2002; 7; d1056–1183.
2.    Klein J, Sato A. The HLA system. N Engl J Med 2000; 343: 702–709; 782–786.
3.    Milner CM, Campbell RD. Genetic organization of the human MHC class III region. Front Biosci. 2001; 6; D914–926.
4.    Dawkins R, Leelayuwat C, Gaudieri S, Tay G, Hui J, Cattley S, Martinez P, Kulski J. Genomics of the major histocompatibility complex: haplotypes, duplication, retroviruses and disease. Immunol Rev 1999; 167; 275–304.
5.    Price P, Witt C, Allcock R, Sayer D, Garlepp M, Kok CC, French M, Mallal S, Christiansen F. The genetic basis for the association of the 8.1 ancestral haplotype (A1, B8, DR3) with multiple immunopathological diseases. Immunol Rev 1999; 167; 257–274.
6.    Candore G, Lio D, Colonna-Romano G, Caruso C. Pathogenesis of autoimmune diseases associated with 8.1 ancestral haplotype, effect of multiple gene interactions. Autoimmunity Reviews 2002; 1; 29–35.
7.    Caruso C, Candore G, Colonna Romano G, Lio D, Bonafè M, Valensin S, Franceschi C: Immunogenetics of Longevity. Is Major Histocompatibility Complex Polymorphism Relevant to the Control of Human Longevity? A Review of Literature Data. Mech Ageing Dev 2001;122; 445–462.
8.    Salter-Cid L, Peterson PA, Yang Y. The major histocompatibility complex-encoded HFE in iron homeostasis and immune function. Immunol Res 2000; 22: 43 9.
9.    Candore G, Mantovani V, Balistreri CR, Lio D, Colonna-Romano G, Cerreta V, Carru C,

Deiana L, Pes G, Menardi G, Perotti L, Miotti V, Bevilacqua E, Amoroso A, Caruso C. Frequency of the HFE Gene Mutations in Five Italian Populations. Blood Cells Mol Dis 2002; 29; 267–273.

10. Balistreri CR, Candore G, Almasio P, Di Marco V, Colonna-Romano G, Craxì A, Motta M, Piazza G, Malaguarnera M, Lio D, Caruso C: Analysis of haemochromatosis gene mutations in the Sicilian population: implications for survival and longevity. Archs Gerontol Geriatrics 2002; S8: 35–42.

11. Hajeer AH, Hutchinson IV. Influence of TNF-alpha gene polymorphisms on TNF-alpha production and disease. Hum Immunol. 2001; 62; 1191–1199.

12. Lio D, Candore G, Colombo A, Colonna Romano G, Gervasi F, Marino V, Scola L, Caruso C: A genetically determined high setting of TNF-α influences immunological parameters of HLA-B8,DR3 positive subjects: implications for autoimmunity. Human Immunol 2001; 62; 705–713.

13. Candore G, Colonna-Romano G, Lio D, Caruso C. Immunological and Immunogenetic markers of successful and unsuccessful ageing. Advances in Cell Aging and Gerontology vol 13, 29–45, 2003.

14. Walport MJ. Complement. N Engl J Med 2001; 344: 1058–1066; 1140–1144.

15. Candore G, Modica MA, Lio D, Colonna-Romano G, Listì F, Grimaldi MP, Russo M, Triolo G, Accardo-Palumbo A, Cuccia M, Caruso C. Pathogenesis of autoimmune diseases associated with 8.1 ancestral haplotype: a genetically determined defect of C4 influences immunological parameters of healthy carriers of the haplotype. Biomed Pharmacother, 2003;57:274–277.

16. Soti C, Csermely P. Molecular chaperones and the aging process. Biogerontology. 2000;1; 225–233.

17. Rea IM, McNerlan S, Pockley AG. Serum heat shock protein and anti-heat shock protein antibody levels in aging. Exp Gerontol 2001; 36:341–352.

18. Caruso C, Candore G, Colonna Romano G, Lio D, Bonafè M, Valensin S, Franceschi C: HLA, Aging and Longevity: A Critical Reappraisal. Human Immunol 2000;61;942–949.

19. Lagaay AM, D'Amaro J, Ligthart GJ, Schreuder GM, van Rood JJ, Hijmans W. Longevity and heredity in humans. Association with the human leucocyte antigen phenotype. Ann. NY Acad Sci 1991; 621, 78–89.

20. Izaks GJ, Remarque EJ, Schreuder GMT, Westendorp RGJ, Ligthart GJ. The effect of geographic origin on the frequency of HLA antigens and their association with ageing. Eur J Immunogenet 2000; 27; 87–92.

21. Ivanova R, Hénon N, Lepage V, Charron D, Vicaut E, Schächter F.. HLA-DR alleles display sex-dependent effects on survival and discriminate between individual and familial longevity. Hum Mol Genet 1998; 7; 187–194.

22. Henon N, Busson M, Dehay Martuchou C, Charron D, Hors J. Familial versus sporadic longevity and MHC markers. J Biol Regulat Homeost Agent 1999;13; 27–3.

23. Proust J, Moulias R, Fumeron F, Bekkhoucha F, Busson M, Schmid M, Hors J. HLA and longevity. Tissue Antigens 1982; 19; 168–173.

24. Rea IM, Middleton D. Is the phenotypic combination A1B8Cw7DR3 a marker for male longevity? J. Am Geriatr Soc 1994; 42, 978–983.

25. Papasteriades C, Boki K, Pappa H, Aedonopoulos S, Papasteriadis E, Economidou J. HLA phenotypes in healthy aged subjects. Gerontology 1997; 43, 176–181.

26. Franceschi C, Motta L, Valensin S, Rapisarda R, Frantone A, Berardelli M, Motta M, Monti D, Bonafe M, Ferrucci L, Deiana L, Pes GM, Carru C, Desole MS, Barbi C, Sartoni

G, Gemelli C, Lescai F, Olivieri F, Marchigiani F, Cardelli M, Cavallone L, Gueresi P, Cossarizza A, Troiano L, Pini G, Sansoni P, Passeri G, Lisa R, Spazzafumo L, Amadio L, Giunta S, Stecconi R, Morresi R, Viticchi C, Mattace R, De Benedictis G, Baggio G. Do men and women follow different trajectories to reach extreme longevity? Italian Multicenter Study on Centenarians. Aging (Milan) 2000; 12, 77–84.

27. Izaks GJ, van Houwelingen HC, Schreuder GM, Ligthart GJ. The association between human leucocyte antigens (HLA) and mortality in community residents aged 85 and older. J Am Geriatr Soc 1997;45;56–60.

28. Lio D, Pes M G, Carru C, Listi; F, Ferlazzo V, Candore G, Colonna-Romano G, Ferrucci L, Deiana L, Baggio G, Franceschi C, Caruso C. Association between the HLA-DR alleles and longevity: a study in Sardinian population. Exp Gerontol 2003; 38; :313–318.

29. Ross OA, Curran MD, Rea IM, Hyland P, Duggan O, Barnett CR, Annett K, Patterson C, Barnett YA, Middleton D. HLA haplotypes and TNF polymorphism do not associate with longevity in the Irish. Mech Ageing Dev 2003;124:563–567.

30. Lio D, Balisteri CR, Colonna-Romano G, Motta M, Franceschi C, Malaguarnera M, Candore G, Caruso C. Association between the MHC class I gene HFE polymorphisms and longevity: a study in Sicilian population. Genes and Immunity 2002; 3; 20–24.

31. Carru C, Pes GM, Deiana L, Baggio G, Franceschi C, Lio D, Balistreri CR, Candore G, Colonna-Romano G, Caruso C. Association between the HFE mutations and longevity: a study in Sardinian population. Mech Ageing Dev 2003; 124:529–532.

32. Candore G, Licastro F, Chiappelli M, Franceschi C, Lio D, Rita Balistreri C, Piazza G, Colonna-Romano G, Grimaldi LM, Caruso C. Association between the HFE mutations and unsuccessful ageing: a study in Alzheimer's disease patients from Northern Italy. Mech Ageing Dev 2003; 124:525–528.

33. Candore G, Balistreri CR, Lio D,Mantovani V, Colonna-Romano G, Chiappelli M, Tampieri C, Licastro F, Branzi A, Averna M, Caruso M, Hoffmann E, Caruso C. Association between HFE mutations and acute myocardial infarction: a study in patients from Northern and Southern Italy. Blood Cells Mol Dis 2003; 31:57–62.

34. Bathum L, Christiansen L, Nybo H, Ranberg KA, Gaist D, Jeune B, Petersen NE, Vaupel J, Christensen K. Association of mutations in the hemochromatosis gene with shorter life expectancy. Arch Intern Med 2001;;161;2441–2444.

35. Willis G, Wimperis JZ, Smith KC, Fellows IW, Jennings BA. Haemochromatosis gene C282Y homozygotes in an elderly male population. Lancet 1999;354;221–222.

36. Van Aken MO, De Craen AJ, Gussekloo J, Moghaddam PH, Vandenbroucke JP, Heijmans BT, Slagboom PE, Westendorp RG. No increase in mortality and morbidity among carriers of the C282Y mutation of the hereditary haemochromatosis gene in the oldest old: the Leiden 85-plus study. Eur J Clin Invest. 2002; 32;750–754.

37. Wang XY, Hurme M, Jylha M, Hervonen A. Lack of association between human longevity and polymorphisms of IL-1 cluster, IL-6, IL-10 and TNF-alpha genes in Finnish nonagenarians. Mech Ageing Dev 2001;123:29–38.

38. Lio D, Scola L, Crivello A, Colonna-Romano G, Candore G, Bonafè M, Cavallone L, Marchigiani F, Olivieri F, Franceschi C, Caruso C. Inflammation, genetics and longevity: further studies on the protective effects in men of IL-10 -1082 promoter SNP and its interaction with -308 TNF- promoter SNP. J Med Genet 2003; 40; 296–299.

39. Kramer J, Fulop T, Rajczy K, Nguyen AT, Fust G. A marked drop in the incidence of the null allele of the B gene of the fourth component of complement (C4B*Q0) in elderly subjects: C4B*Q0 as a probabile negative selection factor for survival. Hum Genet 1991;

86:595–598.
40. Bellavia D, Frada G, Di Franco P, Feo S, Franceschi C, Sansoni P, Brai M. C4, BF, C3 allele distribution and complement activity in healthy aged people and centenarians. J Gerontol A Biol Sci Med Sci 1999;54;B150–153.
41. Ross OA, Curran MD, Crum KA, Rea IM, Barnett YA, Middleton D. Increased frequency of the 2437T allele of the heat shock protein 70-Hom gene in an aged Irish population. Exp Gerontol. 2003;38; 561–565.
42. Altomare K, Greco V, Bellizzi D, Berardelli M, Dato S, DeRango F, Garasto S, Rose G, Feraco E, Mari V, Passarino G, Franceschi C, De Benedictis G. The allele (A)(-110) in the promoter region of the HSP70–1 gene is unfavourable to longevity in women. Biogerontology 2003;4:215–220.
43. Matsuzaka Y, Makino S, Nakajima K, Tomizawa M, Oka A, Bahram S, Kulski JK, Tamiya G, Inoko H. New polymorphic microsatellite markers in the human MHC class III region. Tissue Antigens. 2001;57:397–404.

# III.   AGEING OF THE ENDOCRINE SYSTEM

*The Neuroendocrine Immune Network in Ageing*
Edited by R.H. Straub and E. Mocchegiani

# Neuroplasticity in the Human Hypothalamus During Ageing

MICHEL A. HOFMAN and DICK F. SWAAB

*Netherlands Institute for Brain Research, Amsterdam, The Netherlands*

ABSTRACT

The various cell groups in the human hypothalamus show different patterns of ageing, which are the basis for changes in biological rhythms, hormone production, immune processes, autonomic functions and behavior. The suprachiasmatic nucleus (SCN), the clock of the brain, exhibits circadian and seasonal rhythms in peptide synthesis that are disrupted later in life. Furthermore, the age-related sexual differences in the number of vasoactive intestinal polypeptide neurons in this nucleus reinforces the idea that the SCN is not only involved in the timing of circadian rhythms but also in the temporal organization of reproductive functions. The sexually dimorphic nucleus of the preoptic area (SDN-POA), or intermediate nucleus, is twice as large in men as in women, a difference that arises between the ages of 2–4 years and puberty. During ageing a dramatic, sex-dependent decrease in cell number occurs, leading to values which are only 10–15% of the cell number found in early childhood. The vasopressin and oxytocin producing cells in the supraoptic nucleus (SON) and paraventricular nucleus (PVN) are examples of neuron populations that seem to stay perfectly intact in old age. The synthesis of vasopressin in the SON is even increased in elderly women and remains stable in elderly men. Other examples of neuronal activation during ageing are found in the parvocellular corticotropin-releasing-hormone (CRH) containing neurons in the PVN, as indicated by the increase in the number of CRH expressing neurons and by their co-expression of vasopressin. The nucleus basalis of Meynert (NMB), a major source of cholinergic innervation to the neocortex, shows a progressive degeneration of cholinergic neurons in the course of ageing, along with a decreased neuronal metabolic activity, severe cytoskeletal alterations and a loss of high and low affinity neuroptrophin receptors. On the other hand, the protein synthetic activity of the ventromedial nucleus (VMN), a hypothalamic area involved the organization of sex-specific functions and sexual behavior, depends on the age and gender of the subject. The size of the Golgi apparatus (GA) relative to the cell size in the VMN, for example, is larger in young women than in young men and larger in elderly men than in young men. In addition, the GA/cell size ratio is positively correlated with age in men and not in women. The arcuate nucleus of the hypothalamus (ARH), or tubero-infundibular nucleus, contains hypertrophic neurons in postmenopausal women. These hypertrophied neurons contain neurokinin-B, substance P and estrogen receptors and probably act on LHRH neurons as interneurons. The lateral tuberal nucleus (NTL), probably involved in feeding behavior and energy metabolism, does not show any neuronal loss in senescence. To sum up, each cell group of the human hypothalamus has its own sex-specific pattern of ageing. In fact, some hypothalamic nuclei show a dramatic functional decline with ageing, whereas others seem to become more active later in life.

## 1.    INTRODUCTION

The human hypothalamus, which measures only 4 cm³, is confined anteriorly by the lamina terminalis, posteriorly by the midbrain tegmentum, and superiorly by the hypothalamic sulcus [1,2]. Although the complex cellular arrangement and multitude of afferent and efferent projections have made analysis of hypothalamic organization difficult, most authors distinguish three major regions: (i) the *preoptic or chiasmatic region* – containing the suprachiasmatic nucleus, sexually dimorphic nucleus, supraoptic nucleus and paraventricular nucleus. In addition, the nucleus basalis of Meynert, the diagonal band of Broca and the bed nucleus of the stria terminalis are considered in connection with the preoptic region; (ii) the *tuberal region* – containing the ventromedial and dorsomedial hypothalamic nuclei, along with the arcuate (or tubero-infundibular) nucleus, lateral tuberal nucleus and tubero-mamillary nucleus, and (iii) the *posterior or mamillary region*, which is dominated by the mamillary bodies that abut the midbrain tegmentum (Figure 1) [for reviews, see 1–5].

Before 1900 there were only vague intimations of the function of the brain surrounding the third ventricle and these were based primarily on various pathological and assorted clinical observations [6]. Since then a large body of experimental evidence has been derived implicating that this part of the brain contains the control systems which are critically involved in a wide range of homeostatic and rheostatic regulatory processes [see e.g., 4,7,8]. Among these are the control of water balance, food ingestion and energy metabolism, sleep, body temperature and neuroendocrine secretion and the regulation of reproduction and various emotional-affective states. Increasing age and a variety of diseases may impair many of these functions and may have deleterious effects on the morphology of hypothalamic structures [see, e.g., 2,5,9–11]. Some hypothalamic cell groups, on the other hand, remain functionally intact or even show clear signs of neuroplasticity by becoming more active later in life. In the present review, recent data on a number of prominent hypothalamic nuclei in the preoptic and tuberal regions are discussed in relation to sexual differentiation, ageing and some neuropathological conditions.

## 2.    PREOPTIC OR CHIASMATIC REGION

### 2.1.    Suprachiasmatic nucleus

The suprachiasmatic nucleus (SCN) is the principal component of the mammalian biological clock, which is responsible for generating and coordinating many physiological, endocrine, and behavioral circadian rhythms [for reviews, see 12,13]. Consistent with its role in the temporal organization of circadian processes, investigations in rodents and non-human primates suggest that the SCN is also involved in the seasonal control of reproduction, sexual behavior and energy metabolism [14,15].

The human SCN is a small collection of parvocellular neurons in the basal part of the anterior hypothalamus, just dorsal to the optic chiasm on either side of the third ventricle (Figure 1). It is usually elongated in the anterior-posterior direction. The external boundaries of the SCN are often difficult to discern in conventionally thionin-stained sections, but by labelling the SCN

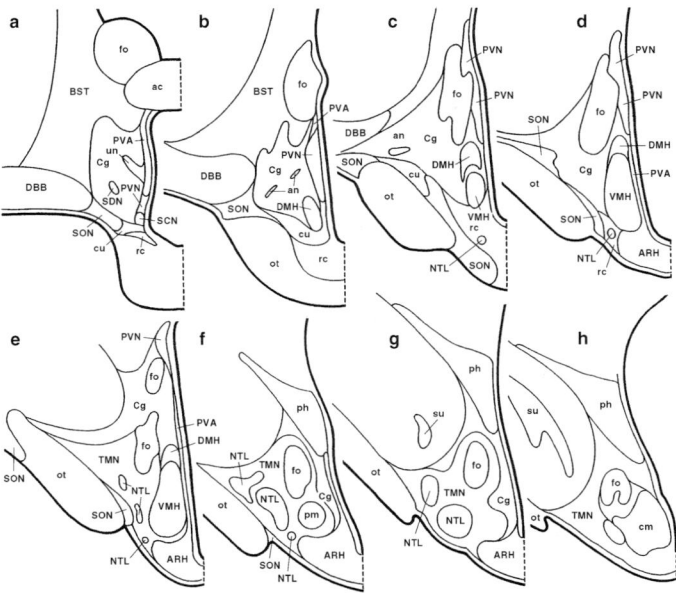

Figure 1. Preoptic and tuberal region of the hypothalamus in the adult human brain. The diagram shows the main land-marks and nuclear grays that are encountered as coronal sections and traced antero-posteriorly (a-h). Each of the sections is spaced apart by 800 μm. AC, anterior commissure; AN, accessory neurosecretory nucleus; ARH, arcuate nucleus; BST, bed nucleus of the stria terminalis; CG, chiasmatic gray; CM, corpus mamillare; CU, cuneate nucleus; DBB, nucleus of the diagonal band (of Broca); DMH, dorsomedial nucleus; FM, fasciculus mamillo-thalamicus; FO, fornix; NTL, lateral tuberal nucleus; OT, optic tract; PH, posterior hypothalamic nucleus; PM, posteromedial nucleus; PVA, periventricular area; PVN, paraventricular nucleus; RC, retrochiasmatic nucleus; SCN, suprachiasmatic nucleus; SDN, sexually dimorphic nucleus (or intermediate nucleus; INAH-1); SON, supraoptic nucleus; SU, subthalamic nucleus; TG, tuberal gray; TMN, tuberomamillary nucleus; UN, unicate nucleus; VMH, ventromedial nucleus. Based upon Braak and Braak [3], with permission.

immunocytochemically with antibodies against neuropeptides, such as arginine-vasopressin (AVP), vasoactive intestinal polypeptide (VIP), neuropeptide-Y, or neurotensin, it becomes clearly recognizable [16,17]. Neurons that are immunoreactive to these antisera appear to form distinct subpopulations of cells with different, chemically specific, connections. Our recent find-ings in humans support the notion of the SCN being the principal neural substrate that organizes biological rhythms, as the biosynthesis of particular neuropeptides in this nucleus, such as AVP and VIP, exhibit marked diurnal as well as seasonal fluctuations [15,18]. It appears that photic information may have a synchronizing effect on the clock mechanism of the SCN by inducing changes in the functional activity of certain groups of neurons.

With advancing age, however, the circadian timing system is progressively disturbed, both in humans and other mammals, as is clearly demonstrated by a reduced amplitude and period length of circadian rhythms and an increased tendency towards internal desynchronization [for reviews see, 19,20]. The functional changes in the temporal organization of older individuals may be due to a gradual age-related deterioration in the activity of specific cell groups in the SCN, and presumably other parts of the clock, leading to disruptions of the circadian timing

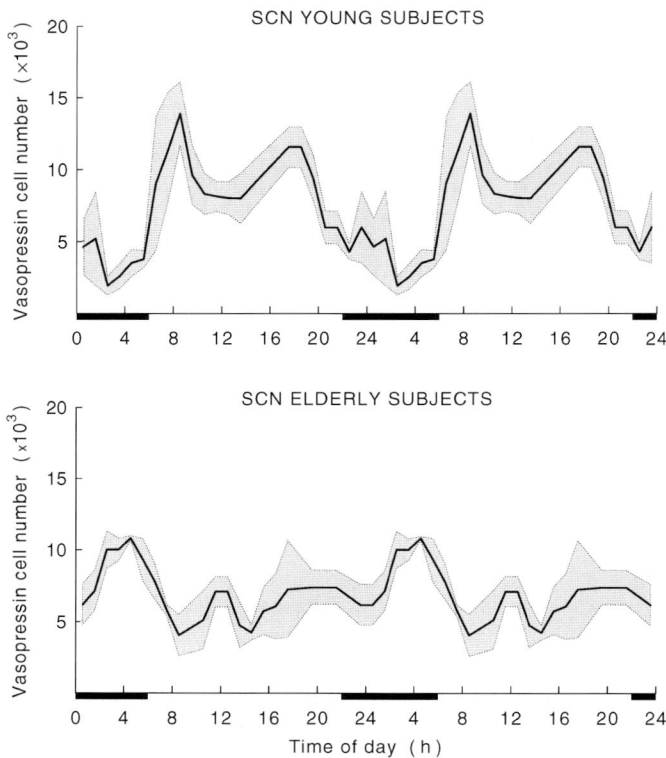

Figure 2. Circadian rhythm in the number of vasopressin-containing neurons in the human suprachiasmatic nucleus (SCN) of (A) young subjects (< 50 years of age) and (B) elderly subjects (≥ 50 years of age). The black bars indicate the night period (22.00–06.00 h). The general trend in the data is enhanced by using a smoothed double plotted curve and is represented by mean ± S.E.M. values. Note the circadian rhythm in the SCN of young people with low values during the night period and peak values during the early morning. From Hofman and Swaab [21], with permission.

system. Part of the neural substrate for the disrupted circadian rhythms with ageing might be the population of AVP neurons in the SCN. It was shown that the number of AVP-expressing neurons in the human SCN exhibits a marked diurnal oscillation in young (up to 50 years of age), but not in elderly people (over 50 years of age) [21,22]. Whereas in young subjects low numbers of AVP-expressing neurons were found during the night period and peak values during the early morning, the SCN of elderly people showed a reduced amplitude and a tendency towards a reversed diurnal pattern with high instead of low values during the night (Figure 2). Similar age-related decrements have been reported for the seasonal timing system [23].

Neuroanatomical studies, furthermore, provide evidence that degenerative alterations in the human SCN occur at a later phase in life than the reported functional changes in circadian organization. More frequent and prolonged awakenings and shorter sleep periods have already been found in 50- to 60-year-old subjects [24,25], whereas a reduction in SCN volume and number of AVP-expressing neurons are only present from the age of 80 years onwards [9,11,26]. Thus, the observed loss of AVP-expressing neurons in the SCN of very old people may only be a relatively late correlate of functional changes in the biological clock appearing much earlier. Many findings indicate that changes in the SCN and other parts of the circadian timing system underlie some of the sleep disturbances among elderly people [20,22,27]. Based on the hypothesis that increased stimulation of the brain can improve or even restore the decreased neuronal activity [28], it has been demonstrated in aged rats that both overt sleep-wake rhythms [29] and cell function in the SCN [30] can be restored by enhancing the stimuli the circadian timing system normally uses for synchronization of the sleep-wake rhythm. There is evidence that the same mechanism may be effective in humans. Improvement of the sleep-wake rhythm of elderly people has been demonstrated by application of a variety of potent modulators of the circadian timing system, like bright light, melatonin and physical activity [for reviews see, 20,31]. Contrary to treatment with hypnotics, the improvement of sleep following these treatments is without adverse effects and even results in improvement of mood, performance, daytime energy, and quality of life.

## 2.2. Sexually dimorphic nucleus (Intermediate Nucleus, INAH-1)

The sexually dimorphic nucleus of the preoptic area (SDN-POA) was first described in the rat brain by Gorski et al. [32]. In rats sexual differentiation of the SDN-POA has been reported to occur during the first ten days postnatally [33] due to differences in perinatal steroid levels [34]. On the basis of lesion experiments in rodents, it was found that the SDN-POA is involved in aspects of male sexual behavior, i.e., mounting, intromission and ejaculation. However, the effects of lesions on sexual behavior are only slight, so it may well be that the major functions of this hypothalamic nucleus are still unknown at present.

The SDN-POA in the young adult human brain is twice as large in males (0.2 mm$^3$) as in females (0.1 mm$^3$) and contains twice as many cells [35,36]. The SDN-POA is located between the supraoptic (SON) and the paraventricular nucleus (PVN), at the same rostrocaudal level as the SCN (Figure 1). The SDN-POA is identical to the intermediate nucleus described by Braak and Braak [1], and to the INAH-1 of Allen et al. [37]. A sex difference does not occur until the 4th postnatal year, when cell numbers start to decrease in girls, whereas in men the cell numbers in the SDN-POA remain stable until approximately the age of 50 years, after which they decline rapidly [36] (Figure 3). In the period from 45 to 60 years we found that the cell number diminished at a rate of 3% per year. After the age of 60, the cell number in the male SDN-POA does not decrease any further. In women, a second phase of cell loss sets in after the age of about 50 years. After a period of relative stability, from 70 years onwards, a more dramatic reduction in cell number could be detected (Figure 3). In the period between 75 and 85 years the cell number in the SDN-POA of females decreases at a rate of 4.8% per year, leading to values which are only 10–15% of the peak value found at 2–4 years postnatally. The sharp decrease in cell numbers in the SDN-POA later in life might be related to the dramatic hormonal changes and decreasing sexual activity which accompany senescence [38,39]. It is not clear whether the age-related alterations in gonadal function are cause or effect of the observed cell loss in this nucleus. Cell numbers in the SDN-POA of Alzheimer's disease patients were found to be within the normal range for age and sex [40].

An intriguing consequence of this sex-dependent pattern of ageing is that the sexual dimorphism of the human SDN-POA is by no means constant throughout adult life. As can be seen in Figure 3, the largest discrepancies in SDN-POA cell number between the sexes are found around the age of 30 years and in people older than 80 years, whereas the sexual dimorphism is least around the age of 60 years.

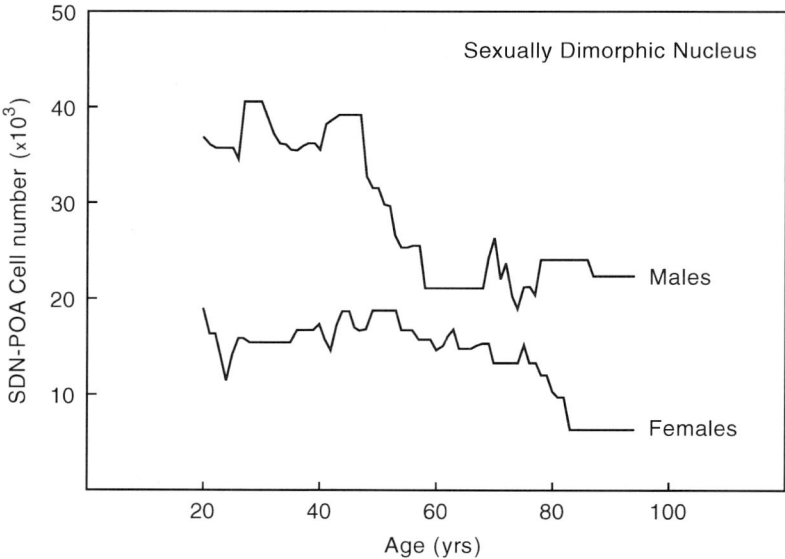

Figure 3. Age-related changes in the total cell number of the sexually dimorphic nucleus of the preoptic area (SDN-POA) in the human hypothalamus. The general trend in the data is enhanced by using smoothed growth curves. Note that in men the SDN-POA cell number steeply declines between the ages of 50 and 60 years, whereas in women, from the age of about 50 years onwards, a more gradual cell loss is observed, which continues up to old age. These curves demonstrate that the reduction in cell number in the human SDN-POA in senescence is a non-linear, sex-dependent process. From Hofman and Swaab [36], with permission.

## 2.3.    Supraoptic and paraventricular nuclei

The large neurosecretory cells of the hypothalamic supraoptic nucleus (SON) and paraventricular nucleus (PVN) produce the neuropeptides arginine vasopressin (AVP) and oxytocin (OT) which are released into the bloodstream in the neurohypophysis. This hypothalamo-neurohypophyseal system is known to be implicated in the regulation of diuresis, vascular resistance, parturition and lactation [4,41]. Parvocellular neurons of the PVN project to brain areas outside the hypothalamus and influence central and autonomic processes [42]. Oxytocin has an effect on, for example, food intake, affiliation, and maternal and sexual behavior [43,44]. In males, oxytocin might also be involved in sexual arousal and ejaculation [45].

The neurons of the human SON and PVN form a population of stable cells in normal ageing and in Alzheimer's disease; no loss of neurons was observed [46–49]. The observation that no cytoskeletal alterations were found in Alzheimer patients with several antibodies in the SON [50] is in accordance with this stability. Although in the PVN of Alzheimer patients some neuronal and dystrophic neurite staining is observed with cytoskeletal antibodies [50], the number of peptidergic neurons in the PVN of these patients is not affected.

Various observations provide evidence for the hypothesis that an enhanced activity of neurons may interfere with the process of ageing, and thus prolongs the lifespan of neurons or restores their function [28]. Many cell groups in the human SON and PVN are not only metabolically highly active throughout life, especially the magnocellular neurons, but they even seem to be more active in elderly subjects than in young adults, as can be derived from the increase in size of the vasopressin- and oxytocin-containing perikarya [51,52], nucleoli [53] and Golgi apparatus [54,55] in senescence. Also the plasma levels of vasopressin are enhanced in elderly subjects [56]. Furthermore, cell counts in the human SON suggest that a substantial proliferation of glial cells takes place in this nucleus with advancing age [57].

The enhanced neuronal activity of the human SON during ageing, however, was only found in women, but not in men. It has therefore been hypothesized that the low-affinity neurotrophin receptor p75 (p75$^{NTR}$) might be involved in the mechanism of activation of the protein synthesis in postmenopausal women, since this receptor has been identified in the PVN and SON of a few aged women, and because estrogens were found to downregulate p75$^{NTR}$ expression [58]. Therefore, we investigated whether the drop of estrogens in the postmenopausal period influences the expression of p75$^{NTR}$ in the AVP neurons of the human SON [59]. The area of p75$^{NTR}$ immunoreactivity per cell was found to be positively correlated with age and with Golgi apparatus size only in women but not in men (Figure 4), suggesting that the neurotrophin receptor p75 is indeed involved in postmenopausal activation of vasopressinergic neurons in the human SON.

In contrast to the SON, the PVN does not only contain magnocellular AVP and OT neurons, but also parvocellular ones that project to brain areas outside the hypothalamus, and to the median eminence. Examples of this type of parvocellular neurons are the corticotropin-releasing hormone (CRH) neurons. In the human PVN they are not located in a well-defined subnucleus as they are in the rat, but are spread all over the PVN, except for the most rostral part where they are absent. Another property of CRH neurons in the PVN is that they may coexpress AVP when activated. This occurs, for example, in the process of ageing, and their activation may be causally related to symptoms of depression [60].

2.4.    Nucleus basalis of Meynert

The nucleus basalis of Meynert (NBM) is the major source of cholinergic innervation to the neocortex [61]. The human NBM includes a series of clusters of magnocellular neurons and scattered perikarya extending from the medial septum and diagonal band of Broca rostrally, through the substantia innominata to the most caudal part of the global pallidus [2,62]. The cholinergic activity of NBM neurons is diminished in elderly subjects, as a result of the reduced choline-acetyltransferase (ChAT) and acetylcholinesterase (AChE) activity [63]. Moreover, the nucleolar volume [64] and Golgi apparatus size [65] of NBM neurons are smaller in elderly subjects. Estimates of changes in neuronal numbers in the human NBM with ageing have yielded similar results. The NBM is affected in a number of cognitive disorders, such as Down's syndrome, Parkinson's disease and especially in Alzheimer's disease, where a progressive degeneration of cholinergic neurons has been observed [2], along with a decreased neuronal metabolic activity [65,66], severe cytoskeletal alterations [50] and a loss of high and low affinity neuroptrophin receptors [67–69].

There is some evidence suggesting a beneficial effect of estrogens on the risk and course of Alzheimer's disease [70,71], although the potential role of estrogen replacement therapy (ERT) is certainly not without controversy. Estrogens may interact with the cholinergic system, nerve growth factor and its high affinity trk receptors [72–74]. In the human NBM both estrogen receptor alpha (ERα) and estrogen receptor beta (ERβ) are present [62], suggesting that estrogens may have a direct effect on the NBM. ERα was expressed to a higher degree than ERβ and was localized mainly in the cell nucleus, while ERβ was mainly confined to the cytoplasm. A significant positive correlation between age and the percentage of ERα nuclear positive neurons

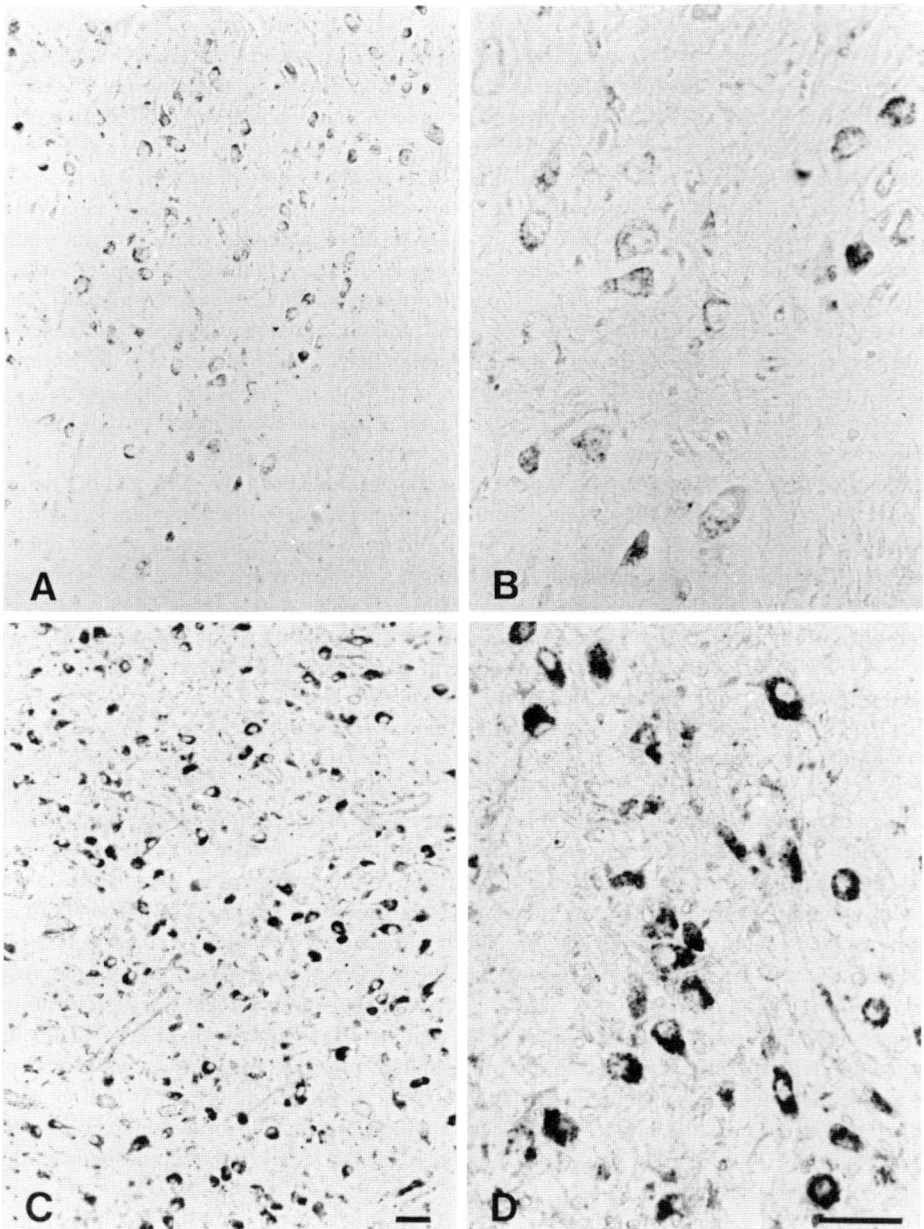

Figure 4. Neurotrophin receptor p75 immunoreactivity in the human supraoptic nucleus (SON) of a young woman (**A,B**) and an elderly woman (**C,D**). Note the difference in the intensity of the staining which is more prominent in the elderly woman. Bar = 50 μm. From Ishunina et al. [59], with permission.

was found only in men and not in women, whereas the proportion of ERβ cytoplasm positive cells was not related with age in either of the sexes [62]. In Alzheimer's disease the proportion of neurons showing nuclear staining for both ERα and ERβ and cytoplasmic staining for ERβ was markedly increased, providing a neural substrate for estrogen replacement therapy in treating cognitive decline in postmenopausal women with Alzheimer's disease.Taken together, these findings demonstrate that there are hardly any changes in ER-immunoreactivity in the human NBM during normal ageing, whereas a pronounced upregulation of both ERs and, in particular, nuclear ERα is observed in Alzheimer patients.


3.     TUBERAL REGION

3.1.     Ventromedial and dorsomedial nuclei

The human ventromedial nucleus (VMN), also known as the nucleus of Cajal, is a pear-shaped structure in the tuberal region of the hypothalamus (Figure 1) with a cell density that is higher in its peripheral region than in the center of the nucleus. It features a narrow, cell-sparse zone surrounding the nucleus, which facilitates its delineation from adjoining nuclear grays [3]. Two main parts are distinguished in the VMN: the largest part is the ventrolateral part (vl-VMN), while a smaller dorsomedial part is situated closer to the fornix [5]. In humans the VMN contains a dense network of somatostatin cells and fibers. In addition, thyrotropin-releasing hormone fibers, substance P, preproenkephalin and NADPH diaphorase were found in this nucleus [for reviews, see 2,75]. Oxytocin binding was demonstrated in the human VMN by autoradiography, suggesting the presence of oxytocin receptors in this region, but they did not show any correlation with age or sex of the subjects [76].

The VMN is interconnected with many neighbouring areas but is also reciprocally connected to the magnocellular nuclei of the basal forebrain, suggesting that the VMN influences higher cortical functions and behavior via these pathways. Animal studies clearly indicate that the VMN is involved in the organization of sex specific functions, such as female reproductive behavior, gonadotropin secretion, feeding, aggression and sexual orientation [2]. PET scan studies in humans suggest that the VMH in our brain is involved in the gender specific integration of pheromonal input [77]. By measuring the size of the Golgi apparatus (GA) and the volume of individual neurons in the human VMN we found the protein synthetic activity of the neurons to be age- and sex dependent [75]. The size of the Golgi apparatus relative to the cell size appeared to be larger in young women than in young men and to be larger in elderly men than in young men. In addition, the GA/cell size ratio correlated significantly with age in men and not in women. Moreover, androgen receptors are clearly expressed in the male VMN and not in the female VMN [78]. These data suggest an inhibitory effect of androgens on the activity of VMN neurons in humans, which is possibly mediated via androgen receptors. Indeed, in men, plasma testosterone levels tend to decline with age [39]. However, the possibility of a stimulatory effect of estrogens on the VMN neurons via estrogen receptors should also be considered.

It should be noticed that the activity of the human VMN neurons was completely opposite to that observed in the SON and PVN. The biosynthesis of vasopressin in the human SON and PVN was higher in young men than in young women and was higher in elderly women than in young women, suggesting an inhibitory effect of estrogens on the activity of these neurons [52,55]. This inhibitory effect in the SON and PVN is most probably mediated via estrogen receptor β, whereas the complex interaction of sex and age differences in the bioactivity of SON

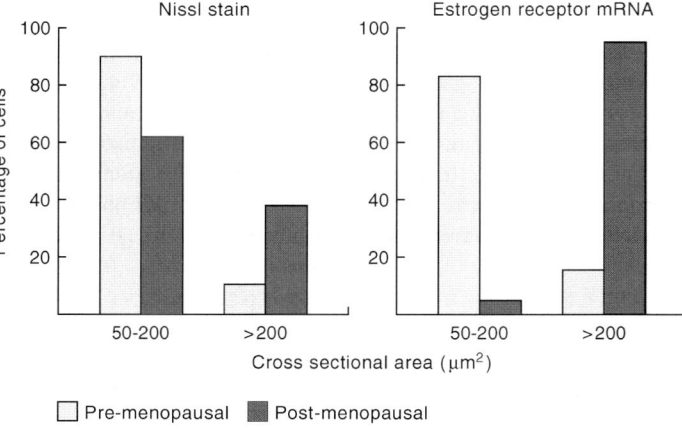

Figure 5. Percentages of neurons less than and exceeding 200 μm cross-sectional area in the arcuate nucleus of the hypothalamus (ARH) of pre-menopausal and post-menopausal women. Left: histogram of all neurons stained with cresyl violet; right: histogram of neurons expressing estrogen receptor gene transcripts. Note the distinct increase in cell size in the estrogen receptor mRNA expressing neurons in post-menopausal women, indicating that the hypertrophy occurs in a subpopulation of neurons. Based upon Rance et al. [92], with permission.

neurons might be explained by the differential expression of both estrogen receptors, ERα and ERβ [59].

The dorsomedial nucleus (DMN) is poorly differentiated in the human brain and covers the anterior and superior poles of the ventromedial nuclues. Large numbers of cells which, according to their cytological features, belong to the hypothalamic gray, invade peripheral portions of the nucleus. The medium-sized nerve cells of the DMN are markedly richer in lipofuscin deposists than those of the VMN [3].

## 3.2.    Arcuate (infundibular) nucleus

The horseshoe-shaped arcuate (or infundibular) nucleus of the hypothalamus (ARH) surrounds the lateral and posterior entrance of the infundibulum (Figure 1) and contains a large number of neurotransmitters and neuropeptides, such as somatostatin, neuropeptide-Y and neurotensin [5]. Various neuroendocrine cell types that are involved in the hypothalamo-pituitary-gonadal (HPG) axis are also located in this nucleus.

The menopause represents a dramatic alteration in the function of the HPG axis in women. As a consequence of decreased levels of sex hormones, neuronal hypertrophy has been reported in the ARH of postmenopausal women [79]. The mean profile area of arcuate neurons of postmenopausal women, but also of older men [80], was significantly larger than that of young subjects (Figure 5). The density of these hypertrophied neurons was also increased. Arcuate neuronal hypertrophy has also been described in chronically ill, hypogonadal men and in patients suffering from starvation and gonadal atrophy [for a review, see 79].

The hypertrophied neurons in the ARH of postmenopausal women contain increased amounts of neurokinin B, substance P, and estrogen receptor mRNAs, indicating an increased neuronal activity as well [79]. Luteinizing hormone-releasing hormone (LHRH) neurons are also found in this nucleus, but the hypertrophied neurons themselves do not contain this peptide. The sensitivity of the luteinizing hormone to LHRH does not seem to decrease in elderly men [81]. On the other hand, a decreased sensitivity to sex steroid feedback was found in elderly (80 years) postmenopausal women as compared to younger postmenopausal women (55 years) [82]. No data on changes in the exact number of LHRH expressing neurons in relation to ageing in the human hypothalamus are available at present.

In addition to LHRH neurons, the human ARH contains a large population of growth hormone-releasing hormone (GHRH) neurons [10]. Together with the inhibiting somatostatin neurons that are situated in the periventricular nucleus, they regulate the secretion of growth hormone (GH), which is not only necessary for normal infant and childhood growth, but also for the regulation of normal body composition and metabolism in adulthood [83]. A current hypothesis states that a defective GH secretion may be one of the pacemakers of ageing [84]. In line with this is the observation that in elderly people the response of GH to GHRH is significantly reduced [85]. It has even been reported that GH responses to GHRH decline after 30 to 40 years in men, and after menopause in women [86]. Although it has been suggested that a decreased GHRH or increased somatostatinergic activity are the main events underlying the age-related decline of GH secretion in mammals, no direct evidence is available about the changes of GHRH or somatostatin neurons in the human hypothalamus with ageing.

### 3.3.    Lateral tuberal nucleus

The lateral tuberal nucleus (nucleus tuberalis lateralis, NTL) is an evolutionary new hypothalamic structure that can only be recognized in anthropoid primates, including man. There are indications that the primate NTL is homologous to the calcitonin gene-related peptide (CGRP) containing the terete nucleus of the rat [87].The most characteristic feature of the human NTL is an extremely great variability of its shape and segmentation and its population of somatostatin expressing neurons [9,48]. In addition, some galanin and LHRH immunoreactive fibers have been found in this nucleus [88,89]. Because, under different conditions, the pathology of this area is accompanied by cachexia, the NTL is hypothesized to play a role in feeding behavior and energy metabolism.

In adulthood the human NTL contains about 60,000 neurons, a number that does not seem to change dramatically during lifespan. In contrast to this stable neuron number, a considerable increase in the neuron density and a decrease in the size of cell nuclei were observed in the NTL of subjects over 60 years old [48]. The NTL is affected in a variety of human neurodegenerative diseases, including Huntington's disease. In fact, neuronal loss in the NTL may be a good marker of the severity of this disease [90]. Pathological changes in the NTL have also been described in depression and some other neurological diseases [2]. In Alzheimer's disease, on the other hand, the number of NTL neurons did not differ from that of controls [91]. In fact, the NTL seems to represent a brain area in which Alzheimer's disease affects the neurons in a limited way, without further progress to the classical changes of silver-staining of tangles and neuronal cell loss.

### 4.    CONCLUDING REMARKS

The main conclusion of the present review is that the ageing process causes differential changes in the human hypothalamus. Declining neuronal activity and cell death may already start around the age of 40 to 50 years in some peptidergic systems (e.g., the decreased number of VIP neurons in the SCN of middle-aged men and the sharp decline in the number of neurons of the SDN-POA after 50 years). On the other hand, some hypothalamic cell groups remain functionally intact and are sometimes even activated in the course of ageing (e.g., the vasopressin-producing neurons of the PVN and SON). The selective vulnerability of neurons in the human brain argues against the notion that normal ageing is accompanied by a general decline in cellular activity. Moreover, the typical sex-specific patterns of ageing, characteristic of many hypothalamic cell groups, suggest that steroid hormones play a key role in the process of ageing and neuroplasticity. It can be concluded that the various hypothalamic nuclei are involved in a great number of functions and that the multitude of changes in this region of the human brain during ageing forms the basis of many physiological, endocrine and behavioral alterations in elderly people.

## ACKNOWLEDGEMENTS

The authors are grateful to Mr. H. Stoffels for drawing most of the figures. Brain material was obtained from the Netherlands Brain Bank, Amsterdam (coordinator Dr. R. Ravid).

## REFERENCES

1.  Braak H, Braak E. The hypothalamus of the human adult: chiasmatic region. Anat Embryol 1987; 176: 315–330.
2.  Swaab DF. Neurobiology and neuropathology of the human hypothalamus. In: Bloom FE, Björklund A, Hökfelt T, editors. Handbook of Chemical Neuroanatomy. The Primate Nervous System, Part1, vol. 13. Amsterdam: Elsevier Science Publishers, 1997; 39–137
3.  Braak H, Braak E. Anatomy of the human hypothalamus (chiasmatic and tuberal region). Prog Brain Res 1992; 93: 3–16.
4.  Swanson LW. The hypothalamus. In: Björklund A, Hökfelt T, Swanson LW, editors. Handbook of Chemical Neuroanatomy, vol. 5. Amsterdam: Elsevier Science Publishers, 1987; 1–124.
5.  Saper CB. Hypothalamus. In: Paxinos G, editor. The Human Nervous System. San Diego, CA: Academic Press, 1990; 389–413.
6.  Morgane PJ, Historical and modern concepts of hypothalamic organization and function. In: Morgane PJ, Panksepp J, editors. Anatomy of the Hypothalamus. New York: Marcel Dekker, 1979; 1–64.
7.  Brooks C McC. The history of thought concerning the hypothalamus and its functions. Brain Res Bull 1988; 20: 657–667.
8.  Mrosovsky N. Rheostasis. The Physiology of Change. New York: Oxford University Press; 1990.
9.  Swaab DF, Hofman MA, Lucassen PJ, Purba JS, Raadsheer FC, Van de Nes JAP. Functional neuroanatomy and neuropathology of the human hypothalamus. Anat Embryol 1993; 187: 317–330.
10. Zhou J-N, Swaab DF. Age-related changes in neuropeptidergic neurons in the human hypothalamus. In: Macieira-Coelho A, editor. Molecular Basis of Aging. Boca Raton, CA: CRC Press, 1995; 527–552.
11. Hofman MA. Lifespan changes in the human hypothalamus. Exp Gerontol 1997; 32: 559–575.
12. Klein DC, Moore RY, and Reppert SM, editors. Suprachiasmatic Nucleus; The Mind's

Clock. New York: Oxford University Press; 1991.

13. Van Essenveldt LE, Lehman MN, Boer GJ. The suprachiasmatic nucleus and the circadian time-keeping system revisited. Brain Res Rev 2000; 33: 44–77.

14. Turek, F.W. and Van Cauter, E. Rhythms in reproduction. In: Knobil E, Neill JD, editors. Physiology and Reproduction 2nd ed. New York: Raven Press, 1994; 487–540.

15. Hofman MA, Swaab DF. A brain for all seasons: cellular and molecular mechanisms of photoperiodic plasticity. Prog Brain Res 2002; 138: 255–280.

16. Moore RY. The organization of the human circadian timing system. Prog Brain Res 1992; 93: 101–117.

17. Hofman MA, Zhou J.-N, Swaab DF. Suprachiasmatic nucleus of the human brain: an immunocytochemical and morphometric analysis. Anat Rec 1996; 244: 552–562.

18. Hofman MA, Swaab DF. Diurnal and seasonal rhythms of neuronal activity in the suprachiasmatic nucleus of humans. J Biol Rhythms 1993; 8: 283–295.

19. Touitou Y, Haus, E. Biological rhythms and aging. In: Touitou Y, Haus E, editors. Biological Rhythms in Clinical and Laboratory Medicine. Berlin: Springer Verlag, 1992; 188–207.

20. Van Someren EJW, editor. Circadian Rhythms and Sleep in Human Aging. Chronobiol Inter. (Spec issue) 2000; vol. 17(3): 233–432.

21. Hofman MA, Swaab DF. Alterations in circadian rhythmicity of the vasopressin-producing neurons of the human suprachiasmatic nucleus (SCN) with aging. Brain Res 1994; 651: 134–149.

22. Hofman MA. The human circadian clock and aging. Chronobiol Int 2000; 17: 245–259.

23. Hofman MA, Swaab DF. Influence of aging on the seasonal rhythm of the vasopressin-expressing neurons in the human suprachiasmatic nucleus. Neurobiol Aging 1995; 16: 965–971.

24. Myers BL, Badia P. Changes in circadian rhythms and sleep quality with aging: mechanisms and interventions. Neurosci Biobehav Rev 1995; 19: 553–571.

25. Prinz PN, Bailey SL, Woods DL. Sleep impairments in healthy seniors: roles of stress, cortisol, and interleukin-1 beta. Chronobiol Int 2000; 17: 391–404.

26. Swaab DF, Fliers E, Partiman TS. The suprachiasmatic nucleus of the human brain in relation to sex, age and senile dementia. Brain Res 1985; 342: 37–44.

27. Dijk D-J, Duffy JF, Czeisler, CA. Contribution of circadian physiology and sleep homeostasis to age-related changes in human sleep. Chronobiol Int 2000; 17: 285–311.

28. Swaab DF. Brain aging and Alzheimer's disease: "wear and tear" versus "use it or lose it". Neurobiol Aging 1991; 12: 317–324.

29. Witting W, Mirmiran M, Bos NP, Swaab DF. Effect of light intensity on diurnal sleep-wake distribution in young and old rats. Brain Res Bull 1993; 30: 157–162.

30. Lucassen PJ, Hofman MA, Swaab DF. Increased light intensity prevents the age related loss of vasopressin-expressing neurons in the rat suprachiasmatic nucleus. Brain Res 1995; 693: 261–266.

31. Van Someren EJW, Riemersma RF, Swaab DF. Functional plasticity of the circadian timing system in old age: light exposure. Prog Brain Res 2002; 138: 205–231.

32. Gorski RA, Gordon JH, Shryne JE, Southam AM. Evidence for a morphological sex difference within the medial preoptic area of the rat brain. Brain Res 1978; 148: 333–346.

33. Jacobson CD, Shryne JE, Shapiro F, Gorski RA. Ontogeny of the sexually dimorphic nucleus of the preoptic area. J Comp Neurol 1980; 193: 541–548.

34.  Kawata M. Roles of steroid hormones and their receptors in structural organization in the nervous system. Neurosci Res 1995; 24: 1–46.

35.  Swaab DF, Fliers E. A sexually dimorphic nucleus in the human brain. Science 1985; 228: 1112–1115.

36.  Hofman MA, Swaab DF. The sexually dimorphic nucleus of the preoptic area in the human brain: a comparative morphometric study. J Anat 1989; 164: 55–72.

37.  Allen LS, Hines M, Shryne JE, Gorski RA. Two sexually dimorphic cell groups in the human brain. J Neurosci 1989; 9: 497–506.

38.  Sonntag WE. Hormone secretion and action in aging animals and man. Rev Biol Res Aging 1987; 3: 299–335.

39.  Vermeulen A. Androgens and male senescence. In: Nieschlag E, Behre HM, editors. Testosterone. Action, Deficiency, Substitution. Berlin: Springer Verlag, 1990; 261–276.

40.  Swaab DF, Gooren LJG, Hofman MA. The human hypothalamus in relation to gender and sexual orientation. Prog Brain Res 1992; 93: 205–219.

41.  Hatton GI. Emerging concepts of structure-function dynamics in adult brain: the hypothalamo-neurohypophysial system. Prog Neurobiol 1990; 34: 437–504.

42.  Buijs RM, Kalsbeek A. Hypothalamic integration of central and peripheral clocks. Nature Reviews/ Neuroscience 2001; 2: 521–526.

43.  Carter LS. Oxytocin and sexual behavior. Neurosci Biobehav Rev 1992; 16: 131–144.

44.  Insel TR. Oxytocin - a neuropeptide for affiliation: evidence from behavioral, receptor autoradiographic, and comparative studies. Psychoneuroendocrinology 1992; 17: 3–35.

45.  Murphy MR, Seckl JR, Burton S, Checkley SA, Lightman SL. Changes in oxytocin and vasopressin secretion during sexual activity in men. J Clin Endocrinol Metab 1987; 65: 738–741.

46.  Goudsmit E, Hofman MA, Fliers E, Swaab DF. The supraoptic hypothalamus in relation to sex, age and Alzheimer's disease. Neurobiol Aging 1990; 11: 529–536.

47.  Wierda M, Goudsmit E, Van der Woude PF, Purba JS, Hofman MA, Bogte H, Swaab DF. Oxytocin cell number in the human paraventricular nucleus remains constant with aging and in Alzheimer's disease. Neurobiol Aging 1991; 12: 511–516.

48.  Morys J, Dziewiatkowski J, Switka A, Sadowski M, Narkiewicz O. Morphometric parameters of some hypothalamic nuclei: age-related changes. Folia Morphol (Warz) 1994; 53: 221–229.

49.  Van der Woude PF, Goudsmit E, Wierda M, Purba JS, Hofman MA, Bogte H, Swaab, DF. No vasopressin cell loss in the human hypothalamus in aging and Alzheimer's disease. Neurobiol Aging 1995; 16: 11–18.

50.  Swaab DF, Grundke-Iqbal I, Iqbal K, Kremer HPH, Ravid R, Van de Nes JAP. τ and ubiquitin in the human hypothalamus in aging and Alzheimer's disease. Brain Res 1992; 590: 239–249.

51.  Fliers E, De Vries GJ, Swaab DF. Changes with aging in the vasopressin and oxytocin innervation of the rat brain. Brain Res 1985; 348: 1–8.

52.  Ishunina TA, Swaab DF. Vasopressin and oxytocin neurons of the human supraoptic and paraventricular nucleus; size changes in relation to age and sex. J Clin Endocrinol Metab 1999; 84: 4637–4644.

53.  Hoogendijk JE, Fliers E, Swaab DF, Verwer RWH. Activation of vasopressin neurons in the human supraoptic and paraventricular nucleus in senescence and senile dementia. J Neurol Sci 1985; 69: 291–299.

54.  Lucassen PJ, Ravid R, Gonatas NK, Swaab DF. Activation of the human supraoptic and

paraventricular nucleus neurons with aging and in Alzheimer's disease as judged from increasing size of the Golgi apparatus. Brain Res 1993; 632: 105–113.

55.  Ishunina TA, Salehi A, Hofman MA, Swaab DF. Activity of vasopressinergic neurones of the human supraoptic nucleus. J Neuroendocrinol 1999; 11: 251–258.

56.  Frolkis VV, Golovchenko SF, Medved VI, Frolkis RA. Vasopressin and cardiovascular system in aging. Gerontology 1982; 28: 290–302.

57.  Hofman MA, Goudsmit E, Purba JS, Swaab DF. Morphometric analysis of the supraoptic nucleus in the human brain. J Anat 1990; 172: 259–270.

58.  Miranda RC, Sohrabji F, Toran-Allerand D. Interactions of estrogen with the neurotrophins and their receptors during neural development. Horm Behav 1994; 28: 367–375.

59.  Ishunina TA, Salehi A, Swaab DF. Sex- and age-related expression of p75 neuroptrophin receptor in the human supraoptic nucleus. Neuroendocrinology 2000; 71: 243–251.

60.  Raadsheer FC, Van Heerikhuize JJ, Lucassen PJ, Hoogendijk WJ, Tilders FJH, Swaab DF. Corticotropin-releasing hormone mRNA levels in the paraventricular nucleus of patients with Alzheimer's disease and depression. Am J Psychiatry 1995; 152: 1372–1376.

61.  Mesulam M-M, Mufson EJ, Levey AI, Wainer BH. Cholinergic innervation of cortex by the basal forebrain: cytochemistry and cortical connections of the septal area, diagonal band nuclei, nucleus basalis/substantia innominata and hypothalamus in the rhesus monkey. J Comp Neurol 1983; 214: 170–197.

62.  Ishunina TA, Swaab DF. Increased expression of estrogen receptor  and  in the nucleus basalis of Meynert in Alzheimer's disease. Neurobiol Aging 2001; 22: 417–426.

63.  Sparks DL, Hunsaker JC, Slevin JT, DeKosky ST, Kryscio RJ, Markesbery WK. Monoaminergic and cholinergic synaptic markers in the nucleus basalis of Meynert (NBM): normal age-related changes and the effect of heart disease and Alzheimer's disease. Ann Neurol 1992; 31: 611–620.

64.  Mann DMA, Yates PO, Marcyniuk B. Changes in nerve cells of the nucleus basalis of Meynert in Alzheimer's disease and their relationship to ageing and the accumulation of lipofucsin pigment. Mech Aging Dev 1984; 25: 189–204.

65.  Salehi A, Lucassen PJ, Pool CW, Gonatas NK, Ravid R, Swaab DF. Decreased neuronal activity in the nucleus basalis of Meynert in Alzheimer's disease as suggested by the size of the Golgi apparatus. Neuroscience 1994; 59: 871–880.

66.  Swaab DF, Lucassen PJ, Salehi A, Scherder EJA, Van Someren EJW, Verwer RWH. Reduced neuronal activity and reactivation in Alzheimer's disease. Prog Brain Res 1998; 117: 343–377.

67.  Mufson EJ, Li J-M, Sobreviela T, Kordower JH. Decreased trkA gene expression within basal forebrain neurons in Alzheimer's disease. NeuroReport 1996; 8: 25–29.

68.  Salehi A, Verhaagen J, Dijkhuizen PA, Swaab DF. Co-localization of high affinity neurotrophin receptors in nucleus basalis of Meynert neurons and their differential reduction in Alzheimer's disease. Neuroscience 1996; 75: 373–387.

69.  Salehi A, Ocampo M, Verhaagen J, Swaab DF. P75 neurotrophin receptor in the nucleus basalis of Meynert in relation to age, sex and Alzheimer's disease. Exp Neurol 2000; 161: 245–258.

70.  Van Duijn CM. Hormone replacement therapy and Alzheimer's disease. Maturitas 1999; 15: 201–205.

71.  Waring SC, Rocca WA, Petersen RC, O'Brien PC, Tangalos EG, Kokmen E. Postmenopausal estrogen replacement therapy and risk of AD: a population-based study. Neurology 1999; 52: 965–970.

72. Schneider LS, Farlow MR, Henderson VW, Pogoda J. Effects of estrogen replacement therapy on response to tacrine in patients with Alzheimer's disease. Neurology 1996; 46: 1580–1584.

73. Gibbs RB, Aggarwal P. Estrogen and basal forebrain cholinergic neurons: implications for brain aging and Alzheimer's disease-related cognitive decline. Horm Behav 1998; 34: 98–111.

74. Mufson EJ, Cai WJ, Jaffer S, Chen E-Y, Stebbins G, Sendera T, Kordower JH. Estrogen receptor immunoreactivity within subregions of the rat forebrain: neuronal distribution and association with perikarya containing choline acetyltransferase. Brain Res 1999; 849: 253–274.

75. Ishunina TA, Unmehopa UA, Van Heerikhuize JJ, Pool CW, Swaab DF. Metabolic activity of the human ventromedial nucleus neurons in relation to sex and ageing. Brain Res 2001; 893: 70–76.

76. Loup F, Tribollet E, Dubois-Dauphin M, Dreifuss JJ. Localization of high affinity binding sites for oxytocin and vasopressin in the human brain. An autoradiographic study. Brain Res 1991; 555: 220–232.

77. Savic I, Berglund H, Gulyas B, Roland P. Smelling of odorous sex hormone-like compounds causes sex-differentiated hypothalamic activations in humans. Neuron 2001; 31: 661–668.

78. Fernandez-Guasti A, Kruijver FPM, Fodor M, Swaab DF. Sex differences in the androgen receptor distribution in the human hypothalamus J Comp Neurol 2000; 425: 422–435.

79. Rance NE. Hormonal influences on morphology and neuropeptide gene expression in the infundibular nucleus of post-menopausal women. Prog Brain Res 1992; 93: 221–236.

80. Rance NE, Uswandi SV, McMullen NT. Neuronal hypertrophy in the older men. Neurobiol Aging 1993; 14: 337–342.

81. Kaufman JM, Giri M, Deslypere JM, Thomas G, Vermeulen A. Influence of age on the responsiveness of the gonadotrophs to luteinizing hormone-releasing hormone in males. J Clin Endocrinol Metab 1991; 72: 1255–1263.

82. Rossmanith W, Reichelt C, Scherbaum W. Neuroendocrinology of aging in humans: attenuated sensitivity to sex steroid feedback in elderly postmenopausal women. Neuroendocrinology 1994; 59: 355–362.

83. Christiansen JS, Jorgensen JO, Pedersen SA, Moller J, Jorgensen J, Kakkebaek NE. Effects of growth hormone on body composition in adults. Horm Res 1990; 33 (suppl. 4): 61–64.

84. Rudman D, Feller AG, Nagraj HS, Gergans, GA, Lalitha PY, Goldberg AF, Schlenker RA, Cohn L, Rudman I, Mattson DE. Effect of human growth hormone in men over 60 years old. New Engl J Med 1990; 323: 1–6.

85. Coiro V, Volpi R, Bertoni P, Finzi G, Marcato A, Caiazza A, Colla R, Giacalone G, Rossi G, Chiodera P. Effect of potentiation of cholinergic tone by pyridostigmine on the GH response to GHRH in elderly men. Gerontology 1992; 38: 217–222.

86. Iranmanesh A, Lizarralde G, Veldhuis JD. Age and relative adiposity are specific negative determinants of the frequency and amplitude of growth hormone (GH) secretory bursts and the half-life of endogenous GH in healthy men. J Clin Endocrinol Metab1991; 73: 1081–1088.

87. Lantos TA, Gorcs TJ, Palkovits M. Immunohistochemical localization of calcitonin gene-related peptide in the terete nucleus of the rat hypothalamus. Neurobiology 1996; 4: 73–84.

88. Gai WP, Geffen LB, Blessing WW. Galanin immunoreactive neurons in the human hypothalamus: colocalization with vasopressin-containing neurons. J Comp Neurol 1990; 298: 265–280.
89. Najimi M, Chigr F, Jordan D, Leduque P, Block B, Tommasi M, Rebaud P, Kopp N. Anatomical distribution of LHRH-immunoreactive neurons in the human infant hypothalamus and extrahypothalamic regions. Brain Res 1990; 516: 280–291.
90. Kremer HPH, Roos RAC, Dingjan G, Marani E, Bots GTM. Atrophy of the hypothalamic lateral tuberal nucleus in Huntington's disease. J Neuropathol Exp Neurol 1990; 49: 371–382.
91. Kremer HPH. The hypothalamic lateral tuberal nucleus: normal anatomy and changes in neurological diseases. Prog Brain Res 1992; 93: 249–261.
92. Rance NE, McMullen NT, Smialek JE, Price DL, Young WS III. Postmenopausal hypertrophy of neurons expressing the estrogen receptor gene in the human hypothalamus. J Clin Endocrinol Metab 1990; 71: 79–85.

# The Role of Growth Hormone Signaling in the Control of Ageing

A. BARTKE[1], M. HEIMAN[2], D. TURYN[3], F. DOMINICI[3] and J. J. KOPCHICK[4,5]

*[1]Geriatric Research, Department of Medicine, School of Medicine, Southern Illinois University, Springfield, Illinois; [2]Lilly Research Labs, Corporate Center, Indianapolis, Indiana; [3]Instituto de Quimica y Fisicoquimica Biologicas, Facultad de Farmacia y Bioquimica, Buenos Aires, Argentina; [4]Edison Biotechnology Institute, Konneker Research Laboratories, Ohio University, Athens, Ohio; [5]Department of Biomedical Sciences, College of Osteopathic Medicine, Ohio University, Athens, Ohio*

ABSTRACT

Physiological functions of growth hormone (GH) and insulin-like growth factor 1 (IGF-1), the key mediator of GH actions, include effects on ageing and longevity. Mutant mice with GH resistance, GH deficiency or partial IGF-1 resistance exhibit symptoms of delayed ageing and live significantly longer than normal (wild type) animals. Homologous pathways regulate ageing in invertebrates. Mechanisms of GH action on ageing are not known but probably include alterations in metabolism, insulin signaling, and generation of reactive oxygen species. The release of GH gradually declines during adult life and this may have a role in protecting the organism from cancer and other age-related diseases. However, reduced GH levels have also been linked to age-related changes in body composition, muscle and brain function, and are suspected of contributing to the deterioration of quality of life in the elderly. In spite of great interest in GH treatment as an anti-ageing therapy, the risks and benefits of its use are not well characterized and the concept of GH replacement during "somatopause" is controversial. It is suggested that the relationship of GH signaling to ageing is not the same at different stages of life history and that GH actions may mediate some of the apparent trade-offs between growth, reproduction and longevity.

## 1. INTRODUCTION

Ageing is associated with progressive alterations in all organ systems. Age-related changes in the function of the hypothalamic-pituitary axis and the peripheral endocrine glands have been studied in considerable detail and decline in the levels of various hormones, including growth hormone (GH) in ageing animals and humans is well documented. There is also a considerable amount of information on the mechanisms responsible for reduced GH release during ageing and on the consequences of the resulting reduction in circulating GH levels. However, experimental findings obtained during the last decade indicate that the intricate relationship of ageing to endocrine function is not limited to alterations in hormone levels that accompany ageing. There is increasing evidence that hormonal signaling plays an important role in the control of ageing, and that endocrine function during early stages of life history is important in determining the rate of ageing and life expectancy. These complex relationships which we are only beginning to understand are well illustrated by considering the proposed roles of GH in ageing. As was mentioned above, the rate of GH release is profoundly influenced by age. There is considerable evidence that physiological actions of this hormone somehow promote ageing and thus contribute to life expectancy in endocrinologically normal individuals being considerably shorter than the achievable maximum. Curiously, physiological levels of GH signaling are apparently also responsible also for preventing age-related functional decline in several organ systems, and for maintaining youthful body composition.

In this review, we will summarize evidence for these seemingly opposing actions of GH and attempt to place them in the context of genetic and environmental control of ageing. We will also relate GH actions to the features of life histories and reproductive strategies that impact longevity, and to the prospects of devising anti-ageing interventions. We will discuss findings concerning both GH and insulin-like growth factor 1 (IGF-1) because, in mammals, IGF-1 mediates most of the known actions of GH, and GH stimulates IGF-1 production in various organs and acts as the principal regulator of IGF-1 levels in peripheral circulation.

## 2.   STUDIES OF LONGEVITY GENES LINK IGF SIGNALING TO THE CONTROL OF AGEING

Microscopic soil-dwelling round worm, *Caenorhabditis elegans (C. elegans)* and a common fruit fly, *Drosophila melanogaster* are widely used for studies of developmental biology and genetics. Elegant studies in the laboratories of Riddle, Johnson, Kenyon, Ruvkun, Hekimi, Guarente, Tatar, Larsen, Partridge and others [reviewed in: 1,2] provided evidence that natural mutations or targeted disruption (the so called knock-out) of individual genes in these organisms can produce a significant and, in some cases quite dramatic extension of life span. Interestingly, most of these "longevity genes" were shown to be involved in the control of identifiable signaling pathways which can be activated by environmental stimuli and influence life history traits related to stress resistance, diapause, reproduction and life expectancy. In 1997, Kimura et al. [3] reported that daf-2, a key gene in the control of ageing in *C. elegans* exhibits homology to mammalian IGF-1 and insulin receptors and suggested that the genetic control of ageing may be conserved in taxonomically distant groups of animals. This important concept and the involvement of GH and IGF-1 signaling, the so-called *somatotropic axis*, in the control of ageing in mammals received strong support from the reports of significantly prolonged life span in house mice (*Mus musculus*) with spontaneous or experimentally induced loss of function mutations that affect peripheral GH and/or IGF-1 levels and action. In mice, significant extension of longevity was reported in three spontaneous mutants and four types of animals with targeted gene disruption, the so-called gene knock-out (KO) [Table I; Refs 4–13]. Three of these seven types of long lived mice are GH deficient and one is GH resistant. Of the remaining three, one exhibits partial IGF-1 resistance. Thus reduced somatotropic axis signaling is present in five of seven known instances of loss of function of a single gene leading to prolonged longevity in mammals.

Concerning the role of GH in the control of ageing, data obtained in GHR-KO mice are particularly compelling. In these animals, physiological actions of GH during their entire life span are prevented by disruption of the GH receptor [9] and thus all of the phenotypic characteristics of these animals, including delayed ageing [14] and prolonged longevity [10,15] reflect either direct or secondary consequences of the absence of GH signaling. It should be mentioned that

Table I    Spontaneous mutations and gene knock-outs that prolong longevity in mice.

| | Primary effect | % increase in mean life-span | References* |
|---|---|---|---|
| Ames dwarf Prop1[df] | GH, PRL & TSH deficiency | 40–65% | Schaible & Gowen, 1961 [4]<br>Brown-Borg et al., 1996 [5] |
| Snell dwarf Pit1[dw] | GH, PRL & TSH deficiency | ~42% | Snell, 1929 [6]<br>Flurkey et al., 2001 [7] |
| Little GHRHR[lit] | GH deficiency | 24% on low fat diet | Eicher & Beamer, 1976 [8]<br>Flurkey et al., 2001 [7] |
| GHR/GHBP –/– | GH resistance | 40–55% | Zhou et al, 1997 [9]<br>Coschigano et al., 2000 [10] |
| P66[shc] –/– | Stress resistance | 30% | Migliaccio et al., 1999 [11] |
| IGF1R +/– | Partial IGF-1 resistance | 33% (females) | Holzenberger et al., 2003 [12] |
| FIRKO –/– | Adipose-specific insulin resistance | 18% | Blüher et al., 2003 [13] |

*When two references are listed, the first is the original description of the mutation (or the knock-out) and the second is the report of the effect on longevity.

in *C. elegans* and *D. melanogaster*, multiple insulin-like molecules signal through a common receptor and a signaling pathway which corresponds to both IGF-1 and insulin signaling pathways in mammals [2]. Early in evolution of chordate ancestors of vertebrates, these pathways became separated and in mammals they involve three major ligands (insulin, IGF-1 and IGF-2) acting primarily via distinct insulin and IGF-1 receptors. Mutations affecting GH and IGF-1 signaling in mice lead to secondary alterations in insulin signaling and thus involvement of both pathways in producing the long-lived phenotype in these animals can be suspected.

The effect of mutations of the various longevity genes on life expectancy can be quite dramatic with increases in the average life span well over 100% in several *C. elegans* mutants and well over 50% (depending on gender, diet and genetic background) in some mutant mice. It should also be emphasized that these increases of the average life span are accompanied by significant, generally comparable increases in maximal life span and by physiological, histopathological, and behavioral indices of delayed ageing [7,14,16,17]. This provides evidence that longevity genes can influence the fundamental biological process of ageing rather than preventing disease or some pathological changes that lead to early mortality. Another important implication of these findings is that life extension in these mutants does not represent a prolonged period of senility, but rather a longer "health-span" i.e. postponement of age-related pathology and functional decline. We will return to the issue of distinguishing between gains in the average vs. the maximal life span, and between ageing per se and age-related disease later in this review. Although difficult to separate both conceptually and experimentally, age-related disease and ageing are believed to be distinct biological processes.

3.    SEARCH FOR MECHANISMS THAT LINK REDUCED GH AND IGF-1 SIGNAL-
      ING WITH DELAYED AGEING

Putative mechanisms that link somatotropic axis to the control of longevity can be proposed on the basis of known actions of GH and IGF-1, and phenotypic characteristics shared by several long-lived mouse mutants, as well as some of the long-lived invertebrate mutants. Additional leads can be obtained from comparing the phenotype of long lived mutants to the characteristics of genetically normal (wild type) animals that have been subjected to caloric restriction (CR). Caloric restriction is usually accomplished by feeding the animals 50–70% of the food they would have consumed under conditions of unlimited access (ad libitum, AL) and is well documented to prevent disease, delay ageing, and prolong life in laboratory populations of mice and rats [reviewed in: 18,19]. Consequently, characteristics shared by long-lived mutant and CR mice emerge as possible mechanisms of delayed ageing. Obviously, association of these characteristics with prolonged longevity can not prove cause/effect relationships and it is possible that they represent markers rather than causes of delayed ageing. However, known actions of some of the hormones affected by both "longevity genes" and CR and available experimental and epidemiological findings allow developing some mechanistic hypothesis or, at the very least, speculations. Mechanisms suspected of mediating the effects of reduced GH signaling on longevity are listed below.

### 3.1. Activity of the somatotropic axis, growth and adult body size

In mice, small body size correlates with increased life expectancy. This relationship is supported by studies of lines produced by selection for differential growth rate [20,21], as well as various strains, mutants, knock-outs and transgenics [22] and, more recently by relating individual differences in adult body size in a genetically heterogeneous population of mice to life span [23]. In some of these comparisons, and in studies of wild-derived vs. domesticated mice, low-plasma IGF-1 levels correlate with small size and prolonged longevity [23,24]. Association of small body size and increased life expectancy is well documented in domestic dogs [25] and size in this species correlates with circulating IGF-1 levels [26]. There is considerable evidence that negative association of body size and longevity applies also to other species, including the human [27].

It is unclear what mechanisms are responsible for the apparent "cost" of growth in terms of longevity, particularly when viewed in light of evidence that in the comparisons between, rather than within species, it is the larger animals that generally live longer, presumably due to lower metabolic rate. However, epidemiological [28,29] and experimental [30] evidence for IGF-1 being a risk factor for cancer suggests at least one possibility. Pathological elevations of peripheral GH and IGF-1 levels in individuals affected with gigantism or acromegaly and in transgenic mice overexpressing GH are associated with significantly reduced life expectancy [31–33]. However, relating these observations to the role of GH and IGF-1 in normal ageing is complicated by various disease processes in acromegalic humans and giant transgenic mice [31,34,35]

The relationship of extended longevity of mutant and knock-out mice to alterations in their body composition is unclear. Obesity is a risk factor for age-related disease in the human and important role of trunkal obesity in the control of responsiveness to insulin is strongly supported by epidemiological data [36] and by results of surgical removal of abdominal fat depots in the rat [37]. Prolonged longevity of lean FIRKO mice [13] adds important evidence for this relationship. However, long lived GHR-KO mice have increased adiposity (Table II) resembling humans with Laron dwarfism [38]. Snell dwarf and Ames dwarf mice are frequently obese, but average proportion of body fat and plasma leptin levels in middle aged and old Ames dwarfs are generally lower than the values measured in normal mice of the same chronological age [39].

Table II    Body composition in adult growth hormone resistant GHR-KO mice in comparison to normal animals from the same stock.

|  | Females | | Males | |
|---|---|---|---|---|
|  | Normal | GHR-KO | Normal | GHR-KO |
| Number of animals | 27 | 23 | 25 | 22 |
| Body wt. (g) | 21.5 ± 0.8 | 13.0 ± 0.6 * | 32.9 ± 1.5 | 16.1 ± 0.6 * |
| % lean | 80.1 ± 1.9 | 73.9 ± 2.4 * | 73.8 ± 2.0 | 52.7 ± 2.6 * |
| % fat | 10.5 ± 1.8 | 14.5 ± 2.4 | 18.8 ± 2.0 | 34.6 ± 3.1 * |
| Bone mineral density | 6.92 ± 0.16 | 6.18 ± 0.14 | 8.00 ± 0.57 | 6.43 ± 0.12 * |

Determined by DEXA using procedures described by Heiman et al. [39].

*Significantly different from normal mice of the same sex; $P < 0.05$.

## 3.2.    Reduced insulin levels and increased sensitivity to insulin

Long-lived dwarf mice and GHR-KO mice have reduced plasma insulin levels [40–43]. This is not surprising in light of well documented anti-insulin actions of GH and stimulatory effects of GH, IGF-1 and PRL on the insulin-producing cells in the pancreas [44–46]. Concomitant reduction in plasma insulin and glucose levels indicates increased sensitivity to insulin. In support of this conclusion, responses to acute insulin stimulation in the liver (including phosphorylation of insulin receptor and its substrates) are significantly enhanced in these animals [42,43]. However, relationship of GH and IGF-1 actions to insulin signaling is complex and suppression of hepatic IGF-1 action in organ-specific knock-outs as well as in dwarf and GHR-KO mice is associated with symptoms of insulin resistance in skeletal muscle [47,48]. It should be emphasized that reduction in plasma insulin levels is among the earliest and most consistent responses to CR in rodents, subhuman primates and humans [18,19,49,50].

We hypothesize that reduced insulin release and improved sensitivity to insulin represent key mechanisms linking reduced activity of the somatotropic axis and improved life expectancy. We also believe that the same mechanisms may be critical for determination of human longevity. In support of this hypothesis, insulin resistance is a major risk factor for most age-related diseases in the human [51]. Plasma insulin levels are inversely related to survival in the Baltimore Ageing study [52], Italian centenarians are characterized by exceptional sensitivity to insulin [53], and incidence of diabetes in Japanese centenarians is 1/10 of the incidence in people in their 80s in the same population [54]. Separation of the effects of disease from alterations in biological ageing on the basis of demographic data is exceedingly difficult. However, there is indirect evidence that centenarians represent a segment of the population that is genetically predisposed to delayed ageing and thus findings in these exceptional individuals may be particularly relevant in this regard.

## 3.3    Body temperature, metabolic rate, and oxidative stress

Body core temperature (Tco) is markedly reduced in dwarf mice [55] resembling the findings in CR mice [18,19]. Smaller but significant reduction in Tco was detected also in GHR-KO mice [56]. Reduced release of thyroid hormones, GH and insulin is likely to be responsible for these findings since each of these hormones is thermogenic. It is unclear whether reduced Tco represents a reduction in metabolic rate. Although absolute levels of food consumption, oxygen consumption, and metabolism are reduced in dwarf and CR mice in comparison to control animals (wild type [WT] and AL, respectively), metabolic rate per unit of lean body weight may not be affected [57]. Interpretation of these findings is very complicated because long-lived mutant, KO and CR animals are much smaller than the corresponding controls and therefore have less favorable surface: mass ratio and face a greater challenge when housed in standard laboratory conditions i.e. much below thermoneutral temperature. Moreover, their body composition is different from normal (wild-type) mice with relatively greater weight of the brain, which is very metabolically active, and relatively smaller amount of fat which has low metabolic rate. The comparisons are further complicated by differences in average daily Tco between these groups and by the ability of small rodents to enter a state of torpor with a major further reduction in Tco.

Wild-type rodents subjected to CR resemble long lived Snell and Ames dwarf and GHR-KO mice in that they are small and hypothermic with reduced plasma levels of GH, IGF-1, insulin, and glucose, and variably suppressed reproductive functions. There is evidence in rats that long term CR does not alter metabolic rate expressed per unit of lean body mass [57]. However, there are some controversies concerning analysis and interpretations of data on metabolic rate of CR animals [58] and findings in non-human primates and in the human have been interpreted as evidence that CR reduces metabolism [59]. These complications notwithstanding, it is reasonable to propose that the slow growing, hypothyroid animals with reduced Tco may have alterations in energy metabolism that lead to reduced generation of reactive oxygen species (ROS).

Reduced ROS generation [60], combined with improved antioxidant defenses [61–63] may lead to diminished oxidative stress and reduced oxidative damage to DNA and other cell components. Protein oxidative damage in the liver and brain, and lipid peroxidation in the brain are reduced in Ames dwarf, in comparison to normal mice [60]. Moreover, oxidative damage to mitochondrial DNA in the brain of Ames dwarf mice is markedly reduced [64]. Inasmuch, as oxidative stress and damage are believed to be the key mechanisms of ageing at the cellular level, these alterations are likely to represent the final, common effector by which reduced somatotropic signaling and secondary changes in insulin action prolong longevity.

Possible contributions of delayed sexual maturation and reduced gonadal function [65,66] to prolonged longevity in dwarf, GHR-KO and CR mice are probably minor and will not be discussed in this article. Results obtained in IGF1R +/– mice provide evidence that reduced IGF-1 signaling can increase life span without inhibiting reproductive functions [12].

## 4.   CONFLICTING FINDINGS

Evidence for the involvement of the somatotropic axis in the control of ageing is compelling. However, evidence for the causative role of reduced GH and/or IGF-1 actions in extending longevity in mammals is largely correlative and is viewed by some as controversial. Reducing plasma IGF-1 levels by expression of a GH antagonist in transgenic mice does not influence life span [67], selective GH deficiency in GH releasing hormone receptor (GHRHR) mutant (GHRHR[lit]) mice is associated with prolonged longevity only when obesity is prevented by a low fat diet [7], and in transgenic "minirats" life span is prolonged in heterozygotes with moderate reduction in IGF-1 but shortened in homozygotes with a more profound suppression of IGF-1 [68]. Injections of GH were reported to shorten life span in cardiomyopathic hamsters [69], to extend it in mice [70], and have no effect on lifespan in rats [71]. Data on the impact of congenital GH deficiency on ageing and longevity in the human are extremely limited and difficult to interpret. Hypopituitarism caused by mutation of the Prop1 gene (homologous to Ames dwarfism in mice) and GH resistance due to mutations of the GH receptor are apparently compatible with reaching an advanced age [72,73] but the number of patients is too small to allow meaningful comparisons with endocrinologically normal individuals from the same populations. A recent case report [74] comments on a youthful appearance of a middle aged patient with isolated GH deficiency. Findings most difficult to reconcile with the anti-ageing actions of reduced activity of the somatotropic axis concern the normal decline in GH release during ageing and the beneficial effects of GH replacement in elderly individuals. These important issues will be discussed below.

## 4.1.  GH levels decline with age

It is well documented that concentration of GH in peripheral circulation declines during ageing. Detailed studies in the rat, in the dog and in the human provided evidence that this decline is due primarily to reduced release of GH releasing hormone (GHRH) from the hypothalamus [75–78] and entails suppression of secretory bursts of GH from the pituitary [75,78]. Age related suppression of GH release is very pronounced. In the human, it starts soon after attainment of adult height, with secretory rate of GH declining by approximately one half every seven years [78]. This alteration in endocrine function is not without consequences. Studies in individuals who at some point in their adult life lost the ability to secrete GH (most commonly as a result of pituitary tumors and their treatment) lead to the realization of multiple roles of GH after linear growth is completed and to description of the syndrome of adult GH deficiency (GHD). Individuals afflicted with GHD are characterized by obesity, loss of muscle mass, and increased cardiovascular disease risk factors, as well as reduced energy level, reduced sexual interest, psychological alterations, more frequent need to seek medical attention, and other symptoms collectively described as reduced quality of life [79,80]. These patients improve dramatically when treated with GH [79,80]. Since GH levels in the elderly are very low and many symptoms of ageing resemble symptoms of GHD, it was proposed that age-related changes in body composition and various measures of the quality of life are caused by reduced activity of the somatotropic axis, the so-called somatopause. This proposal received strong support from the landmark study of Rudman et al. [81], who reported in 1990 that six months of treatment with GH in elderly men who had very low IGF-1 levels resulted in reduced adiposity, increased muscle mass and increased skin thickness. Growth hormone treatment also improved bone density on one of the examined sites of the skeleton [81]. These findings lead to great interest in GH treatment in the management of symptoms of ageing. This very interesting and controversial issue will be briefly discussed in the next section of this article.

## 4.2.  Potential use of GH in anti-ageing medicine

Following publication of results obtained by Rudman et al. [81], several groups examined effects of GH therapy in unselected elderly subjects. The ability of GH to reduce percent body fat and to increase lean body mass (primarily muscle volume) was confirmed, but beneficial effects were generally not as clear as they are in younger patients with adult GHD [82]. Numerous side effects were noted including athralgias, carpal tunnel syndrome, joint pain and development of insulin resistance that, in a few cases appeared to have progressed to diabetes [82,83]. Moreover, it was pointed out that GH induced increase in muscle mass does not necessarily imply improved strength [83]. Most investigators concluded that in the elderly GH should be used with caution, that more studies are needed to evaluate the risk-benefit ratio of GH therapy in this group of patients, and that at present, advanced age alone does not constitute an indication for GH treatment [82,84].

In spite of these reservations, GH, GH releasers and numerous ill-defined "GH-related" products are aggressively promoted as anti-ageing agents [85] and advertisements directed at potential consumers promise reversal of virtually every accompaniment of ageing including graying of the hair and baldness. These claims are usually supported by quoting beneficial effects of GH replacement in middle aged patients with GHD. However, most of the evidence for problems associated with untreated adult GHD (including cardiovascular risk factors and increased mortality) and for the benefits of GH in these patients is derived from studies in subjects who lost the ability to secrete GH as a result of pituitary tumors and/or their treatment [86]. Ethiology of GH deficiency in this type of patients represents a serious confounding factor in any attempt to extrapolate to normal elderly subjects. Some endocrinologists believe that undesirable side effects of GH treatment in the elderly can be avoided by use of lower doses and also by starting with a very low dose and increasing it gradually during early stages of treatment.

Multiple lines of evidence for neuroprotective actions of IGF-1 [87–89] support the concept of potential benefits of GH therapy. Moreover, GH treatment of ageing rats was shown to stimulate brain vascularity and regional blood flow [90] as well as cognitive function [91] thus reducing or reversing the effects of ageing on these parameters.

Clinical information available to date can be summarized by stating that the benefits of GH therapy in the elderly are probably proportional to the degree of suppression of the somatotropic axis in these individuals, that doses comparable to those used in young GHD patients should be avoided, that risks of long-term GH therapy in the elderly are not known and include potentially serious problems, and that there is no data on the possible effects of GH replacement in elderly humans on life expectancy. Detailed discussion of clinical, ethical and philosophical issues related to replacement therapy aimed at correcting physiological change in hormonal function is outside the scope of this article. Extrapolation from the data and opinions concerning gonadal hormone replacement therapy (HRT) probably applies to many aspects of the use of GH in the elderly.

## 5.    POSSIBLE PROTECTIVE ROLE OF AGE-RELATED DECLINE IN SOMATO-TROPIC ACTIVITY

In view of delayed ageing and prolonged longevity of GH resistant and GH deficient mice, it is interesting to consider the possibility that evolution may have favored a decline in GH release after somatic growth is completed or nearly completed. In the preceding section(s) of this article, we have discussed the effects of age-related changes in GH release on fat deposition, muscle mass, brain vascularity, neuronal survival, and CNS function. None of these changes appears to be beneficial or adaptive. However, reduced somatotropic activity may have a role in delaying or preventing major age-related diseases and disease risk factors including cancer, insulin resistance and diabetes.

Epidemiological studies suggest that plasma IGF-1 levels and other indices of GH action constitute risk factors for neoplasms including breast, prostate and colon cancer [29,92,93]. Stimulatory effects of IGF-1 on the growth of cancer cells [94] support this concept. Tumor development is delayed and severity of tumors reduced in long-lived dwarf mice with combined GH,

PRL and TSH deficiency [17]. Dwarf mice as well as dwarf rats with isolated GH deficiency are protected from chemically induced carcinogenics [30,95] and GH treatment promotes development of mammary tumors in response to dimethlybenzenthracene (DMBA) in dwarf rats [30].

Anti-insulinemic actions of GH are well documented and GH treatment as well as pathological elevation of GH due to pituitary tumors or overexpression of GH in transgenic animals can lead to hyperinsulinemia and insulin resistance [34,96]. Incidence of diabetes is increased in acromegalic patients [34] and a recent study suggested that GH treatment of elderly subjects may constitute risk factor for diabetes [82]. However, in patients with GHD, GH replacement may reduce, rather than exacerbate insulin resistance by reducing abdominal fat deposits [80]. If age-related decline in GH should prove to have a role in maintaining sensitivity to insulin, this would have important implications for survival. Evidence reviewed earlier in this chapter suggests that insulin signaling plays an important role in the control of ageing in many organisms and that enhanced responses to insulin are found in exceptionally long-lived humans [53] as well as in long-lived mutant mice [40,42,43].

6. RELATIONSHIP OF GH ACTIONS DURING DIFFERENT STAGES OF LIFE HISTORY TO THE REGULATION OF AGEING AND LONGEVITY; EVOLUTIONARY CONSIDERATIONS

The mechanisms responsible for negative association of the activity of somatotropic axis with longevity, for the possible protective role of reduced GH release in adults, and for the relationship of low GH levels to functional decline during ageing remain to be clearly identified. It is a matter of speculation how these complex and somewhat counterintuitive relationships may have evolved. During pre- and peri-pubertal period, high levels of GH, and unrestricted food intake promote somatic and reproductive development but may not be optimal for prolonged survival. The existence of trade-offs between sexual maturation, fertility, reproductive effort, and longevity is supported by considerable amount of evidence. This complex topic will not be discussed here due to limitations of space. The putative mechanisms linking GH actions with ageing might include increased production of reactive oxygen species related to anabolic and thermogenic effects of GH, as well as increased nonenzymatic glycation [97], and detrimental effects of insulin [98] related to GH induced reduction in sensitivity to insulin and consequent stimulation of insulin release.

If the gradual decline in GH release after sexual maturation is indeed protective, as discussed earlier in this review, it may have been favored by the impact of prolonged adult survival on the length of the reproductive period. Extended period of parental care of the young in primates may have provided additional evolutionary advantage to individuals less likely to succumb to cancer, diabetes, or other diseases related to insulin resistance. The apparent costs of such protective effects would include detrimental consequences of reduced GH levels on skeletal muscles, fat deposition, brain vascularity and brain function in old individuals. These consequences of age-related reduction in GH levels may not have been selected against because they emerge mainly in the post-reproductive period when the force of natural selection is greatly reduced if not eliminated.

The assumptions concerning the relationship of GH secretion at different stages of life history to the regulation of longevity are highly speculative. However, they appear plausible when viewed in the light of information on the effects of caloric intake on longevity. Caloric restriction (CR) started before sexual maturation, in early adulthood, or in mid-life prevents age-related disease and prolongs life, with early onset CR being most effective [18,19]. However, in ageing individuals, reduced food intake is detrimental rather than beneficial and survival benefits can be produced by improved nutrition and increased food consumption [99,100]. The proposed highly complex interplay between GH and age-related changes in susceptibility to disease, in body composition, and in physical and mental function may constitute one of the mechanisms of the apparent trade-offs between growth and reproduction and life expectancy.

## 6.1.   Implications for possible anti-ageing strategies

It is interesting to consider whether any of the information discussed in this review might allow development of novel and effective anti-ageing interventions. The suspected role of altered insulin signaling in long lived GH resistant and GH deficient mice suggests that reducing insulin and glucose levels in addition to its well documented role in preventing age-related disease may also slow the biological process of ageing. This approach to "anti-ageing medicine" seems particularly attractive, since reduction in peripheral insulin and glucose levels in the human can be safely and effectively achieved by diet and exercise. Pharmacological interventions aimed at improving sensitivity to insulin, or perhaps selectively interfering with insulin actions in the adipose tissue [13] or in particular fat depots [37] emerge as additional, more remote possibilities.

It is also possible that a low dose regimen of GH replacement in selected elderly subjects will be proven in future studies to be beneficial in preventing or reversing frailty and in improving the quality of life. In this context, it is interesting to mention that in the human, physical exercise increases GH release [101, 102] and benefits of exercise in preventing and reversing the age-related functional decline are documented beyond any doubt. Importantly, this includes evidence from studies in very old subjects [103].

## ACKNOWLEDGEMENTS.

Studies in our laboratory and preparation of this article were supported by NIH, AG19899, Illinois C-FAR and SIU Geriatric Medicine and Research Initiative. We apologize to those whose work relevant to the discussed issues was not cited due to limitations of space or our inadvertent omission.

## REFERENCES

1.    Guarente L, Kenyon C. Genetic pathways that regulate ageing in model organisms. Nature 2000; 408: 255–262.
2.    Clancy DJ, Gems D, et al. Extension of life-span by loss of CHICO, a *Drosophila* insulin receptor substrate protein. Science 2001; 292: 104–106.
3.    Tatar M, Bartke A, et al. The endocrine regulation of aging by insulin-like peptides: translating molecular homology into function. Science 2003; 299: 1346–1351.
4.    Schaible R, Gowen JW. A new dwarf mouse. Genetics 1961; 46: 896.
5.    Brown-Borg HM, Borg KE, et al. Dwarf mice and the ageing process. Nature 1996; 384: 33.
6.    Snell GD. Dwarf, a new mendelian recessive character of the house mouse. Proc Natl Acad Sci U S A 1929; 15: 733–734.

7.   Flurkey KJ, Papaconstantinou, et al. Lifespan extension and delayed immune and collagen aging in mutant mice with defects in growth hormone production. Proc Natl Acad Sci U S A 2001; 98: 6736–6741.

8.   Eicher EM, Beamer WG. Inherited ateliotic dwarfism in mice. Characteristics of mutation, little, on chormosome 6. J Hered 1976; 67: 87–91.

9.   Zhou Y, Xu BC, et al. A mammalian model for Laron syndrome produced by targeted disruption of the mouse growth hormone receptor/binding protein gene (The Laron mouse). Proc Natl Acad Sci U S A 1997; 94:13215–13220.

10.   Coschigano KT, Clemmons D, et al. Assessment of growth parameters and life span of GHR/BP gene-disrupted mice. Endocrinology 2000; 141: 2608–2613.

11.   Migliaccio E, Giorgio M, et al. The p66she adaptor protein controls oxidative stress response and life span in mammals. Nature 1999; 402: 309–313.

12.   Holzenberger M, Dupont J, et al. IGF-1 receptor regulates lifespan and resistance to oxidative stress in mice. Nature 2003; 421: 182–187.

13.   Bluher M, Kahn B, et al. Extended Longevity in Mice Lacking the Insulin Receptor in Adipose Tissue. Science 2003; 299: 572.

14.   Kinney BA, Coschigano KT, et al. Evidence that age-induced decline in memory retention is delayed in growth hormone resistant GH-R-KO (Laron) mice. Physiol Behav 2001; 72: 653–660.

15.   Bartke A, Chandrashekar V, et al. Consequences of growth hormone (GH) overexpression and GH resistance. Neuropeptides 2002; 36: 201–208.

16.   Kinney BA, Meliska CJ, et al. Evidence that Ames dwarf mice age differently from their normal siblings in behavioral and learning and memory parameters. Horm Behav 2001; 39: 277–284.

17.   Ikeno Y, Bronson RT, et al. The delayed occurrence of fatal neoplastic diseases in Ames dwarf mice: Correlation to extended longevity. J Gerontol Biol Sci Med Sci 2003; 58: B291–B296.

18.   Weindruch R, Walford RL. The retardation of aging and disease by dietary restriction. Springfield, IL, Charles C. Thomas 1988.

19.   Masoro EJ. Dietary Restriction: An Experimental Approach to the Study of the Biology of Aging. Handbook of the Biology of Aging (Fifth Edition). San Diego, Academic Press 2001; 396–420.

20.   Roberts RC. The lifetime growth and reproduction of selected strains of mice. Heredity 1961; 16: 369–381.

21.   Eklund J, Bradford CE. Longevity and lifetime body weight in mice selected for rapid growth. Nature 1977; 265: 48–49.

22.   Rollo CD. Growth negatively impacts the life span of mammals. Evol Dev 2002; 4: 55–61.

23.   Miller RA, Harper JM, et al. Big mice die young: early life body weight predicts longevity in genetically heterogeneous mice. Aging Cell 2002; 1: 22–29.

24.   Miller RA, Dysko R, et al. Mouse (Mus musculus) stocks derived from tropical islands: new models for genetic analysis of life history traits. J Zool 2000; 250: 95–104.

25.   Patronek GJ, Waters DJ, et al. Comparative longevity of pet dogs and humans: implications for gerontology research. J Gerontol 1997; 52A: B171–B178.

26.   Eigenmann JE, Amador A, et al. Insulin-like growth factor I levels in proportionate dogs, chondrodystrophic dogs and in giant dogs. Acta Endocrinol 1988; 118:105–108.

27.   Samaras TT, Elrick H. Is height related to longevity? Life Sci 2003; 72: 1781–1802.

28. Tretli S. Height and weight in relation to breast cancer morbidity and mortality; a prospective study of 570,000 women in Norway. Int J Cancer 1989; 44: 23–30.

29. Yu H, Rohan T. Role of the insulin-like growth factor family in cancer development and progression. J Natl Cancer Inst 2000; 20: 1472–1489.

30. Ramsey M, Ingram R, et al. Growth hormone-deficient dwarf animals are resistant to Dimethylbenzanthracine (DMBA)-induced mammary carcinogenesis. Endocrinology 2002; 143: 4139–4142.

31. Orme SM, McNally RJQ, et al. Mortality and cancer incidence in acromegaly: a retrospective cohort study. J Clinic Endocrinol Metab1998; 83: 2730–2734.

32. Wolf E, Kahnt E, et al. Effects of long-term elevated serum levels of growth hormone on life expectancy of mice: lessons from transgenic animal models. Mech Ageing Dev 1993; 68: 71–87.

33. Cecim M, Bartke A, et al. Expression of human, but not bovine growth hormone genes promotes development of mammary tumors in transgenic mice. Transgenics 1994; 1: 431–437.

34. Jadresic A, Banks LM, et al. The acromegaly syndrome. Quart J Med 1982; 202: 189–204.

35. Wanke R, Wolf E, et al. The GH-transgenic mouse as an experimental model for growth research: clinical and pathological studies. Horm Res 1992; 37: 74–87.

36. Cefalu W, Wang Z, et al. Contribution of visceral fat mass to the insulin resistance of aging. Metabolism 1995; 44: 954–959.

37. Gabriely I, Ma X, et al. Removal of visceral fat prevents insulin resistance and glucose intolerance of aging. Diabetes 2002; 51: 2951–2958.

38. Rosenfeld RG, Rosenbloom AL, et al. Growth hormone (GH) insensitivity due to primary GH receptor deficiency. Endocr Rev 1994; 15: 369–390.

39. Heiman M, Tinsley F, et al. Body composition of prolactin-, growth hormone-, and thyrotropin-deficient ames dwarf mice. Endocrine 2003; 20:149–154.

40. Borg KE, Brown-Borg HM, et al. Assessment of the primary adrenal cortical and pancreatic hormone basal levels in relation to plasma glucose and age in the unstressed Ames dwarf mouse. Proc Soc Exp Biol Med 1995; 210: 126–133.

41. Hsieh CC, DeFord JH, et al. Effects of the *Pit 1* mutation on the insulin signaling pathway: implications on the longevity of the long-lived Snell dwarf mouse. Mech Ageing Dev 2002; 123: 1244–1255.

42. Dominici FP, Hauck S, et al. Increased insulin sensitivity and upregulation of insulin receptor, insulin receptor substrate (IRS)-I and IRS-2 in liver of Ames dwarf mice. J Endocrinol 2002; 173: 81–94.

43. Dominici FP, Diaz GA, et al. Compensatory alterations of insulin signal transduction in liver of growth hormone receptor knockout mice. J Endocrinol 2000; 166: 579–590.

44. Garcia-Ocana A, Vasavada R, et al. Using beta cell growth factors to enhance human pancreatic islet transplantation. J Clin Endocrinol Metab 2001; 86:984–988.

45. Nielsen J, Svensson C, et al. Beta cell proliferation and growth factors. J Mol Med 1999; 77: 62–66.

46. Parsons JA, Bartke A, et al. Number and size of islets of langerhans in pregnant, human growth hormone-expressing transgenic, and pituitary dwarf mice: effect of lactogenic hormones. Endocrinology 1995; 136: 2013–2021.

47. Yakar S, Liu JL, et al. Liver-specific IGF-I gene deletion leads to muscle insulin insensitivity. Diabetes 2001; 50: 1110–1118.

48.  Dominici, F, et al. unpublished data, 2003.
49.  Roth GS, Ingramn DK, et al. Calorie restriction in primates: will it work and how will we know? J Am Geriatr Soc 1999; 47: 896–903.
50.  Hursting S, Lavigne J, et al. Calorie restriction, aging, and cancer prevention: mechanisms of action and applicability to humans. Annu Rev Med 2003; 54: 131–152.
51.  Facchini FS, Hua N, et al. Insulin resistance as a predictor of age-related diseases. J Clin Endocrinol Metab 2001; 86: 3574–3578.
52.  Roth GS, Lane M, et al. Biomarkers of caloric restriction may predict longevity in humans. Science 2002; 297: 811
53.  Paolisso G, Gambardella A, et al. Glucose tolerance and insulin action in healthy centenarians. Am J Physiol 1996; 270: E890–E894.
54.  Hirose T. 2002, personal communication.
55.  Hunter WS, Croson WB, et al. Low body temperature in long-lived Ames dwarf mice at rest and during stress. Physiol Behav 1999; 67: 433–437.
56.  Hauck SJ, Hunter WS, et al. Reduced levels of thyroid hormones, insulin, and glucose, and lower body core temperature in the growth hormone receptor/binding protein knockout mouse. Ex Biol Med 2001; 226: 552–558.
57.  McCarter RJ, Palmer J. Energy metabolism and aging: a lifelong study of Fischer 344 rats. Am J Physiol 1992; 263: E448–E452.
58.  Poehlman E. Editorial: Reduced metabolic rate after caloric restriction-can we agree on how to normalize the data? J Clin Endo Metab 2003; 88: 14–15.
59.  Blanc S, Schoeller D, et al. Energy expenditure of rhesus monkeys subjected to 11 years of dietary restriction. J Clin Endo Metab 2003; 88: 16–23.
60.  Brown-Borg HM, Johnson W, et al. Mitochondrial oxidant generation and oxidative damage in Ames dwarf and GH transgenic mice. Am Aging Assoc 2001; 24: 85–96.
61.  Brown-Borg HM, Rakoczy SG. Catalase expression in delayed and premature aging mouse models. Exp Gerontol 2000; 35: 199–212.
62.  Hauck S, Bartke A. Effects of growth hormone on hypothalamic catalase and Cu/Zn superoxide dismutase. Free Radic Biol Med 2000; 28: 970–978.
63.  Brown-Borg HM, Rakoczky S, et al. Effects of growth hormone and insulin-like growth factor-1 on hepatocyte antioxidative enzymes. Exp Biol Med 2002; 227: 94–104.
64.  Sanz A, Bartke A, et al. Long-lived Ames dwarf mice: oxidative damage to mitochondrial DNA in heart and brain. J Am Aging Assoc 2002; 25: 119–122.
65.  Bartke A. Delayed aging in Ames dwarf mice. Relationships to endocrine function and body size. In: The Molecular Genetics of Aging. Z. Hekimi (Ed.) Springer-Verlag; 2000: 29:181–202.
66.  Bartke A, Chandrashekar V, et al. Consequences of growth hormone (GH) overexpression and GH resistance. Neuropeptides 2002; 36: 201–209.
67.  Bartke A, Brown-Borg HM, et al. Growth hormone and aging. J Am Aging Assoc 2000; 23: 219–225.
68.  Shimokawa I, Yoshikazu H, et al. Life span extension by reduction in growth hormone-insulin-like growth factor-1 axis in a transgenic rat model. Am J Pathol 2002; 160: 2259–2265.
69.  Marleau S, Lapointe N, et al. Effect of chronic treatment with bovine recombinant growth hormone of cardiac dysfunction and lesion progression in UM-X7.1 cardiomyopathic hamsters. Endocrinology 2002; 143: 4846–4855.
70.  Khansari DN, Gustad T. Effects of long-term, low-dose growth hormone therapy on

immune function and life expectancy of mice. Mech Ageing Dev 1991; 57: 87–100.

71. Kalu, DN, Orhii PB, et al. Aged-rodent models of long-term growth hormone therapy - lack of deleterious effect on longevity. J Gerontol A Biol Sci Med Sci 1998; 53: B452–B463.

72. Krzisnik C, Kolacio Z, et al. The "Little People" of the island of Krk – revisited; etiology of hypopituitarism revealed. J Endocr Genet 1999; 1: 9–19.

73. Kopchick JJ, Laron Z. Is the Laron mouse an accurate model of Laron Syndrome? (Minireview). Mol Genet Metab 1999; 68: 232–236.

74. Den Ouden DT, Kroon M, et al. Clinical case seminar: a 43-year-old man with untreated panhypopituitarism due to absence of the pituitary stalk: from dwarf to giant. J Endocinol Metab 2002; 87: 5430–5434.

75. Sonntag WE, Hylka VW, et al. Growth hormone restores protein synthesis in skeletal muscle of old male rats. J Gerontol 1985; 40: 689–694.

76. Müller E, Cella S, et al. Aspects of the neuroendocrine control of growth hormone secretion in ageing mammals. J Reprod Fertil Suppl 1993; 46: 99–114.

77. Rudman D. Growth hormone, body composition, and aging. J Am Geriatr Soc 1985; 33: 800–807.

78. Giustina A, Veldhuis JD. Pathophysiology of the neuroregulation of growth hormone secretion in experimental animals and the human. Endocr Rev 1998; 19: 717–797.

79. Baum HBA, Katznelson L, et al. Effects of a physiological growth hormone (GH) therapy on cognition and quality of life in patients with adult-onset GH deficiency. J Clin Endocr Metab 1998; 83: 3184–3189.

80. Vance ME, Mauras N. Growth hormone therapy in adults and children. N Engl J Med 1999; 341: 1206–1216.

81. Rudman D, Feller AG, et al. Effects of human growth hormone in men over 60 years old. N Engl J Med 1990; 323: 1–6.

82. Blackman MR, Sorkin JD, et al. Growth hormone and sex steroid administration in healthy aged women and men. JAMA 2002; 288: 2282–2292.

83. Papadakis MA, Grady D, et al. Growth hormone replacement in healthy older men improves body composition but not functional ability. Ann Intern Med 1996; 124: 708–716.

84. von Werder K. The somatopause is no indication for growth hormone therapy. J Endocrinol Invest 1999; 22(suppl 5): 137–141.

85. Klatz R. Grow Young With HGH: The Amazing Medically Proven Plan to Reverse Aging. New York, New York, Harper Collins,1999.

86. Rosen T, Bengtsson BA. Premature mortality due to cardiovascular disease in hypopituitarism. Lancet 1990; 336: 285–288.

87. Garcia-Segura LM, Cardona-Gomez GP, et al. Insulin-like growth factor-I receptors and estrogen receptors interact in the promotion of neuronal survival and nuroprotection. J Neurocytol 2000; 29: 425–437.

88. Carro E, Trejo JL, et al. Circulating insulin-like growth factor I mediates the protective effect of physical exercise against brain insults of different etiology and anatomy. J Neurosci 2001; 21: 5678–5684.

89. Frago LM, Paneda C, et al. Growth hormone (GH) and GH-releasing Peptide-6 increase brain insulin-like growth factor-1 expression and activate intracellular signaling pathways involved in neuroprotection. Endocrinology 2002; 143: 4113–4122.

90. Sonntag WE, Lynch C, et al. The effects of growth hormone and IGF-I deficiency on

cerebrovascular and brain ageing. J Anat 2000; 4: 575–585.

91.  Markowska AL, Mooney M, et al. Insulin-like growth factor-I (IGF-I) ameliorates age-related behavioral deficits. Neuroscience 1998; 87: 559–569.

92.  Burroughs KD, Dunn SE, et al. Insulin-like growth factor-1: a key regulator of human cancer risk? J Natl Cancer Inst 1999; 91: 579–581.

93.  Rosen CJ, Pollak M. IGF-I and aging: A new perspective for a new century. Trends Endocrinol Metab 1999; 10: 136–142.

94.  Khandwala H, McCutcheon I, et al. The effects of IGFs on neoplastic growth. Endocr Rev 2000; 21: 215–244.

95.  Bielschowsky F, Bielschowsky M. Carcinogenesis in the pituitary of dwarf mouse. The response to dimethylbenzanthracene applied to the skin. Br. J. Cancer 1961;15: 257–262.

96.  Dominici FP, Cifone D, et al. Loss of sensitivity to insulin at early events of the insulin signaling pathway in the liver of growth hormone-transgenic mice. J Endocrinol 1999; 161: 383–392.

97.  Baynes JW, Monnier VM. The maillard reaction in aging, diabetes and nutrition. New York, Alan R. Liss, 1989.

98.  Parr T. Insulin exposure and aging theory. Gerontology 1997; 43: 182–200.

99.  Sullivan D, Walls R. The risk of life-threatening complications in a select population of geriatric patients: the impact of nutritional status. J Am Coll Nutr 1995; 14: 29–36.

100. MacIntosh C, Morley J, et al. The anorexia of aging. Nutrition 2000; 16: 983–995.

101. Felsing N, Brasel J, et al. Effects of low and high intensity exercise on circulating growth hormone in men. J Clin Endo Metab 1992; 75: 157–162.

102. Weltman AL, Weltman JY, et al. Endurance training amplifies the pulsatile release of growth hormone: effects of training intensity. J Applied Physiology 1992; 76: 2188–2196.

103. Miszko TA, Cress ME, et al. Effect of strength and power training on physical function in community-dwelling older adults. J Gerontol A Biol Sci Med Sci 2003; 58A: 171–175.

# Ageing and the Adrenal Cortex

V. LAMOUNIER-ZEPTER and S.R. BORNSTEIN

*University Hospital of Duesseldorf, Department of Endocrinology, Moorenstr. 5, 40225, Duesseldorf, Germany*

## ABSTRACT

Human ageing is associated with a clear dissociation of the adrenocortical hormone secretion pattern: circulating levels of adrenal androgens significantly decrease with age in both sexes, while cortisol and aldosterone secretion remains almost unchanged. The exact mechanism to account for this dissociation is still unclear. The regulation of adrenocortical steroidogenesis in the elderly involves diverse extra- and intra-adrenal factors, including oxidative stress, apoptosis and the interaction of these with immunological factors. Here we review the mechanisms proposed for the adrenal senescence and the potential role of the age-related changes in adrenocortical secretion in pathophysiological ageing.

## 1.   INTRODUCTION

Adult human ageing is associated with a number of important changes in physiological function and regulation to which the adrenal function may contribute. The changes in the adrenal cortex that occur with ageing in humans, and how such changes may have an impact on important physiological and pathophysiological processes is the focus of this chapter.

The adrenal gland consists of two endocrine tissues of different embryonic origin: the primarily steroid-producing adrenocortical tissue and the catecholamine-producing medullary chromaffin cells. The adult adrenal cortex is functionally divided in three distinct zones, characterized by the expression of specific steroidogenic enzymes and the ability to produce a unique steroid hormone profile in response to stimulation by specific peptides hormones: outer zona glomerulosa (ZG), central zona fasciculata (ZF) and inner zona reticularis (ZR). The ZG is the unique source of aldosterone. The ZF produces mainly the glucocorticoids, cortisol and corticosterone, whereas the ZR is the main source of adrenal androgens, dehydroepiandrosterone (DHEA) and dehydroepiandrosterone sulfate (DHEAS) [1].

The human and mouse adrenal presents a transient developmental zone between the cortical and the adrenal medulla, so called fetal or X zone, respectively [2–3]. The human fetal zone secretes large amounts of DHEAS, which act as a source of $C_{19}$ steroids for placental estrogen production, thus being crucial in establishing the estrogenic milieu during pregnancy [2].

The chromaffin cells of the adrenal zona medullaris produce mainly the catecholamines, epinephrine and norepinephrine. Chromaffin vesicles contain additionally numerous transmitters, neuropeptides and proteins, which may be released together with the catecholamines.

140

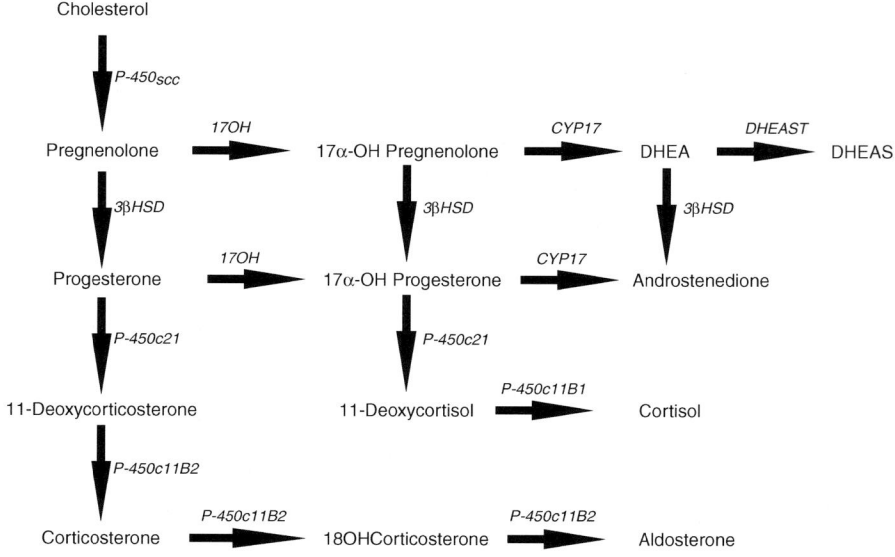

Figure 1. Steroidogenesis in the human adrenal cortex. Abbreviations: DHEA, dehydroepiandrosterone; DHEAS, dehydroepiandrosterone sulfate; P450scc, side-chain cleavage enzyme; 17OH, 17α-hydroxylase; CYP17, 17,20-desmolase; DHEAST, dehydroepiandrosterone sulfotransferase; 3βHSD, 3β-hydroxysteroid dehydrogenase; P-450c21, 21-hydroxylase; P-450c11B2, 11β-hydroxylase; P-450c11B1, 11β1-hydroxylase.

There is a close interaction of the adrenal cortex with the adrenal medulla. In addition, there is an intricate neuroendocrine immune network within the adrenal modulating both adrenocortical steroidogenesis and adrenomedullary catecholamine production [1]. Adrenal senescence is primarily associated with a decline in adrenal androgen production. Therefore we will focus on adrenal androgen secretion and ageing.

## 2.  ANDRENAL ANDROGEN SECRETION AND AGEING

Adrenal androgens are primarily produced in the inner zona reticularis of the adrenal cortex. A small amount of DHEA and its sulfate is produced by the testes and the central nervous system (CNS) [4–5]. DHEA is synthesized from the precursor pregnenolone, derived from cholesterol after side-chain cleavage by cytochrome P450scc enzyme (CYPscc) (Figure 1). Cytochrome P450-17α (CYP17), a 17α-hydroxylase with 17,20-desmolase activity, converts the C-21 steroid pregnenolone to the C-19 steroid DHEA. Hydroxysteroid sulfotransferase (DHEAST) converts DHEA to DHEAS, which is the most abundant circulating steroid hormone in humans [6]. DHEAS is highly expressed in the human adrenal cortex and liver. Its expression is tissue-specific regulated by steroidogenic factor 1 in the adrenal and liver receptor homologue in the liver [7].

DHEA can also be converted to androstenedione by the microsomal non-P450 enzyme 3β-hydroxysteroid dehydrogenase (3βHSD). In the periphery, DHEA and androstenedione can serve as substrates for the formation of testosterone and aromatization to estrogens [8].

## 2.1. Regulation of androgen production

Besides the well established stimulatory effect of adrenocorticotropin (ACTH), the regulation of adrenocortical steroidogenesis involves a complex interaction of several intra- and extraadrenal factors. This intricate intraadrenal regulatory system may explain the discrepancy in cortisol and androgen release in several clinical situations, and especially in the age-related changes in androgen secretion [8].

## 2.2. Extra-adrenal regulation of adrenal androgens

The extra-adrenal factors believed to modulate adrenal androgens secretion include gonadal androgens, insulin, growth hormone (GH), prolactin, and pituitary peptides.

Based on the clinical observation of increased pubic hair growth in girls with gonadal dysgenesis treated with estrogens, it has been postulated that estrogens participate in the regulation of adrenal androgen secretion [8]. Indeed, experiments with human adrenal cell culture demonstrated an increased DHEA/DHEAS production after stimulation with estradiol [9–10]. Similarly, the adrenal response to stimulation with ACTH increases after testosterone treatment *in vivo* [11].

Insulin infusion has been found to cause a decrease in circulating DHEA/DHEAS levels [12]. In addition, in primary cultures of human adrenocortical cells, insulin and insulin-like growth factor 1 (IGF-1) increase the mRNA levels for 17$\alpha$-hydroxylase and type II 3$\beta$HSD in the absence of cyclic AMP or ACTH, and also potentiate the effect of cyclic AMP and ACTH in all steroidogenic enzyme mRNA expression [13].

Studies of growth hormone effects on adrenal steroidogenesis have been inconclusive. Patients with GH deficiency seem to present decreased concentrations of adrenal androgens [14–15], but treatment with human GH as such does not have any effect on the adrenal androgen secretion [15].

*In vitro* studies showed that prolactin stimulates steroidogenesis in guinea pig and human adrenals [16–17]. Moreover, human adrenals express receptors for prolactin [17]. However, during the characteristic age-related changes in adrenal androgens concentration there are no corresponding changes in prolactin concentrations [8].

Besides ACTH, other peptides derived from proopiomelacocortin may be involved in the regulation of steroidogenesis. Concentrations of $\beta$-endorphin and $\beta$-lipotropin have been reported to increase during puberty, suggesting a possible role of both opioids in the development of adrenarche [18]. In fact, *in vitro* studies with guinea pig adrenal cells showed an increase in DHEA production after stimulation with $\beta$-lipotropin [19]. Similar results were obtained with stimulation of human adrenal cells with $\beta$-endorphin *in vitro* [20]. However, if this mechanism functions *in vivo* is still unclear.

At least, the corticotropin-releasing hormone (CRH) may also directly influence the adrenal androgen production. In human fetal adrenal CRH increases DHEAS production as effective as ACTH, whereas at stimulating cortisol production CRH was only 30% as potent as ACTH [21]. On the other hand, ACTH does not elevate cortisol secretion in the absence of CRH action as documented in CRH-type 1 receptor knockout mouse model [22]. Consistent with this preferential effect on DHEAS production, an increased expression of CYP17 mRNA, but not of 3$\beta$HSD mRNA, was observed after stimulation with CRH [21].

2.3.    Intra-adrenal regulation of adrenal androgens

Besides systemic factors, the adrenal androgen secretion is regulated by complex intra-adrenal interactions, involving the adrenal vascular supply, its neural input, the immune system, and local growth factors [1].

The adrenal gland is highly vascularized and its blood flow is finely controlled by a complex network of neural and hormonal mechanisms [1]. Anderson suggested that, particularities of adrenal blood flow would explain the increased androgen secretion during adrenarche [23]. In the adrenal blood flows from cortex to medulla, thus exposing the inner-zone cells to high levels of cortisol. Gradually the reticularis cells would respond to this high cortisol levels by undergoing morphological and functional changes, such as induction of the enzyme 17,20-desmolase and perhaps partial loss of 3βHSD, resulting in an increase in androgen production [23].

In addition to medullary catecholamines and neuropeptides to which adrenocortical cells are exposed in response to activation of the sympatho-adrenal axis, the adrenal cortex receives a direct innervation, which seems to modulate its function. The innervation of the adrenal cortex originates from two different sources: one is the adrenal medulla that is regulated by splanchnic nerve activity. Other neurons have their cell bodies outside the adrenal gland and reach the cortex together with blood vessels [1]. *In vivo* studies reported a regulation of adrenocortical sensitivity to ACTH through splanchnic nerve activity: dissection of the splanchnic nerves decreases adrenocortical response to ACTH [24], while stimulation enhances it [25]. Similarly, perfusion of intact pig adrenals with epinephrine or stimulation of the splanchnic nerves showed an increased release of androstenedione [26]. Besides the release of catecholamines, other possible pathways for the neural activation of adrenal cortex include the release of neuropeptides. For example, the vasoactive intestinal peptide (VIP) has been identified in nerves and cells within the adrenal cortex in various mammals [27]. In our studies carried out on human adrenal primary cell culture, VIP stimulated the release of DHEA, testosterone, and androstenedione significantly [28].

Components of the immune system are located within the adrenal cortex under normal conditions, indicating the existence of an intra-adrenal immune system [29–30]. Macrophages were found primarily in the zona reticularis, expressing different cytokines such as interleukin-1 (IL-1), interleukin-6 (IL-6) and tumor necrosis factor alpha (TNFα) when activated [31]. The adrenocortical cells themselves are also able to synthesize several cytokines. *In situ* hybridization and specific immunostaining studies show that cytokines are expressed mainly in steroid-producing cells in the zona reticularis and in cortical cells located in the adrenal medulla [31]. This intra-adrenal immune system has direct effects on the adrenocortical function, including a regulatory role on adrenal growth and on steroidogenesis [1]. IL-6 receptor is expressed with high density in the zona reticularis and inner zona fasciculata of human adrenal gland and IL-6 stimulates the release of androgens both via ACTH and directly [32–33]. In human fetal adrenals TNFα alters the P450 enzyme expression with a consequent shift to androgen production [34]. Moreover, based on the observation of direct cellular contacts between lymphocytes and reticularis cells, as well as on *in vitro* experiments with HLA-matched lymphocytes and NCI-H295 cells, it has been found that lymphocytes may directly stimulate DHEA secretion [35]. The T lymphocyte infiltration in adrenals increases with age, suggesting that lymphocytes may be involved in the alterations of adrenocortical function during ageing [36].

Various peptide growth factors may also be involved in the regulation of adrenal androgens secretion. Epidermal growth factor (EGF) and fibroblast growth factor (FGF) stimulate growth of bovine and human adrenocortical cells *in vitro* [37–38], but only EGF also stimulates DHEAS secretion in fetal zone cells [39]. Transforming growth factor β reduces DHEAS synthesis by increasing mRNA accumulation of 3βHSD and decreasing those of CYP17 [40–41]. Both Insulin-like growth factors peptides and their receptors have been identified in the adrenal cortex [42–43] and a regulatory effect on androgen biosynthesis has been reported in human adrenal primary cultures [44]. In addition, the treatment of adrenal cells with IGF-I or II leads to increased CYP17 and 3βHSD mRNA expression without changes in the expression of CYPscc [45].

## 3.     ADRENARCHE/ANDROPAUSE

In contrast to the minimal changes of the cortisol and aldosterone secretion occurring with age, the adrenal androgens production in humans follows a characteristic age-associated pattern [46]. DHEA is the main secretory product of the fetal adrenal, leading to high circulating levels of DHEAS at birth. After birth, the DHEAS serum concentrations decrease to almost undetectable levels, in parallel with the involution of the fetal zone in the first year of life [4]. During childhood, DHEA concentrations remain low, but gradually increase again between the sixth and eighth year of age, a phenomenon called adrenarche [47], and reach peak levels between the ages of 20 and 30 [46]. Thereafter, a steady decline of androgen levels is observed throughout adult life, so that only 20–30% of those in young adults are found by the ages of 70 to 80 years. The greatest decline occurs by the ages of 50 to 60, which is called adrenopause [48].

### 3.1.     Regulation of adrenopause

While alterations of adrenal blood flow may explain certain aspects of hormone profile changes during adrenarche, the mechanisms to account for the decline in the androgen secretion in ageing may be different [49]. It has been proposed that a reduction of 17,20-desmolase enzymatic activity may be responsible for the decrease in DHEA synthese [50]. However, no changes in the gene expression of 17α-hydroxylase and other enzymes reponsible for steroidogenesis could be related to ageing [51]. Thus, it is probable that structural alterations in the adrenal cortex other than changes in the enzymatic properties of the zonal cells could explain the decrease in adrenal androgen synthesis evaluated in ageing.

In fact, morphological rearrangements in the adrenal cortex with ageing have been observed [52]. In spite of similar total cortical thickness between young and older men, a significant decrease in the zona reticularis width was observed with ageing. Moreover, when the zona reticularis width was evaluated in proportion to outer cortical zone thickness, a 2-fold increase in the ratio of the width of zona fasciculata and zona glomerulosa to that of zona reticularis has been observed. In paralel, the border between zona reticularis and zona fasciculata becomes irregular [52]. This remodeling of the cortical zones may explain the modifications in the adrenal steroidogenesis observed in the elderly. In contrast to the zone fasciculata, which has abundant 3βHSD, the human zona reticularis contains high levels of DHEAST and little, if any 3βHSD [51,53]. These intra-adrenal differences in the steroidogenic pathway involve a preferential synthesis of DHEA/S but only a small amount of cortisol in the zona reticularis. Thus, a reduced mass of the zona reticularis in ageing but without reduction in the mass of zona fasciculata may result in reduced DHEA synthesis without affecting the cortisol synthese pathway [52].

It remains unclear how exactly these morphological rearrangements in the adrenal cortex occur and whether these observed morphological changes impact on adrenocortical function and

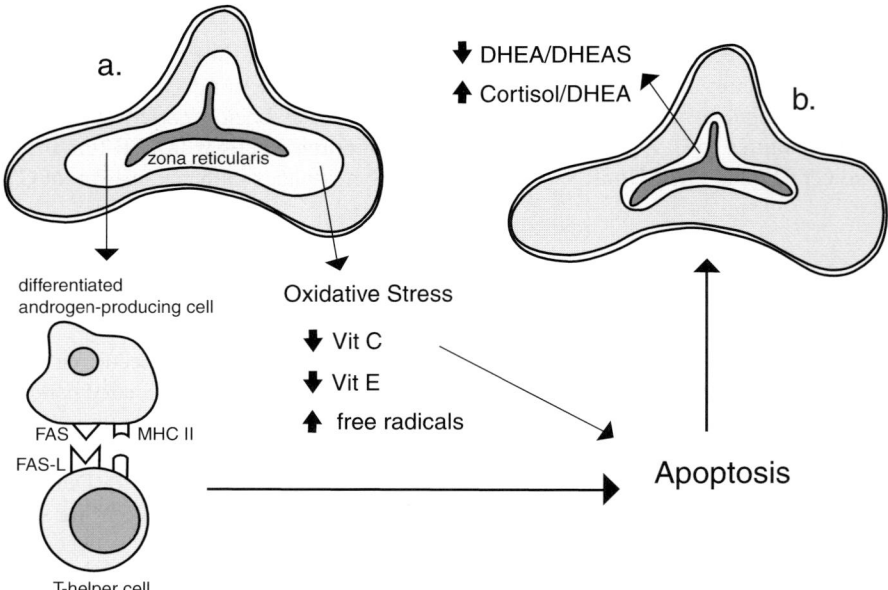

Figure 2. Proposed mechanisms for adrenopause. a, Adult adrenal gland with differentiated androgen-producing cell, expressing MHC class II and FAS. FAS-L expressing T helper cells induce apoptosis in immunocompetent reticularis androgen-producing cells. Oxidative stress also contributes to apoptosis, leading to decrease of zona reticularis width (b) and downregulation of adrenal androgens synthesis.

regulation. We have proposed a multifactorial model (Figure 2), where the interaction between extra-adrenal and intra-adrenal hormonal and immunological factors leads to apoptosis of the cortical androgen-producing cells [54]. The highly differentiated androgen-producing reticularis cells express major histocompatibility (MHC) class II antigens, as well as FAS and FAS ligand [55–57]. FAS is a cell surface molecule of the tumor necrosis factor (TNF)/nerve growth factor receptor families [58–59], whereas FAS ligand (Fas-L) is a type II membrane protein also belonging to the TNF family [58,60]. Both molecules are involved in cellular immune response and programmed cell death [55,61]. There is also evidence of MHC class II mediated programmed cell death [62]. T lymphocytes expressing CD4 and CD8 antigens were present within human adrenal zona reticularis by immunohistochemical subtyping, and electron microscopic analyses demonstrated direct cell-cell contact between T lymphocytes and adrenocortical cells *in situ* [35]. Thus, Fas-L expressing T helper cells may induce apoptosis in immunocompetent androgen-producing cells expressing Fas and MHC class II antigens [63]. A functional link between the MHC class complex, Fas/FasL and apoptosis has already been demonstrated: ligation of MHC class II antigens leads to increased sensitivity of lymphocyte cell lines to Fas-mediated apoptosis [64–65]. A similar mechanism seems to occur in the human adrenal cortex [66] and would explain the Fas/MHC class II-mediated apoptosis of differentiated reticularis cells.

In addition to these immune-endocrine interactions, oxidative damage could contribute to the adrenocortical cellular senescence. The antioxidants ascorbate (vitamin C) and alpha-tocopherol (vitamin E) are concentrated in the adrenal gland to very high levels compared to other organs [67]. It has been proposed that vitamin C and E protect the adrenal membranes from the deleterious effects of lipid peroxidation, thereby playing a protective role in steroidogenesis [68–69].

Interestingly, in the guinea pig as well as in the human adrenal cortex the ZR is the site of greatest lipofuscin accumulation, an end product of lipid peroxidation [70], and is also the region of greatest peroxidative activity *in vitro* [71]. Thus, the accumulation of free radicals from both exogenous sources and metabolites of catecholamines produced by medullary chromaffin cells, together with a decrease in antioxidants, would be responsible for increased oxidative damage of the reticularis cells and promote their apoptosis. The apoptosis of ZR cells with a consequent decrease in CYP17 activity and DHEA production, but with a maintained cortisol secretion, would create the imbalance between androgen and cortisol secretion observed in ageing.

## 3.2. Functional significance

The decrease in the androgen production observed during adrenopause may be an important link between endocrinosenescence and immunosenescence. Several epidemiological studies have reported a negative correlation of serum DHEA and DHEAS with IL-6 [72–73]. Circulating IL-6 increases with ageing [74–75], suggesting an important role of this cytokine in several age-dependent diseases [76–78]. Additionally, *in vitro* evidence has been presented for DHEA-induced inhibition of IL-6 production by human peripheral mononuclear blood cells [73]. Thus, the age-dependent decrease in DHEA may partly explain the increase in serum IL-6 concentrations observed in the elderly. In fact, the treatment of ageing mice with DHEA or DHEAS reduces up-regulated IL-6 production [79]. A similar effect was verified in the serum levels of IL-10, which also is increased in aged mice [80].

Whether DHEA has direct immunomodulatory effects on target cells or whether these effects are indirectly mediated by its downstream steroids, e.g. testosterone, remains unclear. *In vitro* studies showed DHEA conversion to physiological amounts of downstream steroid hormones in macrophages, suggesting a relevant immunomodulatory effect for the downstream hormones of DHEA [81]. In addition, gene profiling studies, which analyzed the effects of dexamethasone, testosterone, and DHEA on gene expression by lymphocytes, suggested a direct effect on gene expression, which may be mediated by a receptor specific for DHEA [82].

It has also been suggested that the balance between the adrenal androgen and glucocorticoids may influence the differentiation of T helper cells into Th1 and Th2 [83]. The Th1 response promotes cellular immunity, whereas the Th2 profile through IL-3, IL-4, IL-5, IL-10 and IL-13 secretion initiates humoral immunity and counteracts the Th1 response [83]. It seems that corticosteroids induce a shift from Th1 to Th2 response [84–85]. Thus, changes in the hormone balance with a relative predominance of glucocorticoids observed in the adrenopause would alter the immune balance, with a switch toward a Th2 cytokine pattern. This dysregulation of cytokine production is in accordance with the predominance of Th2 cytokines in the ageing immune system and may explain the propensity to disease observed in aged people.

Another potential role of adrenal androgens in the pathophysiological ageing refers to the effects of DHEA and DHEAS on the central nervous system. DHEA and DHEAS may be synthesized *de novo* in the CNS, and thus belong to the class of neurosteroids [5]. They act at two major receptors: the γ-aminobutyric acid (GABA) receptor and the N-methyl-D-aspartate (NMDA) receptor [86]. Through their interactions with these neurotransmitters, DHEA and DHEAS are supposed to be involved in complex processes such as learning, memory, neuroprotection and neocortical organization. As a GABA receptor antagonist, DHEA and DHEAS may enhance the neuronal and glial survival in mouse embryo brain cells. Furthermore, central DHEA injection can improve learning and memory capacities in mice [87]. *In vitro* experiments with primary cultures of mouse embryonic neocortical neurons indicate that DHEA and DHEAS may affect axonal and dendritic growth, thus suggesting an organizational role of DHEA and DHEAS in the developing CNS [88]. This neurotrophic effect of DHEA was mediated by the NMDA receptor. In addition, in rat primary hippocampal neurons culture, DHEAS was able to enhance the neuronal resistance to stress relevant to both acute and chronic neurodegenerative conditions. The beneficial effects of DHEAS on neuronal survival was correlated with an activation of a κB-binding transcription factor [89], which appears to protect against the neurotoxicity associated with oxidative stress, calcium and β-amyloid [90]. DHEA also opposes the action of glucocorticoids on the brain, antagonising the neurotoxicity of corticosterone on the hippocampal rat cells [91]. Thus, it has been proposed that the age-related decline of DHEAS may play a role in the occurrence of neurodegenerative diseases, including Alzheimer's disease.

Whether the centrally synthesized DHEAS only or changes in plasma DHEAS levels can also affect CNS functions remains unclear. In fact, the data concerning the correlation of peripheral DHEAS secretion and cognitive function are still conflicting. In a larger study of normal older women, no significant relationship between DHEAS levels in serum and cognitive function has been found [92]. Similarly, in elderly people no correlation has been demonstrated between cognitive impairment and serum DHEAS levels [93]. The studies on DHEAS levels and Alzheimer's disease are also contradictory. Some found that Alzheimer's disease patients had lower DHEAS levels than controls [94–95], but this could not be confirmed by others [96]. When DHEA is expressed in relation to cortisol concentrations, however, an increased cortisol:DHEA ration seems to be associated with cognitive impairment in elderly people [97]. Cortisol has a central neurotoxic activity, whereas DHEAS antagonizes the glucocorticoid action both centrally and peripherally. Thus, not exactly the decrease of adrenal androgens but the dysbalance of adrenocortical secretion during ageing may be involved in the age-related diseases such as neurodegenerative disorders and others.

4.    CONCLUSION

Ageing in healthy people is associated with a clear dissociation of the adrenocortical hormone secretion pattern. These changes in adrenal hormone levels may be an important link between immune-, endocrine- and neuronal-senescence, thereby with a widespread clinical significance. On this basis, it has been questioned whether ageing represents, in part, an androgen deficiency syndrome and is therefore reversible by DHEA supplementation. In fact, increasing use of DHEA as anti-ageing drug has been observed in the USA, where it is considered to be a food supplement and available without prescription. However, the advantage of such therapy in the elderly remains unclear. Especially, the conversion of exogenously administrated DHEA to potent androgens and estrogens in excess may increase the risk of development of sex hormone-dependent neoplasias. Thus, the risk *vs.* beneficial effects of DHEA supplementation in elderly remain to be elucidated before a widespread use of DHEA in advanced age is recommended.

REFERENCES

1.    Ehrhart-Bornstein M, Hinson JP, Bornstein SR, Scherbaum WA, Vinson GP. Intraadrenal interactions in the regulation of adrenocortical steroidogenesis. Endocr Rev 1998;19: 101–143.
2.    Mesiano S, Jaffe RB. Developmental and functional biology of the primate fetal adrenal

cortex. Endocr Rev 1997;18:378–403.

3.	Nussdorfer GG. Cytophysiology of the adrenal cortex. Int Rev Cytol 1986;98:1–405.

4.	De Peretti E, Forest MG. Pattern of plasma dehydroepiandrosterone sulfate levels in humans from birth to adulthood: evidence for testicular production. J Clin Endocrinol Metab 1978;47:572–577.

5.	Baulieu EE, Robel P. Dehydroepiandrosterone (DHEA) and dehydroepiandrosterone sulfate (DHEAS) as neuroactive neurosteroid. Proc Natl Acad Sci USA 1998;95:4089–4091.

6.	Adams JB. Control of secretion and function of C19-delta 5-steroids of the human adrenal gland. Mol Cell Endocrinol 1985;41:1–17.

7.	Saner KJ, Mayhew BA, Carr BR, Rainey WE. Tissue-specific regulation of dehydroepiandrosterone sulfotransferase expression by steroidogenic factor (SF1) in the adrenal and liver receptor homologue (LRH) in the liver. Molecular Steroidogenesis Workshop. United Kingdom, 2003;p13 (Abstract).

8.	Parker LN. Control of adrenal androgen secretion. Endocrinol Metab Clin North Am 1991;20:401–421.

9.	Gell JS, Oh J, Rainey WE, Carr BR. Effect of estradiol on DHEAS production in the human adrenocortical cell line, H295R. J Soc Gynecol Investig 1998;5:144–148.

10.	Fujieda K, Faiman C, Feyes FI, Winter JS. The control of steroidogenesis by human fetal adrenal cells in tissue culture. IV. The effect of exposure to placental steroids. J Clin Endocrinol Metab 1982;54:89–94.

11.	Polderman KH, Gooren LJ, van der Veen EA. Testosterone administration increases adrenal response to adrenocorticotropin. Clin Endocrinol (Oxf) 1994;40:595–601.

12.	Nestler JE, Usiskin KS, Barlascini CO, Welty DF, Clore JN, Blackard WG. Suppression of serum dehydroepiandrosterone sulfate levels by insulin: an evaluation of possible mechanisms. J Clin Endocrinol Metab 1989;69:1040–1046.

13.	Kristiansen SB, Endoh A, Casson PR, Buster JE, Hornsby PJ. Induction of steroidogenic enzyme genes by insulin and IGF-I in cultured adult human adrenocortical cells. Steroids 1997;62:258–265.

14.	Cohen HN, Wallace AM, Beastall GH, Fogelman I, Thomson JA. Clinical value of adrenal androgen measurement in the diagnosis of delayed puberty. Lancet 1981;1:689–692.

15.	Ilondo MM, Vanderschueren-Lodeweyckx M, Vlietinck R, Pizarro M, Malvaux P, Eggermont E, Eeckels R. Plasma androgens in children and adolescents. Part II. A longitudinal study in patients with hypopituitarism. Horm Res 1982;16:78–95.

16.	O'Connell Y, McKenna TJ, Cunningham SK. The effect of prolactin, human chorionic gonadotropin, insulin and insulin-like growth factor 1 on adrenal steroidogenesis in isolated guinea-pig adrenal cells. J Steroid Biochem Mol Biol 1994;48:235–240.

17.	Glasow A, Breidert M, Haidan A, Anderegg U, Kelly PA, Bornstein SR. Functional aspects of the effect of prolactin (PRL) on adrenal steroidogenesis and distribution of the PRL receptor in the human adrenal gland. J Clin Endocrinol Metab 1996;81:3103–3111.

18.	Genazzani AR, Facchinetti F, Petraglia F, Pintor C, Bagnoli F, Puggioni R, Corda R. Correlations between plasma levels of opioid peptides and adrenal androgens in prepuberty and puberty. J Steroid Biochem 1983;19:891–895.

19.	O'Connell Y, McKenna TJ, Cunningham SK. Effects of pro-opiomelanocortin-derived peptides on adrenal steroidogenesis in guinea-pig adrenal cells in vitro. J Steroid Biochem Mol Biol 1993;44:77–83.

20.	Clarke D, Fearon U, Cunningham SK, McKenna TJ. The steroidogenic effects of beta-endorphin and joining peptide: a potential role in the modulation of adrenal androgen production. J Endocrinol 1996;151:301–307.

21. Smith R, Mesiano S, Chan EC, Brown S, Jaffe RB. Corticotropin-releasing hormone directly and preferentially stimulates dehydroepiandrosterone sulfate secretion by human fetal adrenal cortical cells. J Clin Endocrinol Metab 1998;83:2916–2920.

22. Timpl P, Spanagel R, Sillaber I, Kresse A, Reul JM, Stalla GK, Blanquet V, Steckler T, Holsboer F, Wurst W. Impaired stress response and reduced anxiety in mice lacking a functional corticotropin-releasing hormone receptor 1. Nat Genet 1998;19:162–166.

23. Anderson DC. The adrenal androgen-stimulating hormone does not exist. Lancet 1980;2: 454–456.

24. Edwards AV, Jones CT, Bloom SR. Reduced adrenal cortical sensitivity to ACTH in lambs with cut splanchnic nerves. J Endocrinol 1986;110:81–85.

25. Engeland WC, Gann DS. Splanchnic nerve stimulation modulates steroid secretion in hypophysectomized dogs. Neuroendocrinology 1989;50:124–131.

26. Ehrhart-Bornstein M, Bornstein SR, Guse-Behling H, Stromeyer HG, Rasmussen TN, Scherbaum WA, Adler G, Holst JJ. Sympathoadrenal regulation of adrenal androstenedione release. Neuroendocrinology 1994;59:406–412.

27. Holzwarth MA. The distribution of vasoactive intestinal peptide in the rat adrenal cortex and medulla. J Auton Nerv Syst 1984;11:269–283.

28. Bornstein SR, Haidan A, Ehrhart-Bornstein M. Cellular communication in the neuro-adrenocortical axis: role of vasoactive intestinal polypeptidde (VIP). Endocr Res 1996;22: 819–829.

29. Hume DA, Halpin D, Charlton H, Gordon S. The mononuclear phagocyte system of the mouse defined by immunohistochemical localization of antigen F4/80: macrophages of endocrine organs. Proc Natl Acad Sci U S A 1984;81:4174–4177.

30. González-Hernández JA, Bornstein SR, Ehrhart-Bornstein M, Geschwend JE, Adler G, Scherbaum WA. Macrophages within the human adrenal gland. Cell Tissue Res 1994;278: 201–205.

31. Bornstein SR, Ehrhart-Bornstein M, Scherbaum WA. Morphological and functional studies of the paracrine interaction between cortex and medulla in the adrenal gland. Microsc Res Tech 1997;36:520–533.

32. Päth G, Bornstein SR; Spath-Schwalbe E, Scherbaum WA. Direct effects of interleukin-6 on human adrenal cells. Endocr Res 1996;22:867–873.

33. Päth G, Bornstein SR, Ehrhart-Bornstein M, Scherbaum WA. Interleukin-6 and the interleukin-6 receptor in the human adrenal gland: expression and effects on steroidogenesis. J Clin Endocrinol Metab 1997;82:2343–2349.

34. Jäättelä M, Carpén O, Stenman U-H, Saksela E. Regulation of ACTH-induced steroidogenesis in human fetal adrenals by rTNF-α. Mol Cell Endocrinol 1990;68:R31-R36.

35. Wolkersdörfer GW, Lohmann T, Marx C, Schroder S, Pfeiffer R, Stahl HD, Scherbaum WA, Chrousos GP, Bornstein SR. Lymphocytes stimulate dehydroepiandrosterone production through direct cellular contact with adrenal zona reticularis cells: a novel mechanism of immune-endocrine interaction. J Clin Endocrinol Metab 1999;84:4220–4227.

36. Hayashi Y, Hiyoshi T, Takemura T, Kurashima C, Hirokawa K. Focal lymphocytic infiltration in the adrenal cortex of the elderly: immunohistological analysis of infiltrating lymphocytes. Clin Exp Immunol 1989;77:101–105.

37. Schweigerer L, Neufeld G, Friedman J, Abraham JA, Fiddes JC, Gospodarowicz D. Basic

fibroblast growth factor: production and growth stimulation in cultured adrenal cortex cells. Endocrinology 1987;120:796–800.

38. Mesiano S, Mellon SH, Gospodarowicz D, Di Blasio AM, Jaffe RB. Basic fibroblast growth factor is regulated by corticotropin in the human fetal adrenal: a model for adrenal growth regulation. Proc Natl Acad Sci USA 1991;88:5428–5432.

39. Crickard K, Jaffe RB. Control of proliferation of human fetal adrenal cells in vitro. J Clin Endocrinol Metab 1981;53:790–796.

40. Lebrethon MC, Jaillard C, Naville D, Begeot M, Saez JM. Effects of transforming growth factor β1 on human adrenocortical fasciculata-reticularis cell differentiated functions. J Clin Endocrinol Metab 1994;79:1033–1039.

41. Stankovic AK, Dion LD, Parker Jr CR. Effects of transforming growth factor-beta on human fetal adrenal steroid production. Mol Cell Endocrinol 1994;99:145–151.

42. Hansson HA, Nilsson A, Isgaard J, Billig H, Isaksson O, Skottner A, Andersson IK, Rozell B. Immunohistochemical localization of insulin-like growth factor 1 in the adult rat. Histochemistry 1988;889:403–410.

43. Pillion DJ, Arnold P, Yang M, Stockard CR, Grizzle WE. Receptors for insulin and insulin-like growth factor-1 in the human adrenal gland. Biochem Biophys Res Commun 1989;165:204–211.

44. Fottner C, Engelhardt D, Weber MM. Regulation of steroidogenesis by insulin-like growth factors (IGFs) in adult human adrenocortical cells: IGF-I and, more potently, IGF-II preferentially enhance androgen biosynthesis through interaction with the IGF-I receptor and IGF-binding proteins. J Endocrinol 1998;158:409–417.

45. L'Allemand D, Penhoat A, Lebrethon MC, Ardevol R, Baehr V, Oelkers W, Saez JM. Insulin-like growth factors enhance steroidogenic enzyme and corticotropin receptor messenger ribonucleic acid levels and corticotropin steroidogenic responsiveness in cultured human adrenocortical cells. J Cllin Endocrinol Metab 1996;81:3892–3897.

46. Orentreich N, Brind JL, Rizer RL, Vogelman JH. Age changes and sex differences in serum dehydroepiandrosterone sulfate concentrations throughout adulthood. J Clin Endocrinol Metab 1984;59:551–555.

47. Parker LN. Adrenarche. Endocrinol Metab Clin North Am 1991;20:71–83.

48. Kroboth PD, Salek FS, Pittenger AL, Fabian TJ, Frye RF. DHEA and DHEAS: a review. J Clin Pharmacol 1999;39:327–348.

49. Hornsby PJ. Aging of the human adrenal cortex. Ageing Res Rev 2002;1:229–242.

50. Liu CH, Laughlin GA, Fischer UG, Yen SS. Marked attenuation of ultradian and circadian rhythms of dehydroepiandrosterone in postmenopausal women: evidence for a reduced 17,20-desmolase enzymatic activity. J Clin Endocrinol Metab 1990;71:900–906.

51. Endoh A, Kristiansen SB, Casson PR, Buster JE and Hornsby PJ. The zona reticularis is the site of biosynthesis of dehydroepiandrosterone and dehydroepiandrosterone sulfate in the adult human adrenal cortex resulting from its low expression of 3 beta-hydroxysteroid dehydrogenase. J Clin Endocrinol Metab 1996;81:3558–3565.

52. Parker CRJ, Mixon RL, Brissie RM, Grizzle WE. Aging alters zonation in the adrenal cortex of men. J Clin Endocrinol Metab 1997;82:3898–3901.

53. Kennerson AR, McDonald DA, Adams JB. Dehydroepiandrosterone sulfotransferase localization in human adrenal glands: a light and electron microscopic study. J Clin Endocrinol Metab 1983;56:786–790.

54. Alesci S, Bornstein SR. Intraadrenal mechanisms of DHEA regulation: a hypothesis for adrenopause. Exp Clin Endocrinol Diabetes 2001;109:75–82.

55. French LE, Hahne M, Viard I, Radlgruber G, Zanone R, Becker K, Muller C, Tschopp J. Fas and Fas-ligand expression in embryos and adult mice: Ligand expression in several immuneprivileged tissues and coexpression in adult tissues characterized by apoptotic cell turnover. J Cell Biol 1996;133:335–343.

56. Marx C, Bornstein SR, Wolkersdorfer GW, Peter M, Sippell WG, Scherbaum WA. Relevance of major histocompatibility complex class II expression as a hallmark for the cellular differentiation in the human adrenal cortex. J Clin Endocrinol Metab 1997;82: 3136–3140.

57. Khoury EL, Greenspan JS, Greenspan FS. Adrenocortical cells of the zona reticularis normally express HLA-DR antigenic determinants. Am J Pathol 1987;127:580–591.

58. Smith CA, Farrah T, Goodwin RG. The TNF receptor superfamily of cellular and viral proteins: Activation, costimulation, and death. Cell 1994;76:959–962.

59. Itoh N, Yonehara S, Ishii A, Yonehara M, Mizushima S, Sameshima M, Hase A, Seto Y, Nagata S. The polypeptide encoded by the cDNA for human cell surface antigen Fas can mediate apoptosis. Cell 1991;66:233–243.

60. Suda T, Takahashi T, Golstein P, Nagata S. Molecular cloning and expression of the Fas ligand, a novel member of the tumor necrosis factor family. Cell 1993;75:1169–1178.

61. French LE, Tschopp J. Thyroiditis and hepatitis: Fas on the road to disease. Nature Med 1997;3:387–388.

62. Truman JP, Ericson ML, Choqueux-Seebold CJ, Charron DJ, Mooney NA. Lymphocyte programmed cell death is mediated via HLA class II DR. Int Immunol 1994;6:887–896.

63. Marx C, Wolkersdörfer GW, Bornstein SR. A new view on immune-adrenal interactions: role for Fas and Fas ligand? Neuroimmunomodulation 1998;5:5–8.

64. Yoshino T, Cao L, Nishiuchi R, Matsuo Y, Yamadori I, Kondo E, Teramoto N, Hayashi K, Takahashi K, Kamikawaji N, et al. Ligation of HLA class II molecules promotes sensitivity to CD95 (Fas antigen, APO-1)-mediated apoptosis. Eur J Immunol 1995;25:2190–2194.

65. Truman JP, Choqueux C, Tschopp J, Vedrenne J, Le Deist F, Charron D, Mooney N. HLA class II-mediated death is induced via Fas/Fas ligand interactions in human splenic B lymphocytes. Blood 1997;89:1996–2007.

66. Wolkersdörfer GW, Ehrhart-Bornstein M, Brauer S, Marx C, Scherbaum WA, Bornstein SR. Differential regulation of apoptosis in the normal human adrenal gland. J Clin Endocrinol Metab 1996;81:4129–4136.

67. Hornsby PJ, Crivello JF. The role of lipid peroxidation and biological antioxidants in the function of the adrenal cortex. Part 2. Mol Cell Endocrinol 1983;30:123–147.

68. Chakraborty S, Nandi A, Mukhopadhyay M, Mukhopadhyay CK, Chatterjee IB. Ascorbate protects guinea pig tissues against lipid peroxidation. Free Radical Biol Med 1994;16: 417–426.

69. Bahr V, Mobius K, Redmann A, Oelkers W. Ascorbate and alpha-tocopherol depletion inhibit aldosterone stimulation by sodium deficiency in the guinea pig. Endocr Res 1996;22:565–600.

70. Ito T. Histology and histogenesis of the adrenal cortex in the guinea pig. Folia Anat Jap 1951;24:269–291.

71. Staats DA, Colby HD. Regional differences in microsomal lipid peroxidation and antioxidant levels in the guinea pig adrenal cortex. J Steroid Biochem 1987;28:637–642.

72. Young DG, Skibinski G, Mason JI, James K. The influence of age and gender on serum dehydroepiandrosterone sulphate (DHEA-S), IL-6, IL-6 soluble receptor (IL-6-sR) and transforming growth factor beta 1 (TGF- β1) levels in normal healthy blood donors. Clin

Exp Immunol 1999;117:476–481.

73.  Straub RH, Konecna L, Hrach S, Rothe G, Kreutz M, Schölmerich J, Falk W, Lang B. Serum dehydroepiandrosterone (DHEA) and DHEA sulfate are negatively correlated with serum interleukin-6 (IL-6), and DHEA inhibits IL-6 secretion from mononuclear cells in man *in vitro*: possible link between endocrinosenescence and immunosenescence. J Clin Endocrinol Metab 1998;83:2012–2017.

74.  Hager K, Machein U, Krieger S, Platt D, Seefried G, Bauer J. Interleukin-6 and selected plasma proteins in healthy persons of different ages. Neurobiol Aging 1994;15:771–772.

75.  Wie J, Xu H, Davies JL, Hemmings GP. Increase of plasma IL-6 concentration with age in healthy subjects. Life Sci 1992;51:1953–1956.

76.  Manolagas SC, Jilka RL. Bone marrow, cytokines and bone remodeling. Emerging insights into the pathophysiology of osteoporosis. N Engl J Med 1995;332:305–311.

77.  Blum-Degen D, Müller T, Kuhn W, Gerlach M, Przuntek H, Riederer P. Interleukin-1 beta and interleukin-6 are elevated in the cerebrospinal fluid of Alzheimer's and *de novo* Parkinson's disease patients. Neurosci Lett 1995;202:17–20.

78.  Kawano M, Hirano T, Matsuda T, Taga T, Horii Y, Iwato K, Asaol H, Tang B, Tanabe O, Tanaka H et al. Autocrine generation and requirement of BSF-2/IL-6 for human multiple myelomas. Nature 1988;332:83–85.

79.  Daynes RA, Araneo BA, Ershler WB, Maloney C, Li G-Z, Ryu S-Y. Altered regulation of IL-6 production with normal aging: possible linkage to the age-associated decline in dehydroepiandrosterone and its sulfated derivative. J Immunol 1993;150:5219–5230.

80.  Spencer NF, Norton SD, Harrison LL, Li GZ, Daynes RA. Dysregulation of IL-10 production with aging: possible linkage to the age-associated decline in DHEA and its sulfated derivative. Exp Gerontol 1996;31:393–408.

81.  Schmidt M, Kreutz M, Löffler G, Schölmerich J, Straub RH. Conversion of dehydroepiandrosterone to downstream steroid hormones in macrophages. J Endocrinol 2000;164:161–169.

82.  Maurer M, Trajanoski Z, Frey G, Hiroi N, Galon J, Willenberg HS, Gold PW, Chrousos GP, Scherbaum WA, Bornstein SR. Differential gene expression profile of glucocorticoids, testosterone, and dehydroepiandrosterone in human cells. Horm Metab Res 2001;33: 691–695.

83.  Marx C, Ehrhart-Bornstein M, Scherbaum WA, Bornstein SR. Regulation of adrenocortical function by cytokines-relevance for immune-endocrine interaction. Horm Metab Res 1998; 416–420.

84.  Ramírez F, Fowell DJ, Puklavec M, Simmonds S, Mason D. Glucocorticoids promote a Th2 cytokine response by CD4+ T cells in vitro. J Immunol 1996;156:2406–2412.

85.  Elenkov IJ, Papanicolaou DA, Wilder RL, Chrousos GP. Modulatory effects of glucocorticoids and catecholamines on human interleukin-12 and interleukin-10 production: clinical implications. Proc Assoc Am Physicians 1996;108:374–381.

86.  Majewska MD. Neuronal actions of dehydroepiandrosterone. Possible roles in brain development, aging, memory and affect. Ann N Y Acad Sci 1995;774:111–120.

87.  Roberts E, Bologa L, Flood JF, Smith GE. Effects of dehydroepiandrosterone and its sulfate on brain tissues in culture and on memory in mice. Brain Res 1987;406:357–362.

88.  Compagnone NA, Mellon SH. Dehydroepiandrosterone: a potential signalling molecule for neocortical organization during development. Proc Natl Acad Sci USA 1998;95: 4678–4683.

89.  Mao X, Barger SW. Neuroprotection by dehydroepiandrosterone-sulfate: role of an NFkB-

like factor. NeuroReport 1998;9:759–763.

90. Barger SW, Horster D, Furukawa K, Goodman Y, Krieglstein J, Mattson MP. Tumor necrosis factors and protect neurons against amyloid ß-peptide toxicity: evidence for involvement of a B-binding factor and attenuation of peroxide and Ca2+ accumulation. Proc Natl Acad Sci USA 1995;92:9328–9332.

91. Kimonides VG, Spillantini MG, Sofroniew MV, Fawcett JW, Herbert J. Dehydroepiandrosterone antagonizes the neurotoxic effects of corticosterone and translocation of stress-activated protein kinase 3 in hippocampal primary cultures. Neuroscience 1999;89:429–436.

92. Yaffe K, Ettinger B, Pressman A, Seeley D, Whooley M, Schaefer C, Cummings S. Neuropsychiatric function and dehydroepiandrosterone sulfate in elderly women: a prospective study. Biol Psy 1998;43:694–700.

93. Ravaglia G, Forti P, Maioli F, Sacchetti L, Nativio V, Scali CR, Mariani E, Zanardi V, Stefanini A, Macini PL. Dehydroepiandrosterone-sulfate serum levels and common age-related diseases: results from a cross-sectional Italian study of a general elderly population. Exp Gerontol 2002;37:701–712.

94. Näsman B, Olsson T, Backstrom T, Eriksson S, Grankvist K, Viitanen M, Bucht G. Serum dehydroepiandrosterone sulfate in Alzheimer's disease and in multi-infarct dementia. Biol Psychiatry 1991;30:684–690.

95. Rudman D, Shetty KR, Mattson DE. Plasma dehydroepiandrosterone sulfate in nursing home men. J Am Geriatr Soc 1990;38:421–427.

96. Legrain S, Berr C, Frenoy N, Gourlet V, Debuire B, Baulieu EE. Dehydroepiandrosterone sulfate in a long-term care aged population. Gerontology 1995;41:343–351.

97. Kalmijn S, Launer LJ, Stolk RP, deJong FH, Pols HAP, Hofman A, Breteler MMB, Lamberts SWJ. A prospective study on cortisol, dehydroepiandrosterone sulfate and cognitive function in the elderly. J Clin Endocrinol Metab 1998;83:3487–3492.

*The Neuroendocrine Immune Network in Ageing*
Edited by R.H. Straub and E. Mocchegiani

# Hormonal Changes in Ageing Men

EUGEN PLAS[1], STEPHAN MADERSBACHER[2] and PETER BERGER[3]

[1]*Department of Urology and Andrology, LBI for Urology and Andrology, Krankenhaus Lainz, Vienna;* [2]*Department of Urology and Andrology, LBI for Urological Oncology, Donauspital, Vienna;* [3]*Institute for Biomedical Ageing Research, Austrian Academy of Sciences, Innsbruck, Austria*

## ABSTRACT

The hypothalamic-pituitary-gonadal hormone axis undergoes profound changes with ageing in both sexes. While there is a rather abrupt cessation of sex steroid production in women with age (menopause), alterations in men are more subtle. Recent cross-sectional and longitudinal studies have identified a number of age-related changes in men that can ultimately lead to androgen deficiency. A number of symptoms and disorders frequently seen in elderly men (e.g. loss of body mass, sexual dysfunction, osteoporosis, depression) have been linked to these endocrine changes. Up to one third of men beyond the age of 60 have low serum testosterone levels. The role of androgen supplementation for these men is controversial due to the lack of long-term, randomised clinical trials with defined clinical endpoints. The aim of this review is to summarize age-related changes of the hypothalamic-pituitary-axis in men, determine their clinical relevance with particular reference to androgens, and discuss advantages and disadvantages of androgen supplementation.

## 1.    INTRODUCTION

Mean life expectancy continuously increases worldwide and is currently 78 years in Western Europe, 76 in North America, 69 in Latin America, 65 in Asia and 49 years in Sub-Saharan Africa. Life-style changes and technologic advances to reduce infant and maternal mortality, as well as hygienic, medical, surgical and occupational improvements, all contribute to this increase in mean life expectancy. It has been estimated that by the year 2050, 16% of the human population will be older than 65 years in industrialized countries [1]. These demographic changes confront societies with new, emerging problems involving social, political and economical issues.

Ageing affects all organ systems, resulting in metabolic, cardiovascular, intestinal, neurological, orthopaedic, genitourinary and endocrine changes, which can ultimately lead to a state of multi-morbidity. Rising co-morbidities with advanced age are associated with increasing burdens to the health care system, with 80% of medications being prescribed within the last 10 years of life. This emphasizes and justifies the importance of scientific inquiry in the ageing population to establish, improve or develop healthy ageing.

## 2. GENDER RELATED DIFFERENCES OF THE ENDOCRINE SYSTEM DURING AGEING

Menopause in women is characterized by a cessation of ovarian cyclic function endocrinologically resulting in a sharp decrease of estradiol serum levels below 20pg/ml and, through feedback mechanisms, a compensatory rise of follicle stimulating hormone (FSH >35IU/l) and luteinizing hormone (LH >10IU/l) [2]. Usually, menopause occurs around the end of the 5th decade of life. Postmenopausal women lack menstrual cyclicity, that can be accompanied by urinary urgency, vaginal dryness, hot flushes, depression, anxiety, irritability and skin alterations.

In men, age-related endocrine changes generally occur later in life than in women, and these changes are more subtle, since there usually is no complete cessation of testicular function in healthy ageing men like that of the ovaries in women. Therefore the term "andropause", defined as "indefinite syndrome composed of several constellations of physical, sexual, and emotional symptoms brought about by a complex interaction of hormonal, psychological, situational and physical factors" [3] is a misconception and not truly analogous to female menopause [4]. Usually, there is no endocrine discontinuity in male reproductive function, since persistent testicular function enables elderly men to reproduce even above 65 years [5]. Despite the gradual decline of testicular function with age, only about one third of men will become hypogonadal as defined by low serum testosterone levels, which may lead to the development of clinical signs and symptoms.

## 3. HYPOGONADISM IN ELDERLY MEN

Werner published the first report of clinical symptoms related to hypoandrogenism in 1939 [6]. He proposed the term "male climacteric" for a syndrome comprising nervousness, irritability, psychological depression, impaired memory, fatigue, insomnia, hot flushes, periodic sweating and loss of sexual vigour. Later, the terms "andropause" or "male climateric" were popularised and, more recently, the terms "PADAM" (partial androgen deficiency of the ageing male) and then "ADAM" (androgen deficiency of the ageing male) were created to describe all androgenic changes during the male ageing process.

The International Society for The Ageing Male defines hypogonadism as a biochemical syndrome associated with age _and_ a deficiency in serum androgens that may result in alterations affecting quality of life and adversely affect function of multiple organ systems [7]. Currently, there is consensus that serum-testosterone (T) levels of less than 320ng/dl or 12nmol/l are defined as hypogonadism that may require treatment. The prevalence of men with serum testosterone levels in the hypogonadal range increases linearly with age [8]. While only 7% of men less than 60 years fall into this category, this percentage increases to 25% in those aged 60–80 years and to further 30–40% of men above 80 years but this does not mean that these men necessarily develop other symptoms of hypogonadism.

There is no consensus about the endocrine causative mechanisms of the rather modest age-associated decline in T the origin of which may be at all three regulatory and organ levels of the hypothalamo-pituitary-testicular axis [4]. Decreased numbers of Leydig cells, impaired testicular perfusion and a blunted release of T upon stimulation by chorionic gonadotropin point at a partial primary testicular failure (primary hypogonadism). Elevated LH serum levels are common in elderly men, but these levels in response to the age-related decline of T levels of primary cause are lower than those observed in androgen-deficient younger men. Hypothalamic

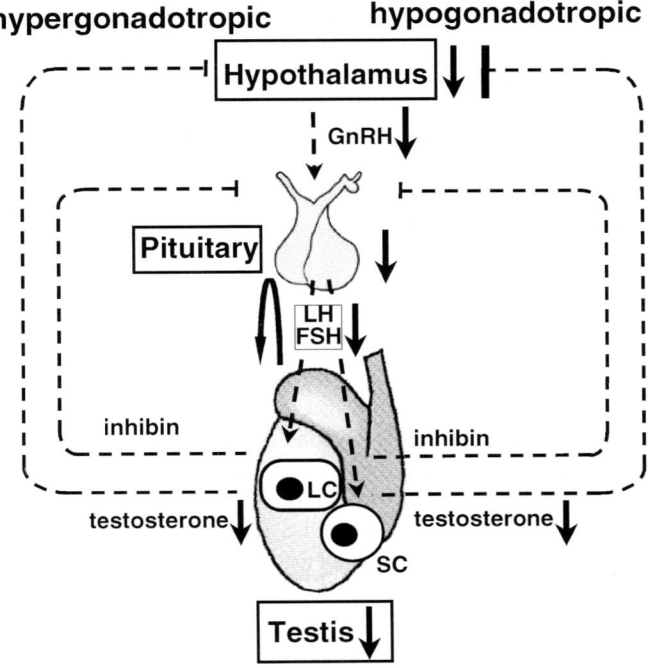

Figure 1. Hypogonadism of the ageing male.

In contrast to young male adults where often clearly defined disorders such as anorchia, Klinefelter syndrome and pituitary deficiency are the cause of primary (testicular) or secondary (hypothalamic, pituitary) hypogonadism, in elderly men age-related gradual regulatory changes of the HPG axis result in a mixture of partial dysfunctions at all organ levels (hypothalamus, pituitary, testis). Primary testicular failure leading to partial hypergonadotropic hypogonadism with increasing levels of follicle stimulating hormone (FSH and luteinizing hormone (LH) (left side), is intermingled with central defects of the hypothalamic-pituitary unit usually causing only mild secondary gonadal failure or hypogonado-tropic hypogonadism (right side). Age-associated changes at the testicular level are reductions in the numbers of Leydig (LC) and Sertoli cells (SC), thickening and hernia-like protrusions of the basal membrane of the seminiferi tubules, development of vacuolization and accumulation of lipofuscin within the Leydig cells and a decrease in the T production capacity upon stimulation with chorionic gonadotropin (CG); Additionally, SHBG levels in serum are upregulated with age: The sum of all these changes is meant to result in diminished levels of bioavailable T. This in turn should lead to full compensatory upregulation of the hypothalamic-pituitary unit but concomitant age-related changes, i.e. failure of the hypothalamus to generate appropriate amplitudes of the pulsatile secretion bursts of gonadotropin-releasing hormone (GnRH), higher sensitivity to the suppressive effects of T, and smaller amounts of GnRH released at each pulse, seem to be responsible for age-associated blunted pituitary response of FSH and LH and consequently inadequate T production and availability. Moreover, despite declining opioidergic suppression of GnRH secretion, old men fail to adequately restore GnRH and LH synchronous pulsatility and amplitudes to compensate for mostly mild primary hypoganadism reflected by too low T serum levels (reviewed in [21] and [22]).

Dashed arrows … regulatory mechanisms; solid arrows … changes in organic functions and of hormone levels with age impairment of stimulating LH, and subsequently androgen secretion is inferred from the loss of circadian rhythm of LH and T levels. Moreover, the nycthemeral variations in plasma T levels are significantly decreased in elderly men. The reduced number of spontaneous high amplitude LH pulses in elderly men does not seem to be a consequence of decreased sensitivity of the gonadotrophs to gonadotropin releasing hormone (GnRH), but of the release of smaller amounts of GnRH at each pulse. These findings speaks in favour of additional functional changes in the hypothalamic-pituitary unit (secondary hypogonadism). Above the age of 80, elevated LH levels may also be explained by the loss of the increased opioid tone of middle aged men [for review see 4,21,22] (Figure 1).

## 4.    LONGITUDINAL ENDOCRINE CHANGES IN AGEING MEN

Although age-related endocrine changes have been repeatedly analysed in cross-sectional data [9], there are only few longitudinal studies available [10–12]. Recently, Feldman et al. reported a follow-up of the Massachusetts Male Ageing Study [13]. In this community-based trial 1.709 men aged 40–71 years were initially investigated in 1991. Most of these men were reinvestigated 7–10 years later (n=1.156) and the data analysed cross-sectionally at baseline and follow-up, as well as longitudinally. It was observed that total testosterone (T: –1.6%/year), bioavailable testo-sterone (BAT: –2.8%/year) and albumin-bound testosterone 2.5%/year) longitudinally declined significantly with age. These longitudinal changes are slightly higher than those of the cross-sectional data at follow-up (0.8%/year, 1.7%/year, 2.0%). Dehydroepiandrosterone (DHEA) and dehydroepiandrosterone sulphate (DHEA-S) longitudinally declined in parallel by 1.4% and 5.2% annually, whereas dihydrotestosterone (DHT), follicle stimulating hormone (FSH) and luteotropic hormone (LH) increased by 3.5%, 3.1%, and 0.9%, respectively (Figure 2). These results are consistent with Harmann et al. who investigated longitudinal changes of T and the free testosterone index (FTI) in healthy men aged 22–91years. They reported an average decline of T by 0.124 nmol/L/year and FTI by 0.0049/year [10]

## 5.    DECLINE IN PULSATILE HORMONE SECRETION WITH AGE

In addition to changes in absolute hormone and hormone metabolite serum levels, the ageing hypothalamic-pituitary-gonadal axis is also characterised by a decline in diurnal pulsatile secre-tion. LH and FSH synthesis and secretion are differentially regulated by the frequency of GnRH pulses. A low frequency of GnRH stimulation preferentially increases FSH, whereas LH is maximally stimulated at higher GnRH pulse frequencies [14,15]. Basal LH pulse frequency in ageing men is comparable to young men, but frequency of LH pulses with amplitudes greater than 2IU/L decreases. Mean LH pulse amplitude, maximal LH pulse amplitude and pulse areas decline with ageing resulting in a loss of circadian changes of T serum concentration [16,17]. In young men, LH and T peak between 5.00–8.00 am and decline in the late afternoon and early evening (5.00–7.00pm) [18].

To address the issue of the necessity of multiple serum sampling due to diurnal hormone variations, we recently investigated 101 healthy men with a median age of 56 years [19]. Three morning serum samples were obtained within 90 minutes, separately analysed and compared to results of three pooled samples. If serum T was greater than 400 ng/dl, single analysis varied by a maximum of ±20% from pooled data, supporting the conclusion that due to the lack of pulsa-

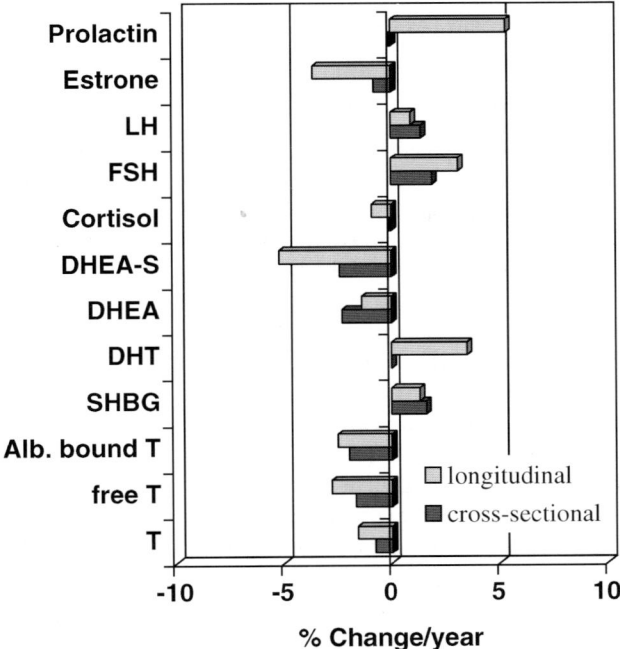

Figure 2. Annual endocrine changes of ageing men in cross sectional and longitudinal studies from the same cohort [9,13].

tility in elderly men, a single blood sample analysis seems to be sufficient. Men with a serum T of less than 400ng/dl, however, require repeated investigations for exclusion of hypogonadism [19].

6.      IMPACT OF HEALTH STATUS ON SERUM ANDROGEN LEVELS

In sharp contrast to the situation in young men, where hypoandrogenism is usually caused by clear pathomechanisms (e.g. Klinefelter syndrome, pituitary deficieny, anorchia etc.), ADAM is a multifactorial disorder. Factors like psychological stress, acute or chronic illness, physical activity, obesity, malnutrition, drug abuse or medication have a significant influence on the hypothalamic-pituitary-gonadal axis. Therefore, it is difficult to determine whether these age-related endocrine changes are caused by the ageing process *per se*, or are a consequence of underlying disorders [13,20]. A poor health status must, therefore, be considered a major risk factor for developing hypogonadism.

## 7.    ANDROGEN SUPPLEMENTATION IN AGEING MALES

### 7.1.    General considerations

First, it must be emphasised that data from large, prospective, randomised trials on androgen supplementation in ageing men are rare [21]. A second limitation is the definition of clinical study
endpoints. Considering the fact that ageing is associated with a tremendous increase in health
care costs, primary endpoints should be improvements of well being, quality of life and healthy
ageing, parameters that are not easy to assess. Prolongation of life by modulation of the ageing
process *per se* is an unrealistic endpoint. To date, only a few trials have studied the improvement
of quality of life. Clinicians should, therefore, be encouraged to not only study changes of serological parameters, but also improvements in quality of life and general well being.

A number of symptoms and syndromes accompanying the ageing process of men, such as
loss of muscle mass and muscle strength, changes in body composition, osteopenia, reduction
of bone density, sexual dysfunction, regression of secondary sexual characteristics, decreased
haematopoiesis, lethargy, decreased cognitive function and decreased general well-being, are
also observed in androgen-deficient young men. This led to speculations that lower androgen
levels may be responsible for some of these changes in ageing men, and that androgen replacement therapy may prevent, retard or even reverse some of these age-related clinical features.
But, since most of the clinical features accompanying the ageing process are non-specific and
of multifactorial origin there are no significant correlations to androgen levels. Nevertheless, it
has been assumed that androgen deficiency or insensitivity is involved, providing a rationale for
androgen supplementation in elderly men [for review see 21].

### 7.2.    Indications for androgen supplementation

Common indications for androgen supplementation for ADAM are decreased musculosceletal
mass, increased adipose tissue, loss of libido and reduced hematocrit, as these symptoms partially can be reversed by androgen replacement in adult men with frank hypogonadism due to
known causes [22,23]. Preliminary results after T replacement in elderly hypogonadal (BAT
<70 ng/dl) men (69–89 years) have shown beneficial effects on libido, density lipoprotein levels
and a significant increase in bone density and lean body mass [24,25]. In a double blind study in
which 108 men over 65 years old were randomized to wear either a T patch or a placebo patch
for 36 months, fat mass decreased and lean mass increased significantly in the T-treated group.
Muscle strength of knee extension and flexion was, however, not significantly different in the
two groups [26]. Although testosterone-treated men reported improved fitness and general well
being [27], the long term outcome remains unclear.

There are only a few studies investigating the effect of hypogonadism on age-associated
physical and cognitive function. In a placebo controlled trial, 15 hypogonadal men (T <60 ng/
dL) with a mean age of 68 received placebo, and 17 hypogonadal men (mean age 65) received
biweekly 200 mg T-cyprionate for 12 months [28]. The primary outcome was improved strength,
increased haemoglobin, and lower leptin levels in treated individuals as compared to placebo
but no significant difference in cognitive function was found. Seventeen percent of the patients
receiving T had to be withdrawn due to abnormally elevated hematocrits.

Although supraphysiological doses of T, especially when combined with strength training,
increased fat-free mass, muscle size and strength in men aged 19–40 years [29], the long term
substitution of supraphysiologic doses of T in frail elderly men cannot be recommended yet.

## 7.3. Depression and androgen supplementation

One of the major psychiatric problems of ageing is depression, which has been repeatedly linked to serum androgen levels. Recently, Perry et al. studied the effect of T-cyprionate on 16 elderly eugonadal males with major depressive disorder who received either 100 mg/week or 200 mg/week for 6 weeks [30]. There was a 42% decrease in depression scores irrespective of androgen dosage. This positive short-term effect emphasizes the importance of long-term trials.

## 7.4. Body composition and androgen supplementation

Several studies have determined the impact of androgen supplementation on body composition, BMI and lean body mass. As mentioned above Snyder and co-workers investigated 108 men (age > 65 years) in a double blind study for 36 months. Following T-treatment, lean body mass increased significantly and fat mass decreased [26]. Similar data were reported by Rolf et al. in 61 hypogonadal men (age 20–67 years) [31], who used 24 untreated hypogonadal men aged 19–65 years and 60 healthy eugonadal men (age 24–78 years) as controls. In eugonadal and untreated hypogonadal men, BMI and total fat mass increased consistently with advancing age. Hypogonadal males receiving T supplementation revealed neither an age-dependent change in BMI nor an increase in body fat content. It was concluded that ageing men benefit from T therapy with respect to body composition.

## 7.5. Bone metabolism and androgen supplementation

Although elderly men are less frequently affected by osteoporosis than post-menopausal women, it is still found in approx. 10% of men >65 years. Whether the physiological reduction of gonadal function in ageing men [32,33] leads to the age-related bone loss is still a matter of debate [34,37]. Concomitant with the rising incidence of osteoporosis in ageing men, the risk of osteoporotic fractures increases as well [38]. A positive correlation between the bone mineral content and plasma T, androstendione and estrone were reported in 30 men aged 60–90 years [33]. Androgen substitution was shown to improve bone mineralization and bone density in hypogonadal adult men [25]. Further, long term T-substitution normalized and maintained bone mineral density in this cohort [39]. It was shown by Snyder and co-workers that increasing the serum T concentrations in men over 65 years of age did not generally increase lumbar spine bone density, but increased it in those men with low pretreatment serum concentrations [40]. Nevertheless, it has to be kept in mind that only a few percent of osteoporotic men are hypogonadal.

The development of osteoporosis may be related to secondary estrogen deficiency rather than a primary testosterone deficiency [34]. In postmenopausal women, estrogen stimulates calcitonin secretion and increases calcium resorption [41]. Whether these mechanisms are also present in men remains unclear. Testosterone might also act indirectly, as human bone contains the enzyme 5$\alpha$-reductase, converting testosterone into dihydrotestosterone [42].

## 7.6. Side effects of androgen supplementation

Most of the concerns regarding long-term androgen supplementation centre on the prostate. Despite the importance of androgens for prostatic growth and the fact that benign prostatic hyperplasia (BPH) and prostate cancer require T during puberty, there is presently no evidence that androgen supplementation in hyopogonadal men will lead to BPH or induce prostate cancer

[43]. Only moderate increases in prostate size and prostate specific antigen (PSA) levels were noticed following androgen supplementation [26,44,45]. However, there is a recent case reports on the development of prostate cancer during androgen supplementation [46].

An elevated PSA serum level should be considered a contraindication for T supplementation. If under androgen-supplementation PSA rose more than 0.8 ng/ml/year ("PSA-velocity") and if the absolute PSA level exceeded 4 mg/ml, therapy should be stopped and prostate biopsy performed. Further possible risks of androgen supplementation in elderly men are the development of acne, a rise of hematocrit, worsening of lower urinary tract symptoms, a decline in HDL, sleep apnea and psychosocial issues.

## 8.  SENIEUR PROTOCOL – AN OPTIMISED STUDY POPULATION TO INVESTIGATE MALE AGEING

Reliable studies on age-related endocrine changes must fulfil two criteria: (i) a reproducible and reliable assay system, and (ii) a well characterized, healthy study population. To achieve this goal, the Human Immunology Group of European Community's Concerted Action Programme on Ageing (EURAGE) designed admission criteria for gerontological studies, summarized under the "*SENIEUR* protocol" [47]. The primary aim of this protocol was to define "healthy aged subjects" for immuno-gerontological studies. To reduce the risk of biased results due to a poorly defined study population, the *SENIEUR* protocol excludes any endogenous and exogenous influences on the immune and endocrine system to provide a standardized study population suitable for gerontological studies. Healthy aged people taking any medication known to adversely affect immune and endocrine systems (e.g. anti-inflammatory drugs, hormones, and analgetics) are excluded from the *SENIEUR*-protocol.

The impact of ageing on individuals selected according to the *SENIEUR* protocol with respect to endocrine changes [48], immune variables, lipid profiles, lipoproteins and neopterin has been intensively studied. To investigate the impact of health status on androgen levels in men, we quantified total T, BAT and SHBG in 526 men (aged 20–89 years) participating in a health screening project and compared the respective values to 35 men selected according to the *SENIEUR* protocol. This study has shown that the lowest annual declines in androgen level were observed in the *SENIEUR*-group (T –0.2%; BAT: –0.4%), emphasizing the importance of the health status for endocrine changes. BMI and levels of cholesterol, triglycerides, and glucose correlated negatively ($p < 0.01$) with T and BAT. Based on *SENIEUR*-data, the proposed lower T reference value declined gradually from 3.1 ng/mL (20–29 years) to 1.7 ng/mL ($\geq$70 years) [49].

## 9.  CONCLUSIONS

Longitudinal changes of the endocrine system are now well established in both genders. However, it remains to be definitively determined that preventive endocrine treatment is beneficial for ageing men. There are data available supporting the beneficial role of androgen supplementation in hypogonadal men, but this group is a minority of the ageing population. In eugonadal ageing men, there are only a few, controversial reports available, and androgen supplementation as a "life style drug" is premature. Investigations on endocrine changes or outcome of endocrine treatment require strict inclusion criteria, reproducible assay systems, randomised trials and clear definitions of primary clinical endpoints, that not only focus on short term, but the long-term outcome.

REFERENCES

1.    World Health Organization. Active Ageing: A Policy Framework 2002.
2.    Cobin RH, Bedsoe MB, Futerweit W, Goldzieher JW, Goodman NF, Petak SM, Smith KD, Steinberger E. AACE Medical Guidelines for clinical practice for management of menopause. Endocrine Practice 1999;5:354–366.
3.    Henker FD. Sexual, psychic and physical complaints in 50 middle-aged men. Psychosomatics 1977;18:23–27.
4.    Hermann M, Berger P. Hormone replacement in the aging male? Exp Gerontol 1999;34: 923–933.
5.    Plas E, Berger P, Hermann M, Pflüger H. Effects of aging on fertility. Exp Gerontol 2000;35:543–51.
6.    Werner AA. The male climacteric. JAMA 1939;112:1441–1443.
7.    Morales A, Lunenfeld B. International Society for the Study of the Aging Male Investigation, treatment and monitoring of late-onset hypogonadism in males. Official recommendations of ISSAM. International Society for the Study of the Aging Male. Aging Male 2002;5:74–86.
8.    Vermeulen A, Kaufman JM. Aging of the hypothalmo-pituitary-testicular axis in men. Horm Res 1995;43:25–28.
9.    Gray A, Feldmann A, McKinlay JB, Longcope C. Age, disease and changing sex hormone levels in middle-aged men: results of the Massachusetts Male Aging Study. J Clin Endocrinol Metab 1991;73:1016–1625.
10.   Harman SM, Metter EJ, Tobin JD, Pearson J, Blackman MR. Longitudinal effects of aging on serum total and free testosterone levels in healthy men. J Clin Endocrinol Metab. 2001;86:724–31.
11.   Zmuda JM, Cauley JA, Kriska A, Glynn NW, Gutai JP, Kuller LH. Longitudinal relation between endogenous testosterone and cardiovascular disease risk factors in middle-aged men. A 13-year follow-up of former Multiple Risk Factor Intervention Trial participants. Am J Epidemiol. 1997;146:609–17.
12.   Morley JE, Kaiser FE, Perry HM 3rd, Patrick P, Morley PM, Stauber PM, Vellas B, Baumgartner RN, Garry PJ. Longitudinal changes in testosterone, luteinizing hormone, and follicle-stimulating hormone in healthy older men. Metabolism 1997;46:410–3.
13.   Feldman HA, Longcope C, Derby CA, Johannes CB, Araujo AB, Coviello AD, Bremner WJ, McKinlay JB. Age trends in the level of serum testosterone and other hormones in middle-aged men: longitudinal results from the Massachusetts male aging study. J Clin Endocrinol Metab. 2002;87:589–98.
14.   Knobil E. The neuroendrocrine control of the menstrual cycle. Recent Prog Horm Res 1980;36:53–88
15.   Shupnik MA, Fallest PC. Pulsatile GnRH regulation of gonadotropin subunit gene transcription. Neurosci Biobehav Rev. 1994;18:597–9.
16.   Vermeulen A, Deslypere JP, Kaufman JM. Influence of antiopioids on luteinizing hormone pulsatility in ageing men. J Clin Endocrinol Metab 1989;68: 68–72.
17.   Bremner WJ, Vitiello V, Prinz PN Loss of circadian rhythmicity in blood testosterone

levels with ageing in normal men. J Clin Endocrinol Metab 1983;56:1278–1281.

18. Tenover JS, McLachlan RI, Dahl KD, Burger HG, deKretser DM, Bremner WJ. Decreased serum inhibin levels in normal elderly man: evidence for a decline in Sertoli cell funtion with aging. J Clin Endocrinol Metab 1988;67:455–459.

19. Gallistl H, Plas E, Pflüger H. Assessment of testosterone in the aging male – how many samples do we need ? J Urol Urogynakol 2003; in press.

20. McKinlay JB, Longcope C, Gray A. The questionable physiologic and epidemiologic basis for a male climacteric syndrome: preliminary results from the Massachusetts Male Aging Study. Maturitas 1989;11:103–115.

21. Hermann M, Berger P. Aging of the male endocrine system. Rev Physiol Biochem 1999;139:89–122.

22. Hermann M, Berger P. Hormonal changes in aging men: A therapeutic indication? Exp Gerontol 2001;36:1075–1282.

23. Bhasin S, Bremner WJ. Emerging issues in androgen replacement therapy. J Clin Endocrinol Metab 1997;82:3–8.

24. Katznelson L, Finkelstein JS, Schoenfeld DA, Rosenthal DI, Anderson EJ, Klianski A. Increase in bone density and lean body mass during testosterone administration in men with acquired hypogonadism. J Clin Endocrinol Metab 1996;81:4358–4365.

25. Finkelstein JS, Klibanski A, Neer RM. Increase in bone density during treatment of men with idiopathic hypogonadism. J Clin Endocrinol Metab 1989;69:776–783.

26. Snyder PJ, Peachey H, Hannoush P, Berlin J, Loh L, Lenrow DA, Holmes JH, Dlewati A, Santanna J, Rosen CJ, Strom BL. Effect of testosterone treatment on body composition and muscle strength in men over 65 years of age. J Clin Endocrinol Metab 1999;84: 2647–2653.

27. Tenover JS. Androgen administration to aging men. Endocrinol Metab Clin North Am 1994;23:877–892.

28. Sih R, Morley JE, Kaiser FE, Perry III HM, P Patrick, Ross C. Testosterone replacement in older hypogonadal men: A 12-month randomized controlled trial. J Clin Endocrinol Metab 1997;82:1661–1667

29. Bhasin S, Storer TW, Berman N, Callegari C, Clevenger B, Phillips J, Bunnell TJ, Tricker R, Shirazi A, Casaburi R. The effects of supraphysiologic doses of testosterone on muscle size and strength in normal men. New Engl J Med 1996;335:1–7.

30. Perry PJ, Yates WR, Williams RD, Andersen AE, MacIndoe JH, Lund BC, Holman TL Testosterone therapy in late-life major depression in males. J Clin Psychiatry 2002;63(12): 1096–101.

31. Rolf C, von Eckardstein S, Koken U, Nieschlag. E Testosterone substitution of hypogonadal men prevents the age-dependent increases in body mass index, body fat and leptin seen in healthy ageing men: results of a cross-sectional study. Eur J Endocrinol 2002;146(4): 505–11.

32. Baker HWG, Burger HG, de Kretser DM, Hudson B, O´Connor S, Wang C, Microvacs A, Court J, Dunlop M, Rennie GC. Changes in the pituitary-testicular system with age. Clin Enodcrinol 1976;5:349–354.

33. Foresta C, Ruzza G, Mioni R, Guarneri G, Grimaldo R, Meneghello A, Mastrogiacomo I. Osteoporosis and decline of gonadal function in the elderly male. Horm Res 1984;19: 18–22.

34. Crilly RG, Francis RM, Nordin BEC. Steroid hormones, ageing and bone. Clin Endocrinol Metab 1981,10:115–121.

35. Meier DE, Orwoll ES, Keenan EJ, Fagerstrom RM. Marked decline in trabecular bone mineral content in healthy men with age: lack of association with sex steroid levels. J Am Geriatr Soc 1987;35:189–197.

36. Odell WD, Swerdloff RS. Male hypogonadism. West J Med 1976;124:446–475.

37. Swerdloff RS, Wang C. Androgens and aging in men. Exp Gerontol 1993;28:435–446.

38. Santavirtas, Konttinen YT, Heliovaara M, Knekt P, Luthje P, Aromaa A. Determinants of osteoporotic thoracic vertebral fracture-Screening of 57000 Finnish women and men. Acta Orthoped Scand 1992;63:198–202.

39. Behre HM, Kliesch S, Leifke E, Link TM, Nieschlag E. Long-term effect of testosterone therapy on bone mineral density in hypogonadal men. J Clin Endocrinol Metab 1997;82:2386–2390.

40. Snyder PJ, Peachey H, Hannoush P, Berlin JA, Loh L, Holmes JH, Dlewati A, Staley J, Santanna J, Kapoor SC, Attie MF, Haddad JG, Strom BL. Effect of testosterone treatment on bone mineral density in men over 65 years of age. J Clin Endocrinol Metab 1999;84:1966–1972.

41. Gallagher JC, Lawrence B, De Luca HF. Effect of estrogen on calcium absorption and serum vitamin D metabolites in postmenopausal osteoporosis. J Clin Endocrinol Metab 1980;51:1359–1364.

42. Schweikert IM, Rulf W, Niederle N, Schafer HE, Keek E, Kruck F. Testosterone metabolism in human bone. Acta Endocr 1980;95:258–264.

43. Behre HM, Bohmeyer J, Nieschlag E. Prostate volume in testosterone-treated and untreated hypogonadal men in comparison to age-matched normal controls. Clin Endocrinol 1994;40:341–349.

44. Morley JE, Perry HM, Kaiser FE, Kraenzle D, Jensen J, Houston K, Mattammal M; Perry HM Jr. Effects of testosterone replacement therapy in old hypogonadal males: a preliminary study. J Am Geriatr Soc 1993;41:149–152.

45. Holmäng S, Marin P, Lindtstedt G, Hedelin H. Effect of long-term oral testosterone undecanoate treatment on prostate volume and serum prostate specific antigen in eugonadal middle-aged men. Prostate 1993;23:99–106.

46. Curran MJ, Bihrle W 3rd. Dramatic rise in prostate-specific antigen after androgen replacement in a hypogonadal man with occult adenocarcinoma of the prostate. Urology 1999;53(2):423–424.

47. Ligthart GJ, Corberand JX, Geertzen HGM, Meinders AE, Knook DL, Hijmans W. Necessity of the assessment of health status in human immunogerontological studies: evaluation of the Senieur protocol. Mech Ageing Dev 1990;55:89–105.

48. Madersbacher M, Stulnig T, Huber LA, Schönitzer D, Dirnhofer S, Wick G, Berger P. Serum glycoprotein hormones and their free α-subunit in a healthy elderly population selected according to the SENIEUR protocol. Mech Ageing Dev 1993;71:223–233.

49. Schatzl G, Madersbacher S, Temml C, Krenn-Schinkel, Nader A, Söregi G, Lapin A, Hermann M, Berger P, Marberger M. Serum Androgen levels in Men: Impact of Health Status and Age. Urology 2003;61:629–633.

# Ageing and the Endocrine Circadian System

YVAN TOUITOU[1] and ERHARD HAUS[2]

[1]Department of Biochemistry and Molecular Biology, Faculty of Medicine Pitié-Salpêtrière, 91 boulevard de l'Hôpital, 75634 paris Cedex 13, France; [2]Jackon Street, St Paul, Minn 55101-2525, USA

ABSTRACT

The neuroendocrine systems shows in its central and peripheral components a time structure consisting of rhythms of multiple frequencies ranging from minutes (e.g. pulsatile secretions) and hours (ultradian rhythms) to the prominent about 24-hour (circadian) rhythms, and to lower (infradian) frequencies including seasonal variations or circannual rhythms. Optimal function of the human organism depends critically on the maintenance of these rhythms and their time relations among each other, and in some frequencies upon maintenance of environmental synchronization.

The adrenal cortex shows with ageing a subtle increase in cortisol and more marked decrease in adrenal androgen production. The circadian amplitudes of some adrenal steroids decrease with ageing. The increase in cortisol with ageing was found to be correlated with cardiovascular pathology in the elderly.

Age is the most important factor affecting circulating melatonin concentrations with significant decline in the elderly. This may lead to a loss in circadian time information and in antioxidant functions.

Ageing changes in the hypothalamic-pituitary-thyroid axis show ethnic-geographic differences which in part may be due to environmental conditions (e.g. iodine supply).

In prolactin secretion, circadian rhythmicity continues in the elderly. Catecholamine circadian rhythms centrally as well as peripherally also persist into old age, however, with changes in circadian mean and circadian amplitude.

In general, in healthy subjects rhythmicity is maintained until very old age. However, some characteristic changes like, e.g., a decrease in the circadian amplitude of numerous variables, occur also in the clinically healthy elderly. With disease states developing with advancing age, circadian as well as rhythms in other frequencies may be markedly disturbed and the altered time structure may contribute to ill health and senility.

## 1.    INTRODUCTION

Living matter displays an organization in space, its molecular and anatomic structure, and an organization in time or time structure. The human time structure consists of a spectrum of rhythms of different periods, ranging from milliseconds to years, which are superimposed upon one another and upon the time-dependent changes that accompany growth, development and ageing (for review see [1–2]). In the neuroendocrine system, rhythmicity is required for optimal function and rhythm disturbances or loss of rhythmicity in some frequencies may lead to malfunction and disease.

Many biological rhythms are genetically fixed with specific genes and gene products now identified in several mammalian species. The prominent circadian (about 24-hour) rhythms, which are ubiquitous in human cells, tissues and organs, are maintained by a multiple gene mechanism consisting of interacting positive and negative transcription/translation feedback loops. These rhythms are directed by a central master clock located in the paired suprachiasmatic nuclei in the hypothalamus which is kept entrained with the solar day-night cycle (or artificial lighting regimen) through non-vision-related ganglion cells in the retina, which serve as slow-acting photoreceptors. Circadian peripheral oscillators, which possess a molecular composition like that of the master clock are found in many peripheral tissues and typically cycle with a 6 to 8 hour delay with references to the central pacemaker (for review see [3]). The genetic mechanisms interact with endocrine factors, like cortisol in the maintenance of rhythmicity in some peripheral oscillators [4].

In some animal models, ageing was found to lead to a loss in rhythmicity at the level of the central oscillator [5]. In others, ageing seemed to affect the rhythms in some peripheral oscillators, but not in all tissues and seemed to act primarily on interaction among central and peripheral circadian oscillators with loss of circadian synchronization between them [6]. Further identification of oscillator genes in the circadian and other frequency ranges may open the way for environmental (light) and pharmacologic manipulation of rhythm disturbances encountered in the aged.

In the study of the neuroendocrinology of ageing in human populations, it has to be realized that the neuroendocrine system in our classic concept is only part of the complex regulatory web which including the central and peripheral nervous system, the immune system, the endocrine glands and peripheral endocrine tissues like adipose and gastro-intestinal tissues. Any separate discussion of parts of this system is by necessity artificial and has to be understood against the background of other presentations in this issue.

## 2.    ADRENAL HORMONES CIRCADIAN RHYTHM

The increasing number of subjects reaching the age of 65 and older thanks to medical progress and improvement of life conditions has raised a growing interest in the study of the ageing process and age-related changes in biologic rhythms [7]. Adrenal function and ageing has been the subject of special care in recent years. Throughout life, the adrenal cortex exhibits dramatic steroidogenic changes, among which a subtle increase in cortisol with age. These steroidogenic changes are considered to worsen some age-related diseases, although their exact biological significance is not completely understood [7–12].

The circadian profile of plasma cortisol is conserved in the elderly but with higher plasma levels at night during the trough time (nadir) which results in an elevated cortisol 24-hour mean level and a reduction in the rhythm amplitude [13–15]. These higher plasma cortisol levels at night may play a role in cognitive and metabolic disturbances found in the elderly. Besides, an increase of the active free fraction of plasma cortisol was reported in elderly subjects, probably in relation to a decrease in the concentration of binding proteins and/or to a decrease in binding capacity of these proteins [13,16]. A dramatic reduction of circadian amplitude was also observed on 18-hydroxy-11-deoxycorticosterone (18-OHDOC), an ACTH-dependent steroid [13,16]. Both ACTH and cortisol circadian rhythms are independent of the sleep-wake cycle although small modulatory effects of this cycle have been shown. Cortisol circadian rhythm is a robust rhythm, which does not respond rapidly to minor and transient environmental changes (as in e.g. jet lag) as they are part of our daily life, which makes it a good candidate as a rhythm marker.

A trend for a phase advance in plasma cortisol and plasma 18-hydroxy-11-deoxycorticos-terone was reported in the elderly [13,17] but was not found in all seasons [16]. A similar phenomenon was described by Haus et al. [18] for plasma DHEA-S. Both plasma DHEA and DHEA-S are decreased in elderly men and women, and have lost their ability to be stimulated by adrenocorticotropic hormone [19], most probably in relation with an enzymatic alteration in the biosynthetic pathway. In contrast, the response of cortisol production to stimulation by adrenocorticotropic hormone and the adrenocorticotropic hormone response to the infusion of corticotropin releasing factor appear to be unimpaired in the aged [review in 20].

Recent data on melatonin [21] and cortisol [22] lead to consider a possible outcome of geo-graphic and ethnic origins and/or eating habits on hormone blood patterns. Therefore, we meas-ured morning (08:00) serum cortisol levels in Chinese people from 31 to 110 years old, stratified by 10-year age groups to explore the influences of age and gender. Because of the high incidence of cardiovascular diseases, especially high blood pressure and ischemic heart disease, in ageing we also documented this aspect in relation with serum cortisol concentrations [23].

We showed the positive effects of ageing and cardiovascular diseases on morning (08:00) serum cortisol levels. In comparison with the young controls (< 60s), serum cortisol levels were increased significantly in the elders (60s, 70s, and >80s); after 60 years of age, morning serum cortisol levels increased significantly with ageing and the increase was even more significant in pathological conditions as we found that the whole patient group had higher serum cortisol levels than the whole healthy elders. While a positive correlation has been reported in a longitu-dinal study between systolic blood pressure and the ageing-cortisol slope [24] which is consist-ent with the glucocorticoids-induced hypertension observed in Cushing's syndrome patients, in the elderly subjects of the present study higher serum cortisol levels seemed more related with ischemic heart disease than with high blood pressure. In the 70s group, no significant difference was found between the healthy and patient groups, possibly because approximately one third of the patients of this age group had High Blood Pressure (HBP) alone versus only one fifth in the 60s age group [23]. This divergence may be due to differences in geographic and ethnic origins and/or eating habits as reported for cortisol [22] and other biological parameters [25–28]. Changes in steroidogenesis implicated in ageing and many age-related diseases were reported also in the literature [10–12,29,30] however, the clinical implications of increase in serum corti-sol in ageing and diseases conditions were not completely understood and need further study.

We also compared the circadian rhythms of serum cortisol and DHEAS in four different age-groups of healthy Chinese men (in their 30s, 40s, 50s and 60s) [22]. The circadian rhythmic pattern of serum cortisol and its serum concentrations were similar to those reported in the literature for Caucasian subjects (a peak concentration of about 320 nmol/L and a trough con-centration around 83 nmol/L [16,31,32]. The two populations differed, nonetheless: the peak serum concentrations in Chinese men (31–63 years old) were observed between 04:00 and 06:00 hrs and the minimum concentrations around 18:00–24:00 hrs, while in both young and elderly Caucasian subjects [7,16,31,32], peak plasma cortisol concentrations occur around 06:00–08:00 hrs and the minimum concentration around 23:00–02:00.hrs. Our results thus indicate that both the peak and the trough of serum cortisol levels in Chinese men occur approximately 2 hours earlier than in Caucasians [22]. This difference may well be a consequence of the regular social routine in China, where it is customary to get up and go to sleep early (06:00 ± 01:00 and 22:00 ± 01:00, respectively) which strongly suggest that social activity and living habits play a major role in the synchronization of circadian rhythms in humans [22]. Nevertheless, differences in geographic and ethnic origins and/or eating habits may also be responsible for the observed differences in serum cortisol peak and trough times between healthy Chinese and Caucasian men. Indeed variations according to ethnic origin and/or eating habits have been reported for some biologic parameters [25–28].

The circadian profiles of serum DHEAS were similar in all four age-groups and the circadian rhythmicity of serum DHEAS was validated in all. These results are consistent with the literature [review in 7,8]. In this study, although circadian rhythmicity and serum cortisol levels remained relatively constant in the elderly, serum DHEAS levels decreased significantly among men in their 60s. The amplitudes of the circadian variation of serum DHEAS were very small (10%, range 9–14%), and levels were higher during the day (small peak around 16:00 hrs); these results are similar to that reported by Ferrari et al. [30]. Our results also showed that serum DHEAS levels fell markedly in ageing Chinese men. The group aged 30–39, had the highest 24-h mean serum DHEAS concentrations, and those in their 60s had the lowest – approximately 50% lower than for those in their 30s. Serum DHEAS fell approximately 20% for those in their 40s and 50s, compared with the youngest group. These data are similar to those reported for Caucasians [33]. In view of the great inter-individual variability of serum DHEAS, which may be considered a highly specific individual marker [34,35], the uneven decrease of serum DHEAS in our different age-groups may result from the individual differences among these age-groups.

Ageing appears to affect cortisol and DHEAS secretions differently: we found that cortisol secretion was generally constant across the age-groups or increased slightly with age. We have previously shown in Caucasian men in their 80s the conservation of the circadian profile of plasma cortisol, but higher plasma concentration of this hormone during the trough time (nadir) which results in an elevated cortisol 24-h mean levels [16]. Adrenal production of DHEA and DHEAS showed a clear-cut age-related decline. The circadian variation of molar ratio of cortisol / DHEAS was thus significantly higher in elderly than young subjects [30]. We evaluated this specific circadian variation in age-groups of men in their 30s, 40s, 50s and 60s. This circadian pattern was similar in all these age-groups, with a peak at early morning (04:00–06:00 hrs) and a trough between 18:00–24:00 hrs. This pattern is similar to the circadian variations of serum cortisol. The circadian pattern of the molar ratio of cortisol / DHEAS thus results mainly from circadian variations of the serum cortisol rhythm, since those for serum DHEAS had a very small amplitude. The level of the cortisol / DHEAS molar ratio, however, was significantly higher in the group in their 60s compared with the younger age-groups, obviously because of their significantly lower serum DHEAS levels. Thus, the circadian pattern of the ratio primarily represents the cortisol rhythm, whereas its level represents the change in serum DHEAS levels with age.

## 3.    MELATONIN CIRCADIAN RHYTHM

Melatonin (N-acetyl-5-methoxytryptamine) is the main hormone secreted by the pineal gland with a high amplitude circadian rhythm. Its circadian rhythmicity is comparable in humans and in experimental animals (either diurnal or nocturnal ones) with daily dark span plasma concentrations three to ten times higher than during the light span. Melatonin interacts with other circadian periodic variables and thus indirectly controls or exerts influence upon a wide variety of physiologic functions such as the sleep-wake cycle, thermal regulation, feeding and sexual behavior, and certain cardiovascular functions, and through its interaction with serotonin, it participates in the regulation of the secretion of ACTH, corticosteroids, ß-endorphin, prolactin, renin, vasopressin, oxytocin, growth hormone, and LH [36,37].

Various reports have shown that secretion of plasma melatonin and urinary 6-sulfatoxyme-latonin decline with age in humans [38–42]. More recent data, however, do not support the hypothesis that reduced melatonin levels are a general characteristic of healthy ageing [43,44]. To further assess this hypothesis, we measured serum melatonin concentrations at 02:00 hhrs and 08:00 hrs in a cross-sectional study among 144 subjects aged from 30 to 110 yr [21]. Since we have formerly found higher levels of plasma melatonin in Caucasian elderly women [45], we also explored such a difference in Chinese elderly women. Indeed, differences in melatonin [46,47] and other biological parameters [25–28] have been reported according to ethnic origin and/or eating habits. To determine the age at which the circadian rhythm of serum melatonin concentration changes, we also examined the characteristics of this rhythm in four age-groups – persons in their 30s, 40s, 50s and 60s (cross-sectional study). This issue is important because melatonin is an antioxidant and a free radical scavenger that may be involved in the ageing process and age-related diseases [48–50] and because oxidative damage (from free radicals) is involved in the pathogenesis of various diseases, especially ischemic coronary heart disease. Therefore, we also looked to see if such diseases in elderly subjects might be related to their serum melatonin concentrations.

Our studies confirmed the variation of melatonin levels in all these adult age groups, with night peaks and daytime lows. The 02:00 hr melatonin levels differed significantly according to age and sex, although this consideration might be restricted to women aged 60 years and over [21]. Regardless of age, nighttime serum melatonin levels were higher in women (post menopausal) than in men: this confirms our earlier finding of sex-related differences in blood melatonin concentration [45]. Serum melatonin concentrations at 02:00 hrs differed very significantly by age, between those under 60, and those 70 or older ($P < 0.001$). The nighttime melatonin levels did not decrease further in the oldest group, compared with those in their 70s. To better define the age threshold at which serum melatonin begins to decline we studied serum melatonin levels throughout the day and found that they peaked during the night and were very low during the day. Those in their 60s had a circadian melatonin pattern similar to that among the younger subjects. Although the younger age-groups could not be distinguished by their nighttime melatonin concentrations, the level was significantly lower for those in their 60s than for the younger subjects.

Our results confirm that age was the most important factor affecting nighttime melatonin levels. A decline was observed in the group of patients in their 60s. There was also a significant difference between those in their 60s and those 70s and older. Levels were similar among those in their 70s and those over 80. This may mean that the nighttime melatonin level begins to decline in the early 60s, bottoms out in the 70s, and then remains at that level with no further decline. Our study thus agrees with those by Nair et al. [41] and Sack et al. [42] and conflicts with some other findings. Zeitzer et al. [43] found no negative relation between age and the amplitude of serum melatonin rhythm in subjects following a "constant routine" (except for a small group of elderly subjects), and Kennaway et al. [44] found that the decrease in urinary 6-sulfatoxymelatonin was not significant after the age of 30. The discrepancy with the report of Zeitzer et al. [43] may be due mainly to the study conditions. Constant routine conditions (subjects remaining awake for at least 30 hours in semi-recumbent position under constant dim ambient illumination of 15 lux and receiving hourly equicaloric snacks and fluids) may possibly be suitable to describe the endogenous circadian rhythm of melatonin.Although a group of elderly subjects did have significantly lower amplitudes even with these conditions. These observations are far enough removed from the usual living conditions, especially among the elderly, to bias any assessment of the clinical usefulness of melatonin in the aged. The finding by Kennaway et al. [44] that urinary 6-sulfatoxymelatonin decrease was no longer significant after the age of 30 did involve normal living conditions. The difference between those results and ours may be due to differences in geographic and ethnic origins and/or eating habits as has been established for melatonin [46,47] and to other biological variables [25–28].

Melatonin has recently been shown to be an antioxidant and a free radical scavenger that may be involved in the ageing process and with age-related diseases [48–50]. Although oxidative damage (from free radicals) is involved in the pathogenesis of various diseases, especially the ischemic coronary heart disease, in this cross-sectional study we found no significant relation between serum melatonin levels and these diseases. This may be a consequence of the treatment the subjects received: either the effect of the treatment was to lower melatonin levels greater than in controls and thus these possible greater levels had no role in preventing the disease or the effect of the treatment was to increase melatonin concentrations to levels similar to those in control subjects which has not yet been reported. To our knowledge the only report dealing with an attempt to stimulate melatonin secretion in man with beta-adrenoreceptor agonists [51] did not yield a positive result. Further studies are needed to answer the question of whether physiological concentrations (ten to sixty pg/ml) of melatonin might influence free radicals sufficiently to protect the cardiovascular system.

## 4. HYPOTHALAMIC-PITUITARY-THYROID (HPT) AXIS

### 4.1. Clinical considerations

The hypothalamic-pituitary-thyroid (HPT) axis is part of a complex web of neuroendocrine functions. It shows an intricate time structure with rhythmic variations of multiple frequencies found at all levels of the system from hypothalamic neurons to the cells of the peripheral target tissues. The frequencies observed range from rapid neuronal discharges to pulsatile secretions and ultradian rhythms to circadian and circannual rhythms. The time-dependent rhythmic (and non-rhythmic) variations of the HPT system interact with, and are modulated by similar time-dependent variations of other neuroendocrine, metabolic and immune functions. The rhythmic variations are superimposed upon trends like child development and ageing. A description of the chronobiology of the ageing changes in the HPT axis is quite complex, with some apparent differences between observations obtained in different laboratories at different geographic locations.

During the process of ageing, alterations of the hypothalamic-pituitary-thyroid axis have to be considered which are not directly age related, but caused by thyroidal or non-thyroidal illness, the incidence of which increases with age [52]. Overt or occult thyroid disease in the elderly interacts with the physiologic changes due to the ageing process. Ethnic-geographic differences also have to be taken into account, including climate, daylight exposure, and diet. Sleep deprivation and sleep fragmentation as seen frequently in the aged lead to a marked decrease in the mean 24-hour TSH secretion [53–56]. In iodine deficiency regions, the prevalence of goiter is estimated at between 30 and 50 percent increasing in frequency with advancing age [57–62].

Prolonged iodine deficiency leads to nodular goiter, often with alteration in thyroid function.

Prevalence of primary hypothyroidism in elderly populations determined by elevated TSH concentration ranges in the U.S.A. and in Norway between 9.6 and 20.3%, with a 3–5 fold preponderance in women [63–66].

In contrast in Germany (a country with iodine deficiency), the incidence of hypothyroidism, in the elderly (including subclinical forms) is only 1–2%, while there is an incidence of 8–10% of hyperthyroidism [52]. Apart from these conditions, subtle changes due to the physiologic ageing process can be found at all levels of the hypothalamic-pituitary-thyroid axis.

## 4.2.  Thyrotropin Releasing Hormone (TRH)

TRH is a tripeptide neurotransmitter with multiple actions in the central nervous system and beyond [67–68]. It is produced also in peripheral tissues, including the immune system [69]. In addition to its capacity to stimulate the release of thyroid-stimulating hormone (TSH) from the anterior pituitary, it also stimulates prolactin. Its hypothalamic content shows a circadian rhythm in rodents, which is in its timing light-dark cycle dependent [70–72]. TRH interacts with other circadian periodic neurotransmitters, e.g., 5-hydroxytryptamine (5-HT), dopamine and norepinephrine [73].

The action of TRH upon the pituitary via a specific membrane receptor leads to de novo synthesis and release of thyroid stimulating hormone (TSH). After binding with TRH, the receptor is only slowly degraded. Therefore, a short pulse of TRH, can lead to a prolonged secretion of TSH from the pituitary cell [74]. The density of TRH receptors can be modulated by other hormones. Thyroid hormones exert a powerful negative feedback control over the pituitary response to TRH [75]. At the TRH receptor level, there is interaction with gonadal and adrenal steroids [76–77]. This heterologous modulation suggests the likelihood of rhythmic interactions between the thyrotropic, gonadotropic and adrenotropic axes.

TRH is the main stimulator of TSH production and release and appears to be the most important factor in the regulation of the pulsatile secretion and of the circadian rhythm of TSH. The pulsatile stimulation of TSH secretion results from an interplay of an apparent hypothalamic oscillator modulated by inhibitory hypothalamic influences mediated by dopamine and somatostatin. Triiodothyronine ($T_3$) secreted by the thyroid or converted from $T_4$ in peripheral tissues inhibits synthesis and secretion of TRH [78] and stimulates the release of inhibitory factors such as dopamine or somatostatin. At the level of the thyrotrope cell in the anterior pituitary $T_3$ inhibits the synthesis of TSH secretion by interaction with specific $T_3$ receptors [79].

## 4.3.  Thyroid Stimulating Hormone (TSH)

Thyroid stimulating hormone is secreted from the pituitary gland in a series of discrete pulses with an average pulse frequency of 9 pulses/24 hours (range 7–12) in normal men and women. These pulses are not equally distributed, but cluster during the evening and night hours when fusion of the pulses and an increase in amplitude leads to the nightly increase of TSH concentration forming the circadian rhythm of this hormone with a maximum in day-night synchronized subjects usually between 02:00 and 04:00 hrs [54,80]. The relatively high peak values of individual TSH pulses during the evening hours have to be kept in mind since the values reached during this time may be slightly above the usually accepted normal range.

The pulse pattern may be necessary for normal thyroid gland function with better response of the thyrotrops in the pituitary to intermittent than to continuous TRH stimulation [81]. Loss of the usual nocturnal variations in TSH and pulse amplitude may be sufficient to cause clinical hypothyroidism [80]. The TSH pulses are statistically significantly concordant with the pulses of LH, FSH and the α-subunit of the glycoprotein hormones suggesting that an underlying unified signal coordinates the pulsatile secretion of both gonadotrops and thyrotrops [80].

Pulse patterns are changed by prolonged fasting and malnutrition. Fasting (e.g., 36 hours) leads in clinically healthy subjects to a decrease in mean pulse amplitude and lowering of the 24-hour mean. The pulse frequency during fasting remains unchanged [82]. The starvation related decrease in TSH could not be prevented by TRH infusions [83] but in animal studies in rodents could be prevented by administration of leptin [84], suggesting an interaction between these agents.

Sleep deprivation and sleep fragmentation as seen frequently in the aged lead to a marked decrease in the mean 24-hour TSH secretion, and lowering of pulse amplitude also without change in peak frequency [53,54,56]. The pituitary response to TRH stimulation is consistently reduced in the elderly [85].

The nocturnal peak of TSH was blunted in the elderly due to a decrease of TSH pulse amplitude [86–89]. However, the temporal structure of the pulsatile TSH secretion and of the circadian rhythm remains constant with a trough in the late afternoon and peak around midnight together with an increased frequency in TSH pulses [55]. While most reports on pulsatile and circadian rhythms in TSH in the elderly show a blunting of the nocturnal peak and a decrease in circadian amplitude [68,90], the circadian mean concentrations may vary according to geographical region and iodine status. When the iodine supply is experimentally decreased, the TSH levels drop in clinically healthy subjects by approximately 50 percent without any effect on the frequency of TSH pulses [91]. In reports from iodine deficiency areas like Germany, a prospective study of mostly elderly subjects in a rehabilitation clinic showed only in 0.8% of subjects elevated TSH serum levels indicating hypothyroidism, but a prevalence of 22.6% of subjects with suppressed TSH [92]. Lower TSH levels in the elderly were reported also, e.g., from Holland [93]; from India [94], and in centenarians in Italy [95,96], in one region, but elevated levels were reported from another [97].

We compared 277 clinically healthy and euthyroid elderly men and woman ($77 \pm 11$ years of age) with 193 children ($11 \pm 1.5$ years of age) living in Eastern Europe (Romania). The activity-rest pattern was synchronized by the routine of a home for the aged or by their school routine: average time of raising 06:30 hrs, time of retiring 22:00 hrs, and three meals per day sampled six times over a 24-hour span. We found that in the elderly the amplitude of the circadian rhythm of TSH was reduced due to an increase of the trough values during daytime with little difference in the peak values during evening and night. The circadian mean concentration was higher in the elderly due to the higher values during daytime.

## 4.4.   Thyroid hormone

Low amplitude circadian rhythms were found in total thyroxine ($TT_4$) and triiodothyronine ($TT_3$), the timing of which varies to some extent between different studies by different investigators and in our laboratory in subjects of different age groups and geographic locations [98]. In most studies, the acrophase of $TT_4$ was found during the forenoon or noon hours with lower values during the night [68,98–102]. The circadian rhythm in free $T_4$ ($fT_4$) showed in elderly subjects a peak during daytime.

The circadian rhythm in $TT_3$ showed in elderly subjects an acrophase in the late afternoon in contrast to a group of children $11 \pm 1.5$ years of age, who showed the acrophase of $TT_3$ and also of free $T_3$ ($fT_3$) during the late night and early morning hours [59,98] similar to a group of subjects reported by Weeke and Gundersen [103]. The elderly subjects in our series showed no statistically significant circadian rhythm in $TT_3$. The acrophase of reverse $T_3$ ($rT_3$) in children was found during the afternoon and in elderly subjects during late afternoon and evening. A comparison of the circadian timing of the HPT axis in children and in elderly subjects showed in the elderly a slight phase delay in the circadian acrophase of TSH and a much more marked phase delay in $TT_3$, $TT_4$, and $rT_3$. This observation suggests a delayed response of the thyroid in the elderly to TSH stimulation occurring at about the same time and with the same TSH concentration as seen in children or young adults [68].

## 4.4. Seasonal variations and/or circannual rhythms of the HPT axis

The HPT axis shows marked seasonal variations, which are to a large degree, but probably not exclusively, a response to changes in environmental temperature [104,105]. Stimulation of TRH secretion in the hypothalamus and of TSH release in response to cold has been reported in human subjects [106–108] leading to an increase in thyroid hormone production [109].

Seasonal variations in TSH in euthyroid subjects were reported by numerous investigators [59,89,99,102,110–112]. The seasonal peak in plasma TSH concentration, however, was quite variable in different studies. The peak was observed Halberg et al. [113] in adult women studied in Minnesota, United States, during winter; and by Lagoguey and Reinberg [112]; in young adult men in France during the spring by Nicolau et al. [59,102] and by Haus et al. [98]; in elderly women and in children with and without endemic goiter in Romania during summer and fall. In the Romanian children the circannual acrophase in TSH did precede the circannual acrophase of the thyroid hormones, suggesting that the seasonal variation of the latter is not a direct result of an increase in plasma TSH concentrations. The circannual TSH peak found in France in spring [111–112] and in Romania during the early fall [59,98,102] also makes a temperature dependence of the circannual rhythm of TSH unlikely.

Most investigators report (irrespective of the yearly temperature average) a seasonal variation in serum or plasma $TT_3$ and $TT_4$ concentrations inversely related with the seasonally changing environmental temperature [98,99,102,114–116]. Studies in a subtropical climate with little or no seasonal temperature variations failed to demonstrate a seasonal variation in circulating $TT_3$ and $TT_4$ concentrations [110,117].

In children [59], the circannual variations in thyroid parameters in moderate climates were highly reproducible. In contrast, elderly subjects at the same geographic location showed no seasonal variation in $TT_4$ but did show a circannual rhythm in TSH with the highest values found during summer and fall [68,98,102,118]. $TT_3$ showed in the elderly a seasonal variation with a peak during winter [98]. The apparent absence of a group synchronized seasonal variation in some thyroid hormones raises the question of a defect in cold adaptation in the elderly, but phenomena like desynchronization of a circannual rhythm within the group or a free-running circannual rhythm cannot be excluded.

In conclusion the hypothalamus-pituitary-thyroid axis shows an intricate multi-frequency time structure, which in clinically healthy subjects is maintained into very old age.

Changes in the HPT axis can occur as consequence of environmental conditions (e.g. iodine deficiency). Over a lifetime developing extra-thyroidal and intra-thyroidal disease (e.g., autoimmune thyroiditis, multinodular goiter with or without autonomy, autoimmune thyroidal disease, etc.) make recognition of changes due to the physiology of normal ageing difficult.

Many observations suggest in the aged a decreased hypothalamic stimulation of thyrotro-

174

pin release. In addition, a delayed and/or decreased response of the thyroid gland may lead to changes in timing of thyroid hormone rhythms and/or a compensating increase in TSH secretions, especially at the time of the trough of the circadian TSH rhythm.

A decreased or absent seasonal variation in HPT axis may indicate absence or alteration of a circannual rhythm or a defect of cold-adaptation in the elderly.

## 5.    PROLACTIN (PRL)

Prolactin plasma concentrations show pulsatile variations superimposed upon ultradian rhythms of an about 8-hour frequency and upon a circadian oscillation. The prolactin 24-hour profile reflects both tonic and intermittent hormone release [119]. The normal secretory pattern of PRL consists of a series of daily pulses, occurring every 2–3 hours, which vary in amplitude. The bulk of the hormone is secreted during rapid eye movement (REM) sleep. In diurnally active human subjects, REM-sleep occurs predominantly during the latter half of the daily sleep phase, so that the highest prolactin plasma concentrations usually occur during the night [120]. In men and non-parous women, REM-sleep is the dominant organizer of prolactin secretion. The mechanism is not known at this time, but it has been shown that conversely PRL infusion increases REM activity in the electroencephalogram [121,122]. In lactating women, the reflex elevation of PRL and oxytocin by nipple stimulation during nursing becomes the predominant controller of circulating prolactin concentrations. Differences in response to nursing as function of circadian stage have thus far not been investigated.

Sleep onset is associated with an increase in PRL secretion also during daytime naps irrespective of the time of the day, but the amplitude of the prolactin rise is usually less than when associated with nocturnal sleep. Conversely modest elevations in prolactin concentration may occur at the time of the usual sleep onset even when the patient remains awake. Thus, PRL plasma concentrations appear to be regulated by a circadian rhythm and the superimposed pulsatile secretions modulated by the sleep-wakefulness pattern with maximal secretion when sleep and circadian rhythmicity are in phase [56,123]. Shallow and fragmented sleep, prolonged awakening and interrupted sleep patterns as frequently seen in the elderly, are associated with a dampening of the nocturnal prolactin rise with decreased amplitude of the nocturnal prolactin pulses [93,124] and decreased prolactin concentrations [125].

PRL secretion in man is normally restrained by the action of dopamine, which is secreted from the hypothalamus. PRL is the only pituitary hormone that is secreted at unrestrained high levels when completely isolated from any trophic influences of the hypothalamus. However, a variety of stimulatory PRL secretagogues have been identified including steroids (estrogen), hypothalamic peptides (thyrotrophin releasing hormone [TRH], VIP and oxytocin) and growth factors such as EGF and FGF-2. Assays of hormone release from single cells have revealed substantial variations in function between cells and marked temporal variations in single cell responses [126]. Numerous medications used in everyday medical practice in the elderly, elevate PRL secretion and can mask the physiologic rhythmicity and occasionally may even lead to symptomatic hyperprolactinemia. These include commonly used antidepressants, antiemetics and narcotics which antagonize dopamine action or elevate serotonin or endorphin bioactivity [127]. Hypnotics like benzodiazepines (e.g. triazolam) and imidazopyridines (e.g. zolpidem) taken at bedtime concordant with the tendency of the daily prolactin rise may lead to substantial raises of serum PRL concentrations leading into the range regarded as abnormal [128,129]. Endogenous estrogens play a role in the differential regulation of prolactin in relation to age and sex. Mean

PRL concentrations, pulse amplitude and pulse frequency are all higher in normally cycling young women than either in postmenopausal women or in men [130]. Also, seasonal variations of serum prolactin concentrations have been found in women but not in men [18,131,132]. Ethnic-geographic differences in the circadian and the circannual amplitude of plasma prolactin concentrations were reported with, e.g., Japanese women showing higher amplitudes in both frequencies as compared to Americans [132].

During pregnancy, serum prolactin concentrations rise but the 24 hour secretion pattern is maintained. Postpartum, the PRL pulses follow suckling episodes and the nocturnal rise becomes manifest only after breastfeeding has ended [133]. In subjects with insulin-dependent diabetes, the circadian and sleep modulation of PRL secretion is preserved, but overall levels are markedly diminished [134,135]. Since prolactin secretion is part of the physiologic response to stress both physically and psychologically stressful events can cause serum PRL to rise [136].

Abnormal 24 hr. prolactin profiles with blunting of the nocturnal rise and inadequate suppression of prolactin concentrations after L-DOPA administration were found in women with narcolepsy and with narcolepsy and sleep apnea [137]. Blunting of the nocturnal rise is not specific and is found also in other conditions, including patients with breast cancer [138].

Circadian rhythmicity of prolactin continues in the elderly [38,98,131,139] with both unchanged and decreased serum concentrations and unchanged or a decreased circadian rhythm amplitude reported [18,131]. The plasma prolactin concentrations in elderly men were reported to be low [134], although in men during the 8th and 9th decade, the MESOR increased but rhythm detection by cosinor was lost during the 9th decade [18].

## 5.1. Hyperprolactinemia

Hyperprolactinemia is a frequent endocrine syndrome, which can occur at any age but is seen most frequently in middle aged and elderly subjects. Chronic use of neuroactive drugs may cause hyperprolactinemia. Any drug affecting dopamine synthesis and release or action at the lactrotroph receptor or post-receptor levels may cause hyperprolactinemia (e.g. antipsychotic drugs such as chlorpromazine, perphenazine, haloperidol or anti-emetics and gastric motility regulators such as metoclopramide and domperidone, etc. increase serum PRL concentrations through dopamine receptor blockage [140].

Hyperprolactinemia caused by hormone secreting tumors is the most common human pituitary disease. In hyperprolactinemia associated with pituitary prolactinomas the regularity of the pulse pattern is decreased and the number of pulses is increased but the nocturnal elevation of prolactin concentration may be preserved [141,142]. In Cushing's disease PRL levels are elevated throughout the 24 hr. cycle while the relative amplitude is reduced [143,144].

## 5.2. Prolactin and the immune system

The effects of prolactin in the human body are many fold. Of importance in the aged is its immune regulatory role. Prolactin receptors are found on a majority of immune-precursor and effector cells in each of the major hematopoietic and lymphopoietic organs (bone marrow, spleen, thymus). However, the action of prolactin upon the immune system is complex and depends upon the stage of both the circadian timing of the prolactin rhythm and its time relations to the circadian rhythms of immune related functions in the target organs [145].

In animal experiments prolactin restores immune competence in hypophysectomized animals [146]. Inhibition of prolactin secretion by bromocriptine results in immunosuppression [147–

149]. Prolactin antagonizes the immunosuppressive effects of glucocorticoids [150]. While lowered prolactin leads to immunodeficiency and exogenous prolactin, short-term experiments produces immunoenhancement persistently elevated prolactin levels due to a variety of conditions are associated with immunosuppression [151–158].

Some of the discordant results on the effects of prolactin on the immune system may be due to the pronounced circadian variation in its immune regulatory action [145] and the marked time dependent difference of immuno cellular responses. In the male Balb/c mouse the immunostimulatory activity of PRL was restricted to only an 8 hr. daily interval from 4–12 hrs. after light on (HALO) in animals kept on a 12 hr. light-dark regimen. PRL administration outside this sensitive interval was occasionally associated with immunosuppressive effects both in the one way mixed lymphocyte reaction (MLR) and in hapten-specific delayed-type hypersensitivity (DTH) responses.

Reducing endogenous levels of prolactin with bromocriptine inhibited immune functions only when the drug was administered during this daily interval of immunoregulatory sensitivity to the hormone [145]. This observation is similar to that of Bernton et al [150] who found that the effect of the prolactin inhibitor Cysteamine (a dopamine beta-hydroxylase inhibitor) on splenocyte mitogenic response was circadian time dependent.

Prolactin effects on immune activity and immunologic disorders were reported in human subjects including lupus erythematosus and the post-partum exacerbation of rheumatoid arthritis [146,151–153,156]. A chronobiologic explanation of the interaction of the rhythms in prolactin secretion and in target cell responsiveness has not been reported in human subjects. It appears that in the immunoregulatory action of prolactin the overall level of plasma PRL is of less importance than the circadian rhythms of PRL and that of the apparently circadian periodic responses of the immunocompetent target cell systems. The phase relation between the circadian rhythm in PRL and that of the rhythm in immunocellular response may be the determining factor for the PRL effect upon the immune system. Circadian rhythm disruption or phase shifts of either of these rhythms may be associated with immunologic dysfunction, which may be of interest for shiftwork, transmeridian flights, circadian rhythm and sleep disturbances in the elderly.

## 6.    CATECHOLAMINES

### 6.1.    Central and tissue catecholamines

The central and peripheral sympathetic nervous system and its humoral messengers show rhythmic variations in various frequencies, which are especially prominent in the circadian range and undergo characteristic changes with ageing (for review see [159]). In the hypothalamus, noradrenaline (NE) concentrations decline with increasing age [160] with a reduced noradrenaline turnover [161]. The age related decline in noradrenergic activity influences the functions of the gonadotropic, somatotropic and thyrotropic axes [162]. The changes in central noradrenergic and dopaminergic regulations may cause changes in the noradrenergic stimulated luteinizing hormone releasing hormone release (LHRH) [163] and the tonic inhibition of prolactin secretion by dopamine (DA) [164]. Since prolactin in turn stimulates DA secretion from dopaminergic neurons the autoregulatory feedback control of prolactin and possibly of TSH appears to be altered in the aged animals through a dopaminergic mechanism [165]. The tuberoinfundibular system shows marked decreases in dopamine concentrations with age in the medial-basal hypothalamus and the median eminence [166]. There is a decrease of about 40% in the number of dopaminergic cells in the substantia nigra up to age 60 [167]. Dopamine receptors in the rat striatum are lost concomitant with an impaired ability to adapt to various stimuli. This loss can be substantially retarded by dietary restriction together with an approximately 40% increased survival time [168]. The higher circadian mean serotonin turnover in the corpus striatum (34% increase) and lower circadian mean of dopamine turnover (69% decrease) in aged rats as compared to their young counterparts could be related to some of the changes in motor function in the aged [169].

The levels and the circadian rhythms of hypothalamic striatal and hypophyseal catecholamines are age-dependent. In young as well as in aged male Wistar rats a circadian rhythm was found in hypothalamic and striatal norepinephrine (NE) content and in hypothalamic DA turnover with acrophase (by 24 hr. cosinor) during the second half of the activity span or the first half of the rest span (under an LD12:12 lighting regimen) [170]. However, the old rats had a lower NE content in the anterior and medial hypothalamus, a smaller circadian amplitude in posterior hypothalamic NE content, a lower DA turnover in the medial and posterior hypothalamus, an increased adeno-hypophysial DA but a decreased NE and DA content in the pituitary neurointermediate lobe [170]. Immunization of the animals with Freund's adjuvant 18 days before study led to circadian rhythm alterations with a difference between aged and young animals. Circadian rhythms in the number of alpha- and beta-receptors were found in the rat forebrain and hypothalamus, which differ in their wave form and timing (acrophase) of their peak. These rhythms persist in the absence of external time cues indicating their endogenous nature but do change in curve form and timing during different seasons of the year suggesting in addition to the circadian, a circannual rhythm [171].

There is a circadian rhythm in norepinephrine (NE) content of the pineal gland in young and in old animals. The old rats had a lower amplitude and lower mean values of their circadian variations in pineal NE content which was only partially restored by melatonin treatment [172]. Circadian variations in cerebrospinal fluid noradrenalin concentrations were found in monkeys and with a 3-point spot check in human subjects [173]. In peripheral tissues of rodents, there is an age-dependent decrease of tissue catecholamine levels [174–175]. Differences in NE metabolism in aged rats were found in the heart [176] and in peripheral tissues [177]. Bruzzone et al [178] found an age-related decrease in sympathetic nerve fibers in rat heart and coronary arteries together with a marked decrease in norepinephrine levels in tissue homogenates of rat heart and coronary vessels. The response of human lymphocyte-adenylate-cyclase to isoproterenol, a beta-adrenergic catecholamine, is decreased in the elderly [179]. The widely demonstrated reduced catecholamine responsiveness in the elderly is thought to be due to a defect in the peripheral beta-receptor linked adenyl-cyclase complex [180].

Brusco et al, [181] studied the effect of age on the circadian rhythms in autonomic nervous system activity by measuring tryosine hydroxylase activity and $^3$H-choline conversion into $^3$H-acetylcholine in 50 day and 18 month old male Sprague-Dawley rats. Circadian rhythms were found in both variables in autonomic ganglia, adrenal medulla and heart muscle in the young animals with peak of tyrosine hydroxylase activity during the daily activity span of the animals except in contrast to cardiac acetylcholine synthesis, which peaked in the second half of the rest (light) span. In acetylcholine activity in the hypogastric ganglion, there were two maxima in the young animals (suggesting a 12 hour rhythm) while there was only one peak in the aged. The old rats exhibited a significant decrease of rhythm amplitude and elevation of the circadian mean in tyrosine hydroxylase activity in autonomic ganglia and adrenal medulla and abolition of the tyrosine hydroxylase activity in the heart. Circadian rhythms in acetylcholine synthesis were impaired or abolished in aged rats. Melatonin in low and high doses given before onset of the daily light span re-established or augmented the amplitude of the rhythms in the aged rats and in high doses also increased the amplitude in the young rats. The timing (acrophase) of these rhythms of biochemical indicators of norepinephrine and acetylcholine synthesis are relatively unmodified by the ageing process.

In the course of sympathetic activation NE is produced at the nerve endings. Although most of it is taken up again locally, some is released into circulation where it represents an indicator of sympathetic activity. Using isotope dilution methods to study catecholamine release to plasma in humans, Esler et al, [182] found with ageing sympathetic activation in the heart, the gut and the liver at rest. Sympathetic nervous responses to stress were augmented in the elderly. Conversely, adrenal medullary release of epinephrine was subnormal at rest and during stress. Circadian variations in human lymphocyte adrenoceptor density were reported, with peak values around noon and a trough around midnight [183]. Ageing leads to an impairment of beta-adrenoceptor density and/or activity [180].

The adrenomedullary and sympathetic system is assumed to reach maturity around the 5th year of life [184]. A progressive increase in catecholamine excretion was reported in children from 1 month to 16 years of age [185], which seems to be proportional to the body surface area [186,187]. Diurnally active clinically healthy human subjects of both sexes only moderately restricted in their activity by the sampling procedures show a circadian rhythm both of plasma epinephrine (E) and of plasma NE with a peak in the late morning and low levels during sleep [188]. This circadian variation consists of a low amplitude circadian rhythm with superimposed episodic fluctuations. With a 20 minute sampling interval 8 to 9 pulses per 24 hour span can be detected. The magnitude of the pulses is greater during the day and larger for epinephrine than for norepinephrine [189]. Further superimposed upon this low amplitude circadian rhythm at rest are the reactive changes due to the subjects' diurnal activity pattern [190,191].

Sleep reversal studies indicate a predominant dependence of the circadian variation of norepinephrine upon sleep-wakefulness and posture [192,193]. In contrast, the circadian rhythm in epinephrine persists for at least 75 hours in sleep deprived subjects kept under as far as feasible constant conditions [192]. No relation of plasma catecholamine concentrations to sleep stages was found [188].

Also the circulating sulfo-conjugated catecholamines show a circadian variation, which shows a marked phase difference from that of the free compounds [191]. Norepinephrine plasma concentrations were greater in elderly as compared with young subjects [194] by 28% during the day (11:00 hr) and by 75% during the night (22:00 – 09:00 hrs) [188].

The changes in catecholamine secretion and metabolism during the ageing process are complex and seem to involve their secretion [195], blood levels [196,197], receptors [198–201], the peripheral tissue content [177], metabolism [202], and possibly clearance [195,203]. The urinary excretion of NE increases from 1 to 59 years of age [204], and then decreases during the 7th to 10th decades of life. Descovich et al, [205] found in elderly subjects (66–99 years of age) no difference in the timing of circadian rhythms in catecholamine excretion and no decrease in the MESOR and amplitude of NE and E. Nicolau et al [206] compared the circadian rhythm and the urinary excretion of free epinephrine, norepinephrine and dopamine between elderly men and women (77 ± 8 years of age) and boys and girls (11 ± 1.5 years of age). The circadian MESOR and the amplitude in E and DA, and to a lesser degree in NE, were higher in the children than in the elderly subjects. The acrophase of catecholamine excretion remained unchanged in spite of a 4 1/2 hour phase shift of the acrophase of the urine excretion in the same elderly subjects into the late night hours [98,207]. The excretion of DA follows in its timing more closely that of the urine volume. In contrast, Lehmann and Keul [208] found in healthy adults norepinephrine excretion to correlate positively with age. The lower excretion of norepinephrine in the elderly found by Nicolau et al. [206] contrasts with reports on higher concentrations of plasma norepinephrine concentrations in the aged [196,209]. An alteration in plasma norepinephrine concentrations during ageing could be the result of changes at several sites. There is evidence of increased norepinephrine production and release in the aged [210] both in supine and standing positions [211] and an age related increased plasma catecholamine rise during exercise [212]. The rate of clearance was found to be the same in young adults and elderly subjects [195]. In contrast, Esler et al [203] suggested a decrease in norepinephrine clearance with ageing and Itskovitz and Wynn [213] reported a large reduction in the excretion of free catecholamines and total dopamine as renal function decreased in the elderly.

The circadian variation in blood pressure parallels that of plasma norepinephrine concentrations. The nocturnal fall in blood pressure was found to be related to a decrease in plasma NE concentrations [214]. This decrease could be abolished in hypertensive patients taking bromocriptine suggesting there is a dopaminergic mechanisms involved in its control [215]. The relation of plasma norepinephrine concentrations to essential hypertension are controversial. Most investigators found no major difference between the age adjusted plasma norepinephrine concentrations in normotensive and hypertensive subjects [216–218]. The urinary excretion of catecholamines in hypertensive individuals was found to be slightly higher during the first half and significantly higher during the second half of life expectancy [208].

Of interest for the circadian periodicity of catecholamines in regards to blood pressure regulation and ageing is the interaction with melatonin. Melatonin blunts noradrenergic activation [219] and decreases catecholamine concentrations and blood pressure [220]. Endogenous melatonin concentrations decrease with ageing, which may contribute to the decrease in extent and percentage of elderly subjects showing nocturnal dipping of their blood pressure.

Children showed a circannual rhythm in systolic and diastolic blood pressure and in norepinephrine excretion with peaks during winter [206]. In elderly subjects, a circannual rhythm in urinary norepinephrine excretion was statistically significant only in women, also with a peak during winter, but was out of phase with the circannual rhythm in blood pressure [221–223]. Absence of a circannual rhythm as a group phenomenon does not necessarily imply absence of such a rhythm in the individual but may be the consequence of a desynchronization within the group. Circannual variations in human lymphocyte adrenoceptor density were found with high values in spring and summer and low values during winter [183,224].

The ubiquitous role of the sympathetic nervous system and its messenger substances in the human body determines the importance and the great variety of physiologic functions shaped or modulated by the multi-frequency rhythms encountered at all levels of the system. The basic rhythmic variations are mostly endogenous in nature but in some variables are greatly modified (masked) by environmental factors and by rhythms of response at the level of the target organs. Ageing changes are found at all levels of the system and are mainly expressed in changes of mean and amplitude rather than the timing of the rhythms concerned.

## 7.   CONCLUSION

Age related changes in the human endocrine time structure occur in different frequency ranges including pulsatile secretions [225], ultradian, circadian and circannual rhythms [7,98,226–227]. Some of these changes are part of the physiologic ageing process, but in some instances may lead to functional impairment, and if exaggerated to senility. A decline in neuroendocrine function has been implicated in the development of several age related diseases.

In clinically healthy, diurnally active human subjects, circadian rhythmicity in many variables may be well maintained into very old age. However, such observations found in transversally studied groups of subjects of different ages are not necessarily representative for the dynamic process of ageing. Longitudinal studies, which are difficult in human subjects, could more correctly represent this process. In comparing subjects transversally of different ages, the upper age groups are biased groups of survivors and do not represent those with genetically determined and/or environmentally induced earlier occurring defects who having experienced an accelerated ageing process have succumbed to a variety of diseases for which advancing age represents a risk factor.

In the circadian range, the most prominent change found in the course of ageing is the reduction in the circadian amplitude, which was found in most variables discussed in this presentation. It may play an important part in the ageing process and related changes. A decrease in amplitude of a physiological variable may be an expression of a functional decline. This may be the case, e.g., in the decrease in the nightly surge in melatonin leading to a decrease in time information provided to peripheral oscillating structures with an ensuring tendency for external and internal desynchronization.

In the circannual range, the disappearance of circannual rhythms or seasonal variations as group phenomenon in elderly subjects may represent a lack of adaptation to the season-dependent environmental stimuli or lack of synchronization of circannual rhythms by these stimuli, both of which may represent adaptive defects which may, e.g., contribute to the high mortality observed among the aged during the winter months.

Alterations in biologic rhythms with internal and external desynchronization and loss of the capability of the organism to maintain under the conditions of daily life a time structure conducive to optimal function may contribute to the dysfunction of the elderly organism and the development of senility. Although the ageing process as such may not be based upon changes in the neuroendocrine system, correction of some of these changes with maintenance or re-establishment of the neuroendocrine time structure may have beneficial effects upon the function of the ageing organism and may contribute to the quality of life during healthy ageing and help to prevent premature senility and other disorders seen in the aged.

## REFERENCES

1.  Touitou Y, Haus E, editors. Biologic Rhythms in Clinical and Laboratory Medicine. New York: Springer Verlag; 1994; pp 730.
2.  Redfern PH, Lemmer Bj, editors. Physiology and Pharmacology of Biologic Rhythms. Berlin, New York: Springer; 1997; pp 668.
3.  Reppert SM, Weaver DR. Molecular analysis of mammalian circadian rhythms. Ann Rev Physiol 2001;63:647–676.
4.  Basalobre A, Brown SA, Marcacci L, Trouche F, Kellendock C, Reichart HM, Schultz G, Schibler V. Resetting of circadian time in peripheral tissues by glucocorticoid signaling. Science 2000;289:2344–2347.
5.  Oster H, Baeriswyl S, Van Der Horst GT, Albrecht U. Loss of circadian rhythmicity in aging mPer 1 -/- mCry 2 -/- mutant mice. Genes Dev 2003;17:1366–1379.
6.  Yamazaki S, Straume M, Tei H, Sakaki Y, Menaker M, Block GD. Effects of aging on central and peripheral mammalian clocks. Proc Natl Acad Sci USA 2002;99:10801–

10806.

7.    Touitou Y, Haus E. Alteration with aging of the endocrine and neuroendocrine circadian system in humans. Chronobiol Int 2000;17:369–390.

8.    Ferrari E, Cravello L, Muzzoni B, Casarotti D, Paltro M, Solerte SB, Fioravanti M, Cuzzoni G, Pontiggia B, Magri F. Age-related changes of the hypothalamic-pituitary-adrenal axis: pathophysiological correlates. Eur J Endocrinol 2001;144:319–29.

9.    Gooren LJG. Endocrine aspects of ageing in the male. Mol Cell Endocrinol 1998;145: 153–159.

10.   Laughlin GA, Barrett-Connor, E. Sexual dimorphism in the influence of advanced aging on adrenal hormone levels: the Rancho Bernardo Study. J Clin Endocrinol Metab 2000;85: 3561–3568.

11.   Dennison E, Hindmarsh P, Fall C, Kellingray S, Barker D, Phillips D, Cooper C. Profiles of endogenous circulating cortisol and bone mineral density in healthy elderly men. J Clin Endocrinol Metab 1999;84:3058–3063.

12.   Ferrari E, Casarotti D, Muzzoni B, Alberteli N, Cravello L, Fioravanti M, Solerte SB, Magri F. Age-related changes of the adrenal secretory pattern: possible role in pathological brain aging. Brain Res Review 2001;37:294–300.

13.   Touitou Y, Sulon J, Bogdan A, Touitou C, Reinberg A, Beck H, Sodoyez JC, Demey-Ponsart E, Van Cauwenberge H. Adrenal circadian system in young and elderly human subjects: a comparative study. J Endocrinol 1982;93:201–210.

14.   Ferrari E, Magri F, Dori D, Migliorati G, Nescis T, Molla G, Fioravanti M, Solerte SB. Neuroendocrine correlates of the aging brain in humans. Neuroendocrinology 1995;61:464–470.

15.   Van Cauter E, Leproult R, Kupfer DJ. Effects of gender and age on the levels and circadian rhythmicity of plasma cortisol. J Clin Endocrinol Metab 1996;81:2468–2473.

16.   Touitou Y, Sulon J, Bogdan A, Reinberg A, Sodoyez JC, Demey-Ponsart E. Adrenocortical hormones, ageing and mental condition: seasonal and circadian rhythms of plasma 18-hydroxy-11- deoxycorticosterone, total and free cortisol and urinary corticosteroids. J Endocrinol 1983;96:53–64.

17.   Sherman X, Wysham C, Pfohl B. Age-related changes in the circadian rhythm of plasma cortisol in man. J Clin Endocrinol Metab 1985;61:439–443.

18.   Haus E, Nicolau G, Lakatua DJ, Sackett-Lundeen L, Petrescu E. Circadian rhythm parameters of endocrine functions in elderly subjects during the seventh to the ninth decade of life. Chronobiologia 1989;16:331–352.

19.   Parker L, Gral T, Perrigo V, Skowksy R. Decreased adrenal androgen sensitivity to ACTH during aging. Metabolism 1981;30:601–604.

20.   Horan MA. Aging, injury and the hypothalamic-pituitary-adrenal axis. Ann NY Acad Sci 1994;719:285–290.

21.   Zhao ZY, Xie Y, Fue YR, Bogdan A, Touitou Y. Aging and the circadian rhythm of melatonin: a cross-sectional study of Chinese subjects 30–110 yr of age. Chronobiol Int 2002;19:1171–1182.

22.   Zhao ZY, Xie Y, Fue YR, Bogdan A, Touitou Y. Circadian rhythms characteristics of serum cortisol and dehydroepiandrosterone sulfate in healthy Chinese men aged 30 to 60 years. A cross-sectional study. Steroids 2003;68:133–138.

23.   Zhao ZY, Lu FH, Xie Y, Fu YR, Bogdan A, Touitou Y. Cortisol secretion in the elderly. Influence of age, sex and cardiovascular disease in a Chinese population. Steroids 2003,

182

sous presse

24.    Lupien S, Roch Lecours A, Schwartz G, Sharma S, Hauger RL, Meaney MJ, Nair NPV. Longitudinal study of basal cortisol levels in healthy elderly subjects: evidence for subgroups. Neurobiol Aging 1996;17:95–105.

25.    Kawasaki T, Ueno M, Uezono K, Matsuoka M, Omae T, Halberg F, Wendt H, Taggett-Anderson MA, Haus E. Differences and similarities among circadian characteristics of plasma renin activity in healthy young women in Japan and the United States. Am J Med 1980;68:91–96.

26.    Woo J, Ho SC, Donnan S, Swaminathan R. Nutritional correlates of blood pressure in elderly Chinese. J Hum Hypertens 1988;1:287–291.

27.    Leung SY, Ng TH, Yuen ST, Lauder IJ, Ho FC. Patterns of cerebral atherosclerosis in Hong Kong Chinese. Severity in intracranial and extracranial vessels. Stroke 1993;24: 779–786.

28.    Woo J, Swaminathan R, Pang CR, MacDonald D, Mak YT, Ho SC, Lau E. Some biochemical indices of bone turnover in elderly Chinese. J Med 1989;20:229–239.

29.    Touitou Y, Bogdan A, Lévi F, Benavides M, Auzéby A. Disruption of the circadian patterns of serum cortisol in breast and ovarian cancer patients: relationships with tumour marker antigens. Br J Cancer 1996;74:1248–1252.

30.    Ferrari E, Arcani A, Gornati R, Pelanconi L, Cravello L, Fioravanti M, Solerte SB, Magri F. Pineal and pituitary-adrenocortical function in physiological aging and in senile dementia. Exp Gerontol 2000;35:1239–1250.

31.    Krieger DT. Rhythms in CRF, ACTH, and Corticosteroids. In: Krieger DT, editor. Endocrine Rhythms. New York: Raven Press; 1979;123–142.

32.    Selmaoui B, Lambrozo J, Touitou Y. Endocrine functions in young men exposed for one night to a 50-Hz magnetic field. A circadian study of pituitary, thyroid and adrenocortical hormones. Life Sci 1997;61:473–486.

33.    Berr C, Lafont S, Debuire B, Dartigues J-F, Baulieu EE. Relationships of dehydroepiandrosterone sulfate in the elderly with functional, psychological, and mental status, and short-term mortality: a french community-based study. Proc Natl Acad Sci USA 1996;93:13410–13415.

34.    Mazat L, Lafont S, Berr C, Debuire B, Tessier J-F, Dartigues J-F, Baulieu EE. Prospective measurements of dehydroepiandrosterone sulfate in a cohort of elderly subjects: relationship to gender, subjective health, smoking habits, and 10-year mortality. Proc Natl Acad Sci USA 2001;98: 8145–8150.

35.    Thomas G, Frenoy N, Legrain S, Sebag-Lanoe R, Baulieu EE, Debuire B. Serum dehydroepiandrosterone sulfate levels as an individual marker. J Clin Endocrinol Metab 1994;79: 1273–1276.

36.    Vanderkar LD, Brownfield MS. Serotonergic neurons and neuroendocrine function. News Physiol Sci 1993;8:202.

37.    Ebadi M, Samejima M, Pfeiffer RF. Pineal gland in synchronizing and refining physiological events. News Physiol Sci 1993;8:30.

38.    Touitou Y, Fèvre M, Lagoguey M, Carayon, A, Bogdan A, Reinberg A, Beck H, Cesselin F, Touitou C. Age and mental health-related circadian rhythm of plasma levels of melatonin, prolactin, luteinizing hormone and follicle-stimulating hormone in man. J Endocrinol 1981;91:467–475.

39.    Touitou Y, Fèvre M, Bogdan A, Reinberg A, de Prins J, Beck H, Touitou C. Patterns of plasma melatonin with ageing and mental condition: stability of nyctohemeral rhythms

and differences in seasonal variation. Acta Endocrinol 1984;106:145–151.

40. Iguchi H, Kato KI, Ibayashi H. Age-dependent reduction in serum melatonin concentrations in healthy human subjects. J Clin Endocrinol Metab 1982;55:27–29.

41. Nair NPV, Hariharasubramanian N, Pilapil N, Isaac I, Thavundayil JX. Plasma melatonin, an index of brain aging in humans? Biol Psychiatry 1986;21:141–150.

42. Sack RL, Lewy AJ, Erb DE, Vollmer WM, Singer CM. Human melatonin production decreases with age. J Pineal Res 1986;3:379–388.

43. Zeitzer JM, Daniels JE, Duffy JF, Klerman EB, Shanahan TL, Dijk DJ, Czeisler CA. Do plasma melatonin concentrations decline with age? Am J Med 1999;107:432–436.

44. Kennaway DJ, Lushington K, Dawson D, Lack L, van-den-Heuvel C, Rogers NJ. Urinary 6-sulfatoxymelatonin excretion and aging: new results and a critical review of the literature. J Pineal Res 1999;27:210–220.

45. Touitou Y, Fèvre-Montange M, Proust J, Klinger E, Nakache JP. Age- and sex-associated modification of plasma melatonin concentration in man. Relationship to pathology, malignant or not, and autopsy findings. Acta Endocrinol 1985;108:135–144.

46. Wetterberg L, Halberg F, Tarquini B, Cagnoni M, Haus E, Griffith K, Kawasaki T, Wallach LA, Ueno M, Uezo K, Matsuoka M, Kuzel M, Halberg E, Omae T. Circadian variation in urinary melatonin in clinically healthy women in Japan and the United States of America. Experientia 1979; 35:416–419.

47. Wetterberg L, Halberg F, Halberg E, Haus E, Kawasaki T, Ueno M, Uezono K, Cornelissen G, Matsuoka M, Omae T. Circadian characteristics of urinary melatonin from clinically healthy young women at different civilization disease risks. Acta Med Scand 1986;220: 71–81.

48. Pieri C, Marra M, Moroni F, Recchioni R, Marcheselli F. Melatonin: a peroxyl radical scavenger more effective than vitamin E. Life Sci 1994;55:271–276.

49. Reiter RJ. The pineal gland and melatonin in relation to aging - A summary of the theories and of the data. Exp Gerontol 1995;30:199–212.

50. Reiter RJ. Oxygen radical detoxification processes during aging: The functional importance of melatonin. Aging Clin Exp Res 1995;7:340–351.

51. Berlin I, Touitou Y, Guillemant S, Danjou P, Puech AJ. Beta-adrenoceptor agonists do not stimulate daytime melatonin secretion in healthy subjects. A double blind placebo controlled study. Life Sci 1995; 56:325–331.

52. Leitolf H, Behrends J, Brabant G. The thyroid axis in ageing. In: Chadwick DJ, Goode J, editors. Novartis Symposium 242 Endocrine Facets of Ageing. New York: Wiley & Sons Ltd; 2002;193–204.

53. Behrends J, Prank K, Dogu E, Brabant G. Central nervous system control of thyrotropin secretion during sleep and wakefulness. Horm Res 1998;49:173–177.

54. Brabant G, Prank K, Ranft U, Schuermeyer T, Wagner TO, Hauser H, Kummer B, Feistner H, Hesch RD, von zur Mühlen A. Physiological regulation of circadian and pulsatile thyrotropin secretion in normal man and woman. J Clin Endocrinol Metab 1990;70: 403–409.

55. Brabant G, Prank K, Hoang Vu C, Hesch RD, von zur Mühlen A. Hypothalamic regulation of pulsatile thyrotropin secretion. J Clin Endocrinol Metab 1991;72:145–150.

56. Spiegel K, Leproult R, van Cauter E. Impact of sleep debt on metabolic and endocrine function. Lancet 1999;354:1435–1439.

57. Berghout A, Wiersinga WM, Smits NJ, Touber JL. Interrelationships between age, thyroid volume, thyroid modularity, thyroid function in patients with sporadic non-toxic goiter.

Am J Med 1990;89:602–608.

58. Gaitan E, Nelson NC, Poole GV. Endemic goiter and endemic goiter disorders. World J Surg 1991;15:205–215.

59. Nicolau GY, Dumitriu L, Plinga L, Petrescu E, Sackett-Lundeen L, Lakatua DJ, Haus E. Circadian and circannual variations of thyroid function in children 11±1.5 years of age with and without endemic goiter. In: Pauly JE, Scheving LE, editors. Advances in Chronobiology: Progress in Clinical and Biological Research. New York: Liss; 1987;227B: 229–247.

60. Hintze G, Burghardt U, Baumert J, Windeler J, Köbberling J. Prevalence of thyroid dysfunction in elderly subjects from the general population in an iodine deficiency area. Aging 1991;3:325–331.

61. Hintze G, Windeler J, Baumert J, Stein H, Köbberling J. Thyroid volume and goiter prevalence in the elderly as determined by ultrasound and their relationships to laboratory indices. Acta Endocrinol (Copenh) 1991;124:12–18.

62. Hampel R, Külberg T, Klein K, Jerichow JU, Pichmann EG, Clausen V, Schmidt I. Goiter incidence in Germany is greater than previously suspected. J Med Klin 1995;90:324–329.

63. Sawin CT, Chopra D, Azizi F, Mannix JE, Bacharach P. The aging thyroid: increased prevalence of elevated serum thyrotropin levels in the elderly. JAMA 1979;242:247–250.

64. Targum SD, Marshall LE, Magac-Harris K, Martin D. TRH tests in a healthy 64.
Targum SD, Marshall LE, Magac-Harris K, Martin D. TRH tests in a healthy elderly population. Demonstration of gender differences. J Am Geriatr Soc 1989;37:533–536.

65. Canaris GJ, Manowitz NR, Mayor G, Ridgway C. The Colorado Thyroid Disease Prevalence Study. Arch Int Med 2000;160:526–534.

66. Brochmann H, Bjoro T, Gaarder PI, Hanson F, Frey HM. Prevalence of thyroid dysfunction in elderly subjects: a randomized study in a Norwegian rural community (Naeroy). Acta Endocrinol (Copenh) 1988;117:7–12.

67. Metcalf G, Jackson IMD, editors. Thyrotropin releasing hormone: biomedical significance. Ann NY Acad Sci 1989;553:1–631.

68. Nicolau GY, Haus E. Chronobiology of the Hypothalamic-Pituitary-Thyroid Axis. In: Touitou Y, Haus E, editors. Biologic Rhythms in Clinical and Laboratory Medicine. New York: Springer Verlag; 1994;330–347.

69. Simard M, Pekary AE, Smith VP, Hershman JM. Thyroid hormones modulate thyrotropin-releasing hormone biosynthesis in tissues outside the hypothalamic-pituitary axis of male rats. Endocrinol 1989;125:529–531.

70. Martino E, Bambini G, Vaudagna G, Breccia M, Baschieri L. Effects of continuous light and dark exposure on hypothalamic thyrotropin-releasing hormone in rats. J Endocrinol Invest 1985;8:31–33.

71. Kerdelhue B, Palkovits M, Karteszi M, Reinberg A. Circadian variations in substance P, luliberin (LH-RH) and thyroliberin (TRH) contents in hypothalamic and extrahypothalamic brain nuclei of adult male rats. Brain Res 1981;206: 405–413.

72. Covarrubias L, Uribe RM, Mendes M, Charli JL, Joseph-Bravo P. Neuronal TRH synthesis: developmental and circadian TRH mRNA levels. Biochem Biophys Res Commun 1988;151:615–622.

73. Bennett GW, Marsden CA, Fone KCF, Johnson JV, Heal DJ. TRH-catecholamine interaction in brain and spinal cord. Ann NY Acad Sci 1989;533:106–120.

74. Hinkle PH. Pituitary TRH receptors. Ann NY Acad Sci 1989;553:176–187.

75. Hinkle RM, Perrone MH, Greer TL. Thyroid hormone action in pituitary cells. Differences

in the regulation of thyrotropin-releasing hormone receptors and growth hormone synthesis. J Biol Chem 1979;254:3907–3911.

76. Tashjian AH Jr, Osborne R, Maina D, Knaian A. Hydrocortisone increases the number of receptors for thyrotropin-releasing hormone on pituitary cells in culture. Biochem Biophys Res Commun 1977;79:333–340.

77. Gershengorn MC, Marcus-Samuels BE, Geras E. Estrogens increase the number of thyrotropin-releasing hormone receptors on mammotropic cells in culture. Endocrinol 1979;105:171–176.

78. Segerson TP, Kauer J, Wolfe HC, Mobtaker H, Wu P, Jackson IMD, Lechan RM. Thyroid hormone regulates TRH biosynthesis in the paraventricular nucleus of the rat hypothalamus. Science 1987;238:78–80.

79. Gurr JA, Kourides IA. Thyroid hormone regulation of thyrotropin alpha- and beta-subunit gene transcription. DNA 1985;4:301–307.

80. Samuels MH, Lillehei K, Kleinschmidt-Demasters BK, Stears J, Ridgway EC. Patterns of pulsatile pituitary glycoprotein secretion in central hypothyroidism and hypogonadism. J Clin Endocrinol Metab 1990;70:391–395.

81. Spencer CA, Greenstadt MA, Wheeler WS, Kletzky OA, Nicoloff JT. The influence of a long-term low dose thyrotropin-releasing hormone infusions of serum thyrotropin and prolactin concentrations in man. J Clin Endocrinol Metab 1980;51:771–775.

82. Romijn JA, Adriannse R, Brabant G, Prank K, Endert E, Wiersinger WM. Pulsatile secretion of thyrotropin during fasting: a decrease of thyrotropin pulse amplitude. J Clin Endocrinol Metab 1990;70:1631–1636.

83. Spencer CA, Lum SM, Wilber JF, Kaptein EM, Nicoloff JT. Dynamics of serum thyrotropin and thyroid hormone changes in fasting. J Clin Endocrinol Metab 1983;56:883–888.

84. Seoane LM, Carro E, Tovar S, Casanueva FF, Dieguez C. Regulation of in vivo TSH secretion by leptin. Regul Pept 2000;92:25–29.

85. Ordene KW, Pan C, Barzel US, Surks MI. Variable thyrotropin response to thyrotropin-releasing hormone after small decreases in plasma thyroid hormone concentrations in patients of advanced age. Metabolism 1983;32:881–888.

86. Barreca T, Franceschini R, Messina V, Boltano L, Rolandi E. 24-hour thyroid stimulating hormone secretory pattern in elderly men. Gerontology 1985;31:119–123.

87. Van Coevorden A, Laurent F, Decoster C, Kerkhofs M, Neve P, van Cauter E, Mockel J. Decreased basal and stimulated thyrotropin secretion in healthy elderly men. J Clin Endocrinol Metab 1989;69:177–185.

88. Greenspan SL, Kilbanski A, Schoenfeld D, Ridgway EC. Pulsatile secretion of thyrotropin in man. J Clin Endocrinol Metab 1986;63:661–668.

89. Greenspan SL, Klibanski A, Rowe JW, Elahi D. Age-related alterations in pulsatile secretion of TSH; role of dopaminergic regulation. Am J Physiol 1991;260:E486-E491.

90. Levy EG. Thyroid disease in the elderly. Med Clin North Am 1991;75:151–167.

91. Brabant G, Bergmann P, Kirsch CM, Köhrel J, Hesch RD, von Zur Mühlen A. Early adaptation of thyrotropin and thyroglobulin secretion to experimentally decreased iodine supply in man. Metabolism 1992;41:1093–1096.

92. Brabant G, Mayr B, Lucke C. Klinische and laborchemische Aspecte des latenten und manifesten Hyperthyreose in höheren Lebensalter. In: Schiddruse. Berlin: de Gruyter; 1995;S 214-S251.

93. Van Coevorden A, Mockel J, Laurent E, Kerkhofs M, L'Hermite-Baleriaux M, Decoster C, Neve P, van Cauter E. Neuroendocrine rhythms and sleep in aging men. Am J Physiol

186

1991;260:E651–661.

94. Chakraborti S, Chakraborti T, Mandal M, Das S, Batabyal SK. Hypothalamic-pituitary-thyroid axis status of humans during development of ageing process. Clinica Chimica Acta 1999;288:137–145.

95. Mariotti S, Barbesino G, Caturegli P, Bartalena L, Sansoni P, Fagnoni F, Monti D, Fagiolo U, Franceschi C, Pinchera A. Complex alteration of thyroid function in healthy centenarians. J Clin Endocrinol Metab 1993;77:1130–1134.

96. Mariotti S, Franceschi C, Cossarizza A, Pinchera A. The aging thyroid. Endocrinol Rev 1995;16:686–715.

97. Ravaglia G, Forti P, Maioli F, Nesi B, Pratelli L, Savarino L, Cucinotta D, Cavalli G. Blood micronutrient and thyroid hormone concentrations in the oldest-old. J Clin Endocrinol Metab 2000; 85:2260–2265.

98. Haus E, Nicolau G, Lakatua DJ, Sackett-Lundeen L. Reference values for chronopharmacology. Ann Rev Chronopharm 1988;4:333–424.

99. Halberg F, Cornelissen G, Sothern RB, Wallach LA, Halberg E, Ahlgren A, Kuzel M, Radke A, Barbosa J, Goetz F, Buckley J, Mandel J, Shuman L, Haus E, Lakatua D, Sackett L, Berg, H, Kawasaki T, Ueno M, Uezono K, Matsuoka M, Omae T, Tarquini B, Cagnoni M, Garcia Sainz M, Perez VE, Griffiths K, Wilson D, Donati L, Tatti P, Vasta M, Locatelli I, Camagna A, Lauro R, Tritsch G, Wetterberg L, Wendt HW. International geographic studies of oncological interest on chronobiologic variables. In: Kaiser HE, editor. Neoplasms comparative pathology of growth in animals, plants and man. Baltimore: Williams and Wilkins; 1981;553–596.

100. Weeke J, Christensen SE, Hansen AP, Laurberg P, Lundbaek K. Somatostatin and the 24-h levels of serum TSH, $T_3$, $T_4$ and reverse $T_3$ in normals, diabetics and patients treated for myxoedema. Acta Endocriinol (Copenh) 1980;94:30–37.

101. Nicolau GY, Haus E, Lakatua D, Bogdan C, Petrescu E, Sackett-Lundeen L, Berg H, Ioanitu D, Popescu M, Chioran C. Endocrine circadian time structure in the aged. Rev Roum Med Endocrinol 1982;20:165–176.

102. Nicolau GY, Lakatua DJ, Sackett-Lundeen L, Haus E. Circadian and circannual rhythms of hormonal variables in clinically healthy elderly men and women. Chronobiol Int 1984;1: 301–319.

103. Weeke J, Gundersen HJG. Circadian and 30 minute variations in serum TSH and thyroid hormones in normal subjects. Acta Endocrinol (Copenh) 1978;89:659–672.

104. Lungu A, Nicolau GY, Cocu F, Teodoru V, Dinu I. Protein-bound iodine variations and spontaneous atmospheric temperature oscillations. Rev Roum Endocrinol 1966;3:279–282.

105. Ogata K, Sasaki T, Murakami N. Central nervous and metabolic aspects of body temperature regulation. Bull Inst Const Med Kumamoto Univ 1966;16(Suppl):1–67.

106. Wilber JF, Baum D. Elevation of plasma TSH during surgical hypothermia. J Clin Endocrinol Metab 1970;31:372–375.

107. Tuomisto J, Mannisto P, Lamberg BA, Linnoila M. Effect of cold exposure on serum thyrotropin levels in man. Acta Endocrinol (Copenh) 1977;83:522–527.

108. Reed HL, Burman KD, Shakir KMM, O'Brian JT. Alterations in the hypothalamic-pituitary-thyroid axis after prolonged residence in Antarctica. Clin Endocrinol (Oxf) 1986;25:55–65.

109. Eastman CJ, Ekins RP, Leith IM, Williams ES. Thyroid hormone response to prolonged cold exposure in man. J Physiol 1974;24:175–181.

110. Halberg F, Lagoguey M, Reinberg A. Human circannual rhythms over a broad spectrum of physiological processes. Int J Chronobiol 1983;8:225–268.

111. Hugues JN, Reinberg A, Lagoguey M, Modigliani E, Sebaoun J. Les rythmes biologiques de la secretion thyréotrope. Ann Med Interne (Paris) 1983;134:83–94.

112. Lagoguey M, Reinberg A. Circadian and circannual changes of pituitary and other hormones in healthy human males: their relationship with gonadal activity. In: Van Cauter E, Copinschi G, editors. Human pituitary hormones: circadian and episodic variations. The Hague: Nijhoff; 1981;261–278.

113. Halberg F, Tarquini B, Lakatua D, Halberg E, Seal U, Haus E, Cagnoni M. Circadian and circannual plasma TSH rhythm, human mammary and prostatic cancer and step toward chrono-oncoprevention. Lab J Res Lab Med 1981;8:251–257.

114. Smals AGH, Ross HA, Kloppenborg PWC. Seasonal variation in serum $T_3$ and $T_4$ levels in man. J Clin Endocrinol Metab 1977;44:998–1001.

115. Perez PR, Lopez JG, Mateos IP, Escribano AD, Sanchez MLS. Seasonal variations in thyroid hormones in plasma. Rev Clin Esp 1980;156:245–247.

116. Oddie TH, Klein AH, Foley TP, Fisher DA. Variation in values for iodothyroinine hormones, thyrotropin and thyroxine-binding globulin in normal umbilical cord serum with season and duration of storage. Clin Chem 1979;25:1251–1253.

117. Rastogi RD, Sawhney RC. Thyroid function in changing weather in a subtropical region. Metabolism 1976;25:903–908.

118. Nicolau GY, Haus E. Chronobiology of the endocrine system. Endocrinol 1989;27:153–183.

119. Veldhhuis JD, Johnson ML, Lizarralde G, Iranmanesh A. Rhythmic and nonrhythmic modes of anterior pituitary secretion. Chronobiol Int 1992;9:371–379

120. Sassin JF, Frantz AG, Kapen S, Weitzman ED. The nocturnal rise of human prolactin is dependent on sleep. J Clin Endocrinol Metab 1973;37:436–440.

121. Obál F Jr, Payne L, Kacsoh B, Opp M, Kapas L, Grosvenor CE, Krueger JM. Involvement of prolactin in the REM sleep promoting activity of systemic vasoactive intestinal peptide (VIP). Brain Res 1994;645:143–149.

122. Rocky R, Obál F Jr, Valatx JL, Bredow S, Fang J, Pagano LP, Krueger JM. Prolactin and rapid eye movement sleep regulation. Sleep 1995;18:536–542.

123. Waldstreicher J, Duffy JF, Brown EN, Rogacz S, Allan JS, Czeisler CA. Gender differences in the temporal organization of prolactin (PRL) secretion: Evidence for a sleep-independent circadian rhythm of circulating PRL levels - A clinical research center study. J Clin Endocrinol Metab 1996;81:1483–1487.

124. Greenspan SL, Klibanski A, Rowe JW, Elahi D. Age alters pulsatile prolactin release: Influence of dopamine inhibition. Am J Physiol 1990;258:E799–804.

125. Spiegel K, Luthringer R, Follenius M, Schaltenbrand N, Macher JP, Muzet A, Brandenberger G. Temporal relationship between prolactin secretion and slow wave electroencephalic activity during sleep. Sleep 1995;18:543–548.

126. Castano JP, Kineman RD, Frawley LS. Dynamic fluctuations in the secretory activity of individual lactotropes as demonstrated by a modified sequential plague assay. Endocrinol 1994;135:1747–1752.

127. Ben-Jonathan N. Regulation of prolactin secretion. In: Imura H, editor. The Pituitary gland, Ed 2. New York: Raven Press; 1994; p 261.

128. Copinschi G, Van Onderbergen A, L'Hermite-Baleriaux M, Szyper M, Caufriez A, Bosson D, L'Hermite M, Robyn C, Turek FW, Van Cauter E. Effects of the short-acting

benzodiazepine triazolam, taken at bedtime, on circadian and sleep-related hormonal profiles in normal men. Sleep 1990;13:232–244.

129. Copinschi G, Akseki E, Moreno-Reyes R, Leproult R, L'Hermite-Baleriaux M, Caufriez A, Vertongen F, Van Cauter E. Effects of bedtime administration of Zolpidem on circadian and sleep-related hormonal profiles in normal women. Sleep 1995;18:417–424.

130. Katznelson L, Riskind PN, Saxe VC, Klibanski A. Prolactin pulsatile characteristic in postmenopausal women. J Clin Endocrinol Metab 1998;83:761–764.

131. Touitou Y, Carayon A, Reinberg A, Bogdan A, Beck H. Differences in the seasonal rhythmicity of plasma prolactin in elderly human subjects. Detection in women but not in men. J Endocrinol 1983;96:65–71.

132. Haus E, Lakatua DJ, Halberg F, Halberg E, Cornelissen G, Sackett LL, Berg HG, Kawasaki T, Ueno M, Uezono K, Matsouka M, Omae T. Chronobiological studies of plasma prolactin in women in Kyushu, Japan and Minnesota, USA. J Clin Endocrinol Metab 1980;51:632–640.

133. Tay CC, Glasier AF, McNeilly AS. Twenty-four hour pattern of prolactin secretion during lactation and the relationship to suckling and the resumption of fertility in breast-feeding women. Hum Reprod 1996;11:950–955.

134. Iranmanesh A, Veldhuis JD, Carlsen EC, Vaccaro VA, Booth RA Jr, Lizarralde G, Asplin CM, Evans WS. Attenuated pulsatile release of prolactin in men with insulin-dependent diabetes mellitus. J Clin Endocrinol Metab 1990;71:73–78.

135. Drejer JC, Hendriksen C, Nielsen LM, Binder C, Hagen C, Kehlet H. Diurnal variations in plasma prolactin, growth hormone, cortisol and blood glucose in labile diabetes mellitus. Clin Endorinol 1977;6:57–64.

136. Reavley S, Fisher AD, Owen D, Creed FH, Davis JR. Psychological distress in patients with hyperprolactinaemia. Clin Endocrinol 1997;47:343–348.

137. Clark RW, Schmidt HS, Malarkey WB. Disordered growth hormone and prolactin secretion in primary disorder of sleep. Neurology 1979;29:855–861.

138. Malarkey WB, Schroeder LL, Steven VC, James AG, Lanese RR. Disordered nocturnal prolactin regulation in women with breast cancer. Cancer Res 1977;37:4650–4654.

139. Lakatua DJ, Nicolau GY, Bogdan C, Petrescu E, Sackett-Lundeen L, Irvine PW, Haus E. Circadian endocrine time structure in humans above 80 years of age. J Gerontol 1984;39: 654–684.

140. Faglia G. Prolactinomas and hyperprolactinemia syndrome. In: DeGroot LJ, Jameson JL, editors. Endocrinology. New York: WB Saunders; 2001;329–342.

141. Boyar RM, Kapen S, Finkelstein JW, Perlow M, Sassin JF, Fukushima DK, Weitzman ED, Hellman L. Hypothalamic-pituitary function in diverse hyperprolactinemic states. J Clin Invest 1974;53:1588–1598.

142. Groote, Veldman R, Van den Berg G, Pincus SM. Increased episodic release and disorderliness of prolactin secretion in both micro- and macro-prolactinomas. Eur J Endocrinol 1999;140:192–200.

143. Caufriez A, Désir D, Szyper M, Robyn C, Copinschi G. Prolactin secretion in Cushing's disease. J Clin Endocrinol Metab 1981;53:843–846.

144. Krieger DT, Howanitz PJ, Frantz AG. Absence of nocturnal elevation of plasma prolactin concentrations in Cushing's disease. J Clin Endocrinol Metab 1976;42:260–272.

145. Cincotta AH, Kniseley TL, Landry RJ, Miers WR, Gutierrez PJA, Esperanza P, Meier AH. The immunoregulatory effects of prolactin in mice are time of day dependent. Endocrinology 1995;136(5):2163–2171.

146. Gala RR. Prolactin and growth hormone in the regulation of the immune system. Proc Soc Exp Biol Med 1991;198:513–527.

147. Berczi I. Immunoregulation by neuroendocrine factors. Dev Comp Immunol 1989;13: 329–341.

148. Bernton EW, Meltzer MS, Holaday JW. Suppression of macrophage activation and T-lymphocyte function in hypoprolactinemic mice. Science 1988;239:401–404

149. Hiestand PC, Mekler P, Nordmann R, Grieder A, Permmongkol C. Prolactin as a modulator of lymphocyte responsiveness provides possible mechanism of action for cyclosporine. Proc Natl Acad Sci USA 1986;83:2599–2603.

150. Bernton E, Bryant H, Holaday J, Jitendra D. Prolactin and prolactin secretagogues reverse immunosuppression in mice treated with cysteamine, glucocorticoids or cyclosporin-A. Brain Behav Immunol 1992;6:394–408.

151. Vidaller A, Liorente L, Larrea F, Mendez JP, Alcocer-Varela J, Alarcon-Segovia D. T-cell dysregulation in patients with hyperprolactinemia: effect of bromocriptine treatment. Clin Immunol Immunopathol 1986;38:337–343.

152. Gerli R, Riccardi C, Nicoletti I, Orlandi S, Cernetti C, Spinozzi F, Rambotti P. Phenotypic and functional abnormalities of T lymphocytes in pathological hyperprolactinemia. J Clin Immunol 1987;7:463–470.

153. Nicoletti I, Gerli R, Orlandi S, Migliorati G, Rambott P, Ricaardi C. Defective natural-killer cell activity in puerperal hyperprolactinemia. J Reprod Immunol 1989;15:113–121.

154. Karmali R, Lauder I, Horrobin DF. Prolactin and the immune response. Lancet 1974;2: 106–107.

155. Vidaller A, Guadarrama F, Llorente L, Mendez JP, Larrea F, Villa AR, Alarcon-Segovia D. Hyperprolactinemia inhibits natural killer (NK) cell function in vivo and its bromocriptine treatment not only corrects it but makes it more efficient. J Clin Immunol 1992;12:210–215.

156. Lavalle C, Loyo E, Paniagua R, Bermodez JA, Hererra J, Graef A, Gonzalez-Barcena D, Fraga A. Correlation study between prolactin and androgens in male patients with SLE. J Rheumatol 1987;14:268–272.

157. McMurray R, Keisler D, Kanuckel K, Izui S, Walker SE. Prolactin influences autoimmune disease activity in the female B/W mouse. J Immunol 1991;147:3780–3787.

158. Jungers P, Dougados M, Pelissier C, Kuttenn F, Tron F, Lesavre P, Bach J-F. Lupus nephropathy and pregnancy report of 104 cases in 36 patients. Arch Intern Med 1982;142: 771–776.

159. Rehman HU, Masson EA. Neuroendocrinology of ageing. Age and Ageing 2001;30: 279–287.

160. Dekosky ST, Palmer AM. Neurochemistry of aging. In: Albert M, Knoefel JE, Clinical Neurology of Aging. New York: Oxford University Press; 1994;79–101.

161. Joseph JA, Berger RE, Engle BT, Roth GS. Age-related changes in the nigrostriatum: a behavioural and biochemical analysis. J Gerontol 1978;33:643–649.

162. Meites J. Aging: hypothalamic catecholamines, neuroendocrine-immune interactions and dietary restriction. Pro Soc Exp Biol Med 1990;195:304–311.

163. Kalra SP, Kalra PS. Neural regulation of leutinizing hormone secretion in the rat. Endocrinol Rev 2001; 4:311–351.

164. Sider LH, Hucke Erica ETS, Florio JC, Felicio LF. Influence of time of day on hypothalamic monoaminergic activity in early pregnancy: effect of a previous reproductive experience. Psychoneuroendocrinology 2003;28:195–206.

165. Demarest KT, Riegle GD, Moore KE. Characteristics of dopaminergic neurons in the aged male rat. Neuroendocrinol 1980;31:222–227.

166. Demarest KT, Moore KE, Riegle GD. Responsiveness of tuberoinfundibular dopamine neurons in the aged female rat to the stimulatory actions of prolactin. Neuroendocrinology 1987; 45:227–232.

167. Kish SJ, Shannak K, Rajput A, Deck JH, Hornykiewicz O. Aging produces a specific pattern of striatal dopamine loss: implications for the etiology of idiopathic Parkinson's disease. J Neurochem 1992;58:642–648.

168. Levin P, Janda JK, Joseph JA, Ingram DK, Roth GS. Dietary restriction retards the age-associated loss of rat striatal dopaminergic receptors. Science 1981;214:561–562.

169. Esquifino AI, Cano P, Chacon F, Reyes Toso CF, Cardinali DP. Effect of aging on 24-hour changes in dopamine and serotonin turnover and amino acid and somatostatin contents of rat corpus striatum. Neurosignals 2002;11:336–344.

170. Cano P, Cardinali DP, Chacon F, Castrillon PO, Reyes Toso CA, Esquifino AI. Age-dependent changes in 24-hour rhythms of catecholamine content and turnover in hypothalamus, corpus striatum and pituitary gland of rats injected with Freund's adjuvant. BMC Physiology 2001;1:14.

171. Kafka MS, Wirz-Justice A, Naber D. Circadian and seasonal rhythms in alpha- and beta-adrenergic receptors in the rat brain. Brain Research 1981;207:409–419.

172. Pazo D, Cardinal DP, Cano P, Reyes Toso CA, Esquifino AI. Age-related changes in 24 hour rhythms of norepinephrine content and serotonin turnover in rat pineal gland: Effect of Melatonin Treatment. Neurosignals 2002;11:81–87.

173. Ziegler MG, Lake CR, Wood JH, Ebert MH. Circadian rhythm in cerebrospinal fluid noradenaline of man and monkey. Nature 1976;264:656–658.

174. Amenta F, Mione MC. Age-related changes in the noradrenergic innervation of the coronary arteries in old rats: A fluorescent histochemical study. J Auton Nerv Syst 1988;22:247–251.

175. Leger J, Croll RP, Smith FM. Regional distribution and extrinsic innervation of intrinsic cardiac neurons in the guinea pig. J Comp Neurol 1999;407:303–317.

176. Gey KF, Burkhard WP, Pietscher A. Variation of the norepinephrine metabolism of the rat heart with age. Gerontologia 1965;11:1–11.

177. Martinez JL Jr, Vasquez BJ, Messing RB, Jensen RA, Liang KC, McGaugh JL. Age-related changes in the catecholamine content of peripheral organs in male and female F344 rats. J Gerontol 1981;36:280–284.

178. Bruzzone P, Cavallotti C, Mancone M, Tranquilli Leali FM. Age related changes in catecholaminergic nerve fibers of rat heart and coronary vessels. Gerontology 2003;49(2): 80–85

179. Krall JF, Conelly M, Weisbant R, Tuck ML. Age-related elevation of plasma catecholamine concentration and reduced responsiveness of lymphocyte adenylate-cyclase. J Clin Endocrinol Metab 1981;52:863–867.

180. Ebstein RP, Stessman J, Eliakim R, Menczel J. The effect of age on beta-adrenergic function in man: A review. Isr J Med Sci 1985;21:302–311.

181. Brusco LI, Garcia-Bonacho M, Esquifino A, Cardinali DP. Diurnal rhythms in norepinephrine and acetylcholine synthesis of sympathetic ganglia, heart and adrenals of ageing rats: Effect of Melatonin. J of the Autonomic Nervous System 1998;74:49–61.

182. Esler M, Lambert G, Kaye D, Rumantir M, Hastings J, Seals DR. Influence of ageing on the sympathetic nervous system and adrenal medulla at rest and during stress. Biogerontology

2002;3:45–49.

183. Pangerl A, Remien J, Haen E. The number of beta-adrenoceptor sites on intact human lymphocytes depends on time of day, on season and on sex. Ann Rev Chronopharmacol 1986;3:331–334.

184. Nakai T, Yamada R. Urinary catecholamine excretion by various age groups with special reference to clinical value of measuring catecholamines in newborns. Pediatr Res 1983;17: 456–460.

185. DeSchaepdryver AF, Hooft C, Delbeke MJ, den Noortgaete MV. Urinary catecholamines and metabolites in children. J Pediatr 1978;93(2):266–268.

186. Parra A, Ramirez del Angel A, Cervantes C, Sanchez M. Urinary excretion of catecholamines in healthy subjects in relation to body growth. Acta Endocrinol (Copenh) 1980;94:546–551.

187. Kirkland JL, Lye M, Levy DW, Banerjee AK. Pattern of urine flow and electrolyte excretion in healthy elderly people. Br Med J 1983;287:1665–1667.

188. Prinz PN, Halter J, Benedetti C, Raskind M. Circadian variation of plasma catecholamines in young and old men: Relation to rapid eye movement and slow wave sleep. J Clin Endocrinol Metab 1979;49:300–304.

189. Linsell CR, Lightman SL, Mullen PE, Brown MJ, Causon RC. Circadian rhythms of epinephrine and norepinephrine in man. J Clin Endocrinol Metab 1985;60:1210–1215.

190. Cameron OG, Curtis GC, Zelnik T, McCann D, Roth TH, Guire K, Huber-Smith M. Circadian fluctuation of plasma epinephrine in supine humans. Psychoneuroendocrinol 1987;12:41–51.

191. Kuchel O, Buu NT. Circadian variations of free and sulfoconjugated catecholamines in normal subjects. Endocrinol Res 1985;11(1–2): 17–25.

192. Akerstedt T, Levi L. Circadian rhythms in the secretion of cortisol, adrenaline and noradrenaline. Eur J Clin Invest 1978;8:57–58.

193. Lightman SL, James VH, Linsell C, Mullen PE, Peart WS, Sever PS. Studies of diurnal changes in plasma renin activity, and plasma noradrenaline, aldosterone and cortisol concentrations in man. Clin Endocrinol 1981;14:213–223.

194. Ziegler MG, Lake CR, Kopin IJ. Plasma noradrenaline increases with age. Nature 1976;261:333–335.

195. Rubin PC, Scott PJW, McLean K, Reid JL. Noradrenaline release and clearance in relation to age and blood pressure in man. Eur J Clin Invest 1982;12:121–125.

196. Rowe JW, Troen BR. Sympathetic nervous system and aging in man. Endocrinol Rev 1980;1:167–179.

197. Goldstein DS, Lake CR, Chernow B, Ziegler MG, Coleman MD, Taylor AA, Mitchell JR, Kopin IJ, Keiser HR. Age-dependence of hypertension - normotensive differences in plasma norepinephrine. Hypertension 1983;5:100–104.

198. Bertel O, Buhler FR, Kiowski W, Lutold BE. Decreased beta-adrenoreceptor responsiveness as related to age, blood pressure and plasma catecholamines in patients with essential hypertension. Hypertension 1980;2(2):130–138.

199. Conway J, Wheeler R, Sannerstedt R. Sympathetic nervous activity exercise in relation to age. Cardiovasc Res 1971;5:577–581.

200. Lakatta VG, Gerstenblith G, Angeli CS, Shock MW, Weisfeldt ML. Diminished inotropic response of aged myocardium to catecholamines. Circ Res 1975;36(2):262–269.

201. Vestal RE, Wood AJ, Shand DG. Reduced beta-adrenoceptor sensitivity in the elderly. Clin Pharmacol Ther 1979;26(2) 181–186.

202. Bender AD. The influence of age on the activity of catecholamines and related therapeutic agents. J Am Geriatr Soc 1970;18:220–232.

203. Esler M, Skews H, Leonard P, Jackman G, Bobik A, Korner P. Age difference of noradrenaline kinetics in normal subjects. Clin Science 1981;60(2):217–219.

204. Karki NT. The urinary excretion of noradrenaline and adrenaline in different age groups; its diurnal variation and effect of muscular work on it. Acta Physiol Scand 1956;39(Suppl 132):1–96.

205. Descovich GC, Montalbetti N, Kuhl JFW, Rimondi S, Halberg F, Ceredi C. Age and catecholamine rhythms. Chronobiologia 1974;1:163–171.

206. Nicolau GY, Haus E, Lakatua D, Sackett-Lundeen L, Bogdan C, Plinga L, Petrescu E, Ungureanu E, Robu E. Differences in the circadian rhythm parameters of urinary free epinephrine, norepinephrine, and dopamine between children and elderly subjects. Rev Roum Med Endocrinol 1985;23:189–199.

207. Haus E, Lakatua DJ, Sackett-Lundeen L, Swoyer J. Chronobiology in laboratory medicine. In: Reitveld WT, editor. Clinical Aspects of Chronobiology. Baarn: Bakker; 1984;13–82.

208. Lehmann M, Keul J. Urinary excretion of free noradrenaline and adrenaline related to age, sex and hypertension in 265 individuals. Eur J Appl Physiol 1986;55:14–18.

209. Cutler NR, Hodes JE. Assessing the noradrenergic system in normal aging: a review of methodology. Exp Aging Res 1983;9(3):123–127.

210. Wallin BG, Sundlof G. A quantitative study of muscle nerve sympathetic activity in resting normotensive and hypertensive subjects. Hypertension 1979;1:67–77.

211. Vargas E, Rothwell C, Weinkove C, Lye M. The measurement of plasma noradrenaline before and after tilt in young and old healthy subjects. Gerontology 1984;30: 253–260.

212. Fleg JL, Tzankoff SP, Lakatta EG. Age-related augmentation of plasma catecholamines during dynamic exercise in healthy males. J Appl Physiol 1985;59(4):1033–1039.

213. Itskovitz HD, Wynn NC. Renal functional status and patterns of catecholamine excretion. J Clin Hypertens 1985;3:223–227.

214. Morfis L, Howes LG. Nocturnal fall in blood pressure in the elderly is related to presence of hypertension and not age. Blood Press 1997;6:274–278.

215. Sowers JR. Dopaminergic control of circadian norepinephrine levels in patients with essential hypertension. J Clin Endocrinol Metab 1981;53:1133–1137.

216. Lake CR, Ziegler MG, Coleman MD, Kopin IJ. Age-adjusted plasma norepinephrine levels are similar in normotensive and hypertensive subjects. NEJM 1977;296:208–209.

217. Chobanian AV, Gavras H, Melby JC, Gavras I, Jick H. Relationship of basal plasma noradrenaline to blood pressure, age, sex, plasma renin activity and plasma volume in essential hypertension. Clin Science and Molecular Medicine 1978;55:93s-96s

218. Liebau H, Manitius J. Diurnal and daily variations of PRA, plasma catecholamines and blood pressure in normotensive and hypertensive man. Contrib Nephrol 1982;30:57–63.

219. Cagnacci A, Arangino S, Angiolucci M, Maschio E, Melis GB. Influences of melatonin administration on the circulation of women. Am J Physiol 1998;274:R335–338.

220. Arangino S, Cagnacci A, Angiolucci M, Vacca AMB, Longu G, Volpe A. Effects of melatonin on vascular reactivity, catecholamine levels and blood pressure in healthy men. Am J Cardiol 1999;83:1417–1419.

221. Nicolau GY, Haus E, Bogdan C, Plinga L, Robu E, Ungureanu E, Sackett-Lundeen L, Petrescu E. Circannual rhythms of systolic and diastolic blood pressure in elderly subjects and in children. Rev Roum Med Endocrinol 1986;24:97–107.

222. Nicolau GY, Haus E, Popescu M, Sackett-Lundeen L, Petrescu E. Circadian, weekly and

seasonal variations in cardiac mortality, blood pressure and catecholamine excretion. Chronobiology International 1991;8:149–159.

223. Lakatua DJ, Nicolau GY, Bogdan C, Plinga L, Jachimowitz A, Sackett-Lundeen L, Petrescu E. Chronobiology of catecholamine excretion in different age groups. In: Advances in Chronobiology, Part B. Alan R. Liss Inc 1987;31–50.

224. Haen E, Langenmayer I, Pangerl A. Circannual variation in the expression of beta-2-adrenoceptors in human peripheral mononuclear leukocytes. Klin Wscher 1988;66: 579–582.

225. Veldhuis JD, Iranmanesh A, Mulligan T, Bowers CY. Mechanisms of conjoint failure of the somatotropic and gonadal axes in ageing man. In: Chadwick DJ, Goode JA, editors. Endocrine Facets of Ageing. Novartis Foundation Symposium 242. Chichester, West Sussex, England: John Wiley & Sons; 2002:98–124.

226. Touitou Y, Bogdan A, Reinberg A, Haus E. Les rythmes annuels endocriniens chez l'homme. Path Biol 1996;44:654–665.

227. Haus E, Touitou Y. Chronobiology of Development and Aging. In: Redfern PH, Lemmer Bj, editors. Physiology and Pharmacology of Biologic Rhythms. New York: Springer; 1997;95–134.

*The Neuroendocrine Immune Network in Ageing*
Edited by R.H. Straub and E. Mocchegiani

# *Melatonin Rhythms, Melatonin Supplementation and Sleep in Old Age*

RIXT F. RIEMERSMA, CAROLINE A.M. MATTHEIJ, DICK F. SWAAB and
EUS J.W. VAN SOMEREN

*Netherlands Institute for Brain Research, Amsterdam, and VU University Medical Center,
The Netherlands*

## ABSTRACT

The circadian timing system (CTS) allows organisms on earth to synchronize internal rhythms to the environmental 24-hour light-dark cycle and anticipate the body on the forthcoming period of either activity or rest. The central pacemaker of the CTS is the hypothalamic suprachiasmatic nucleus (SCN) regulating most, if not all, circadian rhythms in the body. The plasticity of this system at old age is the subject of this review, with special regard to its major internal stimulus, the pineal hormone melatonin.

The circadian rhythm in melatonin production is regulated by the SCN and results in low daytime circulating levels of melatonin and an increase of circulating melatonin after darkness onset. This rhythm, like other circadian rhythms, is attenuated in elderly subjects. The relation between the age-related change in melatonin levels and sleep is discussed in this paper.

Based on the hypothesis that there is a relationship between the increased prevalence of sleep-disturbances and decreased melatonin levels in elderly, several studies have been performed to investigate the effect of exogenous melatonin supplementation on sleep in elderly and demented subjects. An overview of these findings is presented and the various results are discussed.

## 1. INTRODUCTION

All living organisms on earth are exposed to the daily environmental 24-hour light-dark cycle. The circadian timing system (CTS) is able to synchronize an organism's internal rhythm to that of the environment, to allow the body to anticipate the coming period and to function with maximum efficiency in the given environmental situation. In humans this means that the body is prepared to waken before the light period, and to exhibit optimum physical and mental perform-ance during the day. In the evening the body gets ready for the resting period.

The central coordinating pacemaker of the CTS in mammals is located in the suprachiasmatic nucleus (SCN), situated bilaterally in the anterior hypothalamus, on top of the optic chiasm. The SCN regulates circadian rhythms in body temperature, hormone levels and rest-activity cycle. By receiving environmental stimuli, the so-called Zeitgebers, the SCN synchronizes these rhythms to the 24-hour environmental light-dark cycle. In the absence of Zeitgebers the rhythms generated by the SCN will deviate from the exact 24-hour cycle. The CTS is a flexible system that can adapt to a new light-dark regimen after crossing several time zones.

The functional plasticity of the CTS at old age is the subject of this review. We will focus on a major internal stimulus for the CTS, the pineal hormone melatonin, which is endogenously present as a circadian modulator, but may be supplemented exogenously as well. Other effective stimuli, like bright light, temperature, physical activity and transcutaneous nerve stimulation (TENS) have been reviewed previously [1–6].

In the framework of the question to what extent the CTS and its plasticity are maintained at high age, especially in humans, the following points are subsequently discussed (1) the organization of the CTS, including the pineal gland, (2) age-related changes of the CTS, especially in relation to melatonin synthesis in the pineal gland (3) functional implications of weak or disturbed circadian rhythms, (4) interactions of the pineal melatonin synthesis with drugs frequently used by elderly people and (5) the supplementation of melatonin and its consequences for the CTS.

## 2. THE CIRCADIAN TIMING SYSTEM

### 2.1. Structure of the suprachiasmatic nucleus

The suprachiasmatic nucleus (SCN) is a bilateral structure that is the central pacemaker of the circadian timing system (CTS) and regulates most, if not all, circadian rhythms in the body. Within the SCN several types of peptidergic neurons are found, such as vasopressin (AVP), vasoactive intestinal polypeptide (VIP), neuropeptide Y (NPY) and neurotensin (NT), that each have a specific distribution. The vasopressinergic subnucleus has a volume of 0.25 mm$^3$ per side. The synthesis of vasopressin in the SCN shows both a circadian and seasonal rhythm [7].

### 2.2. Input to the suprachiasmatic nucleus

From the inputs the SCN receives, environmental light is the most effective one and is of direct and indirect importance for the melatonin production and rhythm (see further in this paper). The SCN receives information about the environmental light-dark cycle by a direct projection through the retinohypothalamic tract (RHT) [8], for which glutamate and pituitary adenylate cyclase-activating polypeptide (PACAP) are at present the most likely transmitters [9]. Photoperiodic information from the environment is mediated by melanopsin-containing retinal ganglion cells projecting via the RHT to the SCN [10] (see Figure 1).

In addition to the retinohypothalamic input, the SCN also receives inputs from other hypothalamic nuclei, the raphe nuclei, locus coeruleus, limbic forebrain, and from the hormonal milieu [11].

### 2.3. Output of the suprachiasmatic nucleus innervating the pineal gland

Both the acute suppression of melatonin by environmental light and the circadian modulation of the melatonin level are mediated by the SCN. The SCN-Pineal pathway has been elegantly confirmed by the use of the transneural pseudorabies viral tracer (PRV) [12]. PRV was injected into the pineal gland and labeling was subsequently found in the superior cervical ganglion (SCG), the intermediolateral column of the upper thoracic cord (IML), the autonomic divi-

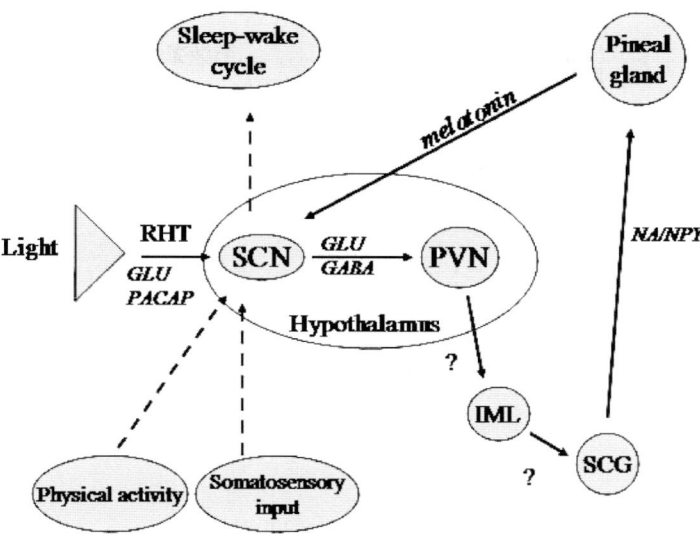

Figure 1. Schematic drawing of that part of the circadian timing system involved in the circadian regulation of melatonin.

Abbreviations: RHT=retinohypothalamic tract, Glu=glutamate, PACAP= pituitary adenylate cyclase-activating polypeptide, SCN=suprachiasmatic nucleus, GABA=gamma-aminobutyric acid, PVN=paraventricular nucleus, SCG=superior cervical ganglion, NA=noradrenalin, NPY=neuropeptide Y

sion of the paraventricular nucleus (PVN), and the SCN. The majority of labeled neurons was found in the dorsomedial position of the SCN. Confocal laser scanning microscopy showed SCN neurons to be double-labeled for PRV and AVP and PRV and VIP. Removal of the SCG resulted in complete absence of the tracer in the SCN – pineal pathway, but not in the pineal gland. Control of the circadian variation of sympathetic input to the pineal is mediated by two signals from the SCN to the PVN, most likely a continuous stimulatory glutaminergic input and a rhythmic inhibitory GABA-ergic input. Without SCN input, a basic stimulatory effect on the pineal gland is maintained by the PVN, which is normally suppressed by the SCN during the light period and enhanced by the SCN during the dark period [13]. The neurotransmitters involved in the pathway from the PVN to the IML, and further from the IML to the SCG, are still unknown [14]. The direct inhibiting effect of nocturnal ocular light on melatonin levels is a result of the inhibition of N-acetyltransferase (NAT) activity (see following section) in the pineal gland [15,16]. In rats, this light-induced inhibition of nocturnal melatonin release is completely prevented by the administration of a GABA-antagonist to the hypothalamic projection areas of the SCN [14]. Within one hour, nocturnal light exposure results in a decrease of melatonin levels. The rate of decline corresponds to the half-life time of melatonin [16], which means that there is a direct effect of light on the production of melatonin. Figure 1 shows schematically the connections between environmental light, the SCN and the pineal gland. For further details about the output pathways of the mammalian suprachiasmatic nucleus we refer to the review of Kalsbeek and Buijs [17]. In humans, the light-induced inhibition of nocturnal melatonin release is fully dependent on light that falls on the eyes, and supposedly mediated by the consequent

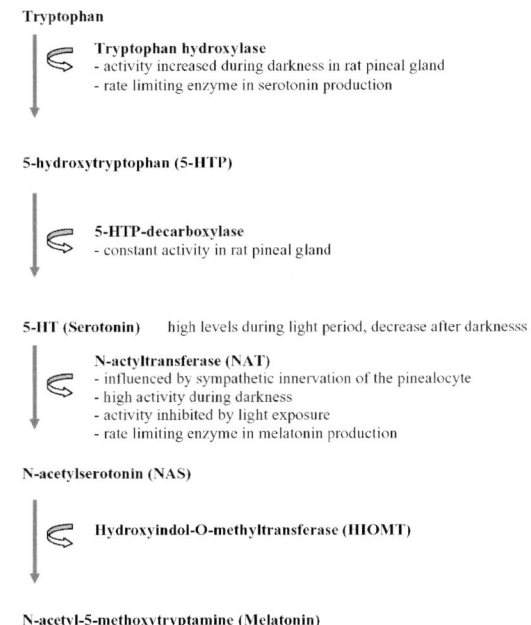

Figure 2. Subsequent steps of melatonin production in the pinealocyte.

retinal signal that reaches the SCN. Since the publication of Campbell et al. in 1998 [18], in which it was stated that extra-ocular light, administered behind the knees, could phase-shift the melatonin rhythm, several researchers have tried to replicate this finding, but without success. Neither did anyone find a suppression of melatonin by extra-ocular light [19,20].

## 2.4.    The pineal gland and its main hormone melatonin

This neuro-endocrine gland is present in many different species, among which all mammals. In humans the pineal gland is localized in the midline of the brain at the posterior level of the third ventricle. The hormone melatonin is the primary product of the pineal gland. Melatonin is synthesized of tryptophan (TRP) [cf. 15]. The subsequent steps are as follows. First, tryptophan hydroxylase (TH) catalyzes the conversion from TRP into 5-hydroxytryptophan (5-HTP). The subsequent oxidation of 5-HTP into 5-HT (serotonin) is catalyzed by 5-HTP-decarboxylase. 5-HT is then N-acetylated by NAT into N-actylserotonin (NAS). NAT is considered the rate-limiting enzyme. Finally hydroxyindol-O-methyltransferase (HIOMT) catalyzes the O-methylation of NAS to form melatonin (N-acetyl-5-methoxytryptamine). These steps are summarized in Figure 2.

The pineal gland is innervated by several neural pathways, as recently reviewed by Moller and Baeres [cf. 21], among which the sympathetic fibres are the most important ones for the circadian regulation of melatonin production. These fibres contain noradrenalin and NPY [cf. 21]. Noradrenalin, by stimulation of β-adrenergic receptors, leads to an increase in NAT [cf. 22]. In addition to the sympathetic innervation via the superior cervical ganglion, by which the SCN regulates its circadian rhythm, the pineal gland is innervated by fibres of parasympathetic nerves, nerves from the trigeminal ganglion, and nerves originating from the brain, entering the pineal gland via the pineal stalk, also called the central innervation [21].

In both nocturnal and diurnal animals, the production of melatonin shows a circadian pattern with high levels after the onset of darkness, and low levels during the light period of the day [23]. The circadian pattern in melatonin levels can be monitored directly, either in serum or saliva, or indirectly, by measuring its main metabolite 6-sulfatoxy melatonin in urine. Especially measurements in saliva are of interest for studies in elderly subjects and demented patients, and in field studies. Saliva sampling is relatively easy and non-invasive, and the correlation with plasma melatonin levels is strong [24]. The dim-light melatonin onset (DLMO) determines the time point at which melatonin starts to rise and rises above a certain threshold. This threshold can be defined in different ways. The DLMO for serum melatonin levels can be defined as the interpolated crossing of the 10 pg/ml threshold during its rise [25]. Other definitions use the time point at which the curve crosses 25% of the peak level [26,27], or the time when the curve exceeds a level calculated from the mean plus two standard deviations of samples at 19:00, 19: 30 and 20:00 h [28].

Its high lipophilicity allows melatonin to diffuse from the pineal into the surrounding tissue with ease and to enter into the bloodstream or the cerebrospinal fluid (CSF). For a long time it has been presumed that melatonin enters the CSF via the bloodstream. Recent studies in sheep, however, indicate that melatonin may also enter the CSF directly from the pineal gland [29].

## 2.5. Melatonin function and sites of action

Melatonin conveys photoperiodic information and thus informs the circadian timing system about the time of day and season [cf. 15,23].

Three pharmacologically distinct melatonin binding sites have been described, the MT-1, MT-2 and MT-3 receptors [cf. 30], of which the MT-1 [31] and MT-2 are G-protein coupled receptors [32], and the MT-3 receptor [33] belongs to the family of quinone reductases. There is an abundant expression of melatonin receptors in the SCN [34]. In rat, the expression of MT-1 mRNA in the SCN exhibits a circadian variation [35], with the highest levels in the early day- and early nighttime periods. The specific binding of melatonin in the SCN of rat is significantly lower during the dark period compared to the light period. This difference disappears in constant darkness [35]. Melatonin decreases the frequency of the spontaneous electrical activity in cultured rat SCN explants [cf. 36]

## 3. AGE-RELATED CHANGES IN THE CIRCADIAN TIMING SYSTEM

### 3.1. Age-related changes in the suprachiasmatic nucleus

In post-mortem studies of the ageing human SCN, the circadian rhythm that was found in young adults in the number of AVP-expressing neurons according to clock-time of death disappeared after the age of 50 years. Also the seasonal fluctuation in AVP-expressing neurons disappeared in the group of elderly subjects. Furthermore, a decline in the number of arginine vasopressin (AVP) expressing neurons was found in subjects over the age of 80 years and even more so, and at a younger age, in Alzheimer patients [cf. 7]. There was also a decrease in the amount of AVP-mRNA in the SCN of Alzheimer patients [37]. The number of VIP-expressing neurons declined only in male subjects and not in female subjects [38].

Table I    Changes of melatonin with age.

| Research group | phase | peak level | mean | overall production | daytime melatonin level | inter-individual variability |
|---|---|---|---|---|---|---|
| Iguichi et al. 1982 [42] | | decrease | decrease | decrease | decrease | |
| Sack et al. 1986 [47] | | | | decrease | | |
| Nair et al. 1986 [43] | delay | decrease | | decrease | | |
| Sharma et al. 1989 [48] | delay | decrease | decrease | | | |
| Touitou et al. 1984 [58] | advance | decrease | decrease | decrease | low | |
| Ferrari et al. 1993 [44] | | decrease | decrease | decrease | | |
| Mishima et al. 1994 [45] | no change | decrease | decrease | decrease | no change | |
| Zhdanova et al. 1998 [50] | advance | decrease | | | | higher |
| Waldhauser et al. 1988 [52] | | decrease | | | | |
| Rodenbeck et al. 1998 [46] | advance | decrease | | | | higher (according to the st. dev.) |
| Zeitzer et al. 1999 [53] | | no change | | no change | | |
| Kennaway et al. 1999 [51] | | decrease (until the age of 30) | | | | |

## 3.2.    Age-related changes in melatonin levels, amplitude and rhythmicity

Age-related changes have been demonstrated throughout the melatonin system, i.e. at the structural level in the pineal gland itself, in the total production level and in the rhythmicity of melatonin levels. Kunz et al. [39] described an increase of the calcified portion of the pineal and suggested the consequent decrease in uncalcified pineal tissue to be responsible for a decrease in aMT6s excretion levels. A post-mortem study showed that, by comparing the pineal glands of subjects that died young (age range 18–54 years) with those of subjects that died at old age (age range 55–92 years), the daily variation in melatonin levels disappears with ageing [40]. The number of beta-adrenergic receptors in the pinealocyte membrane, important for the sympathetic innervation that regulates the circadian rhythmicity of the melatonin production, decreases with age, as does the responsiveness to norepinephrine [41].

A general finding in elderly people is that the nocturnal peak level is attenuated [42–50], resulting in a decrease in the amplitude of the melatonin rhythm (see Table I for an overview of different studies on this topic). This decrease may occur very early in the process of development and ageing. Kennaway [51] found the age-related decrease in urinary excretion of aMT6S, the main metabolite of melatonin, to occur already before the age of 30 years. Waldhauser et al. [52] also found the strongest decline of serum melatonin levels before the age of 20 years. No further decline was found between 20 and 70 years; only after the age of 70 years nighttime levels were lower compared to the younger adult groups. Zeitzer et al. [53] did not find an age-related decline in plasma melatonin levels. One reason for this could be that they included only extremely healthy people in their study. Another reason, suggested by Touitou [54], is the fact that Zeitzer et al. used a constant routine protocol, whereas in other studies subjects remained in their usual living environment. The importance of the illumination levels in the habitual environment has been demonstrated by Mishima et al. [55], who compared groups of young adults, elderly insomniacs and elderly people without sleep-complaints. They found significantly smaller amplitude of the melatonin rhythm in both groups of elderly subjects. An interesting finding then was that the supplementation of midday bright light (2500 lux) in the elderly insomniacs resulted in an increase of the melatonin amplitude to levels comparable with those in young adults [55]. Unfortunately this effect was not tested for the elderly subjects without sleep complaints. The importance of sufficient daytime bright light for the melatonin rhythm was also shown by Baskett et al. [56], who compared a group of hospitalized elderly with a community-based group of elderly. They found an attenuated daytime suppression of melatonin levels in the hospitalized group, together with more variable nighttime melatonin levels. Indeed, such an age-related increase of daytime melatonin level has often been reported in addition to the decrease in nocturnal level, as reviewed in Van Someren et al. [5]. In a study of Ohashi et al. [57], both demented and non-demented hospitalized psychiatric patients showed an increase of daytime melatonin levels compared to a group of non-hospitalized elderly subjects. Two hours of morning bright light (3000 lux) decreased daytime melatonin only in psychiatric non-demented elderly, but did not affect daytime melatonin in the demented group. In relation to this, the season might be an important variable to be taken into account when studying circadian rhythms. Whereas the decrease in amplitude is the most consistent finding on the melatonin rhythm in elderly, some studies have reported a change in the circadian phase of the melatonin rhythm. On this point the findings are less consistent: some studies reported a phase advance [46,50,58] while others found a phase delay [43,48].

## 4.   FUNCTIONAL IMPLICATIONS OF WEAK OR DISTURBED CIRCADIAN RHYTHMS

An extensive review on the implications of weak or disturbed circadian rhythms was recently given by Van Someren [5]. Disturbed circadian rhythms appear to have negative effects on the impact of cardiovascular diseases, mood and cognition. In the present paper we will focus on the effect of changes in melatonin levels.

The relation between melatonin levels and sleep disturbances is not entirely clear. The knowledge that melatonin levels are decreased in elderly subjects, and that the prevalence of sleep disturbances is increased in this group [59], has led to the expectation that there would be a relation between melatonin levels and sleep. Although Haimov et al. [60] found a difference in melatonin levels between elderly insomniacs and elderly without sleep problems, others were unable to confirm this observation, either for melatonin levels, or for its metabolite aMT6S, and sleep parameters [61–64]. Although there is thus not a clear cut relationship between decreased melatonin levels and sleep, there might be a relation between sleep disturbances and the circadian period of the melatonin rhythm. Kripke et al. [61] reported that elderly volunteers with more deviant acrophases of aMT6S slept fewer hours and had more wake time within the sleep period.

## 5.   INTERACTIONS WITH DRUGS FREQUENTLY USED BY ELDERLY

Several studies have addressed the influence of sympathetic agonists and antagonists on melatonin production. Hurlbut et al. [65], for example, investigated the influence of β-agonists and α- and β-antagonists on NAT and melatonin levels in Richardson's ground squirrels. They found that isoproterenol, a β-receptor agonist, stimulated both pineal NAT activity and pineal melatonin content; phentolamine, an α-blocker, partially blocked the rise of NAT. Propranolol, a β-blocker, totally blocked this rise. Melatonin synthesis was not influenced by phentolamine, but administration of propranolol prevented the rise of melatonin during the night. This means that NAT activity is under the influence of both α and β innervation, while melatonin production is under the influence of β-innervation only.

The study of Stoschitzky et al. [66] is interesting in this context, because they investigated the influence of the beta-blocking (S)-enantiomers, and also the non-blocking (R)-enantiomers of propranolol and atenolol in humans. Enantiomers are two related molecules with a different conformation, so that one of them (the S-enantiomer) can bind to a receptor and thus act as an agonist, while the other (R-enantiomer) one cannot bind to the receptor because of its different conformation, and can thus act as a placebo. They found that the (S)-enantiomers decreased the nocturnal excretion of aMT6s, whereas the (R)-enantiomers had no effect. The use of enantiomers makes sure that the β-blocking effect is responsible for the decreased excretion of aMT6s, and thus melatonin production, rather than an unspecific effect of β-blockers.

The role of GABA is an important one. Especially since sleep disturbances are often treated with benzodiazepines, which bind to the GABA receptor. In the pineal, the release of GABA is triggered by noradrenalin, through α1-adrenergic receptors. The effects of GABA are generally inhibitory to the noradrenalin-induced NAT activity, and thus to melatonin production. There is also feedback from GABA to the sympathetic fibres. Pre-synaptically, GABA acts on type A and B receptors, which appear to have opposite functions, respectively facilitating and inhibiting noradrenalin release. The inhibitory function dominates. This circuit offers resistance to the passage of information to the pineal, leading to more balanced responses [cf. 67].

GABA receptors have a number of binding sites for clinically important drugs. Benzodiazepines bind to the GABA-receptor complex and enhance its sensitivity for GABA. In the pineal, the acute and chronic administration of the benzodiazepine diazepam inhibits NAT activity in the rat pineal gland [68], and decreases the content of NAS and melatonin [69]. Chronic administration of diazepam resulted also in decreased nocturnal plasma levels of melatonin. Benzodiazepines may not only directly affect the melatonin rhythm by acting on the pineal innervation, but also by altering the level of activity and consequently the phase and period of the central pacemaker in the suprachiasmatic nucleus [70,71] which is responsible for the melatonin rhythm.

## 6.   MELATONIN SUPPLEMENTATION AND CONSEQUENCES FOR THE CIRCADIAN TIMING SYSTEM

### 6.1.   Pharmacokinetics

Melatonin is quickly absorbed and quickly excreted. Metabolisation takes place in the liver, where 70% is converted in 6-sulphatoxy melatonin (aMT6S). All metabolites are excreted by the kidneys. The half-life time of melatonin varies between 10 and 40 minutes [cf. 15].

Fourtillan et al. [72] described the biological availability of $D_7$ melatonin (melatonin with seven hydrogen atoms replaced by seven deuterium atoms) administered intravenously (i.v.) or orally to twelve young healthy volunteers. Following i.v. administration of 23 μg $D_7$ melatonin, a large difference in peak levels was seen for males and females (mean peak level was 124.8 ± 32.8 pg/ml for males and 169.0 ± 29.6 pg/ml for females). After oral administration of 250 μg melatonin, melatonin was rapidly absorbed with a mean maximum absorption time of 23 minutes for both males and females. Here, the mean peak levels were almost three times higher for females than males (243.7 ± 124.6 pg/ml for males and 623.107 ± 575.1 pg/ml for females). The fraction of melatonin systemically absorbed ranged from 1 to 37% (mean values were 8.6 ±

3.6% for males versus 16.8 ± 12.7% for females). A low bioavailability can either be due to poor oral absorption, a large first pass effect, or a combination of the two.

In addition to the individual and sex-related differences in pharmacokinetics, Zhdanova [50] further reported an increased variance in serum melatonin levels in elderly subjects as compared to young adults after the ingestion of 0.3 mg melatonin.

## 6.2.    Clinical studies on the effects of exogenous melatonin on sleep in the elderly

In Table II an overview is presented of randomized placebo-controlled studies on the effect of melatonin on sleep, performed in elderly.

The two most recent studies, by Serfaty et al. [73] and Baskett et al. [64], did not show a therapeutic effect on actigraphically derived parameters of sleep quality, with the only exception that Baskett found a significant decrease in awakenings in a group of normal sleepers. Serfaty et al. investigated the effect of melatonin on sleep in demented elderly patients without further specification of a clinical diagnosis. Since dementia can be caused by various underlying diseases, this might be a very heterogeneous group of subjects, and a possible effect in a specific disease group might therefore have been missed. Another factor that could play a role in the negative results in the study of Serfaty et al. is the choice of the outcome measures. The actigraphical parameters tested in the study of Serfaty are the median total time asleep, median number of awakenings, and sleep efficiency. Van Someren [74] reported on the improved sensitivity to the effect of bright light therapy on rest-activity rhythms in Alzheimer patients, and demonstrated that the light-induced improvement in coupling of the rest-activity rhythm to the environmental Zeitgeber of bright light is better detected using nonparametric procedures. Worthwhile to mention in this respect is also the finding by Serfaty and co-workers [73] that the carers' reports of sleep problems in demented elderly were not consistent with the objective evidence as obtained by actigrapy. This means that data collected through logs kept by carers should be interpreted with caution.

Baskett et al. [64] looked at elderly insomniacs and normal sleeping elderly, as was also done by Haimov et al. [60] and Zhdanova et al. [63]. Beside differences in dosage, Baskett et al. studied a period of 4 weeks of active treatment, whereas both other studies looked for the effect of only 1 week of active treatment. To measure sleep quality, Haimov and Baskett both made use of actigraphy to measure sleep quality. Haimov found an improvement of sleep efficiency and activity level after 1 week of daily ingestion of 2 mg sustained release melatonin, and a shorter sleep latency after 1 week of daily ingestion of 2 mg fast release melatonin. Baskett found neither an improvement in sleep efficiency, nor in sleep latency after 4 weeks of daily administration of 5 mg melatonin. Zhdanova did show an improvement of sleep efficiency, using polysomnography to measure sleep quality. However, Hughes [27], also using PSG as the outcome measure, found no change in sleep-efficiency. Three studies reported an improvement in sleep latency [27,75,76] without improvement of total sleep time, except for Garfinkel [77], who studied elderly insomniacs using benzodiazepines. In a later study they reported on the ability of melatonin to facilitate discontinuation of benzodiazepine use. Andrade [78] found that melatonin in medically ill hospitalized patients improved their self-rated sleep quality. Objective parameters were not tested in this study.

In general, this overview shows that the various publications show distinct results on the effect of melatonin on sleep in elderly subjects. It is therefore hard to draw any definite conclusions at this moment.

Table II    Randomized placebo-controlled studies on the effect of exogenous melatonin on circadian disturbances in the elderly.

| Research group | Number of subjects | Subject specification | Duration of active treatment | Dosage | Time of suppletion | Main outcome | Side effects |
|---|---|---|---|---|---|---|---|
| Garfinkel et al. 1995 [79] | 12 | Elderly insomniacs | Three weeks | 2 mg controlled release | Two hours before desired bedtime | AG measured improvement of sleep efficiency and wake-after sleep onset period, trend to shorter sleep latency, no effect on total sleep time | Self limiting pruritus in one subject in both placebo and melatonin treated group |
| Haimov et al. 1995 [60] | 51 | Normal sleeping elderly (n=25) and elderly insomniacs (n=26)* | One week (each type of tablet) | 2 mg sustained-release (S-r) or 2 mg fast-release (F-r) | Two hours before desired bedtime | AG measured improvement of sleep efficiency and activity level (S-r) or shorter sleep latency (F-r) | No side effects reported |
| Garfinkel et al. 1997 [75] | 21 | Benzodiazepine treated elderly insomniacs | Three weeks | 2 mg controlled release | Two hours before desired bedtime | AG measured improvement of sleep latency, wake-after sleep onset period, total sleep time, fragmental index and number of awakenings | No side effects reported |
| Hughes et al. 1998 [27] | 16 | Elderly insomniacs | Two weeks | 0.5 mg immediate release or 0.5 mg controlled-release | Thirty minutes before fixed bedtime and/or 4 hours after bedtime | PSG measured decrease of sleep-latency, no effect on PSG measured sleep efficiency or total sleep time and actigraphically measured parameters or subjective ratings of sleep quality | No side effects reported |
| Jean Louis et al. 1998 [76] | 10 | Elderly insomniacs with MCI | Ten days | 6 mg | Two hours before bedtime | AG measured improvement of sleep latency, circadian amplitude, transition from sleep to wakefulness, no effect on total sleep time and wake time after sleep onset, improvement of mood and delayed recall | No side effects reported |

| Research group | Number of subjects | Subject specification | Duration of active treatment | Dosage | Time of suppletion | Main outcome | Side effects |
|---|---|---|---|---|---|---|---|
| Garfinkel et al. 1999 [77] | 34 | Benzodiazepine treated elderly insomniacs | Six weeks | 2 mg controlled release | Two hours before bedtime | Facilitation of discontinuation of benzodiazepine use | Headache in two subjects treated with melatonin and one treated with placebo |
| Andrade et al. 2001 [78] | 33 | Medically ill subjects | 8-16 days | Flexible dose regimen | At night | Improvement of self rated time to fall asleep, sleep quality, sleep depth, freshness on awakening | No side effects reported |
| Zhdanova et al. 2001 [63] | 30 | Both normal sleeping elderly (n=15) and elderly insomniacs (n=15) | One week (each dosage) | 0.1 mg, 0.3 mg or 3 mg | Half an hour before fixed bedtime | PSG measured increase of sleep-efficiency, with the best effect after ingestion of 0.3 mg in the insomniac group, no effect in normal sleepers | 3 mg dose lowered significantly core body temperature |
| Serfaty et al. 2002 [73] | 44 | Demented elderly | Two weeks | 6 mg slow-release | Usual bedtime | No therapeutic effect on either AG derived sleep-parameters nor subjective measures of sleep quality | No side effects reported |
| Baskett et al. 2003 [64] | 40 | Normal sleeping elderly (n=20) and elderly insomniacs (n=20) | Four weeks | 5 mg | At bedtime | Lower number of awakenings in normal sleepers, further no therapeutic effect on AG measured parameters of sleep-quality | Excessive drowsiness in one person only in melatonin condition, in another person both in melatonin and placebo condition |

Abbreviations: PSG=polysomnographically, AG=actigraphically, MCI=mild cognitive impairment

* Of the insomniacs 8 where living independently and 18 where institutionalized elderly.

## 6.3.   Adverse effects and dosage

The studies mentioned in Table II did not report any serious side-effects of melatonin. Only self-limiting pruritus [79], headache and excessive drowsiness [64] were reported.

Various dosages have been studied (see Table II). Zhdanova et al. [63], who studied 0.1 mg, 0.3 mg and 3 mg melatonin found an effect of all three dosages on sleep efficiency in elderly insomniacs, but the best effect was observed after the ingestion of 0.3 mg. This dosage resulted in physiological plasma levels of melatonin, whereas the ingestion of 3 mg melatonin not only resulted in pharmacological plasma melatonin levels but also in a significant decrease in core body temperature.

## 7.   CONCLUSIVE REMARKS

In spite of the effort that has been made in studying the efficacy of melatonin on sleep disturbances in the elderly, still no clear conclusions can be drawn.

Since multiple factors influence circadian rhythmicity, conditions that resulted in clear effects in a laboratory setting might be less effective in a natural setting, where the effect of melatonin is counteracted by other factors influencing the circadian timing system. It seems possible – and may be preferable - to stimulate the endogenous production of melatonin rather than to increase melatonin levels by the supplementation of exogenous melatonin. The studies of Mishima et al., Ohashi et al. and Baskett et al. [55–57] suggest that sufficient daytime bright light will increase night-time melatonin levels and attenuate day-time melatonin levels, and thus improve the circadian rhythmicity of circulating endogenous melatonin. This also means that the season is an important variable in studies regarding melatonin rhythmicity.

It is clear that changes in the circadian system occur with ageing, both at the side of input from the environment as in the endogenous parameters that inform us about the functioning of the SCN. It is also clear that this system is still susceptible to an increased input of the appropriate stimuli and shows flexibility to changes according to the level of input of these stimuli.

Only Zeitzer et al. [53] did not show changes in the peak level of melatonin of a group of very healthy elderly. It would be interesting to know wether the few older subjects that still show a clear circadian rhythm in melatonin levels belong to the subgroup of successful ageing, and if so, what the exact role of melatonin might be in the process of ageing. In that case melatonin supplementation might also be considered as one of the strategies to influence the process of ageing e.g. by acting as an anti-oxidant or immune stimulant [cf. 80].

As concluded by Olde Rikkert in 2001[81], there is the need for larger studies before widespread use of melatonin in sleep disorders can be advocated. These studies should have the power to distinguish patient characteristics as co-variates in the analysis of a therapeutic effect of melatonin, in order to see which subgroup of patients can be expected to show a positive effect of melatonin and which cannot. We hope to answer a number of these questions in the near future, when our 3.5-year follow-up study on the effect of bright light, melatonin or a combination of the two on sleep, mood, behavior, and cognition in demented elderly will be finished.

ACKNOWLEDGEMENTS

Dr. J.Kruisbrink is acknowledged for excellent library assistance, Mrs. W.T.P. Verweij for correction of the English language, Financial support is acknowledged of ZON-MW The Hague (project 28-3003), NWO The Hague (projects SOW 014-90-001, Vernieuwingsimpuls and RIDE), Stichting Centraal Fonds RVVZ, Japan Foundation for Aging and Health, Stichting De Drie Lichten, Cambridge Neurotechnology.

REFERENCES

1.  Van Someren EJ, Kessler A, Mirmiran M, Swaab DF. Indirect bright light improves circadian rest-activity rhythm disturbances in demented patients. Biol Psychiatry 1997;41(9):955–63.
2.  Van Someren EJ, Lijzenga C, Mirmiran M, Swaab DF. Long-term fitness training improves the circadian rest-activity rhythm in healthy elderly males. J Biol Rhythms 1997;12(2): 146–56.
3.  Van Someren EJ, Scherder EJ, Swaab DF. Transcutaneous electrical nerve stimulation (TENS) improves circadian rhythm disturbances in Alzheimer disease. Alzheimer Dis Assoc Disord 1998;12(2):114–8.
4.  Van Someren EJ, Raymann RJ, Scherder EJ, Daanen HA, Swaab DF. Circadian and age-related modulation of thermoreception and temperature regulation: mechanisms and functional implications. Ageing Res Rev 2002;1(4):721–78.
5.  Van Someren EJ, Riemersma RF, Swaab DF. Functional plasticity of the circadian timing system in old age: light exposure. Prog Brain Res 2002;138:205–31.
6.  Van Someren EJW. Thermosensitivity of the circadian timing system. Sleep and Circadian Rhythms 2003;1:55–64.
7.  Swaab DF. Biological rhythms in health and disease: the suprachiamsatic nucleus and the autonomic nervous system. In: Appenzeller O, editor. The autonomic nervous system part I: normal functions: Elsevier; 1999. p. 467–521.
8.  Dai J, Van der Vliet J, Swaab DF, Buijs RM. Human retinohypothalamic tract as revealed by in vitro postmortem tracing. J Comp Neurol 1998;397(3):357–70.
9.  Hannibal J. Neurotransmitters of the retino-hypothalamic tract. Cell Tissue Res 2002;309(1):73–88.
10. Brainard GC, Hanifin JP, Rollag MD, Greeson J, Byrne B, Glickman G, et al. Human melatonin regulation is not mediated by the three cone photopic visual system. J Clin Endocrinol Metab 2001;86(1):433–6.
11. Moore RY. Entrainment pathways and the functional organization of the circadian system. Prog Brain Res 1996;111:103–19.
12. Teclemariam-Mesbah R, Ter Horst GJ, Postema F, Wortel J, Buijs RM. Anatomical demonstration of the suprachiasmatic nucleus-pineal pathway. J Comp Neurol 1999;406(2): 171–82.
13. Perreau-Lenz S, Kalsbeek A, Garidou ML, Wortel J, Van Der Vliet J, Van Heijningen C, et al. Suprachiasmatic control of melatonin synthesis in rats: inhibitory and stimulatory mechanisms. Eur J Neurosci 2003;17(2):221–8.
14. Kalsbeek A, Cutrera RA, Van Heerikhuize JJ, Van Der Vliet J, Buijs RM. GABA release from suprachiasmatic nucleus terminals is necessary for the light-induced inhibition of

nocturnal melatonin release in the rat [In Process Citation]. Neuroscience 1999;91(2): 453–61.

15. Reiter RJ. Pineal melatonin: cell biology of its synthesis and of its physiological interactions. Endocr Rev 1991;12(2):151–80.

16. Lewy AJ, Wehr TA, Goodwin FK, Newsome DA, Markey SP. Light suppresses melatonin secretion in humans. Science 1980;210(4475):1267–9.

17. Kalsbeek A, Buijs RM. Output pathways of the mammalian suprachiasmatic nucleus: coding circadian time by transmitter selection and specific targeting. Cell Tissue Res 2002;309(1):109–18.

18. Campbell SS, Murphy PJ. Extraocular circadian phototransduction in humans. Science 1998;279(5349):396–9.

19. Lockley SW, Skene DJ, Thapan K, English J, Ribeiro D, Haimov I, et al. Extraocular light exposure does not suppress plasma melatonin in humans. Journal of Clinical Endocrinology and Metabolism 1998;83(9):3369–3372.

20. Hebert M, Martin SK, Eastman CI. Nocturnal melatonin secretion is not suppressed by light exposure behind the knee in humans. Neurosci Lett 1999;274(2):127–30.

21. Moller M, Baeres FM. The anatomy and innervation of the mammalian pineal gland. Cell Tissue Res 2002;309(1):139–50.

22. Moore RY. Neural control of the pineal gland. Behav Brain Res 1996;73(1–2):125–30.

23. Arendt J. Mammalian Pineal Rhythms. Pineal Research Reviews 1985;3:161–213.

24. Gooneratne NS, Metlay JP, Guo W, Pack FM, Kapoor S, Pack AI. The validity and feasibility of saliva melatonin assessment in the elderly. J Pineal Res 2003;34(2):88–94.

25. Lewy AJ, Cutler NL, Sack RL. The endogenous Melatonin Profile as a Marker for Circadian Phase Position. Journal of Biological Rhythms 1999;14(3):227–236.

26. Koorengevel KM, Gordijn MC, Beersma DG, Meesters Y, den Boer JA, van den Hoofdakker RH, et al. Extraocular light therapy in winter depression: a double-blind placebo-controlled study. Biol Psychiatry 2001;50(9):691–8.

27. Hughes RJ, Sack RL, Lewy AJ. The role of melatonin and circadian phase in age-related sleep-maintenance insomnia: assessment in a clinical trial of melatonin replacement. Sleep 1998;21(1):52–68.

28. Voultsios A, Kennaway DJ, Dawson D. Salivary Melatonin as a Circadian Phase Marker: Validation and Comparison to Plasma Melatonin. Journal of Biological Rhythms 1997;12(5):457–466.

29. Tricoire H, Malpaux B, Moller M. Cellular lining of the sheep pineal recess studied by light-, transmission-, and scanning electron microscopy: Morphologic indications for a direct secretion of melatonin from the pineal gland to the cerebrospinal fluid. J Comp Neurol 2003;456(1):39–47.

30. Witt-Enderby PA, Bennett J, Jarzynka MJ, Firestine S, Melan MA. Melatonin receptors and their regulation: biochemical and structural mechanisms. Life Sci 2003;72(20):2183–98.

31. Reppert SM, Weaver DR, Ebisawa T. Cloning and characterization of a mammalian melatonin receptor that mediates reproductive and circadian responses. Neuron 1994;13(5): 1177–85.

32. Reppert SM, Godson C, Mahle CD, Weaver DR, Slaugenhaupt SA, Gusella JF. Molecular characterization of a second melatonin receptor expressed in human retina and brain: the Mel1b melatonin receptor. Proc Natl Acad Sci U S A 1995;92(19):8734–8.

33. Nosjean O, Ferro M, Coge F, Beauverger P, Henlin JM, Lefoulon F, et al. Identification of

the melatonin-binding site MT3 as the quinone reductase 2. J Biol Chem 2000;275(40): 31311–7.

34. Dubocovich ML, Benloucif S, Masana MI. Melatonin receptors in the mammalian suprachiasmatic nucleus. Behav Brain Res 1996;73(1–2):141–7.

35. Poirel VJ, Masson-Pevet M, Pevet P, Gauer F. MT1 melatonin receptor mRNA expression exhibits a circadian variation in the rat suprachiasmatic nuclei. Brain Res 2002;946(1): 64–71.

36. Vanecek J. Cellular mechanisms of melatonin action. Physiological Reviews 1998;78(3): 687–721.

37. Liu RY, Zhou JN, Hoogendijk WJ, van Heerikhuize J, Kamphorst W, Unmehopa UA, et al. Decreased vasopressin gene expression in the biological clock of Alzheimer disease patients with and without depression. J Neuropathol Exp Neurol 2000;59(4):314–22.

38. Zhou JN, Hofman MA, Swaab DF. VIP neurons in the human SCN in relation to sex, age, and Alzheimer's disease. Neurobiol Aging 1995;16(4):571–6.

39. Kunz D, Schmitz S, Mahlberg R, Mohr A, Stoter C, Wolf KJ, et al. A new concept for melatonin deficit: on pineal calcification and melatonin excretion. Neuropsychopharmacology 1999;21(6):765–72.

40. Skene DJ, Vivien-Roels B, Sparks DL, Hunsaker JC, Pevet P, Ravid D, et al. Daily variation in the concentration of melatonin and 5-methoxytryptophol in the human pineal gland: effect of age and Alzheimer's disease. Brain Res 1990;528(1):170–4.

41. Greenberg LH. Regulation of brain adrenergic receptors during aging. Fed Proc 1986;45(1): 55–9.

42. Iguichi H, Kato KI, Ibayashi H. Age-dependent reduction in serum melatonin concentrations in healthy human subjects. J Clin Endocrinol Metab 1982;55(1):27–9.

43. Nair NP, Hariharasubramanian N, Pilapil C, Isaac I, Thavundayil JX. Plasma melatonin-- an index of brain aging in humans? Biol Psychiatry 1986;21(2):141–50.

44. Ferrari E, Casale G, Dori D, Solerte SB, Fioravanti F, Magri G, et al. Circadian pattern of melatonin and cortisol secretion in physiological and pathologcal cerebral aging. In: Touitou Y, Arendt J, Pevet P, editors. Melatonin and the pineal gland - From basic science to clinical application: Elsevier Science Publishers B.V.; 1993. p. 379–382.

45. Mishima K, Okawa M, Hishikawa Y, Hozumi S, Hori H, Takahashi K. Morning bright light therapy for sleep and behavior disorders in elderly patients with dementia. Acta Psychiatr Scand 1994;89:1–7.

46. Rodenbeck A, Huether G, Ruther E, Hajak G. Altered circadian melatonin secretion patterns in relation to sleep in patients with chronic sleep-wake rhythm disorders. J Pineal Res 1998;25(4):201–10.

47. Sack RL, Lewy AJ, Erb DL, Vollmer WM, Singer CM. Human melatonin production decreases with age. Journal of Pineal Research 1986;3(4):379–388.

48. Sharma M, Palacios-Bois J, Schwartz G, Iskandar H, Thakur M, Quirion R, et al. Circadian rhythms of melatonin and cortisol in aging. Biol Psychiatry 1989;25(3):305–19.

49. Touitou Y, Haus E. Alterations with aging of the endocrine and neuroendocrine circadian system in humans. Chronobiology International 2000;17(3):369–390.

50. Zhdanova IV, Wurtman RJ, Balcioglu A, Kartashov AI, Lynch HJ. Endogenous Melatonin Levels and the Fate of Exogenous Melatonin: Age Effects. Journal of Gerontology 1998;53A(4):B293-B298.

51. Kennaway DJ, Lushington K, Dawson D, Lack L, Van den Heuvel C, Rogers N. Urinary 6-sulfatoxymelatonin excretion and aging: New results and a critical review of the literature.

J. Pineal Res. 1999;27:210–220.

52. Waldhauser F, Weiszenbacher G, Tatzer E, Gisinger B, Waldhauser M, Schemper M, et al. Alterations in nocturnal serum melatonin levels in humans with growth and aging. J Clin Endocrinol Metab 1988;66(3):648–52.

53. Zeitzer JM, Daniels JE, Duffy JF, Klerman EB, Shanahan TL, Dijk DJ, et al. Do plasma melatonin concentrations decline with age? [In Process Citation]. Am J Med 1999;107(5): 432–6.

54. Touitou Y. Human aging and melatonin. Clinical relevance. Exp Gerontol 2001;36(7): 1083–100.

55. Mishima K, Okawa M, Shimizu T, Hishikawa Y. Diminished melatonin secretion in the elderly caused by insufficient environmental illumination. J Clin Endocrinol Metab 2001;86(1):129–34.

56. Baskett JJ, Cockrem JF, Todd MA. Melatonin levels in hospitalized elderly patients: a comparison with community based volunteers. Age Ageing 1991;20(6):430–4.

57. Ohashi Y, Okamoto N, Uchida K, Iyo M, Mori N, Morita Y. Daily rhythm of serum melatonin levels and effect of light exposure in patients with dementia of the Alzheimer's type. Biol Psychiatry 1999;45(12):1646–52.

58. Touitou Y, Fevre M, Bogdan A, Reinberg A, De Prins J, Beck H, et al. Patterns of plasma melatonin with ageing and mental condition: stability of nyctohemeral rhythms and differences in seasonal variations. Acta Endocrinologica 1984;106(2):145–51.

59. Bliwise DL. Sleep in normal aging and dementia. Sleep 1993;16(1):40–81.

60. Haimov I, Lavie P, Laudon M, Herer P, Vigder C, Zisapel N. Melatonin replacement therapy of elderly insomniacs. Sleep 1995;18(7):598–603.

61. Kripke DF, Elliot JA, Youngstedt SD, Smith JS. Melatonin: marvel or marker? Ann Med 1998;30(1):81–7.

62. Baskett JJ, Wood PC, Broad JB, Duncan JR, English J, Arendt J. Melatonin in older people with age-related sleep maintenance problems: a comparison with age matched normal sleepers. Sleep 2001;24(4):418–24.

63. Zhdanova IV, Wurtman RJ, Regan MM, Taylor JA, Shi JP, Leclair OU. Melatonin treatment for age-related insomnia. J Clin Endocrinol Metab 2001;86(10):4727–30.

64. Baskett JJ, Broad JB, Wood PC, Duncan JR, Pledger MJ, English J, et al. Does melatonin improve sleep in older people? A randomised crossover trial. Age Ageing 2003;32(2): 164–170.

65. Hurlbut EC, King TS, Richardson BA, Reiter RJ. The effects of the light:dark cycle and sympathetically-active drugs on pineal N-acetyltransferase activity and melatonin content in the Richardson's ground squirrel, Spermophilus richardsonii. Prog Clin Biol Res 1982;92:45–56.

66. Stoschitzky K, Sakotnik A, Lercher P, Zweiker R, Maier R, Liebmann P, et al. Influence of beta-blockers on melatonin release. Eur J Clin Pharmacol 1999;55(2):111–5.

67. Drijfhout WJ. Melatonin on-line. Development of trans pineal microdialysis and its application in pharmacological and chronobiological studies. Groningen: Rijksuniversiteit Groningen; 1996.

68. Djeridane Y, Touitou Y. Chronic diazepam administration differentially affects melatonin synthesis in rat pineal and Harderian glands. Psychopharmacology (Berl) 2001;154(4): 403–7.

69. Wakabayashi H, Shimada K, Satoh T. Effects of Diazepam Administration on Melatonin Synthesis in the Rat Pineal Gland in vivo. Chem Pharm Bull 1991;39(10):2674–2676.

70. Joy JE, Losee-Olson S, Turek FW. Single injections of triazolam, a short-acting benzodiazepine, lengthen the period of the circadian activity rhythm in golden hamsters. Experientia 1989;45(2):152–4.

71. Johnson RF, Smale L, Moore RY, Morin LP. Lateral geniculate lesions block circadian phase-shift responses to a benzodiazepine. Proc Natl Acad Sci U S A 1988;85(14):5301–4.

72. Fourtillan JB, Brisson AM, Gobin P, Ingrand I, Decourt JP, Girault J. Bioavailability of melatonin in humans after day-time administration of D(7) melatonin. Biopharm Drug Dispos 2000;21(1):15–22.

73. Serfaty M, Kennell-Webb S, Warner J, Blizard R, Raven P. Double blind randomised placebo controlled trial of low dose melatonin for sleep disorders in dementia. Int J Geriatr Psychiatry 2002;17(12):1120–7.

74. Van Someren EJ, Swaab DF, Colenda CC, Cohen W, McCall WV, Rosenquist PB. Bright light therapy: improved sensitivity to its effects on rest-activity rhythms in Alzheimer patients by application of nonparametric methods. Chronobiol Int 1999;16(4):505–18.

75. Garfinkel D, Laudon M, Zisapel N. Improvement of sleep quality by controlled-release melatonin in benzodiazepine-treated elderly insomniacs. Archives of Gerontology & Geriatrics 1997;24:223–231.

76. Jean-Louis G, Von Gizycki H, Zizi F. Melatonin effects on sleep, mood, and cognition in elderly with mild cognitive impairment. J Pineal Res 1998;25:177–183.

77. Garfinkel D, Zisapel N, Wainstein J, Laudon M. Facilitation of benzodiazepine discontinuation by melatonin: a new clinical approach. Arch Intern Med 1999;159(20):2456–60.

78. Andrade C, Srihari BS, Reddy KP, Chandramma L. Melatonin in medically ill patients with insomnia: a double-blind, placebo-controlled study. J Clin Psychiatry 2001;62(1):41–5.

79. Garfinkel D, Laudon M, Nof D, Zisapel N. Improvement of sleep quality in elderly people by controlled-release melatonin. Lancet 1995;346(8974):541–4.

80. Yu BP. Approaches to anti-aging intervention: the promises and the uncertainties. Mech Ageing Dev 1999;111(2–3):73–87.

81. Olde Rikkert MG, Rigaud AS. Melatonin in elderly patients with insomnia. A systematic review. Z Gerontol Geriatr 2001;34(6):491–7.

# IV.    AGEING OF THE NERVOUS SYSTEM

*The Neuroendocrine Immune Network in Ageing*
Edited by R.H. Straub and E. Mocchegiani

# Age-Related Changes of the Human Autonomic Nervous System

MARCUS W. AGELINK[1], DIRK SANNER[1] and DAN ZIEGLER[2]

[1]*Institute of Biological Psychiatry & Neuroscience, Ruhr-University, Bochum, Germany;*
[2]*The German Diabetes Research Institute, Heinrich-Heine University, Düsseldorf, Germany*

## ABSTRACT

Ageing is accompanied by significant modifications of the cardiovascular system both, structural and functional. Various components of the ANS may be differentially and selectively affected. Considering the ANS control of the heart there is a decrease of cardiovagal modulation with increasing age indicated by a decrease of cardiovagally mediated indices of heart rate variability (HRV) during rest and also during various provocative maneuvers. Sympathetic outflow to the heart is elevated, however, it is not well transformed into a correspondingly enhanced end-organ response of the heart because of a decrease of α-and ß-adrenergic receptor potency. Orthostatic dysregulation is common in the elderly due to several reasons; one major point is that the barore-flexes are impaired with increasing age.

1.  INFLUENCE OF AGE ON AUTONOMIC NERVOUS SYSTEM (ANS) FUNCTION: METHODOLOGICAL ASPECTS AND LIMITATIONS

Although general consensus exists that the ANS is universally affected by age, controversial findings have appeared regarding the effects of the ageing process on defined, individual aspects of ANS function. This is due in part to considerable differences between the cohorts studied: not only do genetic, cultural and psychosocial factors significantly influence ANS function, but also other lifestyle factors such as smoking, alcohol consumption and body weight have an effect. To name one example, an increase in blood pressure with ageing occurs in Western societies but less often or not at all in developing societies [1]. Another major problem in the approach to the biology of ANS ageing is differentiating truly age-dependent alterations from those aris-ing from disease conditions (e.g. latent atherosclerosis, coronary heart disease, cerebrovascular disease, metabolic disease) and / or medication. For example, it is known from autopsy studies that over 60% of patients dying at the age of 60 or older have at least one coronary artery with greater occlusion, and that the majority of older people with significant coronary artery disease are asymptomatic [2]. Ultimately, misleading interpretations frequently arise with the evaluation of age effects on the ANS when extrapolations are made on the basis of individual findings. The age effects on the ANS, however, are in fact of an extremely complex nature that depend on the autonomic feedback system under consideration, the level of the autonomic neuroaxis, and the end-organ tissue, each of which may be differentially and selectively affected. For example, there is convincing evidence that plasma norepinephrine (NE) levels increase with ageing [3–6].

Plasma NE arises from sympathetic postganglionic nerve terminals and is affected by the rate of NE secretion and clearance [7]. Some investigators have concluded that sympathetic function in general increases with ageing [8] without taking into consideration the age-related changes in preganglionic neurons, baroreceptor reflexes and effector organ function. Such an oversimplification may lead to misinterpretations as the following examples illustrate: with preganglionic sympathetic innervation, a progressive reduction in the number of preganglionic sympathetic neurons of the intermediolateral cell column occurs that starts in adult life [9–11]. Concerning postganglionic sympathetic activity, ageing was found to be associated with an elevated cardiac NE spillover rate, a higher muscle sympathetic nerve activity (MSNA), and a reduction in overall heart rate variability, especially regarding the spectral low frequency (LF) power [12]. The microneurographically measured MSNA is the most direct measure of peripheral, postganglionic sympathetic nerve activity. Cardiac NE spillover is related to both sympathetic nerve firing and electrochemical coupling of the neural signal to the heart. The $LF_{HR}$-power reflects the sympathetic and parasympathetic modulation of heart rate [13]. Even though ageing was associated with an elevation of sympathetic outflow to the heart, such an increase in sympathetic nerve activity was obviously not transformed into an enhanced functional end-organ response, since the $LF_{HR}$-power did not reveal the expected increase, but instead a significant reduction with increasing age [12]. This pattern of findings can be understood if one considers that with increasing age a desensitization of beta-adrenoreceptors occurs as well as an impairment of post-receptor signal transduction [14,15]. The example presented emphasizes that various parameters of sympathetic nerve activity such as HRV indices (LF-power, low to high frequency ratio), MSNA or plasma NE describe different aspects of efferent sympathetic nerve activity. In other words: *the MSNA recorded by microneurographic measurements of postganglionic efferents that innervate the skeletal muscle with nerve fibers does not necessarily correspond with the firing rate of cardiac sympathetic nerve fibers* [16]. Also, the LF-power reflects the sum of all the physiological effects that determine the sympathetic nerve activity at the heart; accordingly, this is modified not only by the firing rate of sympathetic neurons, but also by their electrochemical signal transduction, the sensitivity of adrenergic receptors and the cardiovascular reflex pattern. Thus, any study of only single isolated aspects of ANS function might in fact misrepresent the true relationship between autonomic nerve activity and its functional end-organ response.

The presented chapter describes the effects of age on ANS function, focusing on ANS regulation of the cardiovascular system. The influence of age on the results of autonomic functional tests that are of special interest to clinicians evaluating patients with autonomic neuropathy shall be addressed in particular.

## 2.     INFLUENCE OF AGE ON THE HEART AND THE BLOOD VESSELS

Structural and functional differences concerning the heart muscle, the cardiac bioelectrical system and blood vessels exist between younger and older adults. Structural changes in the heart include slightly increased organ weight with a certain degree of left ventricular hypertrophy; functionally, there is a slight decrease in the early left ventricular filling rate [17]. Cell death of the sinoatrial pacemaker cells occurs which results in a reduction in resting heart rate and the risk of an AV-block [2]. Age-conditioned, functional alterations in cardiac function become more manifest during exercise: the effort tachycardia is diminished and the same applies to the various indices of myocardial contractility [18]. With increasing age there is thickening above all of the vascular intima and media; an overall loss in distensibility results from this, which is the

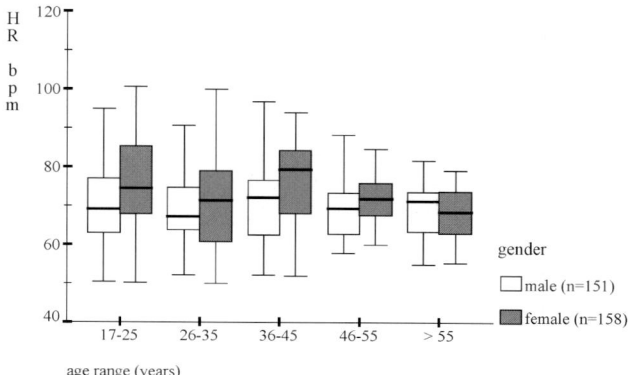

Figure 1 shows the distribution of the resting heart rate (5-minute short term HRV recordings) separated by age and gender. The upper and lower boundaries of the boxes represent the upper and lower quartiles. The box length is the interquartile distance. The line inside the box identifies the group median. The lines emanating from each box extend to the smallest and largest observations less than one interquartile range from the end of the box.

most relevant consequence of ageing in the arterial system. Several important humoral factors involved in circulatory control are affected by ageing. Ageing is accompanied by a significant increase in plasma NE and vasopressin levels and counteractively by an attenuated function of the renin-angiotensin system [18]. Ageing is associated with an increased sympathetic nervous system outflow to the heart, gut and skeletal muscle, but not to the kidney [19–22]. However, the responsiveness to β-adrenergic stimuli is decreased [14,15] and this is in part compensated by an increase in circulating catechol-amines. While on the one hand the heart rate variability is reduced in older compared to younger adults (see also paragraph 4), the overall blood pressure variability might be increased [18].

## 3.    EFFECTS OF AGEING ON RESTING HEART RATE

It was rather consistently reported that the resting heart rate tends to decrease with increasing age [23]; some studies found an inverse correlation between heart rate and age [24,25], although most studies failed to confirm such a linear relationship [26–32]. The measured heart rate is the sum of the intrinsic heart rate and its mutual influencing by sympathetic and parasympathetic modulation, of which vagal influences on resting heart rate predominate in man. Since older people show a reduced cardio-vagal modulation compared to younger adults (see paragraph 4), a higher heart rate in older people compared to younger people should in fact be expected when considering the ANS control of heart rate. However, most studies including our own investigations [26,32], failed to verify any increase in resting heart rate in older compared to younger people. We believe that a reduction in intrinsic sinus node rate with ageing [24,25,33] is responsible for this pattern of findings. Cardiac electrophysiologic studies have demonstrated a progressive decline in sinoatrial conduction and sinus node recovery time with ageing [24]. Remarkably, a number of large studies [28,29,32,34–38] found that healthy, young and middle-aged women had a higher heart rate compared to age-matched healthy men (Figure 1), whereby the reason underlying this finding remains unclear. Hormonal factors were discussed as possible causes for these sex differences in heart rate [32,39].

Table I    Summary of selected HRV-indices obtained from 5-minute HRV short term recordings.

| HRV parameter | interpretation |
|---|---|
| *time domain method* | |
| coefficient of variation (CV) | reflects cardiovagal activity |
| root mean square of successive differences (RMSSD) | reflects cardiovagal activity |
| *frequency domain method* | |
| very low frequency power; VLF (0.003-0.04Hz) | physiological interpretation is less clear; thermoregulatory, vascular and other effects are discussed |
| low frequency power; LF (0.04-0.15 Hz) | depends on sympathetic and parasympathetic modulations |
| high frequency power; HF (0.15-0.4 Hz) | mainly reflects cardiovagal modulation |
| normalized (relative) spectral power LFnu; HFnu | percentage of LF (HF)-power as a proportion of the total power; mainly reflects sympathetic (LFnu), respectively, vagal (HFnu) modulations (Q) |

## 4.    EFFECTS OF AGEING ON CARDIOVAGAL HEART RATE TESTS

Preganglionic vagal motoneurons originate from the nucleus ambiguus and the dorsal vagal nucleus. The afferent supply of these central areas originates from the nucleus of the solitary tract (NTS), the raphe nucleus, the medullary reticular formation, the hypothalamus and the amygdala. The activity of cardiovagal motoneurons is largely controlled by the excitatory effects of afferent impulses from the baroreceptors that are mediated via the NTS, and the inhibitory effects of breath activity [40] .

### 4.1.    Short-term resting heart rate variability (HRV)

The short-term recording of resting heart rate variability (HRV) represents a non-invasive, economic and easily carried out technique of broad applicability and which allows a sufficient quantitative estimation of cardiovagal and (with some restrictions) cardiac sympathetic modulation. Practical guidelines and recommendations for the evaluation and interpretation of findings have recently been standardized by the European Society of Cardiology for the most part [13]. One can distinguish time- (e.g. coefficient of variation, root mean square of successive differences) and frequency-domain HRV indices (very-low frequency, low frequency and high frequency power); selected HRV parameters and their physiological importance are summarized in Table I. The spectral components can be expressed either as absolute or relative values (i.e. "normalized" with respect to the total power). Normalizing is achieved by dividing each of the spectral components by the total power or by the total power minus the VLF component, and by multiplying this value by 100. The latter calculation is based on the fact that the duration of analysis of the HRV should be equivalent to 10 times the duration of the wavelength of the spectral frequency to be measured (which for the VLF component would normally have to last 50 minutes). For this reason interpretation of the VLF component is limited when short-term recordings of HRV are applied [13].

There is general consensus that the high-frequency (HF) spectral components (Table I) mainly reflect the centrally mediated, efferent vagal modulation [13,41,42]. The low-frequency (LF) spectra probably reflect both sympathetic and parasympathetic modulation [13,41,42]. Atropine reduces both the HF as well as LF components in lying probands. The following observations suggest a predominant role of the sympathetic nervous system in affecting the LF- (and particularly the normalized LF-) component; namely 1.) that an increase in the LF component results during mental stress [43] or orthostatic loading [44]; 2.) that a reduction in the LF component results following beta-adrenoreceptor blockade [45], after application of the central alpha-2-receptor antagonist clonidine [46] or after bilateral severance of the stellatum ganglion in dogs [45,47] and; 3.) that at least with sympathetic activation a correlation could be verified between the LF component of heart rate and the LF component of the microneurographically registered MSNA [48,49]. On the basis of these observations, the quotient of the LF and the HF power has been taken by most [45,46,50,51], although not all research groups [52], as an expression of the sympathovagal balance.

Age effects on the time- and frequency-dependent parameter HRV have been evaluated in numerous recent studies [26–32,35–39,53,54]; some newer studies and their most important findings are summarized in Table II. The time-dependent para-meters, particularly the coefficient of variation and the RMSSD decrease with age. HRV short-term recordings appear to be particularly appropriate for evaluating frequency-dependent HRV indices [13]. Most studies revealed consistently that with increasing age the absolute power decreases for all frequency components and therefore also for the total power. There are inconsistent findings regarding the age dependence of the normalized power (see Table II).

We recently applied 5-min short-term HRV recording to a large sample of 309 healthy subjects aged between 18 and 77 years [32]. Recordings were performed after lying for 10 minutes. Measurements and evaluation of findings were based on previously published guidelines [13]. We found a negative correlation between age and the RMSSD (Figure 2) and the spectral total power (Figure 3); the latter aspect concerned the absolute VLF-, LF- as well as HF power. This correlation was found for both sexes; the best correlation with age was found for the absolute HF-power ($r = -0.53$). The normalized HF (nu)-power revealed a negative correlation with age (Figure 4a) for both sexes; in contrast, no significant association was found between the LF(nu)-power (Figure 4b) and age for either sex. Considering that the relative HF-power is the most reliable indicator of parasympathetic modulation amongst the various HRV indices obtained from short-term recordings, the pattern of findings in our study indicates a reduction of cardiovagal modulation with increasing age. Since there is evidence that sympathetic nervous outflow to the heart is elevated in older subjects, one might theoretically expect an increase in the LF(nu)-power with age. However, other components contributing to the measured LF-power are also impeded in function with the ageing process [12]. With increasing age, for example, a decrease in the sensitivity of adrenergic receptors occurs so that the increased sympathetic outflow is not transformed into an enhanced end-organ response. These counteractive processes – increase in sympathetic outflow and reduction in the sensitivity of adrenergic receptors – might produce the net effect that no significant association can be detected between LF(nu) power and age.

Prospective studies on gender effects including HRV short-term recordings have reported inconsistent results regarding the frequency domain HRV indices. For women compared to men, no differences [27] or a lower LF-power have been reported [37–39]. For HF-power, no difference [37] or an increase was observed in women [27,38,39], while for the LF/HF ratio a reduction was observed among women [37,38]. Inconsistencies among these studies may have arisen due to differences between study populations or analytical methods. We recently found that women showed a lower absolute and normalized LF-power and a lower LF/HF ratio compared to men [32]. There were no significant gender differences for the absolute HF-power and

Table II    Summary of larger studies on the age-effects on various time- and frequency domain HRV indices (for definitions see Table I).

| author | year | subjects (n) | age range | methods | major findings |
|---|---|---|---|---|---|
| Ziegler et al. | 1992 | n = 120 | 15–67 years | 5-min supine resting | time- and frequency domain HRV-indices linearly correlated with age (normalized power not calculated); no gender differences; no age-effects on HR |
| Ryan et al. | 1994 | n = 67 | 20–90 years | 8-min supine resting | total power, LFnu, HFnu and LF/HF lower in older compared to younger subjects; women had higher HFnu and total power compared to men; no age-effects on HR |
| Liao et al. | 1995 | n = 1984 | 45–64 years | 2-min supine resting | LF- and HF-powers inversely associated with age even after adjustment for race and sex effects; lower LF-power in women compared to men |
| Jensen-Urstad et al. | 1997 | n = 101 | 20–69 years | 24-hour Holter recording | time- and frequency domain HRV indices inversely correlated with age; LF and LF/HF ratio lower in women compared to men; no age-effects on HR |
| Stein et al. | 1997 | n = 60 | 26–76 years | 24-hour Holter | analysis of 14 HRV-indices; time and frequency domain HRV indices were lower in older compared to younger men; only shorter term indices were lower in older compared to younger women; young women had a lower LF-power compared to young men |
| Umetani et al. | 1998 | n = 260 | 10–99 years | 24-hour Holter recording | analysis restricted to time domain HRV indices; HRV decrease with age; HR declined with age and was higher in younger females compared to males |
| Ramaekers et al. | 1998 | n = 276 | 18–71 years | 24-hour Holter recording | time domain and frequency domain HRV indices inversely correlated with age (except: normalized spectral powers and LF/HF ratio); LF, LFnu and LF/HF lower in women compared to men; no age-effects on HR |
| Sinnreich et al. | 1998 | n = 294 | 35–65 years | 5-min supine resting | VLF, LF, HF, total power and RMSSD declined with age; lower LF- and higher HF-powers in women compared to men |
| Kuo et al. | 1999 | n = 1070 | 40–79 years | 5-min supine resting | inverse correlation between age and absolute or normalized spectral powers except HF(nu); men had higher LF(nu) and LF/HF compared to women; slight decrease of HR with increasing age |
| Fagard et al. | 1999 | n = 424 | 25–89 years | 15-min supine resting | absolute and normalized LF and HF declined with age LF(nu) in women and LF/HF in men; young men had higher LF- and lower HF-power compared to young women; no age-effects on HR |
| Fukusaki et al. | 2000 | n = 373 | 16–69 years | 5-min supine resting | harmonic (HF-, LF- power) and non-harmonic components of HRV decreased with age; no age-effects on HR |
| Agelink et al. | 2001 | n = 309 | 18–77 years | 5-min supine resting | linear decline of time- and frequency domain HRV indices with age (except LF/HF ratio and HR); middle- aged women have lower (higher) LFnu (HFnu)-power compared to middle aged men |

Table III   Summary of larger studies on the age-effects on heart rate responses to deep breathing.

| author | year | subjects (n) | age (range) | methods | major findings |
|---|---|---|---|---|---|
| Smith and Smith | 1981 | n = 174 | 16–89 years | 6 breaths / min; 1 min; body position ? | E-I difference declined linearly with age, no gender differences, no age-effects on HR |
| Wieling et al. | 1982 | n = 133 | 10–65 years | 6 breaths / min; 1 min; supine resting | E-I difference declined with age |
| Pfeifer et al. | 1983 | n = 103 | 19–82 years | 5 breaths / min; 5 min; supine resting | E-I difference linearly correlated with age |
| Oikawa et al. | 1985 | n = 162 | 4–77 years | 6 breaths / min; 2 min; supine resting | HRV log-linear reduction with age |
| Masaoka et al. | 1985 | n = 143 | 20–80 years | 6 breaths / min; 1 min; supine resting | E-I difference linearly declined with age |
| Vita et al. | 1986 | n = 70 | 25–71 years | 6 breaths /min; 1 min; sitting position | E-I difference lineraly declined with age |
| O'Brien et al. | 1986 | n = 310 | 18–85 years | 1 cycle; 10 sec; body position ? | HRV decreased with age |
| Gautschy et al. | 1986 | n = 120 | 22–92 years | 6 breaths / min; 1 min; sitting position | HRV decreased with age |
| Low et al. | 1990 | n = 122 | 10–83 years | 8 breathing cycles; 1.3 min; supine resting | E-I difference linearly declined with age |
| Ingall et al. | 1990 | n = 72 | 5–85 years | 6 breathing cycles; 1min; semisitting | E-I difference declined with age |
| Ziegler et al. | 1992 | n = 120 | 15–67 years | 6 breaths / min; 100 R-R intervals; supine resting | E-I difference and E/I ratio declined linearly with age, no age-effects on HR |
| Piha | 1993 | n = 224 | 21–80 years | 6 breaths / min; duration? supine resting | E/I ratio declined with age in both sexes; no age-effects on HR; HR was higher in middle-aged women compared to men |
| Braune et al. | 1996 | n = 137 | 18–85 years | 6 breaths / min; duration? body position? | E-I difference declined linearly with age; no age-effects on HR |
| Agelink et al. | 2001 | n = 309 | 18–77 years | 6 breaths / min; 100 R-R intervals; supine resting | E-I difference and E/I ratio declined linearly with age; no gender differences |

222

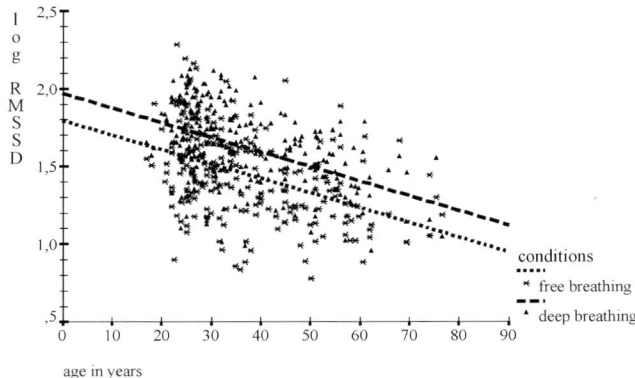

age in years

Figure 2 illustrates the age-dependence of the RMSSD. HRV recordings were performed during free breathing (5-minute supine resting study) and also over 100 consecutive R-R intervals during metronomic, deep breathing cycles (inspiration 6 sec, exspiration 4 sec).

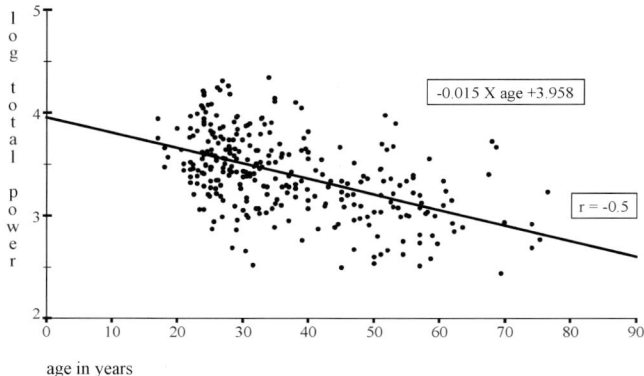

age in years

Figure 3 shows the age dependency of the total spectra power defined as the sum of the LF, HF and VLF spectrum (see also Table I).

the total power. When calculating normalized LF(nu)- and HF(nu)-powers, middle-aged women showed a significantly higher relative HF-power and a significantly lower relative LF-power compared to middle-aged men (Figure 5a,b). Our results are largely compatible with one of the largest studies performed to date (n=1.984) by Liao et al. [37] who also showed a reduction in LF-power in women compared to men and no difference in the absolute HF-power (Table II).

4.2.    HRV during deep respiration

This test is probably the most suitable of all classical HRV tests for evaluating parasympathetic activity, since both afferent and efferent parts of the reflex arch are *mostly vagally* determined [40]. During a respiratory cycle the maximum R-R interval during expiration (E) and the minimum interval during inspiration (I) are measured; the E-I difference ($R-R_{max} - R-R_{min}$) or the E-I quotient ($R-R_{max} / R-R_{min}$) are calculated. The test results mainly depend on breath rate, breath depth and body position [40].

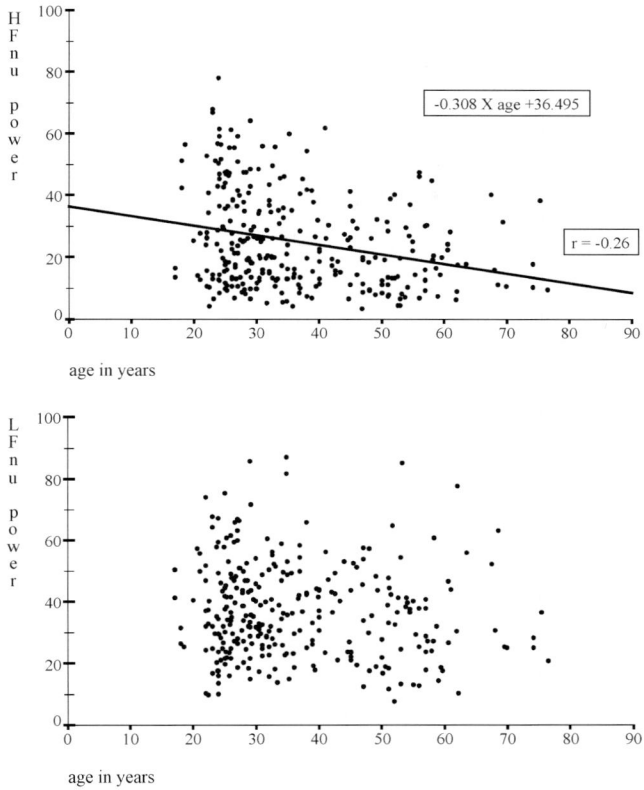

Figure 4 illustrates the relationship between age and the normalized spectral power of the HFnu (fig. 4a) and LFnu (fig. 4b) component. A significant decline with age was only found for the HF(nu) power. Since there were no gender differences in age-effects, values for both sexes were matched.

Numerous studies [26,34,55–65] have examined the influence of age on respiratory sinus arrhythmia; selected large cohort studies are summarized in Table III. All studies involving large cohorts of healthy subjects revealed a progressive reduction in respiratory sinus arrhythmia with age (Table III); most studies demonstrated a linear association between age and the heart rate responses to deep breathing given either in absolute or logarithmic values. We recently studied the heart rate response to deep breathing in 309 healthy subjects aged between 18 and 77 years and found no significant gender differences [32]. There was, however, a significant regression with age for the E-I difference and the E-I ratio (Figure 6). In our study the E-I ratio, unlike the E-I difference, was independent of the resting heart rate.

4.3.    Heart rate response to standing

Changing from lying to standing is followed by an accumulation of blood in the lower limbs, a reduction in central venous pressure, the pressure in the right atrium, and the end-diastolic pressure in the left ventricle; stroke volume and cardiac output decrease transiently. Arterial mean pressure in the carotid artery decreases by around 15–20 mmHg, but aortal pressure shows

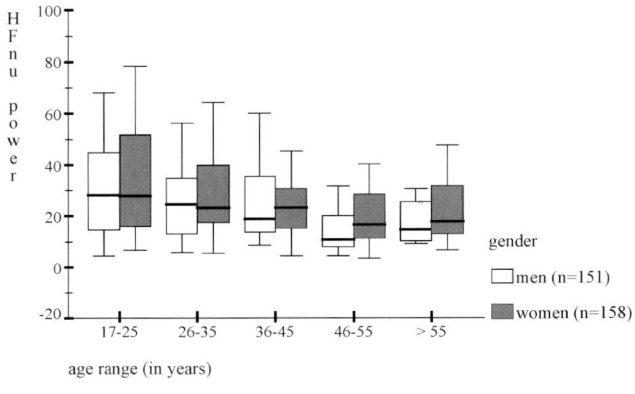

age range (in years)

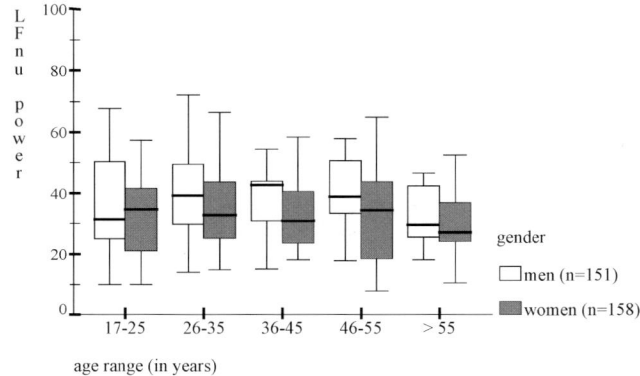

age range (in years)

Figure 5 shows the normalized HFnu (fig. 5a) and LFnu (fig. 5b) powers with respect to age and gender. The upper and lower boundaries of the boxes represent the upper and lower quartiles. The box length is the interquartile distance. The line inside the box identifies the group median. The lines emanating from each box extend to the smallest and largest observations less than one interquartile range from the end of the box. Middle-aged women had a significantly lower LFnu- (F=8.91; p=0.003) and higher HFnu-power (F=5.61, p=0.019) compared to age-matched men (see also 32).

only a small decrease. Upon activation of the baroreceptor reflex there is a counter-regulatory decrease in parasympathetic and an increase in sympathetic modulation together with increases in NE levels, peripheral resistance, venous tone and heart rate [66]. Since reflex tachycardia after orthostatic loading is masked by atropine but not propranolol, this effect is most likely accomplished through a reduction in vagal activity [67]. The increase in heart rate often occurs with a maximum around the 15th heartbeat, followed by a reflex (baroreceptor mediated), overcompensating bradycardia with a maximum around the 30th heartbeat (30:15 ratio). Since the length of the R-R interval varies within certain thresholds [26], it has proved useful to define the quotient within these limits as the R-Rmax(21–45) divided by the R-Rmin(5–25). Although the 30:15 ratio is supposed mainly to assess the vagal regulation [67], it should be considered that the physiological mechanisms underlying the orthostatic reaction are affected by a complicated interplay between baroreceptors, sympathetic and parasympathetic efferents, whereby the reflex response is probably also modulated by interactions with the vestibular and cerebellar systems [66].

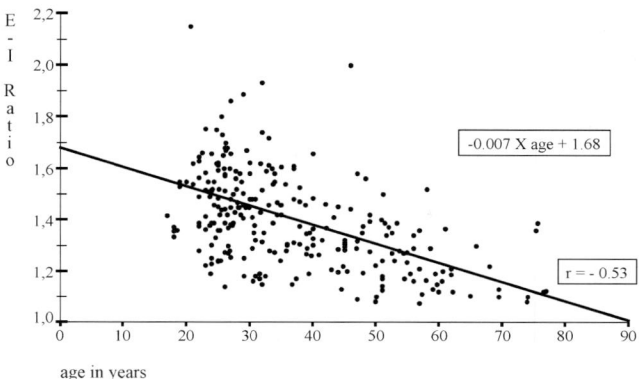

Figure 6 shows the age dependency of the E-I ratio defined as the maximal R-R interval length during exspiration divided by the minimal interval length during inspiration.

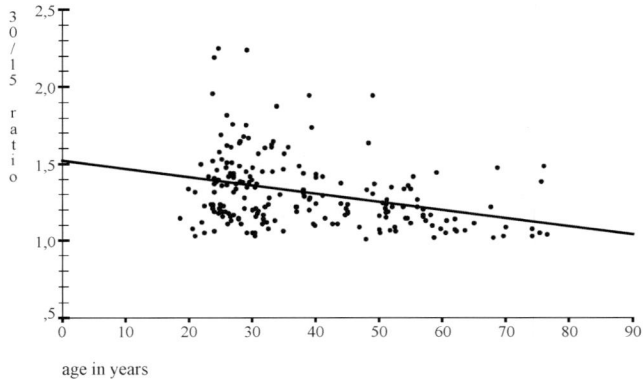

Figure 7. Correlation between age and the "so-called" 30:15 ratio (defined as the quotient of the maximal interval length R-R$_{max(21-45)}$ divided by minimal interval length R-R$_{min(5-25)}$.

A participation of the last-named structures can also be concluded from the observation that the heart rate increases less when changing from a lying to a standing position than it does when changing from a sitting to an upright position [68].

The 30:15 ratio shows a significant age dependence [26,34,56,60–62]. In our own study of 309 healthy subjects [32], no sex differences were observed for the 30:15 ratio, although a significant ($p < 0.001$) correlation with age was seen (Figure 7). The 30:15 ratio did not depend on resting heart rate.

## 4.4.    Valsalva-test

The circulatory responses to the Valsalva maneuver can be subdivided into four phases [66]: after an initial increase of blood pressure (phase 1; probably mechanically mediated) blood pressure declines with continuing pressing due to a reduced blood return to the heart. This was followed by an increase in heart rate (phase 2) mediated by vagal inhibition and probably also sympathetic activation. Immediately upon termination of pressing (phase 3; sympathetically mediated) a short-term fall in blood pressure occurs in combination with a further increase in heart rate. In phase 4 the sympathetically-mediated increase of blood pressure results in a baroreceptor-mediated, transient reduction in heart rate due to an increase in vagal modulation. Although the reflex bradycardia during phase 4 can largely be masked by application of atropine, and therefore the Valsalva test is regarded as a test of cardiovagal modulation, this test in fact involves complex interplays between several populations of afferent autonomic receptors, and between efferents from both, the sympathetic and parasympathetic branch of the ANS [66]. Thus, the valsalva ratio depends on the functional integrity between both sides of the ANS, and does not primarily reflect cardiovagal modulation alone.

Age may affect these various components of the Valsalva maneuver in different ways. Most large cohort studies [26,34,55,61,62,64,69,70], although not all [60,71], were able to verify a reduction of the Valsalva ratio with increasing age. There are also contradictory findings regarding any possible gender differences with the Valsalva ratio [26,34,63].

5.     EFFECTS OF AGEING ON BAROREFLEXES AND BLOOD PRESSURE

The baroreflex is the essential control element of blood pressure regulation. When systemic pulse pressure, or mean arterial pressure is reduced, baroreceptors in the carotid sinus and the aortic arch are activated. Afferents of the baroreceptors influence the activity of cardiovagal motoneurones via the NTS; the activation of the baroreceptors is followed by an increase in heart rate (through vagal inhibition), vasocon-striction of the splanchnic and systemic vasculature, and restoration of blood pressure. The human baroreflex can be tested by assessing heart rate responses to changes in arterial pressure induced by infusions of vasoactive drugs (e.g. phenylephrine, sodium nitroprusside) or to changes in transmural pressure in carotid baroreceptors elicited by the neck chamber technique. Both methods have limitations, which have been discussed elsewhere [72]. The arterial baroreflex is modulated by other reflexes and by respiration: baroreflex stimuli are much more effective in producing vagally mediated heart rate responses when delivered during expiration than during inspiration. There is a positive correlation between the R-R interval and the sensitivity of the baroreflex [66]. Subjects with more sensitive baroreflexes have lower heart rates and vice versa. In addition, the R-R intervals vary more in subjects with increased baroreflex sensitivity. Baroreflexes may be depressed with increasing age [73–77].

We recently compared mean arterial blood pressure between two subgroups of healthy subjects (each n=50) whose age ranged between 21 and of 76 years (cut point: 50 years) and found no significant differences in mean arterial resting blood pressure [Agelink et al.; unpublished data]. Our findings correspond well to the results from others [65]. However, orthostatic hypotension may be a common finding in the elderly, although it may present as mild or even asymptomatic in most cases [78–80]. The tendency towards orthostatic hypotension in older people may have several reasons. First, starting in adulthood, there is a loss of intact, sympathetic preganglionic neurons [9–11]. This becomes symptomatic when about 50% of the neurons are lost [81]. Second, although sympathetic outflow and plasma NE levels become elevated with increasing age, these modifications are probably not followed by a corresponding increase in vascular resistance, since adrenergic $\alpha2$- and probably also $\alpha1$-receptors exhibit reduced potency with ageing [82–85]. Third, with increasing age a desensitization of adrenergic $\beta$-receptors also occurs, which is the reason why cardioacceleration and inotropic responses are reduced in older subjects [14,15]. Fourth, ANS support of blood pressure is altered with age due to the reduced cardiac vagal inhibition of heart rate and cardiac output [86]. Moreover, baroreflexes are impaired with increasing age [73–77].

In summary, ageing is accompanied by significant modifications of the cardio-vascular system both, structural and functional. Various components of the ANS may be differentially and selectively affected. Considering the ANS control of the heart there is a decrease of cardio-vagal modulation with increasing age indicated by a decrease of cardiovagally mediated HRV indices during rest and also during various provocative maneuvers. Sympathetic outflow to the heart is elevated, however, it is not well transformed into a correspondingly enhanced end-organ response because of a decrease of α- and ß-adrenergic receptor potency. Orthostatic dysregulation is common in the elderly due to several reasons; one major point is that the baroreflexes are impaired with increasing age.

# REFERENCES

1.  Kotchen JM, McKean HE, Kotchen TA. Blood pressure trends with aging. Hypertension 1982; 4 (suppl. III): 128–134.
2.  Melvin D, Cheitlin MD. Cardiovascular physiology- changes with aging. Am J Geriatr Cardiol 2003; 12: 9–13.
3.  Ziegler MG, Lake CR, Kopin J J. Plasma noradrenaline increases with age. Nature 1976; 261: 333–335.
4.  Goldstein DS, Lake CR, Chernow B et al. Age-dependence of hypertensive-normotensive differences in plasma norepinephrine. Hypertension 1983; 5: 100–104.
5.  Morrow LA. Age differences in plasma clearance mechanisms for epinephrine and norepinephrine in humans. J Clin Endocrinol Metab 1987; 65: 508–511.
6.  Fleg JL, Tzankoff SP, Lakatta EG. Age-related augmentation of plasma catecholamines during dynamic healthy males. J Appl Physiol 1985; 59: 1033–1039.
7.  Esler M et al. Age-dependent norepinephrine kinetics in normal subjects. Clin Sci 1981; 60: 217–219.
8.  Pfeifer MA et al. Differential changes of autonomic nervous system function with age in man. Am J Med 1983; 75: 249–258.
9.  Low PA, Dyck PJ. Splanchnic preganglionic neurons in man; II. Preganglionic ventral root fibers. Acta Neuropathol 1977; 40: 219–226.
10. Low PA, Dyck PJ. Splanchnic preganglionic neurons in man: III. Morphometry of myelinated fibers of rami communicantes. J Neuropathol Exp Neurol 1978; 37: 734–740.
11. Low PA, Okazaki H, Dyck PJ. Splanchnic preganglionic neurons in man. I. Morphometry of preganglionic cytons. Acta Neuropathol 1977; 40: 55–61.
12. Kingwell BA, Thompson JM, Kaye DM, et al. Heart rate spectral analysis, cardiac norepinephrine spillover, and muscle sympathetic nerve activity during human sympathetic nervous activation and failure. Circulation 1994; 90: 234–240.
13. Task Force of the European Society of Cardiology and the North American Society of Pacing and Electrophysiology. Heart rate variability. Standards of measurement, physiological interpretation, and clinical use. Circulation 1996; 93: 1043–1065.
14. Vestal, RE, Wood AJ, Shand DG. Reduced beta-adrenoreceptor sensitivity in the elderly. Clin Pharmacol Ther 1979; 26: 181–186.
15. Nielson CP, Vestal RE. Alpha-adrenoreceptors, beta-adrenoreceptors, and aging. In: Amery A, Staessen J, eds. Handbook of hypertension, vol12: Hypertension in the elderly. New York/Amsterdam. Elsevier Science Publishing Co Inc.; 1989: 51–62.

228

16. Wallin BG, Esler M, Dorward P, et al. Simultaneous measurements of cardiac noradrenaline spillover and sympathetic outflow to ske-letal muscle in humans. J Physiol 1992; 453: 45–56.
17. Miller TR, Grossman SJ, Schectman KB et al. Left ventricular diastolic filling and its association with age. Am J Cardiol 1986; 58: 531–535.
18. Ferrari AU. Modifications of the cardiovascular system with aging. Am J Geriatr Cardiol 2002; 11: 30–33.
19. Seals D, Esler MD. Human aging and the sympatho-adrenal system. J Physiol 2000; 528: 407–417.
20. Esler MD, Turner AG, Kaye DM et al. Aging effects on human sympathetic neuronal function. Am J Physiol 1995; 268: R278–285.
21. Ng AV, Callister R, Johnson DG, Seals DR. Age and gender influence muscle sympathetic nerve activity at rest in healthy humans. Hypertension 1993; 21: 498–503.
22. Sundlof G, Wallin BG. Human muscle nerve sympathetic activity at rest. Relationship to blood pressure and age. J Physiol 1978; 274: 621–637.
23. Low PA. The effects of aging on the autonomic nervous system. In: Clinical autonomic disorders. Evaluation and management. Low PA (ed)., Little Brown and Company, Boston, Toronto, London, 1993; 685–700..
24. De Marneffe M et al. Variations of normal sinus node function in relation to age: role of autonomic influence. Eur Heart J 1986; 7: 662–672.
25. Jose AD, Collison D. The normal range and determinants of the intrinsic heart rate in man. Cardiovasc Res 1970; 4: 160–167.
26. Ziegler D, Laux G, Dannehl K, et al. Assessment of cardiovascular autonomic function: age-related normal ranges and reproducibility of spectral analysis, vector analysis, and standard tests of heart rate variation and blood pressure responses. Diabet Med 1992; 9: 166–175.
27. Ryan SM, Goldberger AL, Pincus SM, et al. Gender- and age-related differences in heart rate dynamics: are women more complex than men? J Am Coll Cardiol 1994; 24: 1700–1707.
28. Jensen-Urstad K, Storck N, Bouvier F, et al. Heart rate variability in healthy subjects is related to age and gender. Acta Physiol Scand 1997; 160: 235–241..
29. Ramaekers D, Ector H, Aubert AE, et al. Heart rate variability and heart rate in healthy volunteers. is the female autonomic nervous system cardioprotective? Eur Heart J 1998; 19: 1334–1341.
30. Fagard RH, Paerdens K, Staessen JA. Influence of demographic, anthropometric and lifestyle characteristics on heart rate its variability in the population. J Hypertension 1999; 17: 1589–1599.
31. Fukusaki C, Kawakubo K, Yamamoto Y. Assessment of the primary effect of aging on heart rate variability in humans. Clin Autonom Res 2000; 10: 123–130.
32. Agelink MW, Malessa R, Baumann B et al. Standardized tests of heart rate variability: normal ranges obtained from 309 healthy humans, and effects of age, gender and heart rate. Clin Autonom Res 2001; 11: 99–108.
33. Lakatta EG. Ageing: cardiovascular system. In: Masoro EJ (ed). Handbook of physiology. New York, Oxford University Press, 1995; 413–474.
34. Piha SJ. Cardiovascular responses to various autonomic tests in males and females. Clin Auton Res 1993; 3: 15–20..
35. Umetani K, Singer DH, McCraty R, et al. Twenty-four hour time domain heart rate

variability and heart rate: relations to age and gender over nine decades. J Am Coll Cardiol 1998; 31: 593–601.

36. Cowan MJ, Pike K, Burr RL. Effects of gender and age on heart rate variability in healthy individuals and in persons after sudden cardiac arrest. J Electrocardiol 1994; 27 (Suppl): 1–9.

37. Liao D, Barnes RW, Chambless LE, et al. Age, race and sex differences in autonomic cardiac function measured by spectral analysis of heart rate variability – the ARIC study. Atherosclerosis Risk in Communities. Am J Cardiol 1995; 76: 906–912.

38. Sinnreich R, Kark JD, Friedlander Y, et al. Five minute recordings of heart rate variability for population studies: repeatability and age-sex characteristics. Heart 1998; 80: 156–162.

39. Huikuri HV, Pikkujamsa SM, Airaksinen KE et al. Sex-related differences in autonomic modulation of heart rate in middle-aged subjects. Circulation 1996; 94: 122–125.

40. Benarroch EE. (Ed.): Central autonomic network: Functional organization and clinical correlations. Armonk, New York, Futura Publishing Company, 1997

41. Akselrod S, Gordon D, Ubel FA, et al. Power spectrum analysis of heart rate fluctuations: a quantitative probe of beat to beat cardiovascular control. Science 1981; 213: 220–222.

42. Pomeranz B, Macaulay RJB, Caudill MA, et al. Assessment of autonomic function in humans by heart rate spectral analysis. Am J Physiol 1985; 248: H151-H153

43. Moriguchi A, Otsuka A, Kohara K, et al. Spectral change in heart rate variability in response to mental arithmetic before an after the beta-adrenoreceptor blocker, cartelol. Clin Autonom Res 1992; 2: 267–270.

44. Furlan R, Porta A, Costa F, et al. Oscillatory patterns in sympathetic neural discharge and cardiovascular variables during orthostatic stimulus. Circulation 2000; 101: 886–892.

45. Pagani M, Lombardi F, Guzzetti S, et al. Power spectral analysis of heart rate and arterial pressure variabilities as a marker of sympathovagal interaction in man and conscious dog. Circ Res 1986; 59: 178–193.

46. Lazzeri C, LaVilla G, Mannelli M, et al. Effects of acute clonidine administration on power spectral analysis of heart rate variability in healthy humans. J Autonom Pharmacol 1998; 18: 307–312.

47. Rimoldi O, Pierini S, Ferrari A, et al. Analysis of short term oscillations of R-R and blood pressure in conscious dogs. Am J Physiol 1990; 258: H967-H976

48. Saul JP, Rea RF, Eckberg DL, et al. Heart rate and muscle sympathetic nerve variability during reflex changes of autonomic activity. Am J Physiol 1990; 258: H713-H721

49. Pagani M, Montano N, Porta A, et al. Relationship between spectra components of cardiovascular variabilities and direct measures of muscle sympathetic nerve activity in humans. Circulation 1997; 95: 1441–1448.

50. Fei L, Statters DJ, Hnatkova K, et al. Change of autonomic influence on the heart immediately before the onset of spontaneous idiopathic ventricular tachycardia. J Am Coll Cardiol 1994b; 24: 1515–1522.

51. Lanza GA, Pedrotti P, Rebuzzi AG, et al. Usefulness of the addition of heart rate variability to Holter monitoring in predicting in-hospital cardiac events in patients with unstable angina pectoris. Am J Cardiol 1997; 80: 263–267.

52. Eckberg DL. Sympathovagal balance: a critical appraisal. Circulation 1997; 96: 3224–3232.

53. Stein PK, Kleiger RE, Rottman JN. Differing effects on heart rate variability in men and women. Am J Cardiol 1997; 80:302–305.

54. Kuo TB, Lin T, Yang CCH et al. Effect of aging on gender differences in neural control of the heart rate. Am J Physiol 1999; 277: H2233-H2239
55. Smith SE, Smith SA. Heart rate variability in healthy subjects measured with a bedside computer-based technique. Clin Sci 1981; 61: 379–383.
56. Wieling W et al. Reflex control of heart rate in normal subjects in relation to age: a data base for cardiac vagal neuropathy. Diabetologia 1982; 22: 163–166.
57. Pfeifer, MA et al. Differential changes of autonomic nervous system function with age in man. Am J Med 1983; 75: 249–258.
58. Oikawa N et al. Quantitative evaluation of diabetic neuropathy by using heart rate variations – determination of the normal range for the diagnosis of autonomic neuropathy. Tohoku J Exp Med 1985; 145: 233–241.
59. Masaoka S et al. Heart rate variability in diabetes; relationship to age and duration of the disease. Diabetes Care 1985; 8: 64–68.
60. Vita G et al. Cardiovascular reflex tests. Assessment of age-adjusted normal range. J Neurol Sci 1986; 75: 263–274.
61. O'Brien IA, O'Hare P, Corrall RJM. Heart rate variability in healthy subjects: effect of age and the derivation of normal ranges for tests of autonomic function. Br Heart J 1986; 55: 348–354.
62. Gautschy B, Weidmann P, Gnadinger MP. Autonomic function tests as related to age and gender in normal man. Klin Wochenschr 1986; 64: 499–505.
63. Low PA et al. The effect of aging on cardiac autonomic and postganglionic sudomotor function. Muscle Nerve 1990; 13: 152–157.
64. Ingall TJ, McLeod JG, O'Brien PC. The effect of ageing on autonomic nervous system function. Aust N Z J Med 1990; 20: 570–577.
65. Braune S, Auer A, Schulte-Mönting J, et al. Cardiovascular parameters: sensitivity to detect autonomic dysfunction and influence of age and sex in normal subjects. Clin Auton Res 1996; 6: 3–15.
66. Eckberg DL. Parasympathetic cardiovascular control in human disease: a critical review of methods and results. Am J Physiol 1980; 239: H581–593.
67. Ewing DJ, Campbell IW, Murray A, et al. Immediate heart rate response to stan-ding: simple test of autonomic neuropathy in diabetes. Br Med J 1978; 1: 145–147.
68. Doba N, Reis DJ. Role of the cerebellum and vestibular apparatus in regulation of orthostatic reflexes in the cat. Circ Res 1974; 40: 9–18.
69. Clark CV, Mapstone R. Age-adjusted normal tolerance limits for cardiovascular autonomic function assessment in the elderly. Age Ageing 1986; 15: 221–229.
70. Persson A, Solders G. R-R variations, a test of autonomic dysfunction. Acta Neurol Scand 1983; 67: 285–293.
71. Ewing DJ et al. The value of cardiovascular autonomic function tests: 10 years experience in diabetes. Diabetes Care 1985; 8: 491–498.
72. Eckberg DL, Fritsch JM. How should human baroreflexes be tested? News Physiol Sci 1993; 8: 7–12.
73. Duke FC et al. The effects of age on baroreceptor reflex function in man. Can Anaesth Soc J 1976; 23: 111–124.
74. Karemaker JM, Wieling W, Dunning AJ. Aging and the baroreflex. In: Amery A, Staessen J (editors). Handbook of Hypertension, 1989; Vol. 12: Hypertension in the Elderly. Amsterdam: Elsevier Science Publishers BV
75. Gribbin B et al. Effect of age and high blood pressure on baroreflex sensitivity in man. Circ

Res 1971; 29: 424–431.

76. Randall O et al. Determinants of baroreflex sensitivity in man. J Lab Clin Med 1978; 91: 514–519.

77. Shimada K et al. Age-related changes of baroreflex function, plasma norepi-nephrine, and blood pressure. Hypertension 1985; 7: 113–117.

78. Johnson RH. Effect of posture on blood pressure in elderly patients. Lancet 1965;1: 731–733.

79. Caird, FI, Dall JLC, Williams BO. The cardiovascular system. In: Brocklehurst JC (editor). Textbook of Geriatric Medicine and Gerontology, 3rd ed. Edinburgh, 1985; Churchill Livingstone

80. Low PA. Autonomic neuropathy. Semin Neurol 1987; 7: 49–84.

81. Low PA. Quantitation of autonomic responses. In: Dyck PJ et al. (editors). Peripheral Neuropathy, WB Saunders, 2nd ed. Philadelphia 1984; pp. 1139–1165.

82. Docherty JR. Aging and the cardiovascular system. J Auton Pharmacol 1986; 6: 77–84.

83. Elliott HL et al. Effect of age on the responsiveness of vascular alpha-adrenoceptors in man. J Cardiovasc Pharmacol 1982; 4: 388–392.

84. Hyland L, Doeherty JR. An investigation of age-related changes in pre- and post-junctional alpha-adrenoceptors in human saphenous vein. Eur J Pharmacol 1985; 114: 361–364.

85. Docherty JR. Age related changes in adrenergic neuroeffector transmission. Autonom Neurosci 2002; 136: 857–864.

86. Parker Jones P, Shapiro LF, Keisling GA et al. Altered autonomic support of blood pressure with age in healthy men. Circulation 2001; 104: 2424–2429.

*The Neuroendocrine Immune Network in Ageing*
Edited by R.H. Straub and E. Mocchegiani
© 2004 Elsevier B.V. All rights reserved

# Age-Related Alterations in Autonomic Nervous Innervation

DENISE L. BELLINGER[1], KELLEY S. MADDEN[2] and DIANNE LORTON[3]

[1]*Department of Pathology & Human Anatomy, Loma Linda University School of Medicine, Loma Linda, CA 92352;* [2]*Department of Psychiatry, Center for Psychoneuroimmunology Research, University Rochester School of Medicine and Dentistry Rochester, NY 14642;* [3]*Hoover Arthritis Center,Sun Health Research Institute, Sun City, AZ 85351*

## ABSTRACT

Responses of the central nervous system to stressors have modulatory effects on the immune system. Conversely, immune activation in response to antigens and pathogens alter neural activity. Progressive decline in brain and immune functions in the elderly leads to impaired cognition, re-establishment of homeostasis at age-adjusted set points, and immune senescence. The goal of this chapter is to review our understanding of how normal ageing affects cross talk between the autonomic nervous system and the immune system. Age-related changes in the relationship between sympathetic nerves and cells of the immune system have been observed, including differences in the density and compartmentation of sympathetic nerves, norepinephrine turnover, density and coupling of adrenergic receptors, and availability and functional capacity of target lymphoid cells. Strain-dependent alterations in how sympathetic nerves in secondary lymphoid tissues age have been reported, and appear to differentially affect nerve-to-immune signaling. These findings suggest impaired sympathetic regulation of immune function in aged individuals, which may contribute to greater susceptibility to infectious diseases, autoimmunity and cancer in the elderly. It is expected that a better understanding of sympathetic regulation of immunity in aged rodents will be useful in understanding experimental findings from clinical and basic science research. These findings could then be used to develop more optimal strategies for treating diseases, disease prevention and improving overall health in the elderly.

## 1.    INTRODUCTION

Primary and secondary lymphoid organs are innervated by sympathetic nerves that use norepinephrine (NE) as the main neurotransmitter [reviewed in 1]. Release of NE from sympathetic nerves signals cells of the immune system via specific subsets of adrenergic receptors expressed on their cell surface [reviewed in 2]. Interaction of NE with these receptors affects cell biochemistry, and gene expression, to bring about a change in the function of immune cells [reviewed in 3–4]. Whether primary and secondary lymphoid organs from young adult rodents receive parasympathetic innervation is controversial, although cells of the immune system express cholinergic receptors [reviewed in 2], and there is limited information on ageing and the parasympathetic nervous system. Therefore this chapter will focus on ageing and sympathetic nervous system-immune system interactions, and will not consider parasympathetic innervation.

As individuals age, striking changes in the structure and cellularity of lymphoid organs reflect a functional loss of immune competence. Concomitant with the age-associated change in lymphoid microenvironment and immune senescence, remodeling of sympathetic innervation occurs in lymphoid organs, presumably an adaptive response to maintain appropriate contacts with target cells [reviewed in 5]. Additionally, the ability of sympathetic nerves to signal target cells in lymphoid tissue is altered with advancing age. Collectively, these ageing effects change the capacity of the immune system to mount a response to antigen challenge, including responses to pathogens and tumors. This chapter will summarize the evidence for age-related changes in sympathetic noradrenergic (NA) innervation of lymphoid organs and the impact on sympathetic modulation of the immune system.

## 2. BONE MARROW

### 2.1. Age-related changes in bone marrow

Normal ageing alters the morphology of the bone marrow, with increased fat content and myeloid cells, accompanied by decreased lymphoid cells. This age-related change is initiated at the onset of thymic involution (at puberty) [6]. Sustained T cell production in the thymus is dependent on the migration of T cell precursors from the marrow to the thymus [7], and the frequency of bone marrow thymocyte precursors in old mice is reduced by 40-fold. Additionally, bone marrow cells display a reduced potential to colonize thymic organ cultures *in vitro* [8–11] and the thymus *in vivo* [12].

### 2.2. Sympathetic innervation of bone marrow

NA and peptidergic innervation of sinuses and tissue parenchyma has been demonstrated in young adult rodents [13–16]. NA and neuropeptide Y-positive (NPY+) nerves course in small nerve bundles that extend into the bone through the nutrient foramena, forming dense nerve plexuses along the vasculature. From these vascular nerve plexuses, NA/NPY+ nerve fibers extend into the surrounding parenchyma among hematopoietic cells in the marrow [15,17–18].

### 2.3. Age-related changes in sympathetic innervation of bone marrow

Despite age-related changes in the myeloid and lymphoid components of the bone marrow, NA innervation from 21-month-old Fischer 344 (F344) male rats appears to be maintained, as assessed histologically [5]. It is not known whether changes in NE metabolism occurs in bone marrow or whether there are age-associated changes in the ability of sympathetic neurotransmitters to signal target cells.

### 2.4. Functional significance of NA innervation of bone marrow

The functional significance of sympathetic innervation in bone marrow is not entirely clear, but presumed to play a modulatory role in cellular proliferation and differentiation of myeloid and lymphoid cells. In young adult rodents, NE and dopamine concentrations in murine bone marrow display diurnal rhythmicity, peaking at night [19]. This finding may be unique to bone marrow, since diurnal rhythms were not detected in levels of circulating catecholamines, or in thymic or splenic NE [20]. In bone marrow, peak catecholamine concentrations positively correlate with the proportion of bone marrow cells in the $G_2/M$ and S phases of the cell cycle [19,21], suggesting neural regulation of hematopoietic cell proliferation. Chemical sympathectomy with 6-hydroxydopamine (6-OHDA) or administration of the $\alpha_1$-adrenergic antagonist, prazosin, enhanced myelopoiesis and inhibited lymphopoiesis in normal mice and in mice that received syngeneic bone marrow cells [22–23]. NE and the $\alpha_1$-adrenergic receptor agonist, methoxamine, inhibited the growth of granulocyte/macrophage colony forming units (GM-CFU) *in vitro*, which corresponded well with the finding that unfractionated bone marrow cells possess high affinity $\alpha_1$-adrenergic receptors [22]. Chemical sympathectomy in adult mice reduced bone marrow cellularity and increased progenitor cells in the peripheral blood, suggesting that cells were driven out of the bone marrow after 6-OHDA treatment [13]. In sympathectomized mice, progenitor cells and the number of GM-CFU increased, suggesting retention and enhanced proliferation of bone marrow cells. Chemical sympathectomy transiently increased bone marrow cell proliferation *in vivo* [24], suggesting that an intact sympathetic nervous system inhibits spontaneous cell proliferation in bone marrow.

In contrast to these studies, neither neonatal 6-OHDA treatment (which permanently removes NA nerve fibers) nor unilateral surgical denervation of the sciatic nerve in mice had any effect on the percentage of granulocytes or GM-CFU obtained from bone marrow, leading Benestad and colleagues [25] to postulate that NA innervation does not significantly influence differentiation of hematopoietic cells. Differences in the methods of denervation and the age of denervation (neonatal versus adult) most likely account for this discrepancy. No studies to date have examined the age-related changes in sympathetic modulation of bone marrow functions. As we have found that sympathetic modulation of immune function in the spleen is altered with age (see below), studies in bone marrow from aged rodents may reveal the role of sympathetic NA nerves in this immune compartment. Closer examination of the function of sympathetic nerves in bone marrow may provide novel combination therapies to enhance migration of bone marrow precursors into the thymus to increase output of mature naïve T cells. It may also lead to novel therapies that stimulate hematopoiesis after bone marrow transplantation and chemotherapy [26].

## 3.    THYMUS

### 3.1.    Structure and function of the thymus

The thymus consists of incomplete lobules partially separated by septa derived from the connective tissue capsule that envelops the organ. Each lobule consists of the cortex, a peripheral zone of densely packed lymphocytes that surrounds the medulla, a lightly staining central zone. Hassall's corpuscles, several layers of flattened, concentrically arranged epithelial cells, are present in the medulla. These are apparently degenerative bodies that increase in size and number throughout life. The stroma of the thymus consists of dendritic cells, macrophages, and epithelial reticular cells with thin cytoplasmic process that attach to each other by desmosomal contacts.

The thymus is the primary site for differentiation of pre-T cells into mature T cells that recognize and respond appropriately to foreign antigen. Stems cells from the bone marrow enter the thymus in the subcapsular region, where they differentiate from a CD4$^-$CD8$^-$ phenotype (double negative, DN) to an intermediate CD3$^{lo}$CD4$^+$CD8$^+$ (double positive, DP) phenotype [27]. In the thymic cortex, DP thymocytes undergo a selection process requiring T cell receptor interactions with major histocompatibility complex (MHC) molecules on the surface of thymic stromal cells [28]. Positively selected thymocytes up-regulate CD3 expression and down-regulate expression of either CD4 or CD8 to become mature CD3$^{hi}$ CD4$^+$CD8$^-$ or CD3$^{hi}$CD4$^-$CD8$^+$ (single positive, SP) thymocytes. The majority of thymocytes are not selected and die in the thymus via apoptosis. Mature SP thymocytes reside in the medulla of the thymus, and subsequently leave the thymus via blood vessels at the corticomedullary junction to seed secondary lymphoid organs.

## 3.2.   Age-related changes in the thymus

After puberty, the thymus begins to atrophy, but remains a site of T cell selection throughout adulthood. With thymic atrophy, there is a gradual regression in size, weight and cellularity [reviewed in 29]. Deposits of adipose tissue displace and disrupt the thymic stroma, particularly the cortical epithelial network that supports early T cell differentiation and selection [30]. The underlying mechanism(s) of age-related thymic involution is not known, although studies suggest both a genetic basis [31–34] and a role for altered hormone levels [35–36].

Bone marrow-derived stem cell precursor development diminishes with age, but transplanting young bone marrow into old recipients does not reverse decreased T cell regeneration in old thymus, demonstrating that the aged thymic microenvironment itself has a diminished capacity to support thymocyte differentiation [37]. With age, the frequency of CD4$^-$CD8$^-$ DN immature thymocytes increases and CD4$^+$CD8$^+$ DP cells decreases [38–40], indicative of a block in T cell differentiation from the DN to DP phenotype, and leading to reduced naïve T cell output with age. Decreased T cell output contributes to the age-associated dysregulation of the immune system, including reduced cell-mediated immunity, increased autoantibody production, and increased susceptibility to infectious disease [41–42]. Age-related impairment in T cell development in the thymus has been attributed to disruption in T cell receptor rearrangement and β-selection efficiency [32], and in stromal cells defects. Several reports have demonstrated reduced bone marrow precursor and thymocyte interaction with stromal cells [8,43–44]. Other components, such as nerves, endothelial cells, fibroblasts and connective tissue may also be contributory. Von Patay et al. [45–46] have demonstrated that NE stimulated the production of interleukin (IL)-6 by cultured young thymic epithelial cells about 14-fold above baseline in the presence of NE, supporting an indirect role of the sympathetic nervous system in maintaining the internal milieu.

## 3.3.   Sympathetic innervation of the thymus

In the young adult thymus, NA sympathetic nerves from postganglionic neurons in the upper paravertebral ganglia of the sympathetic chain (primarily the superior cervical and stellate ganglia) innervate the thymus [47–49]. NA sympathetic nerves enter the thymus with blood vessels as moderately dense plexuses, and travel in the capsule and interlobular septa, or continue with the vasculature into the subcapsular cortex, where stem cells arrive from the bone marrow. These nerves course along blood vessels in the septa, where they often reside adjacent to mast cells and macrophages. In the capsule and interlobular septa, NA nerves course adjacent to mast cells [14,50], corticotropin-releasing hormone (CRH)-immunoreactive cells [51], and ED1$^+$ and ED3$^+$ macrophages [1,14]. In the cortex, the highest density of fibers occurs near the corticomedullary junction, where T cells undergo differentiation and maturation, and where mature T cells exit the thymus. In the medulla, NA nerves primarily associate with the medullary sinuses, which are continuous with the system of blood vessels in the septa; some nerve fibers extend from these plexuses into the surrounding parenchyma among mature thymocytes and thymic epithelial cells [reviewed in 1].

Thymic nerves containing neurotransmitters other than NE have been demonstrated in young adult rodents, including NPY, tachykinins (substance P, neurokinin A, and neurokinin B, collectively), substance P, and calcitonin gene-related peptide (CGRP), CRH and vasoactive intestinal polypeptide (VIP) [1]. The origin of these fibers is not entirely clear. All of these neurotransmitter-specific nerves have to some extent overlapping distributions in the young adult rat thymus, although this innervation is generally not as robust as seen with sympathetic NA nerves, with the exception of NPY. Chemical sympathectomy with 6-OHDA destroys NPY$^+$ nerves in the rat thymus [52], indicating that NPY co-localizes with NE in sympathetic nerves. Nerve fibers containing CGRP and substance P/neurokinins are presumed to be afferent sensory nerves based on the presence of these neurotransmitters; however, Bulloch et al. [53] have suggested CGRP$^+$ fibers derive from the vagus nerve. Treatment with capsaicin, a neurotoxin selective for small non-myelinated afferent nerves, depletes substance P in the thymus [54] supports derivation from sensory neurons. Based on the concentration of substance P and CGRP in the thymus, Geppetti et al. [55] have suggested that these neurotransmitters are not co-localized, but both CGRP and substance P-containing thymic nerves appear to arise from cervical dorsal root ganglia based on anterograde tracing studies [56]. Little information is available regarding age-related changes in thymic peptidergic innervation. One laboratory reported a similar distribution, but reduced density, of CGRP$^+$ nerves in the aged thymus [53].

3.4.    Sympathetic innervation in the ageing thymus

In a longitudinal study of sympathetic innervation of the F344 rat thymus, we demonstrated a dramatic increase in the density of NA nerve fibers and NE concentration as the thymus involutes with age [57]. Total NE content in the thymus was relatively constant across age despite thymic involution, suggesting that increased nerve density and NE concentration in the ageing thymus likely results from the age-related loss in tissue volume. However, a quantitative study by Zirbes and Novotny [58] demonstrated an increase in the density of sympathetic nerves in the aged rat thymus that exceeded the reduction in thymic volume. If this is indeed the case, it indicates sprouting of sympathetic nerve fibers in the aged rodent thymus. We found age-related increases in thymic sympathetic innervation in other strains of rats and in several mouse strains; the onset occurred with puberty and involution the thymus [5,59]. The thymocytes in the involuting thymus reside among a higher density of NA nerve fibers, suggesting that they are exposed to higher concentrations of NE as the animal ages. Formal substantiation of this hypothesis awaits measurement of NE release and turnover.

3.5.    Functional studies and significance

Most of our knowledge of the role of sympathetic innervation in modulating thymic function derives from studies in young adult rodents. Functional studies, radioligand-binding assays, northern blot analysis and signal transduction studies in young adult rodents provide evidence for heterogeneous expression of $\alpha$- and $\beta$-adrenergic receptors on thymocytes [52,60–64] and thymic epithelial cells ($\beta_1$ and $\beta_2$) [45,65]. Unfractionated thymocytes express very low levels (below the level of detection) of $\beta$-adrenergic receptors compared to mature T cells [3], and mature corticosterone-resistant thymocytes express $\beta$-adrenergic receptors at levels equivalent to peripheral T cells [66–67]. Collectively, these studies suggest that immature T cells express very low levels of β-adrenergic receptors and that β-adrenergic receptor expression is up-regulated later in the differentiation process. These findings are consistent with findings that $\beta_2$-adrenergic receptors localize primarily to the medullary region of the rat thymus [62]. The efficiency of receptor coupling to adenylate cyclase appears to be greater in unfractionated thymocytes than in spleen or lymph node cells [68]. Likewise, unfractionated thymocytes generate an isoproterenol-induced cAMP response that is equivalent to that of splenocytes, even though spleen cells express a far greater number of β-adrenergic receptors [3]. These studies suggest that thymocytes expressing β-adrenergic receptors are exquisitely sensitive to β-adrenergic receptor signaling.

Mice that lack dopamine-β-hydroxylase (*dbh-/-*), an enzyme that converts dopamine to NE, possess a similar number and frequency of thymic subpopulations (CD4+, CD8+, CD4-/CD8-, CD4+/CD8+) compared with age- and sex-matched *dbh+/–* controls [69]. These findings suggest that *dbh–/–* mice do not have intrinsic developmental defects. However, the thymuses of surviving *dbh–/–* mice housed in non-specific pathogen free conditions showed marked involution, and a reduction in the immature CD4+/CD8+ DP thymocytes. Under these conditions, *dbh–/–* mice had higher mortality rates, and impaired resistance to primary and secondary infection. This study supports a role for NA innervation in maintaining a normal microenvironment for T cell development, differentiation and responsiveness to environmental antigens in secondary lymphoid organs.

In studies of young adult thymocytes, exposure of thymocytes *in vitro* to β–adrenergic receptor agonists increased expression of thymocyte Thy-1 and TL antigens, alluding to enhance thymocytes differentiation after β–adrenergic receptor stimulation [70]. Other studies suggest that removal of β–adrenergic receptor signaling may enhance thymocyte proliferation. In one report, more cells were obtained from thymic rudiments implanted into the anterior chamber of mice that had undergone unilateral ablation of the superior cervical ganglia to remove sympathetic innervation [71]. In another study, when NA innervation was removed by 6-OHDA treatment an increase in the number of proliferating cells in the thymic cortex was detected by bromodeoxyuridine uptake [52].

Chemical sympathectomy also increases apoptosis in the thymus, reduce thymic cellularity, and decrease spontaneous thymocyte proliferation *in vitro* [72–73]. Pretreatment with desipramine to inhibit 6-OHDA uptake and prevent NA nerve terminal destruction, blocked the 6-OHDA-induced increase in apoptosis, indicating that ablation of NA nerve fibers was necessary for this effect [73]. However, studies of the thymus using 6-OHDA *in vivo* must be interpreted with caution, since cultured thymocytes incubated in the presence of $10^{-5}$ M 6-OHDA also induced apoptosis, and co-incubation of 6-OHDA with desipramine prevented this effect [73]. These findings suggest that thymocytes possess catecholamine uptake mechanisms, but this awaits further investigation. Furthermore, sympathetic nerves in the thymus are more resistant to the neurotoxic effects of 6-OHDA compared to the spleen [74; personal observations]. Finally, studies need to be performed to rule out the possibility that 6-OHDA-induced effects on thymic apoptosis are due to elevated corticosterone levels secondary to denervation [72,75]. Collectively, these results suggest that NA innervation in the thymus of the young adult may inhibit proliferation and reduce thymocyte apoptosis, but more detailed and direct studies of catecholaminergic influences on specific T cell subsets are necessary.

Madden and Felten [76] assessed thymocyte development and proliferation *in vivo* after chronic ß-adrenoceptor blockade in aged mice. Two- and 18-month-old male BALB/c mice were exposed for 4 weeks to the β-adrenergic receptor antagonist, nadolol, by subcutaneous pellet implantation. Thymocyte differentiation and naïve T cell output were examined by 3-color flow cytometric analysis of T cell populations in the thymus and peripheral blood.

In old, but not young mice, thymocyte CD4/CD8 co-expression was altered by β-adrenergic receptor blockade. Nadolol treatment of aged mice further increased the CD8$^+$ SP thymocytes and reduced the CD4$^+$CD8$^+$ population, in conjunction with an increase in the CD3$^-$CD4$^-$CD8$^-$ population. Thus, nadolol treatment appears to further exacerbate the age-related changes in thymocyte differentiation, suggesting that β-adrenoceptor stimulation counteracts the age-related changes in thymocyte maturation. Because CD3$^{hi}$ expression in the CD4$^+$CD8$^+$ population (the immediate precursors to the SP thymocytes [77]) was not altered by nadolol treatment, increased CD4$^-$CD8$^+$ cells were probably not mediated by more CD4$^+$CD8$^+$ cells undergoing positive selection. Additionally, chronic β-adrenergic receptor blockade did not alter *ex vivo* thymocyte proliferation in either age group, suggesting that changes in thymocyte proliferation did not contribute to the nadolol-induced effects. However, a more detailed analysis of additional thymocytes differentiation markers in combination with such functional measures as proliferation and apoptosis will be needed to further characterize the nadolol-induced changes.

We also wanted to determine whether the nadolol-induced increase in CD8$^+$ T cells in the thymus was due to an inability of these cells to emigrate from the thymus into the circulation. The percentage of CD4$^+$ and CD8$^+$ T cells expressing high and low levels of the adhesion molecule CD44 [78–79] was examined in peripheral blood. In old mice, nadolol treatment increased the percentage of peripheral blood CD4$^+$ and CD8$^+$ T cells expressing the naive (CD44$^{lo}$) phenotype in the absence of changes in the percentages of total CD4$^+$ or CD8$^+$ T cells in peripheral blood [76]. β-adrenergic receptor blockade could increase naïve T cells in the blood by several mechanisms, one of which includes increased emigration of mature T cells from the thymus. Future studies will attempt to examine more directly whether mature thymocyte emigration is increased by β-adrenergic receptor blockade.

These results indicate that β-adrenergic receptor blockade can alter thymocyte differentiation, and suggest that the age-associated increase in sympathetic NA innervation of the thymus modulates thymocyte maturation. The finding that nadolol-induced effects were confined to old animals suggests that the sympathetic nervous system may have a greater influence on thymocyte maturation in aged animals. These results also suggest that the sympathetic nervous system may exert a greater influence on immune function in aged individuals whose immune systems are immunocompromised. Finally, findings that manipulation of sympathetic activity in the aged thymus alters thymocyte differentiation suggests that pharmacological manipulation of NA innervation may provide a novel way of increasing naïve T cell output and improving T cell reactivity to novel antigens with age.

## 4.    SPLEEN

### 4.1.    Structure and function of the spleen

The young adult spleen is a site of antigen-dependent lymphocyte responses and defense against blood-borne pathogens. The parenchyma of the spleen consists of white pulp and red pulp. The red pulp is reticular tissue, with splenic cords of lymphoid cells (cords of Billroth) situated between the sinusoids. Besides reticular cells, the red pulp contains fixed and wandering macrophages, monocytes, lymphocytes, plasma cells, platelets and granulocytes. The white pulp contains lymphatic tissue arranged in sheaths around central arteries and lymphoid nodules appended to the sheaths. The lymphoid cells surrounding the central arteries are mainly T lymphocytes, forming the periarteriolar lymphatic sheath (PALS), while lymphoid nodules or follicles consist primarily of B lymphocytes. Antigen-presenting cells detect blood-borne antigens and present them to PALS-derived T cells for sampling. Antigen-specific T and B cells interact to initiate an immune response. Between the white pulp and red pulp lies a marginal zone, consisting of sinuses and loose lymphoid tissue. In this zone there are few lymphocytes, but many macrophages showing active phagocytosis. This zone harbors blood-borne antigens and thus plays a major role in the immunologic activity of the spleen. The marginal zone also is the site of entry for T and B cells into the spleen before becoming segregated in their specific splenic location.

## 4.2.  Sympathetic innervation of the spleen

Innervation of the rodent spleen has been studied extensively [reviewed in 1]. NA sympathetic nerves enter the spleen as a dense vascular plexus along with the splenic artery at the hilus. These nerves continue to course along arterial and venous branches under the capsule and along the trabeculae. From the trabecular plexuses, NA nerve plexuses continue with the central artery that is surrounded by white pulp. In the white pulp NE-containing nerves extend from the central arterial plexus into the PALS. Double-label immunocytochemistry reveals close association between NA nerves and T lymphocytes in this compartment. NA nerve fibers also are present in the marginal zone, coursing along blood vessels or as individual fibers adjacent to macrophages, and T and B lymphocytes. With electron microscopy, direct contacts were observed between lymphocytes in the PALS and NA nerves with an intervening extracellular space of about 6 nm, a smaller synapse than between sympathetic nerves and more traditional targets, such as smooth muscle cells within the vasculature. In the B cell follicles of the white pulp, only an occasional sympathetic nerve fiber is typically observed.

Nerves containing neurotransmitters other than NE have been demonstrated in the young adult spleen. One of these, NPY, co-localizes with NE as indicated by the loss of NPY-containing nerves after treatment with the NA-selective neurotoxin, 6-OHDA [80]. Nerves containing substance P, CGRP, met-enkephalin, VIP, cholecystokinin, neurotensin, IL-1, CRH and arginine-vasopressin have been described in the rodent spleen [81; review in 1]. These neuropeptide-containing nerves differ in density and distribution from that observed for splenic sympathetic nerves, suggesting that they do not co-localize with these nerves, or perhaps represent a subset of sympathetic nerves that are not NA.

## 4.3.  Sympathetic innervation in the ageing spleen

In contrast to the dramatic age-related increase in sympathetic NA innervation of the thymus, NA innervation of the spleen is diminished markedly in old (>17-month-old) F344 rats [82–84]. This progressive loss in sympathetic innervation is manifested as a decline in the density of nerve fibers in all compartments of the spleen (approximately 75% lower than in young adult rats) and a 50% reduction in splenic NE content by 27 months of age [82,84]. The reduction in splenic NA innervation is associated with degeneration of NA nerve fibers and ultimately, with loss of the cell bodies in the superior mesenteric/celiac ganglionic complex [85, personal observations]. The apparent discrepancy in nerve density and splenic NE concentration results from an increase in NE turnover in age-resistant nerves in the spleen [5]. These age-resistant nerves reside near the hilus with a striking depletion of nerve fibers with increasing distance from the hilus. This pattern of nerve loss resembles a peripheral neuropathy, whereby the longest nerves are most affected, and is suggestive of a metabolic insult. Heightened sympathetic activity in the age-resistant nerves in spleens from aged F344 rats is consistent with evidence for an overall elevation in sympathetic tone in humans with advancing age [86–89]. Conversely, uptake of circulating NE from the circulation into sympathetic nerve terminals may help explain splenic NE concentrations than might be expected from the extensive loss of sympathetic nerves fibers that occur in the aged spleen.

Loss of sympathetic nerves may result from an inability of NA nerves to synthesize enough NE or tyrosine hydroxylase to be visualized by histochemical or immunocytochemical staining, or from an actual loss of nerve fibers. To resolve this issue, alpha-methylnorepinephrine (αMNE), a compound that is uptaken by high affinity carriers into NA nerve terminals and persists because it cannot be catabolized by monoamine oxidase, was administered to aged rats [5,90]. Treatment with αMNE did not reveal additional sympathetic nerve fibers in spleens from old rats. This study supports the actual loss of sympathetic nerves in the age spleen, presuming the high-affinity uptake system for NE in nerve terminals remains functional in aged rats. Studies examining NE uptake in spleen slices from 3- and 21-month-old rats indicate that the uptake of [$^3$H]-NE does remain functional in old animals, and in fact is enhanced [5,90].

We have discovered species and strain differences in how ageing affects sympathetic NA innervation of the spleen [5,59,91]. In aged BALB/c and C57Bl/6 mice, histochemical analysis revealed an increase in the density of splenic NA nerves in the parenchyma of the white pulp, especially near the hilus [5,59]. More distal to the hilus, NA innervation of the white pulp was variable, with some white pulp regions containing a normal density of noradrenergic nerves, and others depleted of NA nerves. Splenic NE concentration was not diminished in aged mice, but in fact was slightly elevated. The variable density of NA nerves in white pulps distal to the hilus suggest that there may be, at least to a limited extent, some loss of nerves in these sites. The increased splenic NE concentration from old mice raises the possibility that NE metabolism is enhanced in these aged mouse strains. Similarly, sympathetic innervation of spleens from aged Brown Norway and Brown-Norway x F344 (F1) rats is equivalent in density to that of young adults, while a striking age-related loss of sympathetic nerves in spleens from Lewis rats progressively occurs with advancing age [91]. These findings indicate that unlike the thymus, age-related changes in splenic NA innervation may be dependent on undefined genetic influences.

We have begun to investigate the mechanisms underlying age-related diminution of splenic NA innervation in F344 rats. In young adults, the integrity of sympathetic nerves critically depends on trophic support provided by target and/or stromal cells in their local microenvironment, and protection from oxidative stress by anti-oxidative enzyme systems. One possibility is that loss of NA innervation in the spleen is related to long-term exposure to antigens. During a primary immune response, several investigators have reported increased NE turnover, that may result in the production of NE and oxidative metabolites that generate free radicals [66,92]. The continuing generation of auto-oxidative catabolites following NE release over time may lead to subsequent autodestruction of NA nerve terminals [reviewed in 93]. In support of this hypothesis, a striking loss of NA nerves in spleens from young adult rodents was observed in rodents chronically-treated with NE [94]. However, in a study where ageing F344 rats were repeatedly immunized with lipopolysaccharide from 12 to 17 months of age, we did not find a significant acceleration in the age-related decline in sympathetic innervation (unpublished data). Other types of antigens that require T cell activation or repeated exposure to superantigens should be examined. Reduced production of neurotrophic factor(s) and/or diminished capacity to handle oxidative stress in the aged spleen also may contribute to diminished splenic NA innervation.

To further test the hypothesis that the age-related reduction in splenic NA innervation is the result of an antigen-driven response, we have begun looking at the possibility that cytokines produced during an immune response can locally regulate sympathetic neurotransmission. We have shown that ganglion cells in the superior mesenteric/celiac complex can synthesize IL-2 receptors [95]. David Snyder has shown that IL-2 can enhance release of NE from synaptosomes (NA nerve terminals) from the spleen, suggesting that IL-2 receptors are present on NA nerve terminals (personal communication). Thus, IL-2 produced locally during an immune response may regulate release of NE from nerve terminals in the spleen. Age-associated changes in IL-2 production and IL-2 receptor expression *in vivo* and the effects of IL-2 on NE release and subsequent alterations in IL-2 signaling of NA nerves with age need to be explored more fully.

### 4.4. Can age-related changes in NA innervation be reversed?

ThyagaRajan and colleagues [96] tested whether the loss of NA innervation in the aged F344 rat spleen was reversible by chronic treatment with L-deprenyl, a selective irreversible monoamine oxidase B (MAO-B) inhibitor. L-deprenyl has been shown to: (1) slow the progression of Parkinson's disease through enhancement of nigrostriatal dopaminergic neurotransmission [97-98]; (2) promote regrowth of transected axons of facial motoneurons at postnatal day 14 [99]; (3) improve survival of rat ganglion cells after damage to optic nerves [100]; and (4) increase neurotrophic factor production and increase activity of superoxide dismutase (SOD) and catalase [99,101] (principal scavenging enzymes that protect neurons from damage caused by free radicals). All these modes of action may serve to maintain NA innervation and promote NA nerve growth in target organs.

Deprenyl partially restored NA innervation in the aged rat spleen in 21-month-old male F344 rats treated with L-deprenyl for 9 weeks [96]. Deprenyl-induced regrowth of NA innervation was associated with increased splenic NE concentration (per mg wet weight) and NE content (per whole spleen), although these levels did not achieve the concentration measured in the young spleen [96]. Furthermore, the age-related decline in splenic natural killer cell activity and concanavalin A (Con A)-induced IL-2 production was partially reversed by deprenyl treatment of aged rats [102]. These results demonstrate plasticity of splenic NA innervation in old animals and indicate that the age-related reduction in NA innervation is not permanent and can be restored, at least partially. Furthermore, they suggest that manipulation of NA innervation may be a novel way to counteract some parameters of age-associated decline in immune function. Long-term treatment with L-deprenyl reduced spontaneous mammary tumor growth in old rats [103]. While L-deprenyl effects on tumor growth are probably mediated via multiple mechanisms, we speculate that at least one of these mechanisms is through drug-induced sympathetically-mediated changes in tumor immunity. Confirmation of this hypothesis awaits further research.

With double-labeled immunocytochemistry, we observed a gradual decline in the density of T lymphocytes in the PALS and ED3+ macrophages in the marginal zone of the F344 rat spleen, concomitant with the loss of NA nerve fibers [83]. Loss of cells in these compartments is most striking with increasing distance from the hilus, similar to the loss of sympathetic nerves. Whether NA nerve loss is causally related to the reduced cellularity of T lymphocytes and macrophages in the spleen is not known. In the spleen, lymphocytes synthesize nerve growth factor (NGF) mRNA and protein [104–105], but macrophages can sequester NGF (in the absence of mRNA) [106]. NGF and other neurotrophins are critical for sympathetic nerve survival in the young [107–108], but neurotrophin production and signaling capacity has not been extensively investigated in aged rodents. Sympathetic nerve distribution in the spleen and thymus of severe combined immunodeficiency (SCID) mice, which are deficient in T and B lymphocytes, is robust in young adults [109, S.Y. Stevens, unpublished data]. Lymphocyte depletion using hydrocortisone or cyclophosphamide in young adult mice at least acutely, did not affect sympathetic nerve integrity [110]. These results suggest that lymphocytes may not provide critical for neurotrophic support to sympathetic nerves in these mice. With advancing age, sympathetic nerves may become more vulnerable to changes in the percentages of lymphocytes subsets and macrophages that populate the aged spleen and changes in their capacity to provide neurotrophic support.

## 4.5.    Age-related changes in the spleen and immune function

Cell-mediated immunity is more impaired in the aged individual compared to humoral or antibody responses [reviewed in 111]. As T cell reactivity towards foreign antigens decline, responses to autoantigen increase with advancing age. Antigen and mitogen-induced proliferation, cytotoxic effector function, natural killer cell activity, cytokine production, IL-2 receptor expression, and signal transduction are commonly reported to be impaired in old T cells [reviewed in 112]. Ageing affects many points in the process of T cell maturation and activation. IL-2 production is generally impaired in ageing, which hampers clonal expansion of T lymphocytes [reviewed in 113]. IL-4 and interferon-$\gamma$ (IFN-$\gamma$) production by T cells appear to be dysregulated in ageing, although the direction of changes reported in aged animals and in humans is variable [reviewed in 114–115]. These conflicting findings can be attributed to species differences, nature of the stimulating agent used *in vitro*, variability in the *in vitro* culture conditions used between laboratories and source of lymphocytes employed. IL-4 and IFN-$\gamma$ are important for regulating humoral- and cell-mediated immune responses, respectively, and are crucial in antimicrobial as well as antineoplastic defenses of the elderly. An increase in IL-5 [114,116] and IL-10 production by T cells from old mice [117] and humans [118], cytokines that promote Th2 immune responses, suggest a possible shift in ageing towards humoral-mediated immune responses [reviewed in 114–115].

Ageing is also associated with a shift in the cell populations that reside in lymphoid compartments. Since the CD4 and CD8 populations contribute differently to the host immune defense, any alteration in their ratio can influence a host's immune response to a biological insult. With advancing age, there is an increase in the CD4/CD8 ratios in human peripheral blood and in murine spleen, and a decrease in CD4/CD8 ratios in murine lymph nodes and peripheral blood [reviewed in 119]. An increase in CD8$^+$CD28$^-$ cells and a reduction in CD8$^+$CD28$^+$ cells is consistent with the age-associated decline in proliferative ability in aged humans, particularly prominent in the CD8 population [120] and hyporesponsiveness to co-stimulation by CD28 in T cells from aged mice [121]. An increase in the ratio of T memory to naïve T cells occurs in lymphoid organs and peripheral blood in humans and in animals [122–124]. The shift towards T memory lymphocytes also contributes to the reduced antigen-induced lymphocyte proliferation and altered patterns of cytokine production with age [116–117].

## 4.6.    Functional studies and significance

In young adult rodents, the role of the sympathetic nervous system in immune regulation has been examined by a variety of experimental approaches. Chemical sympathectomy or $\beta$-adrenergic receptor blockade before immunization reduced a Th2-driven antibody response and decreased the delayed-type hypersensitivity reaction to a contact-sensitizing agent [125–126].

Transgenic mice that lack dopamine-β-hydroxylase (necessary for NE and epinephrine synthesis) had impaired immune reactivity to infectious agents and a protein antigen [69]. Dhabhar and McEwen [127] demonstrated that *in vivo* administration of epinephrine before antigen sensitization augmented the delayed-type hypersensitivity response and increased cellularity in the draining lymph nodes. Collectively, these findings suggest an enhancing effect of the sympathetic nervous system on immune reactivity. In contrast, stressors or agents that elevate sympathetic activity reduced T cell responses, anti-viral immune reactivity, and natural killer cell activity, effects that were prevented by pretreatment with β-adrenergic receptor antagonists or ganglionic blockade [128–131].

Chemical sympathectomy enhanced antigen-induced proliferation and Th1 and Th2 cytokine production *in vitro* [75,132]. Sympathetic denervation before intraperitoneal immunization with a protein antigen also increased serum antibody levels in rats and in C57Bl/6 mice, but not BALB/c mice [75,133]. Differences in sympathetic-induced effects on immune reactivity reported in the literature indicate the importance of a number of factors in determining the outcome of sympathetic manipulation, including genetic background, site of immunization, and the types of immune cells involved in the immune response (i.e., antigen-presenting cell type, type of Th cell). *In vitro* studies with purified cell populations have begun to identify some of the mechanisms underlying sympathetic immune modulation.

Stimulation of β-adrenergic receptors in unfractionated effector cells *in vitro* enhanced alloreactive T lymphocyte cytotoxicity and antibody-secreting cells [134–135]. More recently, highly purified lymphocyte populations have been used to dissect the responses of T and B cells to β-adrenergic receptor stimulation. In this culture system, antigen-specific B cells present antigen and synthesize antibody in the presence of Th clones [136]. Addition of β-adrenergic agonists to this culture system increased antibody production [136] by up-regulating B cell accessory molecule expression and by increasing B cell responsiveness to IL-4 [137]. Conversely, when Th1 clones were cultured in the presence of β-agonists before antigen-induced activation, the number of antibody-forming cells declined concomitantly with lower IFN-γ production [136], suggesting that β-adrenergic receptor stimulation inhibits Th1 responses. However, Th cells also exhibited differential responsiveness to catecholamine stimulation depending on the activating stimulus, and the type of Th cell cultured (Th1 clone vs. naïve CD4+ cells grown in Th1-promoting conditions) [138–139]. Thus, the type of immune cells involved and its activational or maturational state contribute to the complexity of sympathetic interactions with the immune system, and likely reflect the important role of the sympathetic nervous system in "fine-tuning" a response to maintain homeostasis.

In old F344 rats, we wanted to determine if the reduced complement of NA nerve fibers in the aged spleen could maintain functional links with the immune system. We also wanted to explore the possibility that a causal link exists between the age-associated reduction in NA innervation in the spleen and impaired T cell proliferation. Young (3-month-old) and old (17-month-old) F344 rats were chemically sympathectomized, and the functional impact on *in vitro* T cell responses in naïve animals and antigen-specific antibody responses *in vivo* was assessed [5,133,140]. Although young rats were remarkably resistant to sympathectomy-induced changes in T lymphocyte proliferation and cytokine production, splenic T cell activity in aged rats was reduced when splenic NE was depleted, so that the age-related impairment in Con A-induced proliferation was potentiated in sympathectomized animals [140]. In immune animals, serum IgM and IgG anti-keyhole limpet hemocyanin (KLH) antibody responses did not differ between 3- and 17-month-old untreated F344 rats. Sympathectomy increased anti-keyhole limpet hemocyanin (KLH) IgG, IgG1, IgG2b, and IgM antibody responses in young and old rats. Sympathectomy also enhanced KLH-specific proliferation by spleen cells from old, but not young animals [5,133].

If splenocytes in the aged rat were no longer receiving signals from sympathetic nerve terminals, then removal of these nerve fibers by sympathectomy would have no impact on immune reactivity. These results indicate that the sympathetic nervous system can still modulate immune reactivity in aged F344 rats despite the loss of NA nerves in the spleen. The sympathectomy-induced reduction in Con A-induced T cell proliferation suggests that splenic NA innervation in old animals, though diminished, can exert a positive regulatory influence on T lymphocyte function, but it inhibits humoral immune responses in young and old F344 rats.

The differences in sympathectomy-induced alterations in immune function between young and old rats demonstrate that old animals are more susceptible to complete loss of sympathetic NA innervation, perhaps because compensatory mechanisms become limited, while young rats rapidly evoke compensatory mechanism to maintain T lymphocyte activity following NE depletion. Maintenance of sympathetic NA signaling of splenocytes when splenic NE levels are reduced suggests up-regulation of compensatory mechanisms in intact aged rats. Evidence for increased sensitivity to NE in the synapse include: increased spleen cell $\beta$-adrenergic receptor density [5,141], augmented $\beta$-adrenergic receptor signaling (elevated intracellular cAMP) (unpublished data), and elevated NE re-uptake and turnover in the age-resistant NA nerves in the aged F344 rat spleen [5,141]. This evidence suggests that compensatory capacity is approaching maximum in the intact old rat, and further compensation to sympathectomy may be limited. The capacity of young animals to compensate more rapidly and efficiently for loss of sympathetic nerves may negate or at least limit the impact of sympathetic nerve destruction on immune reactivity.

Kruszewska et al. [132], reported similar 6-OHDA-induced enhancement of anti-KLH IgG$_1$ and IgG$_2$ antibody production in the Th1-predominant mouse strain, C57BL/6. This group did not observe significant sympathectomy-induced alterations in serum anti-KLH antibody response in the Th2 predominant strain, BALB/c. Antigen-induced Th1 and Th2 cytokine production *in vitro* was enhanced after sympathectomy in both murine strains, providing no evidence for Th1 versus Th2 dominant cytokine production [75,132]. In F344 rats, there was no evidence for dominance of one Th response over the other based on IgG isotypes. Unfortunately, spleen cells from rats immunized with KLH in the absence of adjuvant do not produce detectable cytokine response to KLH *in vitro*, so we were unable to assess cytokine production in these animals (unpublished observations). The increase in serum antibody levels in young and old animals is directionally opposite to the reports of sympathectomy-induced decreases in cell-mediated and antibody responses in young mice in other studies [126,142] and suppressed antibody responses in mice with SCID reconstituted with an antigen-specific Th2 cell line and antigen-specific B cells [125]. In the latter study, this effect was partially reversed by treatment with NE.

To investigate sympathetic-immune modulation in mice whose splenic NA innervation is maintained relatively well in old age, young and old mice were chemically sympathectomized and immunized with KLH [5]. In contrast to the F344 rat, the anti-KLH antibody response is reduced in NA-intact old mice. Sympathectomy reduced serum anti-KLH antibody titers (IgM, IgG, IgG$_1$, and IgG$_{2a}$), KLH-induced proliferation, and increased IL-2, IL-4, and IFN-$\gamma$ production by KLH-stimulated splenocytes in young, but not old, mice 6 days after immunization with KLH, similar to the effects reported by Kruszewska et al. [132]. This effect was not consistent with previous reports by Kruszewska and colleague [75,132] who reported enhanced KLH-induced proliferation and increased anti-KLH responses in C57BL/6 mice. One major difference between the 2 studies was in housing conditions (single vs. group housed). The inability to detect sympathectomy-induced changes in immune reactivity to KLH in aged mice is consistent with preliminary data from our laboratory demonstrating a striking attenuation of isoproterenol-stimulated cAMP production in splenocytes from old mice. Other investigators have reported impaired peripheral blood lymphocyte responsiveness to NE in elderly subjects [143–144], resulting from a defect in β-adrenergic receptor coupling with adenylate cyclase or the catalytic capacity of adenylate cyclase [145–147]. These findings suggest an age-related block in sympathetic-to-immune signaling in these strains of aged mice. Whether this is an adaptive and/or protective compensatory mechanism to counteract the age-related increase in sympathetic activity awaits further investigation.

Our results demonstrate two types of responses of the sympathetic nervous system to ageing. In the aged F344 rats, splenic NA nerves are markedly diminished, and yet the capacity to respond to sympathectomy is intact, and is associated with an intact ß-adrenergic receptor signal capacity within immune cells. In two strains of aged mice, where splenic NA innervation is largely intact [5,59], ß-adrenergic receptor signaling capacity in lymphocytes from these mice is impaired. Although the biological relevance of differences in the response of the sympathetic nervous system to ageing between aged mice and F344 rats is not known, it is possible that such differences may contribute to individual differences present in aged human populations. It is also possible that individual differences in immune reactivity that occur with age in human populations, for example, in response to vaccination [148] may be related to differences in the sympathetic nervous system response to ageing. The nature of age-related changes in sympathetic NA innervation may also be an influential factor in determining susceptibility to infectious disease and other pathologies of the immune system with age.

## 5.    SUMMARY AND CLOSING REMARKS

NA sympathetic innervation is an important constituent of the microenvironment of lymphoid organs; maintenance of these nerves relies on the secretion of neurotrophins and other neuroprotective mechanisms. The dramatic alterations in NA innervation of lymphoid organs with age indicate a change in the internal milieu with advancing ageing. The role of the sympathetic nervous system in immune modulation is complex, affecting cellular migration, activation and clonal expansion, cytokine production, effector functions and immune cell trafficking. NE signaling of the immune system is neither simplistic nor unidirectional; as a modulatory neurotransmitter it differentially influences multiple cell types at multiple stages during an immune response. Determining the nature and timing of neural-immune interactions is one of the major challenges in this area of study.

Our results suggest that the sympathetic nervous system in some strains of rodents may exert a greater influence on immune responses when the host is immunocompromised, as in ageing, perhaps because of dampened or ineffective feedback mechanisms and/or readjusted homeostatic set points in ageing. Since, the sympathetic nervous system is essential for immune competence, altered sympathetic activity in lymphoid organs with ageing may put the elderly at risk for diseases associated with the immune system, including infectious diseases, autoimmunity, and cancer. This may be particularly true under stressful life events where the elderly display reduced ability to cope with their environment. Investigating the temporal relationship between age-related changes in immune function and NA innervation of lymphoid organs may provide some insight into this issue. Future research in this area may lead to the development of behaviorally based therapies that either reduce stress or improve coping strategies in order to prevent disease or improve disease outcome.

A link between sympathetic dysfunction and immune senescence in ageing would suggest that adrenergic agents may be useful for improving immune responsiveness to vaccines, or be efficacious to the prevention or treatment of certain types of age-related illnesses. It may be possible to develop combination therapies that reduce the doses of cytokines required, when given in combination with adrenergic agents or behavioral approaches, reducing side effects associated with cytokine treatments to which the elderly may be more sensitive. However, before these strategies can be put in place further research is required to better understand the role of the sympathetic nervous system in immune modulation in the elderly. It will be important in these studies to examine strain and species differences, as these differences may lead to a predisposition of certain strains to particular types of illnesses that increase in frequency with advancing age. Clearly, humans display wide variability in how they age, and how they age is likely to be a significant factor in which diseases they become most susceptible to later in life.

## REFERENCES

1.  Bellinger DL, Lorton D, Lubahn C, Felten DL. Innervation of lymphoid organs. Association of nerves with cells of the immune system and their implications in disease. In: Ader R, Cohen N, Felten DL, editors. Psychoneuroimmunology, third edition. New York: Academic Press, 2001;55–111.
2.  Sander VM, Kasprowicz DJ, Kohm AP, Swanson MA. Neurotransmitter receptors on lymphocytes and other lymphoid cells. In: Ader R, Cohen N, Felten DL, editors. Psychoneuroimmunology, third edition. New York: Academic Press, 2001;161–196.
3.  Madden, KS. Catecholamines, sympathetic nerves, and immunity. In: Ader R, Cohen N, Felten DL, editors. Psychoneuroimmunology, third edition. New York: Academic Press, 2001;197–216.
4.  Lorton D, Lubahn C, Bellinger D. Introduction to biological signaling in psychoneuroimmunology. In: Ader R, Cohen N, Felten DL, editors. Psychoneuroimmunology, third edition. New York: Academic Press, 2001; 113–160.
5.  Bellinger DL, Madden KS, Lorton D, Thyagarajan S, Felten DL. Age-related alterations in neural-immune interactions and neural strategies in immunosenescence. In: Ader R, Cohen N, Felten DL, editors. Psychoneuroimmunology, third edition. New York: Academic Press;241–286,
6.  Dominguez-Gerpe L, Rey-Mendez M. Age-related changes in primary and secondary immune organs of the mouse. Immunol Invest 1998;27:153–165.
7.  Donskoy E, Goldschneider I. Thymocytopoiesis is maintained by blood-borne precursors throughout postnatal life. A study in parabiotic mice. J Immunol 1992;148:1604–1612.
8.  Globerson A, Sharp A, Fridkis-Hareli M, Kukulansky T, Abel L, Knyszynski A, Eren R. Aging in the T lymphocyte compartment. A developmental view. Ann NY Acad Sci 1992;673:240–251.
9.  Globerson A. Thymocyte progenitors in ageing. Immunol Lett 1994;40:219–224.
10. Eren R, Zharhary D, Abel L, Globerson A. Age-related changes in the capacity of bone marrow cells to differentiate in thymic organ cultures. Cell Immunol 1988;112:449–455.
11. Eren R, Globerson A, Abel L, Zharhary D. Quantitative analysis of bone marrow thymic progenitors in young and aged mice. Cell Immunol 1990;127:238–246.
12. Tyran ML. Age-related decrease in mouse T cell progenitors. J Immunol 1977;118:846–

851.

13. Afan AM, Broome CS, Nicholls SE, Whetton AD, Miyan JA. Bone marrow innervation regulates cellular retention in the murine haemopoietic system. Br J Haematol 1997;98: 569–577.

14. Müller S, Weihe E. Interrelation of peptidergic innervation with mast cells and ED1-positive cells in rat thymus. Brain Behav Immun 1991;5:55–72.

15. Tabarowski Z, Gibson-Berry K, Felten SY. Noradrenergic and peptidergic innervation of the mouse femur. Acta Histochem 1996;98:453–457.

16. Weihe E, Müller S, Fink T, Zentel HJ. Tachykinins, calcitonin gene-related peptide and neuropeptide Y in nerves of the mammalian thymus: Interactions with mast cells in autonomic and sensory neuroimmunomodulation. Neurosci Lett 1989;100:77–82.

17. Felten SY, Felten DL. The innervation of lymphoid tissue. In: Ader R, Cohen N, Felten DL, editors. Psychoneuroimmunology, second edition. New York: Academic Press, 1991;27–69.

18. Felten DL, Felten SY, Carlson SL, Olschowska JA, Livnat S. Noradrenergic and peptidergic innervation of lymphoid tissue. J Immunol 1985;135:755s–765s.

19. Maestroni GJ, Cosentino, M, Marino F, Togni M, Conti A, Lecchini S, Frigo G. Neural and endogenous catecholamines in the bone marrow. Circadian association of norepinephrine with hematopoiesis? Exp Hematol 1998;26:1172–1177.

20. Kelley SP, Grota LJ, Felten, SY, Madden KS, Felten DL. Norepinephrine in mouse spleen shows minor strain differences and no diurnal variation. Pharmacol Biochem Behav 1996;53:141–146.

21. Cosentino M, Marino F, Bombelli R, Ferrari M, Maestroni GJ, Conti A, Rasini E, Lecchini S, Frigo G. Association between the circadian course of endogenous noradrenaline and the hematopoietic cell cycle in mouse bone marrow. J Chemother 1998;10:179–181.

22. Maestroni GJ, Conti A. Modulation of hematopoiesis via alpha 1-adrenergic receptors on bone marrow cells. Exp Hematol 1994;22:313–320.

23. Maestroni GJ, Conti A, Pedrinis E. Effect of adrenergic agents on hematopoiesis after syngeneic bone marrow transplantation in mice. Blood 1992;80:1178–1182.

24. Madden KS, Felten SY, Felten DL, Hardy CA, Livnat S. Sympathetic nervous system. II. Induction of lymphocyte proliferation and migration in vivo by chemical sympathectomy. J Neuroimmunol 1994;49:67–75.

25. Benestad HB, Strøm-Gundersen, I, Iversen PO, Haug E, Njå A. No neuronal regulation of murine bone marrow function. Blood 1998;91:1280–1287.

26. Maestroni GJ, Togni M, Covacci V. Norepinephrine protects mice from acute lethal doses of carboplatin. Exp Hematol 1997;25:491–494.

27. von Boehmer, H. The developmental biology of T lymphocytes. Ann Rev Immunol 1988;6:309–326.

28. Robey E, Fowlkes BJ. Selective events in T cell development. Ann Rev Immunol 1994;12: 675–705.

29. Bodey B, Bodey B Jr, Siegel SE, Kaiser HE. Involution of the mammalian thymus, one of the leading regulators of aging. In Vivo 1997;11:421–440.

30. George AJ, Ritter MA. Thymic involution with ageing: obsolescence or good housekeeping? Immunol Today 1996;17:267–272.

31. Takeda T, Hosokawa M, Takeshita S, Irino M, Higuchi K, Matsushita T, Tomita Y, Yasuhira K, Hamamoto H, Shimizu K, Ishii M, Yamamuro T. A new murine model of accelerated senescence. Mech Ageing Dev 1981;17:183–194.

32.	Aspinall R. Age-associated thymic atrophy in the mouse is due to a deficiency affecting rearrangement of the TCR during intrathymic T cell development. J Immunol 1997;158: 3037–3045.

33.	Kuro-o M, Matsumura Y, Aizawa H, Kawaguchi H, Suga T, Utsugi T, Ohyama Y, Kurabayashi M, Kaname T, Kume E, Iwasaki H, Iida A, Shiraki-Iida T, Nishikawa S, Nagai R, Nabeshima YI. Mutation of the mouse klotho gene leads to a syndrome resembling ageing. Nature 1997;390:45–51.

34.	Nabarra B, Casanova M, Paris D, Paly E, Toyoma K, Ceballos I, London J. Premature thymic involution, observed at the ultrastructural level, in two lineages of human-SOD-1 transgenic mice. Mech Ageing Dev 1997;96:59–73.

35.	Simpson JG, Gray ES, Beck JS. Age involution in the normal human adult thymus. Clin Exp Immunol 1975;19:261–265.

36.	Bellamy D, Hinsull SM, Phillips JG. Factors controlling growth and age involution of the rat thymus. Age Ageing 1976;5:12–19.

37.	Mackall CL, Punt JA, Morgan P, Farr AG, Gress RE. Thymic function in young/old chimeras: substantial thymic T cell regenerative capacity despite irreversible age-associated thymic involution. Eur J Immunol 1998;28:1886–1893.

38.	Thoman ML. The pattern of T lymphocyte differentiation is altered during thymic involution. Mech Ageing Dev 1995;82:155–170.

39.	Thoman ML. Early steps in T cell development are affected by aging. Cell Immunol 1997;178:117–123.

40.	Thoman ML. Effects of the aged microenvironment on CD4+ T cell maturation. Mech Ageing Dev 1997;96:75–88.

41.	Goidl EA. Aging and the immune response. Cellular and humoral aspects. New York: Marcel Dekker;1987.

42.	Thoman ML, Weigle WO. The cellular and subcellular bases of immunosenescence. Adv Immunol 1989;46:221–261.

43.	Farr AG, Sidman CL. Reduced expression of Ia antigen by thymic epithelial cells of aged mice. J Immunol 1984;133:98–103.

44.	Farr AG, Anderson SK. Epithelial heterogeneity in the murine thymus: fucose-specific lectins bind medullary epithelial cells. J Immunol 1985;134:2971–2977.

45.	von Patay B, Loppnow H, Feindt J, Kurz B, Mentlein R. Catecholamine and lipopolysaccharide synergistically induce the release of interleukin-6 from thymic epithelial cells. J Neuroimmunol 1998;86:182–189.

46.	von Patay B, Kurz B, Mentlein R. Effect of transmitters and cotransmitters of the sympathetic nervous system on interleukin-6 synthesis in thymic epithelial cells. Neuroimmunomodulation 1999; 6:45–50.

47.	Bulloch K, Pomerantz W. Autonomic nervous system innervation of thymic related lymphoid tissue in wildtype and nude mice. J Comp Neurol 1984;228:57–68.

48.	Nance DM, Hopkins DA, Bieger D. Re-investigation of the innervation of the thymus gland in mice and rats. Brain Behav Immun 1987;1:134–147.

49.	Tollefson L, Bulloch K. Dual-label retrograde transport: CNS innervation of the mouse thymus distinct from other mediastinum viscera. J Neurosci Res 1990;25:20–28.

50.	Bellinger DL, Lorton D, Romano TD, Olschowka JA, Felten SY, Felten DL. Neuropeptide innervation of lymphoid organs. Ann NY Acad Sci 1990;594:17–33.

51.	Brouxhon SM, Prasad AV, Joseph SA, Felten DL, Bellinger DL. Localization of CRF in the primary and secondary lymphoid organs of the rat. Brain Behav Immun 1998;12:

107–122).

52. Kendall MD, al-Shawaf AA. Innervation of rat thymus gland. Brain Behav Immun 1991; 5:9–28.

53. Bulloch K, Hausman J, Radojcic T, Short S. Calcitonin gene-related peptide in the developing and aging thymus. An immunocytochemical study. Ann NY Acad Sci 1991;621:218–228.

54. Geppetti P, Maggi CA, Zecchi-Orlandini S, Santicioli P, Meli A, Frilli S, Spillantini MG, Amenta F. Substance P-like immunoreactivity in capsaicin-sensitive structures of the rat thymus. Regul Peptides 1987;18:321–329.

55. Geppetti P, Frilli S, Renzi D, Santicioli P, Maggi CA, Theodorsson E, Fanciullacci M. Distribution of calcitonin gene-related peptide-like immunoreactivity in various rat tissues: correlation with substance P and other tachykinins and sensitivity to capsaicin. Regul Peptides 1988;23:289–298.

56. Elfvin LG, Aldskogius, H, Johansson J. Primary sensory afferent in the thymus of the guinea pig demonstrated with anterogradely transported horseradish peroxidase conjugates. Neurosci Lett 1993;150:35–38.

57. Bellinger DL, Felten SY, Felten DL. Maintenance of noradrenergic sympathetic innervation in the involuted thymus of the age Fischer 344 rat. Brain Behav Immun 1988;2:133–150.

58. Zirbes T, Novotny GE. Quantification of thymic innervation in juvenile and aged rats. Acta Anat (Basel) 1992;145:283–288.

59. Madden KS, Bellinger DL, Felten SY, Snyder E, Maida ME, Felten DL. Alterations in sympathetic innervation of thymus and spleen in aged mice. Mech Ageing Dev 1997;94: 165–175.

60. Durant S. In vivo effects of catecholamines and glucocorticoids on mouse thymic cAMP content and thymolysis. Cell Immunol 1986;102:136–143.

61. Cook-Mills JM, Cohen RL, Perlman RL, Chambers DA. Inhibition of lymphocyte activation by catecholamines: evidence for a non-classical mechanism of catecholamine action. Immunology 1995;85:544–549.

62. Marchetti B, Morale MC, Paradis P, Bouvier M. Characterization, expression, and hormonal control of a thymic beta-2-adrenergic receptor. Am J Physiol 1994;267:E718–731.

63. Morale MC, Gallo F, Batticane N, Marchetti B. The immune response evokes up- and down-modulation of beta-2-adrenergic receptor messenger RNA concentration in the male rat thymus. Mol Endocrinol 1992;6:1513–1524.

64. Morgan JI, Wigham CG, Perris AD. The promotion of mitosis in cultured thymic lymphocytes by acetylcholine and catecholamines. J Pharm Pharmacol 1984;36:511–515.

65. Kurz B, Feindt J, von Gaudecker B, Kranz A, Loppnow H, Mentlein R. Beta adrenoceptor-mediated effects in rat cultured thymic epithelial cells. Br J Pharmacol 1997;120:1401–1408.

66. Fuchs BA, Albright JW, Albright JF. Beta-adrenergic receptors on murine lymphocytes: density varies with cell maturity and lymphocyte subtype and is decreased after antigen administration. Cell Immunol 1988;114:231–245.

67. Radojcic T, Baird S, Darko D, Smith D, Bulloch K. Changes in β-adrenergic receptor distribution on immunocytes during differentiation: an analysis of T cells and macrophages. J Neurosci Res 1991;30:328–335.

68. Bach M-A. Differences in cyclic AMP changes after stimulation by prostaglandins and isoproterenol in lymphocyte subpopulations. J Clin Invest 1975;55:1074–1081.

69. Alaniz RC, Thomas SA, Perez-Melgosa M, Mueller K, Farr AG, Palmiter RD, Wilson CB. Dopamine beta-hydroxylase deficiency impairs cellular immunity. Proc Natl Acad Sci USA 1999;96:2274–2278.

70. Singh U, Owen JJ. Studies on the maturation of thymus stem cells. The effects of catecholamines, histamine and peptide hormones on the expression of T cell alloantigens. Eur J Immunol 1976;6:59–62.

71. Singh U. Lymphopoiesis in the nude fetal mouse thymus following sympathectomy. Cell Immunol 1985;93:222–228

72. Delrue-Perollet C, Li KS, Vitiello S, Neveu PJ. Peripheral catecholamines are involved in the neuroendocrine and immune effects of LPS. Brain Behav Immun 1995;9:149–162.

73. Tsao CW, Cheng JT, Shen CL, Lin YS. 6-Hydroxydopamine induces thymocyte apoptosis in mice. J Neuroimmunol 1996;65:91–95.

74. Kostrzewa RM & Jacobwitz DM. Pharmacological actions of 6-hydroxydopamine. Pharmacol Rev 1974;26:199–288.

75. Kruszewska B, Felten DL, Stevens SY, Moynihan JA. Sympathectomy-induced immune changes are not abrogated by the glucocorticoid receptor blocker RU-486. Brain Behav Immun 1998;12:181–200.

76. Madden KS, Felten DL. Beta-adrenoceptor blockade alters thymocyte differentiation in aged mice. Cell Mol Biol 2001;47:189–196.

77. Petrie HT, Hugo P, Scollay R, Shortman K. Lineage relationships and developmental kinetics of immature thymocytes: CD3, CD4, and CD8 acquisition in vivo and in vitro. J Exp Med 1990;172:1583–1588.

78. Boyd RL, Tucek CL, Godfrey DI, Izon, DJ, Wilson TJ, Davidson NJ, Bean AG, Ladyman HM, Ritter MA, Hugo P. The thymic environment. Immunol Today 1993;14:445–459.

79. MacDonald HR, Budd RC, Cerottini J-C. Pgp-1 (Ly 24) as a marker of murine memory T lymphocytes. Curr Top Microbiol Immunol 1990;159:97–109.

80. Romano TA, Felten SY, Felten DL, Olschowka JA. Neuropeptide-Y innervation of the rat spleen: another potential immunomodulatory neuropeptide. Brain Behav Immun 1991;5:116–131.

81. Bellinger DL, Brouxhon SM, Lubahn C, Tran L, Keng J, Felten DL, Lorton D. Strain differences in the expression of corticotropin-releasing hormone immunoreactivity in nerves that supply the spleen and thymus. Neuroimmunomodulation 2001;9:78–87.

82. Bellinger DL, Felten SY, Collier TJ, Felten DL. Noradrenergic sympathetic innervation of the spleen: IV. Morphometric analysis in adult and aged Fischer 344 rats. J. Neurosci Res 1987;18:55–63, 126–129.

83. Bellinger DL, Ackerman KD, Felten SY, Felten DL. A longitudinal study of age-related loss of noradrenergic nerves and lymphoid cells in the aged rat spleen. Exp Neurol 1992;116:295–311.

84. Felten SY, Bellinger DL, Collier TJ, Coleman PD, Felten, DL. Decreased sympathetic innervation of spleen in aged Fischer 344 rats. Neurobiol Aging 1987;8:159–165.

85. Santer RM, Partanen M, Hervonen A. Glyoxylic acid fluorescence and ultrastructural studies of neurones in the coeliac-superior mesenteric ganglion of the aged rat. Cell Tissue Res 1980; 211:475–485.

86. Esler M, Skews H, Leonard P, Jackman G, Bobik A, Korner P. Age-dependence of noradrenaline kinetics in normal subjects. Clin Sci 1981;60:217–219.

87. Krall JF, Connelly M, Weisbart R, Tuck ML. Age-related elevation of plasma catecholamine concentration and reduced responsiveness of lymphocyte adenylate cyclase. J Clin

Endocrinol Metab 1981;52:863–867.

88. Lake CR, Ziegler MG, Coleman MD, Kopin IJ. Age-adjusted plasma norepinephrine levels are similar in normotensive and hypertensive subjects. New Engl J Med 1977;296: 208–209.

89. Ziegler MG, Lake CR, Kopin IJ. Plasma noradrenaline increases with age. Nature (London) 1976;261:333–335.

90. Bellinger DL, Felten SY, Felten DL. NA sympathetic innervation of lymphoid organs during development, aging and in autoimmune disease. In: Amenta F, editor. Aging of the Autonomic Nervous System. Boca Raton: CRC Press, 1993;243–283.

91. Bellinger DL, Tran L, Kang, JI, Lubahn C, Felten DL, Lorton D. Age-related changes in noradrenergic sympathetic innervation of the rat spleen are strain-dependent. Brain Behav Immun 2002;16:247–261.

92. Besedovsky HO, del Rey A, Sorkin E, Da Prada M, Keller HH. Immunoregulation mediated by the sympathetic nervous system. Cell Immunol 1979;48:346–355.

93. Felten DL, Felten SY, Steece-Collier K, Date I, Clemens JA. Age-related decline in the dopaminergic nigrostriatal system: The oxidative hypothesis and protective strategies. Ann Neurol 1992;32:S133-S136.

94. Felten DL, Felten SY, Bellinger DL, Carlson SL, Ackerman KD, Madden KS, Olschowka JA, Livnat S. Noradrenergic sympathetic neural interactions with the immune system: structure and function. Immunol Rev 1987;100:225–260.

95. Bellinger DL, Brouxhon S, Henderson D, Felten DL, Felten SY. IL-2 increases IL-2R expression in superior mesenteric-celiac ganglion and alters splenic norepinephrine concentration. Soc Neurosci 1996;22:1353.

96. ThyagaRajan S, Felten SY, Felten DL. Restoration of sympathetic noradrenergic nerve fibers in the spleen by low doses of L-deprenyl treatment in young sympathectomized and old Fischer 344 rats. J Neuroimmunol 1998;81:144–157.

97. Birkmayer W, Knoll J, Riederer P, Youdim MB, Hars V, Marton J. Increased life expectancy resulting from addition of L-deprenyl to Mardopar treatment in Parkinson's disease: a long-term study. J Neural Transm 1985;64:113–127.

98. Tedrud JW, Langston JW. The effect of deprenyl (selegiline) on the natural history of Parkinson's disease. Science 1989;245:519–522.

99. Ansari KS, Yu PH, Kruck TP, Tatton WG. Rescue of axotomized immature rat facial motoneurons by R(-)-deprenyl: stereospecificity and independence from monoamine oxidase inhibition. J Neurosci 1993;13:4042–4053.

100. Buys YM, Trope GE, Tatton WG. (-)-Deprenyl increases the survival of rat retinal ganglion cells after optic nerve crush. Curr Eye Res 1995;14:119–126.

101. Seniuk NA, Henderson JT, Tatton WG, Roder JC. Increased CNTF gene expression in process-bearing astrocytes following injury is augmented by R(-)-deprenyl. J Neurosci Res 1994;37:278–286.

102. ThyagaRajan S, Felten SY, Felten DL. L-deprenyl-induced increase in IL-2 and NK cell activity accompanies restoration of noradrenergic nerve fibers in the spleens of old F344 rats. J Neuroimmunol 1998;92:9–21.

103. ThyagaRajan S, Madden KS, Stevens SY, Felten DL. Anti-tumor effect of L-deprenyl is associated with enhanced central and peripheral neurotransmission and immune reactivity in rats with carcinogen-induced mammary tumors. J Neuroimmunol 2000;109:95–104.

104. Santambrogio L, Benedetti M, Chao MV, Muzaffar R, Kulig K, Gabellini N, Hochwald G. Nerve growth factor production by lymphocytes. J Immunol 1994;153:4488–4495.

105. Barouch R, Appel E, Kazimirsky G, Braun A Renz H, Brodie C. Differential regulation of neurotrophin expression by mitogens and neurotransmitters in mouse lymphocytes J Neuroimmunol 2000;103:112–121.

106. Carlson SL, Albers KM, Beiting DJ, Parish M, Conner JM, Davis BM. NGF modulates sympathetic innervation of lymphoid tissues. J Neurosci 1995;15:5892–5899.

107. Yankner BA, Shooter EM. The biology and mechanism of action of nerve growth factor. Annu Rev Biochem 1982;51:845–868.

108. Levi-Montalcini R. The nerve growth factor 35 years later. Science 1987;237:1154–1162.

109. Mitchell B, Kendall M, Adam E, Schumacher U. Innervation of the thymus in normal and bone marrow reconstituted severe combined immunodeficient (SCID) mice. J Neuroimmunol 1997;75:19–27.

110. Carlson SL, Felten DL, Livnat S, Felten SY. Noradrenergic sympathetic innervation of the spleen: V. Acute drug-induced depletion of lymphocytes in the target fields of innervation results in redistribution of noradrenergic fibers but maintenance of compartmentation. J Neurosci Res 1987;18:64–69, 130–131.

111. Pawelec G, Barnett Y, Forsey R, Frasca D, Globerson A, McLeod J, Caruso C, Franceschi C, Fulop T, Gupta S, Mariani E, Mocchegiani E, Solana R. T cells and aging, January 2002 update. Front Biosci 2002;7:d1056–1183.

112. Fulop T Jr. Signal transduction changes in granulocytes and lymphocytes with ageing. Immunol Lett 1994;40:259–268.

113. Pahlavani MA, Richardson A. The effect of age on the expression of interleukin-2. Mech Ageing Dev 1996;89:125–154.

114. Caruso C, Candore G, Cigna D, DiLorenzo G, Sireci G, Dieli F, Salerno A. Cytokine production pathway in the elderly. Immunol Res 1996; 15:84–90.

115. Rink L, Cakman I, Kirchner H. Altered cytokine production in the elderly. Mech Ageing Dev 1998;15:199–209.

116. Hobbs MV, Weigle WO, Noonan DJ, Torbett BE, McEvilly RJ, Koch RJ, Cardenas GJ, Ernst DN. Patterns of cytokine gene expression by CD4+ T cells from young and old mice. J Immunol 1993;150:3602–3614.

117. Hobbs MV, Weigle WO, Ernst DN. Interleukin-10 production by splenic CD4+ cells and cell subsets from young and old mice. Cell Immunol 1994;154:264–272.

118. Castle S, Uyemura K, Wong W, Modlin R, Effros R. Evidence of enhanced type 2 immune response and impaired upregulation of a type 1 response in frail elderly nursing home residents. Mech Ageing Dev 1997;94:7–16.

119. Mackall CL, Gress RE. Thymic aging and T-cell regeneration. Immunol Rev 1997;160: 91–102.

120. Merino J, Martinez-Gonzalez MA, Rubio M, Inoges S, Sanchez-Ibarrola A, Subira ML. Progressive decrease of CD8[high+] CD28[+] CD57[-] cells with ageing. Clin Exp Immunol 1998;112:48–51.

121. Engwerda CR, Handwerger BS, Fox BS. Aged T cells are hyporesponsive to costimulation mediated by CD28. J Immunol 1994;152:3740–3747.

122. Utsuyama M, Hirokawa K, Kurashima C, Fukayama M, Inamatsu T, Suzuki K, Hashimoto W, Sato K. Differential age-change in the numbers of CD4+CD45RA+ and CD4+CD29+ T cell subsets in human peripheral blood. Mech Ageing Dev 1992;63:57–68.

123. Ernst DN, Hobbs MV, Torbett BE, Glasebrook AL, Rehse MA, Bottomly K, Hayakawa K, Hardy RR, Weigle WO. Differences in the expression profiles of CD45RB, Pgp-1, and 3G11 membrane antigens and in the patterns of lymphokine secretion by splenic CD4+ T

254

cells from young and aged mice. J Immunol 1990;145:1295–1302.

124. Jackola DR., Ruger JK, Miller RA. Age-associated changes in human T cell phenotype and function. Aging 1994;6:25–34.

125. Kohm AP, Sanders VM. Suppression of antigen-specific Th2 cell-dependent IgM and IgG1 production following norepinephrine depletion *in vivo*. J Immunol 1999;162:5299–5308.

126. Madden KS, Felten SY, Felten DL, Sundaresan PR, Livnat S. Sympathetic neural modulation of the immune system. I. Depression of T cell immunity *in vivo* and *in vitro* following chemical sympathectomy. Brain Behav Immun 1989;3:72–89.

127. Dhabhar FS, McEwen BS. Enhancing versus suppressive effects of stress hormones on skin immune function. Proc Natl Acad Sci USA 1999;96:1059–1064.

128. Cunnick JE, Lysle DT, Kucinski BJ, Rabin BS. Evidence that shock-induced immune suppression is mediated by adrenal hormones and peripheral β-adrenergic receptors. Pharmacol Biochem Behav 1990;36:645–651.

129. Dobbs CM, Vasquez M, Glaser R, Sheridan JF. Mechanisms of stress-induced modulation of viral pathogenesis and immunity. J Neuroimmunol 1993;48:151–160.

130. Fecho K, Maslonek KA, Dykstra LA, Lysle DT. Evidence for sympathetic and adrenal involvement in the immunomodulatory effects of acute morphine treatment in rats. J Pharmacol Exp Ther 1996;277:633–645.

131. Irwin M, Hauger RL, Brown M, Britton KT. CRF activates autonomic nervous system and reduces natural killer cytotoxicity. Am J Physiol 1988;255:R744-R747.

132. Kruszewska B, Felten SY, Moynihan JA. Alterations in cytokine and antibody production following chemical sympathectomy in two strains of mice. J Immunol 1995;155:4613–4620.

133. Madden KS, Felten SY, Felten DL, Bellinger DL. Sympathetic nervous system-immune system interactions in young and old Fischer 344 rats. Ann NY Acad Sci 1995;771:523–534.

134. Hatfield SM, Petersen BH, DiMicco JA. Beta adrenoceptor modulation of the generation of murine cytotoxic T lymphocytes in vitro. J Pharmacol Exp Ther 1986;239:460–466.

135. Sanders VM, Munson AE. Beta adrenoceptor mediation of the enhancing effect of norepinephrine on the murine primary antibody response in vitro. J Pharmacol Exp Ther 1984;230:183–192.

136. Sander VM, Baker RA, Ramer-Quinn DS, Kasprowicz DJ, Fuchs BA, Street NE. Differential expression of the β$_2$-adrenergic receptor by Th1 and Th2 clones. J Immunol 1997;158:4200–4210.

137. Kasprowicz DJ, Kohm AP, Berton MT, Chruscinski AJ, Sharpe A, Sander VM. Stimulation of the B cell receptor, CD86 (B7–2), and the β2-adrenergic receptor intrinsically modulates the level of IgG1 and IgE produced per B cell. J Immunol 2000;165:680–690.

138. Ramer-Quinn DS, Baker RA, Sanders VM. Activated T helper 1 and T helper 2 cells differentially express the β-2-adrenergic receptor: a mechanism for selective modulation of T helper 1 cell cytokine production. J Immunol 1997;159:4857–4867.

139. Swanson MA, Lee WT, Sanders VM. IFN-γ production by Th1 cells generated from naïve CD4+ T cells exposed to norepinephrine. J Immunol 2001;166:232–240.

140. Madden KS, Stevens SY, Felten DL, Bellinger DL. Alterations in T lymphocyte activity following chemical sympathectomy in young and old Fischer 344 rats. J Neuroimmunol 2000;103:131–145.

141. Ackerman KD, Bellinger DL, Felten SY, Felten DL. Ontogeny and senescence of noradrenergic innervation of the rodent thymus and spleen. In: Ader R, Felten, DL,

Cohen N, editors. Psychoneuroimmunology, second edition. New York:Academic Press, 1991;72–125.

142. Livnat S, Madden KS, Felten DL, Felten SY. Regulation of the immune system by sympathetic neural mechanisms. Prog Neuro-Psychopharmacol Biol Psychiatry 1987;11: 145–152.

143. Feldman RD, Limbird LE, Nadeau J, Robertson D, Wood AJ. Alterations in leukocyte β-receptor affinity with aging. A potential explanation for altered β-adrenergic sensitivity in the elderly. N Engl J Med 1984;310:815–819.

144. O'Hara N, Daul AE, Fesel R, Siekmann U, Bodde OE. Different mechanisms underlying reduced $\beta_2$-adrenoceptor responsiveness in lymphocytes from neonates and old subjects. Mech Ageing Dev 1985;31:115–122.

145. Krall JF, Connelly M, Weisbart R, Tuck ML. Age-related elevation of plasma catecholamine concentration and reduced responsiveness of lymphocyte adenylate cyclase. J Clin Endocrinol Metab 1981;52:863–867.

146. Doyle VM, O'Malley K, Kelly JG. Human lymphocyte beta-adrenoceptor density in relation to age and hypertension. J Cardiovasc Pharmacol 1982;4:738–740.

147. Scarpace PJ, Abrass IB. Decreased beta-adrenergic agonist affinity and adenylate cyclase activity in senescent rat lung. J Gerontol 1983;38:143–147.

148. Webster RG. Immunity to influenza in the elderly. Vaccine 2000;18:1686–1689.

*The Neuroendocrine Immune Network in Ageing*
Edited by R.H. Straub and E. Mocchegiani

# Ageing and the Neuroendocrine System of the Gut

MAGDY EL-SALHY

*Division of Gastroenterology & Hepatology, Department of Molecular and Clinical Medicine, EM-Clinic, University Hospital, SE-581 85 Linköping, Sweden*

## ABSTRACT

Gastrointestinal secretion and absorption of nutrients diminish with ageing, as does gastrointestinal motility, whereas the incidence of gastrointestinal complaints such as reflux esophagitis, earlier satiety, nausea and vomiting, and constipation increases. The gut neuroendocrine system regulates the gastrointestinal functions, which are disturbed in the elderly, namely secretion, absorption and motility. It is therefore conceivable that age-related changes take place, contributing to the development of the gastrointestinal disorders seen in the elderly and to the manifestation of the gastrointestinal symptoms. Age-related changes in the gut neuroendocrine system are both numerous and complicated. It is difficult to distinguish between those that are due to the inevitable process of ageing itself, and those that are secondary (adaptive) to them. Ageing is accompanied by a general loss of neurons of the enteric nervous system and a change in the density of gastrointestinal endocrine cells. These, together with structural and functional changes in the gut and associated glands, may cause the adaptive changes in the enteric nervous system and the gut endocrine cells that have been reported to have a great capacity of plasticity. The accumulated data on the changes in gastrointestinal neuroendocrine system during ageing could prove useful in clinical practice. Thus, a diet that regulates the release of a certain signal substance of this regulatory system could be used to induce a desirable increase or a reduction in this substance. Furthermore, an agonist or antagonist to the gut neuroendocrine system could be utilized.

## 1.    INTRODUCTION

Ageing is associated with several changes in gastrointestinal function such as gastrointestinal secretion, and the intestinal absorption of fat, carbohydrates and probably proteins, which diminish with age [1,2]. Furthermore, gastrointestinal dysmotility with slower gastric emptying and prolonged large intestine transit, has been reported in the elderly [3,13]. Gastrointestinal complaints such as reflux esophagitis, earlier satiety, nausea and vomiting, and constipation have all been reported to increase with ageing [14]. Moreover, gastrointestinal malignancies rise sharply with advancing age [15]. The gut neuroendocrine system is the local regulator of several gastrointestinal functions, such as motility, secretion, absorption, local immune defense, microcirculation of the gut and cell proliferation [16–22]. As these gastrointestinal functions appear to be affected by ageing, it is possible that age-related changes in this regulatory system contribute

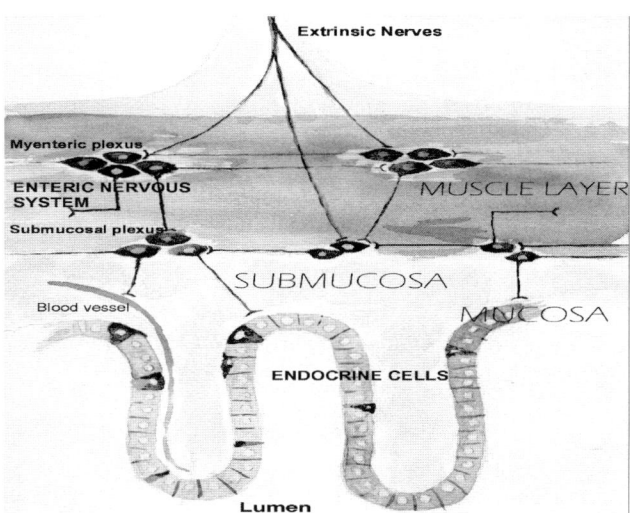

Figure 1. Schematic illustration of the anatomical structure of the gastrointestinal neuroendocrine system. (Supplied kindly by Dr. Olof Sandström, Department of Medicine, University Hospital, Umeå, Sweden).

to the development of gastrointestinal disorders evident in the elderly, and to the manifestation of their symptoms.

The purpose of this chapter is to present an updated review of age-related changes in the neuroendocrine system, to speculate upon the possible role of these changes in the development of gastrointestinal dysfunction in the elderly, and to predict their possible clinical implications.

## 2.  THE NEUROENDOCRINE SYSTEM OF THE GUT

This system consists of two parts: (i) endocrine cells dispersed among the epithelial cells of the mucosa facing the gut lumen, (ii) the peptidergic and serotonergic as well as nitric oxide-containing nerves of the enteric nerve system in the gut wall (Figure 1). This regulatory system comprises a large number of bioactive messengers (Table I). These substances exert their effects by wiring transmission and volume transmission [39]. Wiring transmission is one-to-one transmission including classical synaptic signaling. Volume transmission to distant targets is accomplished by *neurocrine/endocrine* means (releasing into the circulating blood from synapses or endocrine cells), or *by paracrine* mode (by releasing into interstitial fluid to nearby targets). The different parts of this system interact and are integrated both with each other and with afferent and efferent nerve fibers of the central nervous system, especially the autonomic nervous system. The signal substances of the neuroendocrine system may be co-localized in the same cell, or stored separately in mono-expressed cells. Examples of co-localized signal substances are PYY and enteroglucagon in colonic L-cells, serotonin and substance P in enterochromaffin (EC) cells, and VIP, which co-localizes nitric oxide, gastrin releasing peptide (GRP) and enkephalins.

Table I     Overview of the most important neuroendocrine peptides and other signal substances in the neuroendocrine system of the gut.

| Signal substance | Action | Endocrine cells* | Nerve fibres* | Reference |
|---|---|---|---|---|
| Enteroglucagon | Inhibits gastric acid secretion and pancreatic secretion | + | − | Walsh, 1994 |
| Somatostatin | Inhibits intestinal contraction, inhibits gut exocrine and neuroendocrine secretion | + | + | Chiba and Yamada, 1994 |
| Pancreatic polypeptide (PP) | Inhibits pancreatic secretion; stimulates gastric acid secretion; relaxes the gallbladder and stimulates motility of the stomach and small intestine | + | − | Mannon and Taylor, 1994; Walsh, 1994 |
| Peptide YY (PYY) | Delays gastric emptying; inhibits gastric and pancreatic secretion; ileal brake mediator; vasoconstrictor | + | − | Mannon and Taylor, 1994; Walsh, 1994 |
| Neuropeptide Y (NPY) | Inhibits pancreatic and intestinal secretion; decreases gastrointestinal motility; vasoconstrictor - | − | + | Mannon and Taylor, 1994; Walsh, 1994 |
| Secretin | Simulates pancreatic bicarbonate and fluid secretion; inhibits gastric emptying and inhibits contractile activity of small and large intestines | + | − | Leiter et al., 1994; Walsh, 1994 |
| Vasoactive active polypeptide (VIP) | Stimulates gastrointestinal and pancreatic secretion; relaxes smooth muscles in the gut and causes vasodilation | − | + | Dockray, 1994a |
| Gastric inhibitory peptide (GIP) | Incretin; inhibits gastric acid secretion | + | − | Pederson, 1994; Walsh, 1994 |
| Gastrin | Stimulates gastric acid secretion and histamine release; trophic action on gastric mucosa; stimulates contractions of lower oesophageal sphincter (LES) and antrum | + | − | Walsh, 1994 |
| Cholecystokinin (CCK) | Inhibits gastric emptying; stimulates gallbladder contraction and intestinal motility; stimulates pancreatic exocrine secretion and growth; regulates food intake | + | + | Liddle, 1994; Walsh, 1994 |
| Substance P | Stimulates smooth muscle contraction, vasodilator; inhibits gastric acid secretion | + | + | Dockray, 1994b |
| Motilin | Induces phase III MMC (migrating motor complex), stimulates gastric emptying and stimulates contraction of LES | + | − | Poitras, 1994 |
| Neurotensin | Stimulates pancreatic secretion; inhibits gastric secretion; delays gastric emptying and stimulates colon motility | + | + | Shulkes, 1994 |
| Gastrin-releasing peptide (GRP) | Releases gastrin, somatostatin, and some other endocrine peptides; stimulates pancreatic secretion and smooth muscle contraction | − | + | Bunnett, 1994 |
| Enkephalin | Inhibits gastric and pancreatic secretion; delays gastric emptying and intestinal transit | + | + | Dockray, 1994c |
| Galanin | Inhibits gastric, pancreatic and intestinal secretion; delays gastric emptying and intestinal transit; supresses postprandial release of some neuroendocrine peptides | − | + | Rökaeus, 1994 |
| Serotonin | Stimulates gastric antrum and small intestinal and colonic motility; accelerates gastric emptying and both small and large intestinal transit | + | + | Lindberg, 1995;Tally, 1992; Ohe et al.,1994 |
| Nitric oxide (NO) | Smooth muscle relaxation | + | + | Whittle, 1994 |

* present = +; absent = −

3.    AGE-RELATED CHANGES IN THE GUT NEUROENDOCRINE SYSTEM

The age-related abnormalities were mostly observed in studies conducted in animal models, especially rodents. Only a few studies concerning gut endocrine cells were made in humans. The enteric nervous system has been investigated in autopsy material with unknown post-mortems effect. The age-related changes observed in the neuroendocrine system were revealed by either morphological or biochemical methods. Morphological changes are mostly represented by an alteration in the density of cells and/or nerve fibers. These changes signify a change in the anatomical units producing the bioactive signal substances and require time to develop and endure over a period of time. An increase in the density of a certain cell type could indicate an increased amount of the signal substance produced by this particular cell type and *vice versa*. The other types of change observed are expressed as an increase or decrease in the concentration or in coding m-RNA of a signal substance in tissue extracts from gut segments. These changes develop rapidly in response to changes in the physiological conditions but are short-lived. An increase in the level of a particular signal substance implies an increase in its synthesis, which might or might not be accompanied by an increased release. A decrease in a signal substance, on the other hand, could imply either reduced synthesis, or a high rate of synthesis and release.

3.1.    Endocrine cells

In the human stomach, the density of the total endocrine cell population as detected by *chromogranin*, a general marker of endocrine cells, has been found to decline with age [40]. Furthermore, the density of somatostatin cells was also reduced in the elderly in the same study. The plasma gastrin concentration recorded in the same individuals was however higher in women over 55 years of age. The effects of ageing on the endocrine cells of both small and large intestine have been investigated in biopsy material obtained by endoscopy (Figure 2) [41,42]. The endocrine cell population was smaller in the small intestine, as detected by chromogranin A. Among the endocrine cell types that were reduced in number were somatostatin cells, and *GIP cells*. The other endocrine cell types appeared not to be affected by ageing. In the large intestine, the density of both serotonin and somatostatin cells decreased with age. Measurements made in tissue extracts of human large intestine mucosa have revealed that the somatostatin concentration also decreases with age [43].

Age-related changes in the endocrine cells of mice (NMRI) have been investigated systematically in the various segments of the gastrointestinal tract (Figures 3,4) [44–47]. It seems that ageing affects several endocrine cell types in various gut segments. Thus, in the stomach, the density of both gastrin cells and the concentration of gastrin in tissue extracts decreased with age. Whereas the density of somatostatin cells was found to increase in old mice, the concentration of somatostatin in tissue extracts decreased. In the small intestine, the number of serotonin cells increased in old mice. Whereas the density of secretin cells and the secretin concentration in tissue extracts increased with age, those of somatostatin decreased [45,47]. In the colon, the cell density of serotonin (Figure 5), PYY (Figure 6) and enteroglucagon increased with age [46]. Measurements of PYY and somatostatin in colonic tissue extract have shown that the concentration of these two peptides also increases with age [47].

Endocrine cells have been investigated in stomach and small intestine of nude mice (BALB/c-nu) [48]. In this mouse, the total number of endocrine cells, as detected with the Grimelius argyrophilic reaction, increased with increasing age, in both stomach and small intestine. The density of serotonin cells increased in both stomach and small intestine of old mice, whereas the

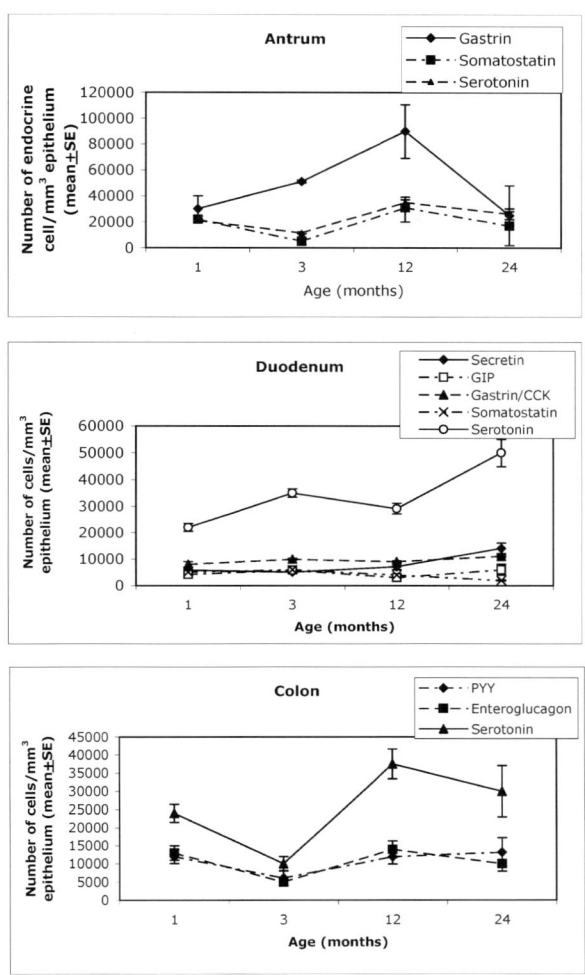

Figure 2. Age-related changes in the densities of different endocrine cell types in the small and large intestine of humans. GIP= gastric inhibitory peptide; CCK= cholecystokinin; PYY= peptide YY; PP= pancreatic polypeptide. [Data from 41 and 42].

gastrin cell population was depleted in the stomach. The authors also found that apoptotic index of epitheliocytes rose with advancing age in these mice. One should bear in mind, however, that nude mice have no thymus and consequently lack cell-mediated immune defense.

It appears that the changes in the gastrointestinal endocrine cells observed in elderly humans differ from those observed in old rodents. This could be due to species-related difference, to the fact that the rodents examined were relatively older than the humans investigated or to a combination of these factors. It is of course important to be cautious when extrapolating data obtained from animal models to humans.

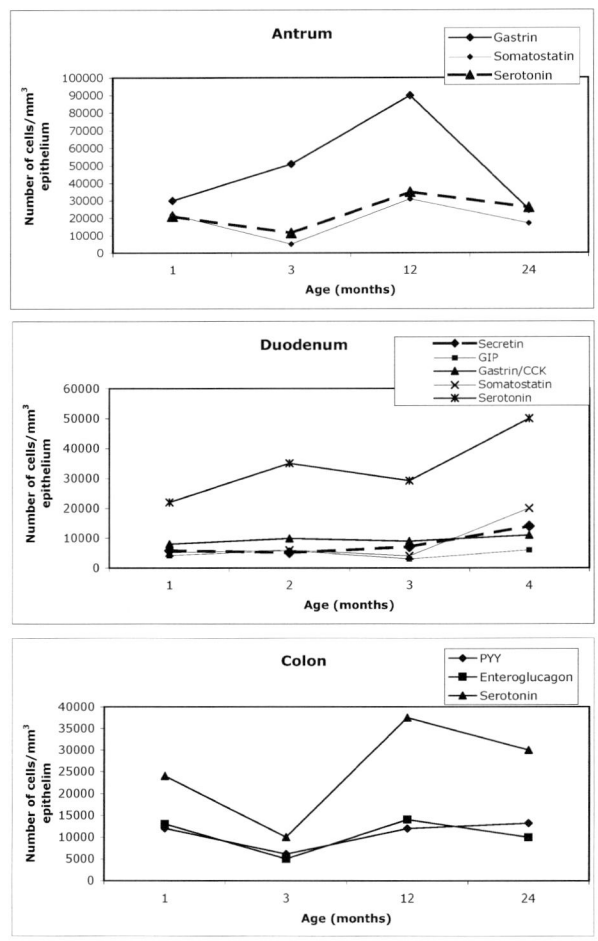

Figure 3. Changes in number of different endocrine cell types with ageing in various gastrointestinal segments of mice. Abbreviations as in Figure 2. [Data from 44–46].

## 3.2.    The enteric nervous system

The number of nerve cells in the myenteric ganglia of the human esophagus, small and large intestine, was found to decrease with ageing [49–51]. Similar results have been reported in the small intestine of guinea pigs [52], in the different segments of the mouse gastrointestinal tract [53], and in the small and large intestine of rat [54,55]. Despite the age-related depletion of neurons in the *myenteric plexus*, the volume density of nerve fibers in different gut segments is not affected in mice [53]. The number of neurons in the submucosal ganglia in different segments of thee gastrointestinal tract of mice decreases with ageing (Figure 7) [53]. However, the volume density of the nerve fibers in the submucosa of the murine gastrointestinal tract is not affected by ageing (Figure 7) [53]. In the wall of the small intestine of senile *vis-á-vis* young rats the number of nerve fibers were depleted [56].

Whereas substance P, VIP, and somatostatin nerve fibers have been reported to decrease in

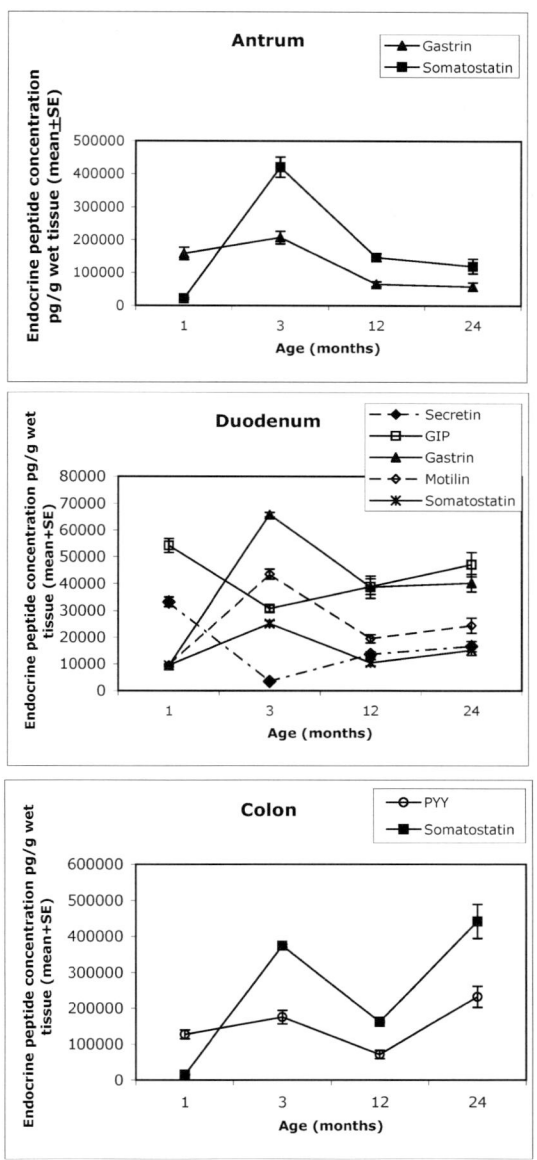

Figure 4. Concentrations of different endocrine peptides in various parts of the gastrointestinal tract at different ages in mice. Abbreviations as in Figure 2. [Data from 47].

small intestine of old rats [56], no age-related changes were found in rat small intestine nerve fibers immunoreactive to substance P, VIP, somatostatin, NPY and enkephalin [55] in the same animals. The number of nitric oxide synthase (NOS)-immunoreactive neurons and their synthesis in the colonic myenteric plexus are reportedly reduced in aged rats [57]. Substance P, and VIP calcitonin gene-related peptide (CGRP) innervations have been investigated in lower esophageal pyloric and ileocecal sphincters of old rats [58]. In the lower esophageal sphincter, substance P

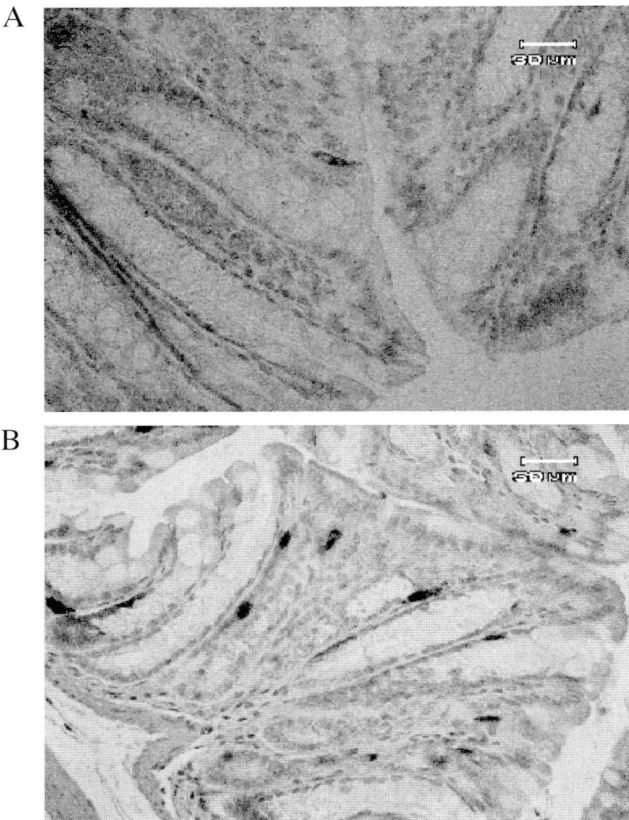

Figure 5. Serotonin-immunoreactive cells in colon of a 3-month-old mouse (A) and of a 24-month-old mouse (B).

and CGRP innervations changed with ageing, whereas VIP-innervation did not. In the pyloric sphincter, there was an increase in substance P- and CGRP- immunoreactive nerve fibers, but a decrease in VIP-fibers. In the ileocecal sphincter, substance P- and VIP-immunoreactive nerve fibers increased with ageing. Radioimmunoassays of tissue extracts of various gastrointestinal segments of the mice have revealed that the neuroendocrine peptide concentrations tend to decrease with ageing (Figure 8), excepting for colonic substance P, which increases in old mice [47].

## 4.  AGE-RELATED CHANGES AND GASTROINTESTINAL DISORDERS IN THE ELDERLY

It was quite evident from the previous presentation that age-related changes in the gut neuroendocrine system do occur. These changes are both numerous and apparently complex. This is not surprising, as the gut neuroendocrine signal substances interact and integrate both with each other and with the endocrine system, as well as with the autonomic and central nervous systems. It is difficult to distinguish between the changes, which are due to the inevitable process of ageing itself, and alterations that are secondary (adaptive) to them.

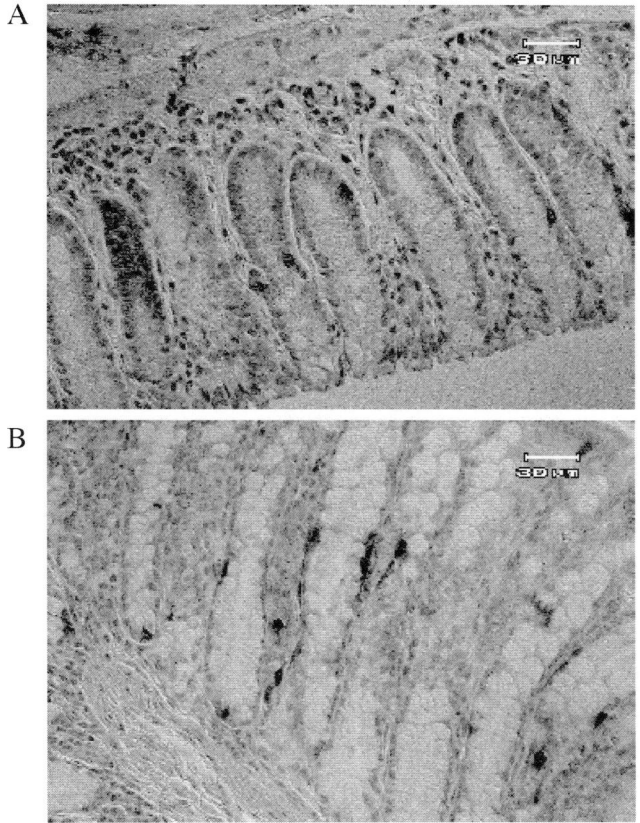

Figure 6. Colonic peptide YY (PYY)-immunoreactive cells in a 3-month-old mouse (A) and of a 24-month old mouse (B).

The process of ageing is accompanied by a general loss of neurons of the enteric nervous system and change in the density of gastrointestinal endocrine cells. Structural and functional changes in the gut, pancreas and biliary pathways have been reported in the elderly human [59]. These together may engender adaptive changes in the enteric nervous system and in the gut endocrine cells, which have been reported to have a great capacity for plasticity [39]. What supports this assumption is the fact that despite the decreased number of neurons in the enteric ganglia with ageing, the density of the nerve fibers remains the same. As the neuroendocrine signal substances of the gut have multi-actions and more than one target organ, adaptive changes that intended to correct a process may provoke a disorder in another process. One example is the increase in secretin, which believed to be secondary to the decline in pancreatic exocrine dysfunction. This increase would inhibit gastric emptying as seen in the elderly. Although this might appear to be highly speculative, it could nevertheless serve as a working hypothesis for further studies.

266

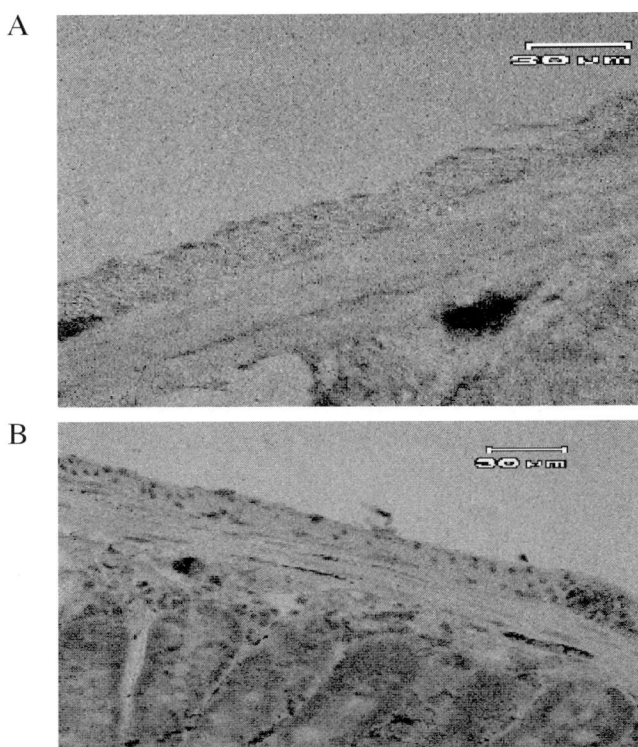

Figure 7. Protein gene-product (PGP) 9.5 -immunoreactive submucosal ganglion (arrow) in a 3-month-old mouse (A) and in a 24-month-old mouse (B). Sumucosal PGP 9.5-immunoreactive nerve fibers (double arrow) in a 24-month-old mouse (C). All are in the duodenum.

## 5.    CLINICAL IMPLICATIONS

No doubt, the gastrointestinal disorders seen in the elderly are multifactorial. Manifestations of diseases to which the elderly are particularly susceptible and side effects of the increased medication in the elderly are important factors. However, age-related alterations in the neuroendocrine system should be considered as one of these. Finding of such changes could be useful and can lead to the use gastrointestinal neuroendocrine signal substances as tool for treatment. The accumulated data concerning the changes in gastrointestinal neuroendocrine signal substances in ageing should therefore be of advantage in clinical practice. Thus, in cases where a gut signal substance increase (or decrease) is desirable, any diet that regulates the synthesis of this substance and its release can well be adapted, or a receptor agonist (or antagonist) can be utilized. (Agonists and antagonists to most neuroendocrine signal substances are available.) Among these substances are serotonin agonists and antagonists that are now undergoing clinical trials in other gastrointestinal motility disorders.

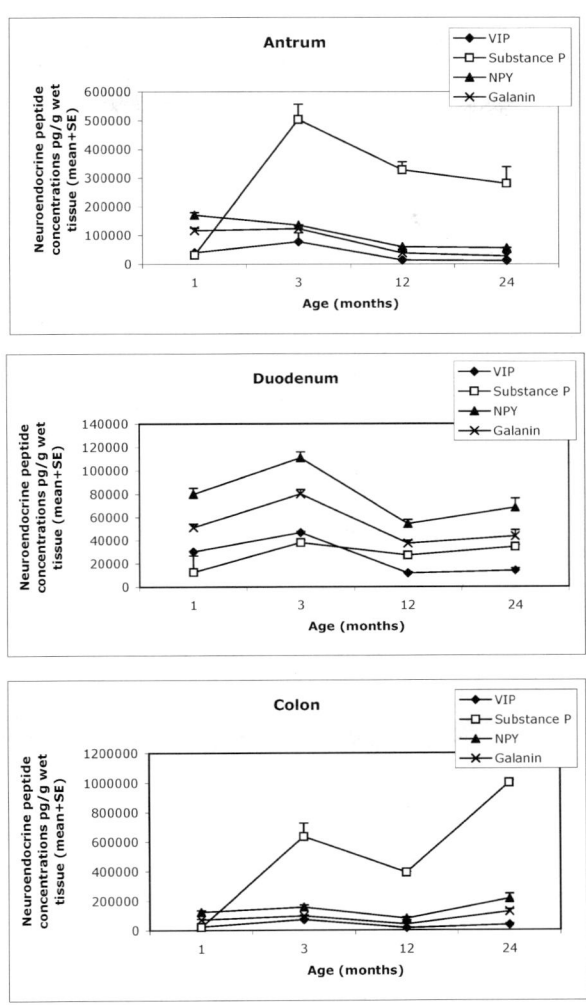

Figure 8. Concentrations of different neuroendocrine peptides in various parts of the gastrointestinal tract at different ages in mice. Abbreviations as in Figure 2. [Data from 47].

## 6. CONCLUSION

Ageing is associated with alteration of the gut neuroendocrine system. These changes appear to be caused by normal ageing processes and by an adaptive response to the age-related changes in the gastrointestinal tract and associated glands. Changes in the neuroendocrine system with ageing, both primary and secondary, could be one of the factors causing gastrointestinal disorders encountered in the elderly. Data regarding changes in gastrointestinal neuroendocrine signal substances in ageing could well be useful in clinical practice and lead to their application for treatment.

268

ACKNOWLEDGEMENT

The author's work cited in this review was supported by grants from The Bengt Ihre Foundation; the Sahlberg Foundation; the Faculty of Medicine, Umeå University Research Funds Foundation, and the Julin Foundation.

REFERENCES

1.  Khalil T, Poston GJ, Thompson JC. Effects of aging on gastrointestinal hormones. Proc Clin Biol Res 1990;326:39–71.
2.  Sandström O. Age-related changes in the neuroendocrine system of the gut. PhD thesis 1999. Faculty of Medicine, Umeå University, Sweden.
3.  Csedes A, Guiraldes E, Bancalari A, Braghetto I, Ayala, M. Relation of gastroesophageal sphincter pressure and esophageal contractile waves and age in man. Scand J Gastroenterol 1978;13:443–447.
4.  Ferrolli, E, Dantas, RO, Olivera, RB, Braga FJNH. The influence of ageing on esophageal motility after ingestion of liquids with different viscosities. Eur J Gastroenterol Hepatol 1996;8:793–798.
5.  Ribeiro AC, Klinger PJ, Hinder RA, DeVault K. Esophageal manometry: a comparison of findings in younger and older patients. Am J Gastroenterol 1998; 93: 706–710.
6.  Evans MA, Triggs EJ, Cheung M, Broe GA. Gastric emptying in the elderly, implications for drug therapy. J Am Geriatr Soc 1981;29:201–205.
7.   Moore JG, Tweedy C, Christian PE, Datz FL. Effect of age on gastric emptying of liquid-solid meals in man. Dig Dis Sci 1983;28: 340–344.
8.  Horowitz M, Maddeun GJ, Chatterton, B.E., Collins PJ, Harding PE, Shaerman DJC. Changes in gastric emptying rates with age. Clin Sci 1984;67:213–218.
9.  Wegener M, Borsch G, Schaffstein J, Luth I, Rickels R, Ricken D. Effect of ageing on the gastro-intestinal transit of lactose-supplemented mixed solid-liquid meals in humans. Digestion 1988;39:40–46.
10. Kupfer RM, Heppell M, Haggith JW, Bateman DN. Gastric emptying and small-bowel transit in the elderly. J Am Geriatr Soc 1985;33:340–343.
11. Gainsborough N, Maskrey VL, Nelson ML, Deating, J. The association of age with gastric emptying. Ageing 1993;22:37–40.
12. Anuras S, Sutherland J. Small intestinal manometry in healthy elderly subjects. J Am Geriatr Soc 1984;32:581–583.
13. Husebye E, Engedal K. The patterns of motility are maintained in the human small intestine throughout the process of aging. Scand J Gastroenterol 1992; 27:397–404.
14. Sandström O, El-Salhy M. Age-related changes in the neuroendocrine system of the gut. A possible role in the pathogenesis of gastrointestinal disorders in the elderly. Minireview based on a doctoral thesis. Upsala J Med Sci 2001;106:81–97.
15. Wallach CB, Kurtz RC. Gastrointestinal cancer in the elderly. Gastroenterol Clin North Am 1990;19:419–228.
16. Allescher H.D. Postulated physiological and pathophysiological roles on motility. In: Daniel EE, editor. Neuropeptides function in gastrointestinal tract. Boca Raton: CRC Press, 1991;309–400.
17. Debas HT, Mulvihill S.J. Neuroendocrine design of the gut. Am J Surg 1991;161:243–249.

18. Ekblad E, Håkanson R, Sundler F. Microanatomy and chemical coding of peptide-containing neurones in the digestive tract. In: Daniel EE, editor. Neuropeptides function in gastrointestinal tract. Boca Raton: CRC Press, 1991;131–180.

19. Rangachari PK. Effects of neuropeptides on intestinal ion transport. In: Daniel EE, editor. Neuropeptides function in gastrointestinal tract. Boca Raton: CRC Press, 1991; 429–446.

20. McConalogue K, Furness JB. Gastrointestinal transmitters. Balliéres Clin Endocrinol Metabol 1994;8:51–76.

21. Goyal PK, Hirano I. Mechanisms of disease: The enteric nervous system. N Engl J Med 1996;334:1106–1115.

22. Walsh JH. Gastrointestinal hormones. In: Johanson LR,editor. Physiology of gastrointestinal tract. 3rd edn. New York: Raven Press, 1994;1–128.

23. Chiba T, Yamada T. Gut somatostatin. In: Walsh JH, Dockray GJ., editors. Gut peptides: Biochemistry and physiology. New York: Raven Press, 1994;123–145.

24. Mannon P, Taylor IL. The pancreatic polypeptide family. In: Walsh JH, Dockray G., editors. Gut peptides: Biochemistry and physiology. New York: Raven Press, 1994; 341–370.

25. Leiter A, Chey W, Kopin A. Secretin. In: Walsh JH, Dockray GJ, editors. Gut peptides: Biochemistry and physiology. New York: Raven Press, 1994;147–173.

26. Dockray G.J, Physiology of enteric neuropeptides. In: Johanson LR,editor. Physiology of gastrointestinal tract, 3rd edn. New York: Raven Press, 1994;169–209.

27. Pederson R. Gastric inhibitory polypeptide. In: Walsh JH, Dockray GJ, editors. Gut peptides: Biochemistry and physiology. New York: Raven Press, 1994;217–259.

28. Liddle R. Cholecystokinin In: Walsh JH, Dockray GJ, editors. Gut peptides: Biochemistry and physiology. New York: Raven Press, 1994;175–216.

29. Dockray G.J. Substance P and other tachykinins. In: Walsh JH, Dockray GJ, editors. Gut peptides: Biochemistry and physiology. New York: Raven Press, 1994;401–422.

30. Poitras P. Motilin In: Walsh JH, Dockray GJ, editors. Gut peptides: Biochemistry and physiology. New York: Raven Press, 1994;261–304.

31. Shulkes A. Neurotensin In: Walsh JH, Dockray GJ, editors. Gut peptides: Biochemistry and physiology. New York: Raven Press, 1994;217–259.

32. Bunnett N. Gastrin-releasing peptide. In: Walsh JH, Dockray GJ, editors. Gut peptides: Biochemistry and physiology. New York: Raven Press, 1994;423–445.

33. Dockray G.J. The opioid peptides. In: Walsh JH, Dockray GJ, editors. Gut peptides: Biochemistry and physiology. New York: Raven Press, 1994;473–491.

34. Rökaeus Å. Galanin. In: Walsh JH, Dockray GJ, editors. Gut peptides: Biochemistry and physiology. New York: Raven Press, 1994;525–552.

35. Lidberg P. On the role of substance P and serotonin in the pyloric motor control. An experimental study in cat and rat. Acta Physiol Scand 1995; Suppl 538:1–69.

36. Tally NJ. Review article: 5-Hydroxytryptamine agonists and antagonists in modulation of gastrointestinal motility and sensation: clinical implications. Aliment. Pharmacol Ther 1992;6:273–289.

37. von der Ohe MR, Hanson RB, Camilleri M. Serotogenic mediation of postprandial colonic tonic and phasic response in humans. Gut 1994;35: 536–541.

38. Whittle B. Nitric oxide in gastrointestinal physiology and pathology. In: Johanson LR,editor. Physiology of gastrointestinal tract, 3rd edn. New York: Raven Press, 1994; 267–294.

39. Giaroni C, De Ponti F, Cosentino M, Lecchini S, Frigo G. Plasticity in the enteric nervous system. Gastroenterology 1999;117:1439–1458.

40. Green M, Bishop AE, Rindin G, Lee FL, Daly MJ, Domin J, Bloom SR, Polak JM. Enterochromaffin-like cell populations in human fundic mucosa: quantitative studies of their variations with age, sex, and plasma gastrin levels. J Pathol 1989;157:235–241.

41. Sandström O, El-Salhy M. Ageing and endocrine cells of the human duodenum. Mech Ageing Dev 1999;108:39–48.

42. Sandström, O, El-Salhy M. Human rectal endocrine cells and ageing. Mech Ageing Dev 1999;108:219–226.

43. Calam J, Ghatei MA, Domin J, Adrian TE, Myszor M, Gupta S, Tait C, Bloom SR. Regional differences in concentrations of regulatory peptides in human colon mucosal biopsy. Dig Dis Sci 1989;34:1193–1198.

44. Sandström O, Madhavi J, El-Salhy M. Age-related changes in antral endocrine cells in mice. Histol Histopathol 1999;14:31–36.

45. Sandström O, El-Salhy M. Duodenal endocrine cells in mice with special regard to age-related changes. Histol Histopathol 2000;15:347–353.

46. Sandström O, Madhavi J, El-Salhy M. Effect of ageing on colonic endocrine cell population in mouse: Gerontology 1998;44:324–330.

47. El-Salhy M, Sandström O. How age changes the content of neuroendocrine peptides in the murine gastrointestinal tract. Gerontology 1999;45:17–22.

48. Kvetnoy I, Popuichiev V, Mikhina L, Anisimov V, Yuzhakov V, Konovalov S, Pogudina N, Franceschi C, Piantanelli L, Rossolini G, Zaia A, Kventania T, Hernandez-Yago J, Blesa JR. Gut neuroenocrine cells: relation to the proliferative activity and apoptosis of mucous epitheliocytes in aging. Neuroendocrinol Lett 2001;22:337–341.

49. Meciano FJ, Carvalho VC, de Souza PR. Nerve cell loss in the myenteric plexus of the human esophagus in relation to age: a preliminary investigation. Gerontology 1995,41:18–21.

50. de Souza PR, Meciano FJ, Borges N, Liberti EA. Age-induced nerve cell loss in the myenteric plexus of the small intestine in man. Gerontology 1993;39:183–188.

51. Gomes OA, de Souza PR, Liberti EA. A preliminary investigation of the effects of aging on nerve cell number in the myenteric ganglia of the human colon. Gerontology 1997;43:210–217.

52. Gabelia G. Fall in the number of myenteric neurons in aging guinea pigs. Gastroenterology 1989;96:1487–1493.

53. El-Salhy M, Sandström O, Holmlund F. Age-induced changes in the enteric nervous system in mouse. Mech Ageing Dev 1999;107:93–103.

54. Santer RM, Baker DM. Enteric neuron numbers and sizes in Auerbach´s plexus in the small and large intestine of adults and aged rats. Auton Nerv Syst 1988;25:59–67.

55. Johmson RJ, Schemann M, Santer RM, Cowen T. The effects of age on the overall population and on sub-populations of myenteric neurones of the rat small intestine. J Anat 1998; 192:479–488.

56. Feher E, Pénzes L. Density of substance P, vasoactive intestinal polypeptide and somatostatin-containing nerve fibers in the aging small intestine of the rats. Gerontology 1987;33:341–348.

57. Takahashi T, Qoubaitary A, Owyang C, Wiley JW. Decreased expression of nitric oxide synthase in colonic myenteric plexus of aged rats. Brain Res 2000;883:15–21.

58. Belai A, Wheeler H, Burnstock G. Innervation of the rat gastrointestinal sphincters:

changes during development and aging. Int J Dev Neurosci 1995;13:81–95.
59. Altman DF. Changes in the gastrointestinal, pancreatic, biliary and hepatic function with aging. Gastoenterol Clin North Amer 1990;19:227–234.

*The Neuroendocrine Immune Network in Ageing*
Edited by R.H. Straub and E. Mocchegiani

# Modulating Effects of Nutrition on Brain Ageing

CARLO BERTONI-FREDDARI, PATRIZIA FATTORETTI, TIZIANA CASOLI, GIUSEP-
PINA DI STEFANO, MORENO SOLAZZI, BELINDA GIORGETTI and MARTA BALIETTI

*Neurobiology of Aging Laboratory, INRCA Research Department, Via Birarelli 8, 60121
Ancona, Italy*

## ABSTRACT

Experimental dietary restriction (DR), vitamin E-deficiency and chronic ethanol administration
were carried out on laboratory animals of different age to assess the role of nutrition on brain
ageing. In DR mice, we measured the ultrastructural features of hippocampal synapses. The
synaptic numeric density (Nv) was significantly increased, while the synaptic average size (S)
was significantly reduced in old DR mice vs. *ad libitum* fed controls. These balanced changes
resulted in a significant increase of the overall synaptic contact area/$\mu m^3$ of tissue, i.e. surface
density (Sv) in DR mice. Smaller contact zones were increased in old DR mice vs. ad libitum
fed controls. Nv, S and Sv reliably report on synaptic plastic rearrangements, thus DR appears to
be able to modulate positively synaptic structural dynamics in the ageing hippocampus. In adult-
vitamin E-deficient rats and normally fed controls, we measured the intranuclear ionic content
by means of X-ray microanalysis. $K^+$ content resulted significantly increased in vitamin E-deficient rats, supporting the idea that the cellular membrane permeability of this ion is markedly
impaired. In the cerebellar cortex and hippocampus of vitamin E-deficient animals we found that
synaptic Nv and Sv were significantly decreased while S was increased: a pattern typical of the
old animals. Computer-assisted morphometry of Purkinje cell mitochondria positive to succinic
dehydrogenase (SDH) activity documented that in vitamin E-deficient animals the organelles'
volume fraction as well as their average volume were significantly reduced. Vitamin E is a well
known biological antioxidant able to quench the lipid peroxidation chain and to protect from
free radical attacks, thus our data support that an increased oxidative stress occurs in adult vitamin E-deficient rats. Chronic alcohol intake in adult rats resulted in a significant increase of $K^+$,
a decrease of $Na^+$ and no change in $Cl^-$. Chronic ethanol intake in old rats vs. controls of the
same age showed no change in $K^+$ content, whereas both $Na^+$ and $Cl^-$ decreased significantly. In
adult ethanol-treated rats vs. their controls, synaptic Nv, S and Sv decreased significantly. In the
old ethanol fed animals vs. age-matched controls, synaptic Nv and Sv decreased, while S was
reduced at a not significant extent. These data confirm that neuronal membranes are markedly
deteriorated by ethanol and that the excessive dietary intake of this alcohol affects their permeability and plasticity. Taken together, the findings of the three experimental dietary manipulations
studied by us support the critical role of nutrition on the occurrence of age-related alterations
in the brain. Namely, dietary components are significantly involved in various stages of the cellular antioxidant defense mechanisms and this appears to be of critical importance for the health
status of the postmitotic nerve cells.

1.    INTRODUCTION

Physiological ageing may be defined as a progressive loss of the capacity of a given organism
or cell to maintain homeostasis. This appears to be particularly true for postmitotic cells, e.g.
neurons. Following neuronal maturation, changes occur with time which underlie an increasing
vulnerability to environmental challenges leading to impair the ability of this type of cells to
survive. Although several theories on the causes of ageing have been proposed on the basis of
reliable experimental data, at present what can be reasonably supported is that in most organisms
senescence occurs because the action of long-term, low intensity stressors are not fully counteracted by protective and repair functions [1]. Oxidative damage due to aerobic metabolism, is
intrinsic to living processes and it is the most investigated cause of stress. Degradation or inactivation of important biological molecules may result from defective control of oxygen reactions,
although prevention of such damage can be carried out by providing molecules that compete for
oxygen, or by lysing, removing, repairing damaged molecules. The antioxidant cell potential,
i.e. all the antioxidant protective mechanisms, involve the complex organisation of intracellular
components, thus the susceptibility to oxidative injury of a given cell or tissue is function of
the overall balance between the degree of oxidative stress and antioxidant defence capabilities.
Despite the important role of oxidative stress in ageing and in the aetiology of some pathological conditions is well documented, much of the research has been focused on intense and acute
stress conditions, while few studies have been carried out on long-term, low intensity oxidative
damage which may occur continuously or frequently throughout the individual lifespan.
    Nutrients are essential for all fundamental cellular processes, thus they may be supposed to
carry out a long lasting modulation on cellular oxidative damage as well as antioxidant defence
mechanisms. To assess the potential role of nutrition in brain ageing processes, we carried out
studies in adult and old rats undergone dietary restriction, vitamin E deficiency and chronic
alcohol administration.

2.    NUTRITIONAL INTERVENTIONS AND METHODS

2.1.    Dietary restriction

2.1.1. Animals and diet

Twenty male CD-COBS rats were divided into four groups each consisting of 5 animals: 1)
14-month-old control rats fed *ad libitum* with the standard diet (14AL); 2) 27-month-old control rats fed *ad libitum* with the standard diet (27AL); 3) 14-month-old rats fed a hypocaloric
diet from weaning (14DR); 4) 27-month-old rats fed a hypocaloric diet from weaning (27DR).
Caloric regimen: at weaning, 50% of the lipids and 35% of carbohydrates of the standard pellet
were replaced by fibres.

2.1.2. Quantitative analysis of synaptic dynamic morphology

Anaesthetised rats were perfused through the left ventricle with saline followed by 2.5% phosphate buffered glutaraldehyde (pH 7.4). Tissue sections from the hippocampi were stained by
means of the ethanol phosphotungstic acid (E-PTA) preferential technique (Figure 1) according to a procedure set up in our laboratory [2-4]. Considering the hippocampal dentate gyrus

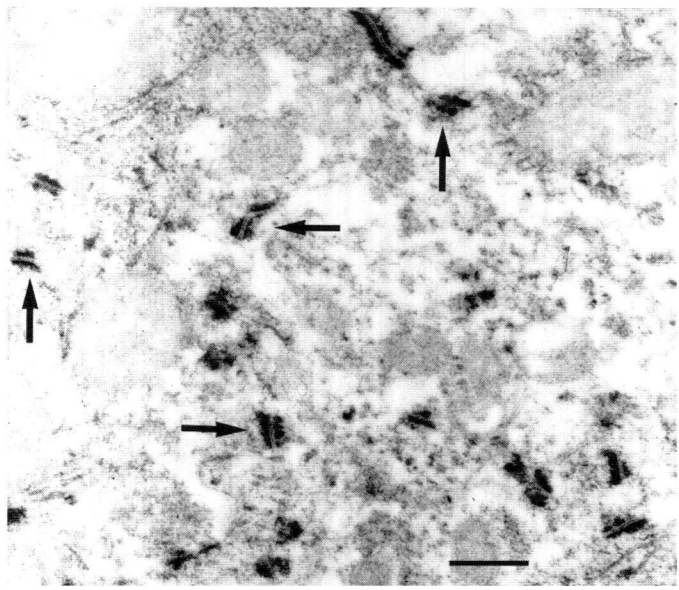

Figure 1. Electron microscopic picture of synaptic contact zones positive to the ethanol phosphotungstic acid (E-PTA) preferential staining procedure: the junctional areas are sharply stained against a pale background (arrows). Bar = 0.5 μm.

granular layer as a reference structure [5], sampling was performed in the neuropil just above the cell bodies. A systematic random sampling was carried out by means of a Kontron KS300 computer-assisted image analysis system equipped with a TV camera directly connected with the electron microscope image plate. The actual number of fields/group of animals was determined by the progressive means method: collection of data was terminated when the standard error of the mean (SEM) for each parameter was less then 5%. On coded samples unknown to the operator, the following parameters were semiautomatically measured by applying currently used morphometric formulas [3,4,6,7]: the number of synapses/$\mu m^3$ (synaptic numeric density: Nv), the average synaptic area (S) and the total area of the synaptic contact zones//$\mu m^3$ of tissue (surface density: Sv).

## 2.2. Vitamin E deficiency

### 2.2.1. Animals and diet

Young rats of 1 month of age were fed for 10 months with a vitamin E-deficient diet [8]. Control littermates, as well as young (3 months) and old (30 months) animals were fed with the normal laboratory chow. The vitamin E-deficient rats and age-matched controls were sacrificed at the age of 11 months.

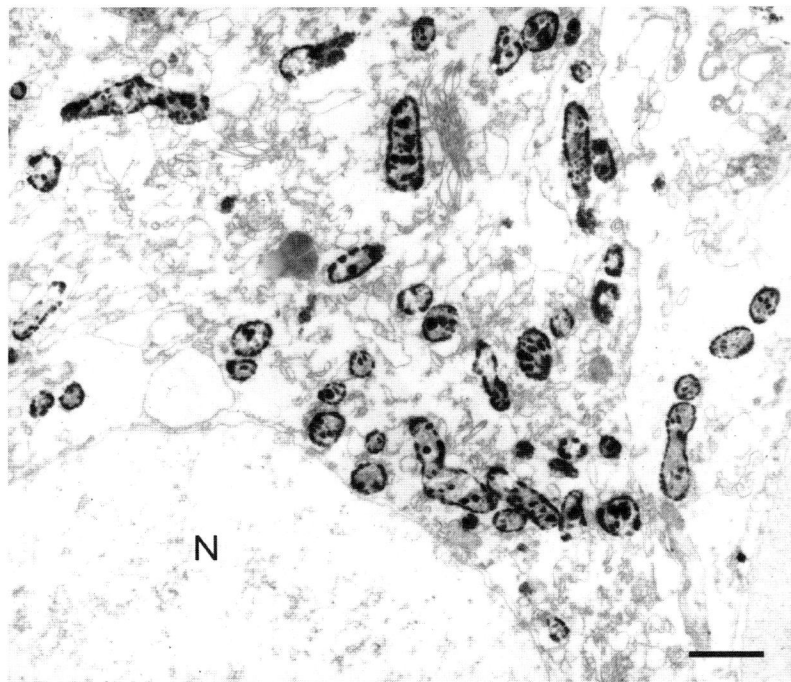

Figure 2. Succinic dehydrogenase (SDH)-positive mitochondria in a Purkinje cell soma. The activity of the enzyme (dark spots inside the organelles) has been evidenced by the copper ferrocyanide method as described in the text. N: nucleus of the Purkinje cell. Bar = 0.5 μm.

### 2.2.2. Quantitative analysis of synaptic dynamic morphology

The ultrastructural features of the synaptic junctional areas in the dentate gyrus supragranular layer were measured in rats of different ages (3, 11, 30-month-old animals) and in vitamin E-deficient rats as described above for DR animals.

### 2.2.3. Quantitative cytochemistry of succinic dehydrogenase (SDH) activity in Purkinje cell mitochondria

SDH, a key enzyme of the respiratory chain, is located at the inner mitochondrial membrane and it is reported to be of critical functional importance when energy request is high [9,10]. Thus, quantitative estimation of its activity may reliably report on the mitochondrial competence to provide adequate amounts of adenosinetriphosphate (ATP). SDH-positive mitochondria were studied in cerebellar Purkinje cell perykaria of 3, 11, 30 month-old animals as well as in 11 month-old vitamin E-deficient rats. Preferential cytochemistry of SDH activity was carried out by the copper-ferrocyanide technique according to Kerpel-Fronius and Hajos [11] (Figure 2). The number of SDH-positive mitochondria/$\mu$m$^3$ of cytoplasm (Numeric density: Nvm), the average volume of the single mitochondrion (V) and the total mitochondrial volume/$\mu$m$^3$ of cytoplasm (Volume density: Vv) were the parameters measured in our study according to currently used morphometric formulas [12].

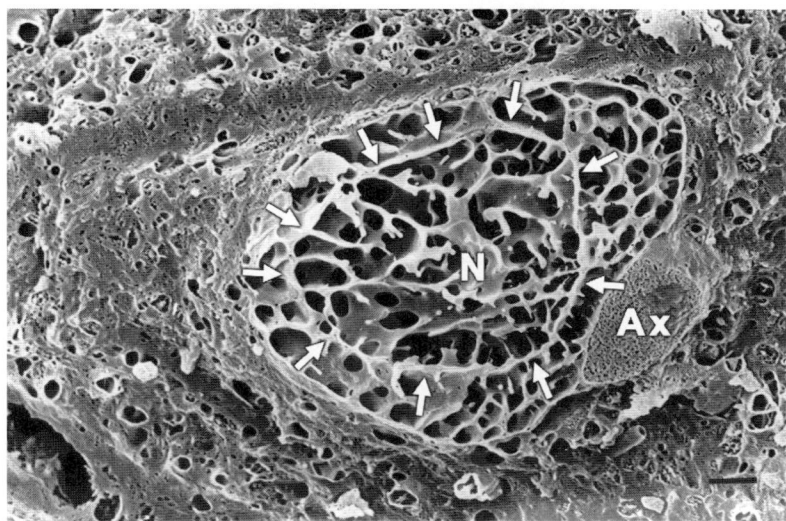

Figure 3. Scanning electron image of a freeze-fractured, dried and carbon sputtered sample of rat hippocampal pyramidal cell CA1 prepared for X-ray microanalysis. The nucleus (N) and the nuclear membrane (arrows) are easily identifiable. Ax: axon Bar = 2 μm.

### 2.2.4.  Intracellular ionic content

The brains of 1, 11 and 24 month-old rats were quickly exposed from the skull after decapitation. The tissue samples were prepared according to a procedure set up in our laboratory [13]. Our microanalysis technique enabled us to measure the content of light elements, i.e. $Na^+$, $Cl^-$, $K^+$, in selected compartments (nucleus and cytoplasm) of dried cells [8]. Although fine ultrastructural details of nerve cells can not be seen by using our procedure, we were able to recognize neuronal nuclei (Figure 3) and to measure their ionic content. We calculated the peak to background ratio for $Na^+$, $Cl^-$, $K^+$ and referred these values as percent of biological dry mass. White counts were taken from 4 to 6 Kev range where no characteristic peak is present [13,14].

### 2.3.  Chronic ethanol administration

### 2.3.1.  Animals and diet

Adult (11 months of age) and old (24 months of age) animals were fed with laboratory chow and 32% ethanol in 25% sucrose solution (w/v) in the drinking water according to Porta et al. [15]. The alcohol-sucrose solution was administered ad libitum to individually housed animals for 4 weeks. Control rats received a solution where ethanol was isocalorically substituted with sucrose. Intake of the drinking water was measured daily and showed an average consumption of 10-12 grams of ethanol/Kg body weight.

### 2.3.2.  Quantitative analysis of synaptic dynamic morphology

As described above for DR animals, we carried out a computer-assisted morphometric study on the ultrastructural features of the synaptic junctional areas in the dentate gyrus supragranular

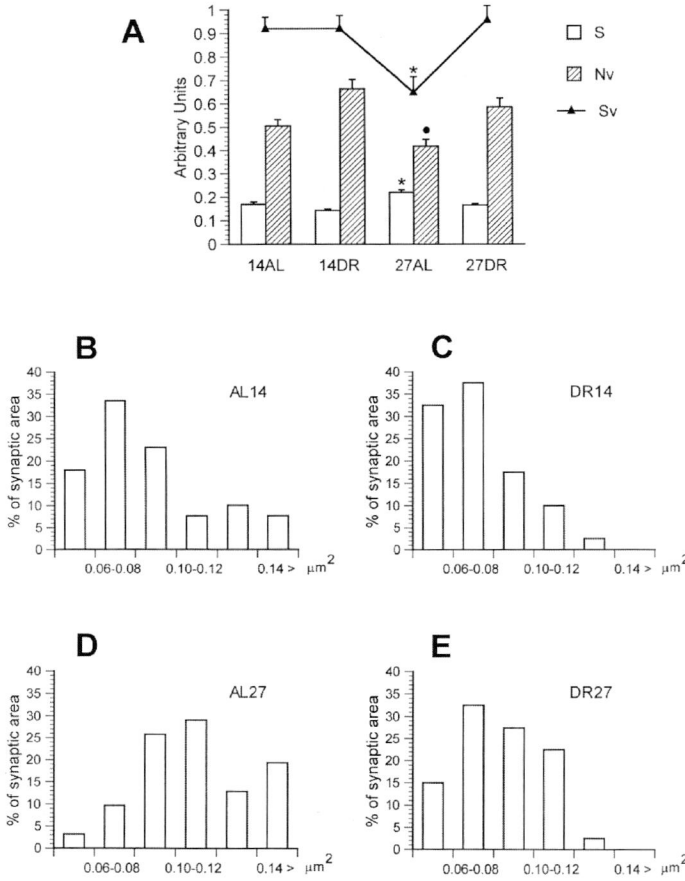

Figure 4. A: summary of the results obtained on synaptic morphometry in dietary restricted (DR) and ad libitum fed (AL) rats of different ages (14 and 27 months of age). To compare all together the three parameters taken into account, we have used arbitrary units (ordinate). * Statistically significant vs. the other groups investigated. • Statistically different from DR14 group.

B–E: percent distribution of synaptic size (S) in ad libitum fed rats and dietary restricted animals. An increased complement of smaller junctional areas can be clearly envisaged in dietary restricted animals.

layer from ethanol treated rats and isocalorically fed animals of the same age.

### 2.3.3. Intracellular ionic content

The right hippocampus was excised from each animal and prepared for X-ray microanalysis as described above. This study was carried out in the nuclei of hippocampal CA1 pyramidal neurons.

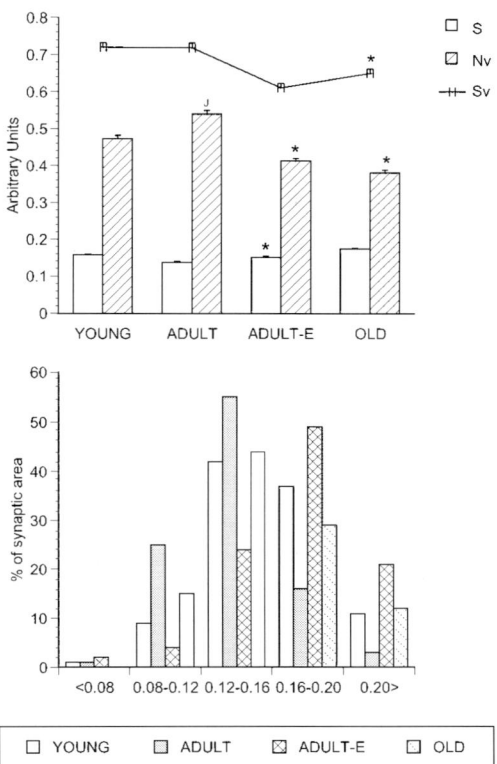

Figure 5. Summary of the results obtained on synaptic morphometry in young, adult and old rats and in adult, vitamin E-deficient animals. * Statistically significant vs. the 11-month-old group. • Statistically significant vs. the other groups investigated. Lower histogram: percent distribution of synaptic size (S) from young, adult and old rats and from adult, vitamin E-deficient animals. Higher percentages of enlarged contact zones are present in vitamin E deficient animals vs. age-matched controls.

## 3.    RESULTS

### 3.1.    Dietary restriction

Figure 4A summarises the results we have obtained on synaptic ultrastructure in adult and old dietary restricted rats. In AL27 rats Sv is significantly lower than in the other groups investigated. No change was found comparing DR14, DR27 and AL14 rats.

Nv decreased by about 18% in AL27 vs. AL14 animals, however this difference was not significant. A significant increase of the synaptic numeric density is clearly evident both in DR14 and DR27 groups compared with AL age-matched controls. No significant difference was envisaged comparing DR14 and DR27 groups of rats. In AL27, Nv was significantly lower than in DR14 animals. S significantly increases with age in AL27 vs. AL14 animals. DR14 and DR27 show a decrease of this parameter vs. their AL littermates, however only in DR 27 rats vs. the AL27 ones the difference was significant. DR14 and DR27 animals showed the same value of S. The percent distributions of S, reported by histograms B-E, show that in DR14 and DR27 rats the junctional areas smaller than 0.08 μm² are markedly increased and account for 72 and

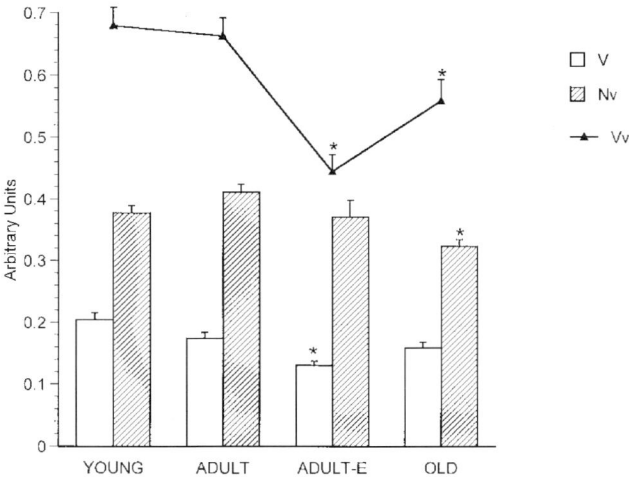

Figure 6. Summary of the results obtained on morphometric studies of SDH-positive mitochondria in Purkinje cell perykarion. Vitamin E deficiency resulted in a marked reductions of the three parameters taken into account. * Statistically significant vs. the 11-month-old control group.

48%, respectively. Conversely, in AL14 and, at a higher extent in AL27, the percent of enlarged contact zones (surface area $> 0.12$ $\mu m^2$) is consistently higher than in the respective age-matched DR groups.

## 3.2. Vitamin E deficiency

Figure 5 reports the results of our morphometric investigation on the synapses of the hippocampal dentate gyrus from rats of different ages and vitamin E-deficient animals. Sv is unchanged between 3 and 11 months of age, but significantly decreases in the old animals. Nv significantly increases in the normally fed adult animals, in contrast with the other age groups analysed. S is significantly smaller at 11 months of age and it is larger in the old rats than in the other groups of age. Vitamin E-deficiency, unlike control condition, resulted in significant decreases of Nv and Sv, while S was not significantly increased. In a comparison with 30-month-old animals, the vitamin E-deficient group showed a significant decrease of Sv and S, while Nv was significantly increased. The lower histogram reports the percent distribution of S: it is clearly evident that, with advancing age, the percentage of larger contact areas markedly increases. Vitamin E-deficient animals vs. control littermates as well as old rats showed higher percentages of larger junctional areas.

The results obtained on SDH-positive mitochondria in Purkinje cell perykaria are shown in Figure 6. Vv is constant between 3 and 11 months of age, but in old rats it is significantly decreased. Nvm increases in adult vs. young animals, but in old rats it is significantly lower than in the other groups of age. V is significantly decreased in adult than in young group and further decreases in old animals. The vitamin E-deficient rats showed no change of Nvm and a marked decrease of Vv in a comparison with all the age groups investigated. In the vitamin E-deficient animals, V is significantly lower than in 3 and 11 month-old normally fed rats, but it is unchanged when compared with the 30 month-old group.

Figure 7 shows the intracellular ionic content of brain cortical cells from rats of different ages

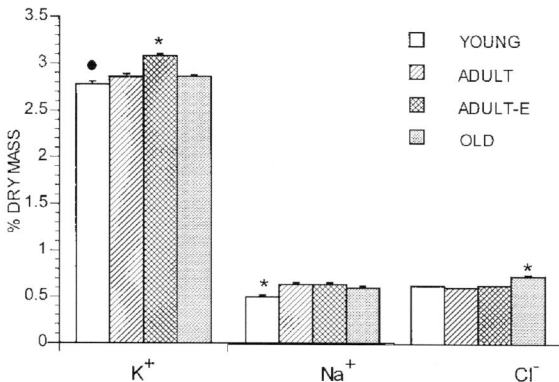

Figure 7. Intracellular monovalent electrolyte content of brain cortex cells in young (1), adult (11) and old (24) rats and in adult, vitamin E-deficient (11-E) animals. * Significantly different from the other groups of age investigated. • Statistically different from 11-month-old rats.

and in adult, vitamin E-deficient animals. The $K^+$ content increases, at a not significant extent, in rats of 11 and 24 months of age. $Na^+$ content is unchanged between 11 and 24 months of age, but it is significantly higher in the 11 month-old group than in the 1 month-old animals. The Cl⁻ content significantly decreases at 11 months of age and is significantly higher in old animals than in the other groups investigated. In vitamin E-deficient rats vs. the other three groups investigated we found an increase of the total intracellular ionic content due to a significant increase of intracellular $K^+$ ions. In a comparison with the old group of animals, the adult vitamin E-deficient rats showed a lower content of Cl⁻, while $Na^+$ content was unchanged.

### 3.3.    Chronic ethanol administration

The effect of alcohol administration on synaptic ultrastructural features are summarized in Figure 8. In ageing, we found a significant decrease of synaptic Nv and Sv, while S was significantly increased. In adult rats, chronic ethanol administration resulted in a significant decrease of all the three parameters taken into account. Old alcohol treated animals showed a significant decrease of Sv and Nv, whereas S decreased at a not significant extent. In Figure 8 the lower histograms B and C show the percent distribution of the synaptic average size (S). It is clearly evident that the complement of smaller contact zones is markedly increased by ethanol administration both in adult and old rats.

Figure 9 reports the results of intracellular ionic content in adult and old animals chronically treated with ethanol. Alcohol administration to adult rats resulted in a significant reduction of $Na^+$ and an increase in $K^+$, while Cl⁻ was almost the same. In old ethanol-treated animals, the intracellular ionic content decreased as compared with age-matched controls and this appeared to be almost due to a significant decrease in $Na^+$ content since $K^+$ and Cl⁻ were not significantly decreased [16]

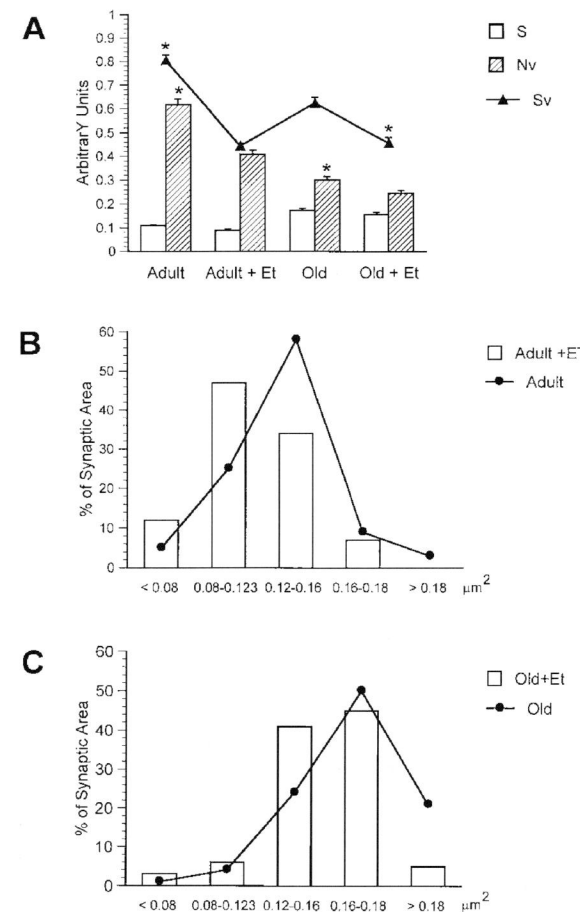

Figure 8. A: summary of the results obtained on synaptic morphometry in adult and old rats chronically treated with ethanol. Both in adult and old rats chronic ethanol administration resulted in significant reductions of number and total area of the contacts. * Statistically significant vs. age-matched ethanol-treated group. B–C: percent distribution of synaptic size (S) from adult and old rats and from ethanol treated animals of the same age. Both in adult and old treated animals an increase in the percent of smaller junctional areas was clearly envisaged.

## 4.    DISCUSSION

### 4.1.    Dietary restriction

#### 4.1.1.  Synaptic dynamic morphology

The main finding of our investigation on dietary restriction is that the reduced food intake resulted in a positive and functional modulation of the synaptic structural dynamics both in adult and old rats. In details, DR14 and DR27 animals vs. their AL controls showed higher Nv values together with an increased complement of junctional areas of smaller size. In ageing, the final outcome of dietary restriction is represented by an increased total area (Sv) in DR27 rats so that

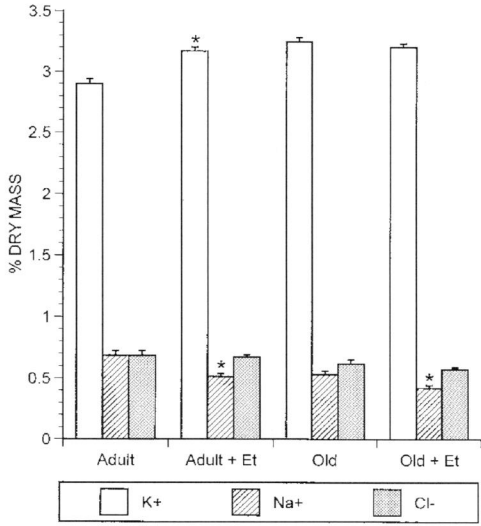

Figure 9. Intracellular monovalent electrolyte content of CA1 hippocampal pyramidal cells in adult and old rats and in ethanol treated animals of the same age. In adult-treated rats, $K^+$ increases and $Na^+$ decreases significantly, respectively, vs. age matched controls. In old-treated rats, $Na^+$ decreases significantly vs. age-matched controls. *Statistically significant vs. age-matched control groups.

the age-related decrease of this parameter is no longer apparent.

The current literature on synaptic structural dynamics supports that number (Nv) and size (S) of the synaptic contact zones are in an inverse relationship which aims at maintaining constant the value of Sv [3,17,18]. Thus, while these three parameters are reporting on particular aspects of synaptic morphology, when considered together per experimental group, they account for the reciprocal rearrangements occurring in the physiological responsive adaptation to environmental stimuli. On the basis of these concepts, it appears that dietary restriction is able to counteract significantly the age-dependent decrease of Sv through a positive balance between Nv and S. The biological significance of this balance is still debated, however higher synaptic densities are associated with better performances of the central nervous system (CNS), while reductions in synaptic number together with enlargement in size of the surviving contacts are reported to occur in adverse and pathological conditions [3,4,17,19,20]. Conceivably, maintenance of a high number of small junctional zones appears to represent a functional task not only for routine CNS functions, such as learning and memory, but also for a proper response to environmental stimuli [21–23].

To give tenable interpretations of the effects of dietary restriction on the synaptic adaptive potential, some current concepts regarding synaptic functional rearrangements must be taken into account. As a consequence of stimulation, synaptic contact zones are reported to: I) enlarge, II) perforate and III) split to yield an increased number of smaller contacts which, in turn, may be reinforced or degraded [22,24]. By improving cell-to-cell communication, this cycle of events is supported to strengthen the synaptic network within the discrete CNS areas undergone stimulation. In this context, the increased complement of smaller synapses in DR animals (Figure 4) suggests that: I) the smaller junctions are better preserved by the reduced caloric intake; II) the cycle of steps purported to remodel the synaptic network is better accomplished in a dietary restricted regimen. These two assumptions do not exclude each other: both may be supposed to act positively on the synaptic potential for plasticity by considering that the many determinants (protein synthesis, energy metabolism, axonal transport etc.) able to modulate synaptic functions may be up regulated by dietary restriction [25–28].

Although the mechanism(s) by which a reduced caloric intake retards many age-related alterations is yet unknown, there is a wide consensus proposing that dietary restriction triggers a multifactorial process responsible of a decreased cellular load of labile and damaged proteins, lipids and DNA [29]. Several experimental data, support that caloric restriction positively modulates free radical metabolism by reducing the generation of reactive oxygen species and thus exerting its anti-ageing effect [30,31]. This action appears to be of very critical functional significance for the postmitotic nerve cells, which are characterised by high lipid content and high oxygen consumption [32]. These concepts support that the higher percent of smaller synapses in DR animals may be the reasonable outcome of the splitting of larger contact zones or of the improved preservation of the persisting junctions. In turn, the present findings document that dietary restriction is capable of a positive modulation of the synaptic dynamic morphology, which may be due to the many reported effects of a reduced caloric intake on free radical generation and/or control [26,30,31,33].

## 4.2. Vitamin E deficiency

### 4.2.1. Synaptic dynamic morphology

Our results document that both in ageing and vitamin E-deficiency the number of synapses (Nv) decreases and, despite a significant increase of the average size of the persisting junctional zones (S), the final outcome is a significant reduction of the total synaptic area (Sv). The distribution of S showed that, from adulthood to old age and in adult-vitamin-E-deficient animals, the percent of enlarged contact zones increases. According to the concepts reported above, a significant synaptic enlargement may be considered a well categorised step of the process(es) responsible of the morphological modulation of the nerve cell terminal areas to environmental stimulations. In this context, an increased oxidative stress, due to the absence from the diet of the biological antioxidant alpha-tocopherol, may be supposed to result in an impaired synaptic resharpening following reactive synaptogenetic processes.

Synaptic junctions are functionally differentiated areas of the neuronal membrane, they are very abundant in unsaturated lipids as clearly demonstrated by the marked osmium tetroxide staining. While the molecular composition of these peculiar sites of the nerve cell membrane entails them a high degree of fluidity, which represents a necessary prerequisite for their proper performances and dynamic plasticity, it also constitutes the structural feature determining their high vulnerability to peroxidations, as it appears to be supported by our present findings.

### 4.2.2. SDH activity in Purkinje cell mitochondria

As shown by Figure 6, SDH activity in perykarial Purkinje cell mitochondria decreases in old rats because of the reduction in number and size of the positive organelles. Also the vitamin E-deficient group showed decreases of the three parameters analysed. SDH molecules are completely synthesised by nuclear genes [34], then they are transported into the mitochondria and metabolically activated by mitochondrial DNA intervention. Conceivably, quantitative estimation of SDH activity, besides providing information on metabolically activated organelles, reports also on interactions between nuclear and mitochondrial genomes. A clear age-related impairment in SDH activity can be envisaged by our findings, which report that a vitamin E-deficient diet is responsible of similar changes in adult animals. SDH molecules are inserted into the mitochondrial inner membrane and their activity is closely coupled with the structural integrity of the surrounding molecular environment where it is located [35]. Thus, an increased deterioration of the mitochondrial membrane may be hypothesised to explain the present findings, although also a defective interaction between nuclear and mitochondrial genomes may contribute to a significant extent.

### 4.2.3. Intracellular ionic content

Both in ageing and vitamin E-deficiency, we found an overall increase of the intracellular ionic concentration. This appeared to be due to the very significant increase of $K^+$ ions. This electrolyte has been reported to contribute at a large extent to the regulation of the intracellular ionic strength, which in turn, may consistently modulate DNA functions. It is well proven that histones neutralise the negative groups of DNA, however within the nucleosomes only 15 to 50% of the charges are neutralised and the result is that chromatin is negatively charged [36]. Neutralisation of chromatin negative groups is very important for DNA stabilisation and it is accomplished also by maintaining, in a physiological range, the intracellular concentration of positive ions [36–38]. On the basis of these concepts, we support that the reported increase of $K^+$ ions may affect the physicochemical properties of chromatin through sterical hindrance of enzymes, displacements of histones and abnormal DNA unfolding. In turn, altered nuclear functions (e.g. template activity) may also impair the ongoing supply of spare parts to the nerve cell terminal regions and contribute to the degeneration of the synaptic junctional areas. In addition to these changes specifically relevant to DNA performances, an increased concentration of ions in the cytoplasm is documented to alter the physicochemical properties of the cellular colloidal system, which results in longer lifetime of the enzyme-substrate complexes [39]. The increased intracellular $K^+$ content documented by the present studies may find a reasonable explanation in the relevant changes of lipid composition of the cellular membranes reported both in ageing and vitamin E-deficiency [40]. As a matter of fact, if the rigidity of the cellular membranes increases, the $K^+$ passive flow to the extracellular space is markedly impaired. In addition, it has been documented that the proper functioning of the $Na^+/K^+$ pump is tightly coupled with the surrounding lipid environment and any decrease in membrane fluidity has been reported to stimulate its activity [41], thus both in ageing and vitamin E-deficiency it may be supposed that $K^+$ ions are pumped into the cells at a high rate, but they are slowly eliminated with the final result of an increased intracellular $K^+$ content.

### 4.3. Chronic ethanol administration

### 4.3.1. Synaptic dynamic morphology

Synaptic ultrastructure is markedly affected both in adulthood and ageing by alcohol administration, as clearly documented by the reduction in number of contacts (Nv) and of the total junctional area (Sv) among nerve cells. The increased complement of smaller synapses observed in the treated animals vs. controls (Figure 8 B and C) regardless of the age of the animals, supports the idea that the synaptic adaptive potential is significantly impaired by alcohol treatment. Namely, considering the cycle of steps of synaptic remodelling mentioned above, the increased percentage of smaller contacts may be interpreted as an insufficient compensation unable to counteract the damaging effects of ethanol.

Despite no clear-cut mechanism(s) of action of ethanol has been demonstrated so far, acute or chronic intake of this alcohol appears to affect CNS membranes structure. Disordering, fluidising, rigidizing effects on plasma membranes have been demonstrated following treatment of laboratory animals [42]. Namely, chronic ethanol intake appears to develop a sort of tolerance which is supposed to be due to modifications in the composition of the lipid moiety of the plasma membrane structure, e.g. increase of the cholesterol/phospholipids mole ratio and decrease of membrane fatty acid unsaturation [43,44]. In agreement with these biochemical results, we interpret our present findings to represent age-related adaptive changes of the synaptic membranes to ethanol intake. As mentioned above, synaptic membranes are very fluid zones of the neuronal membrane because of their very high content in polyunsaturated fatty acids [45], thus it can be hypothesised that ethanol administration may impair synaptic plastic potential by decreasing the unsaturation of nerve cell terminal zones. In physiological ageing, a condition where the chemical composition of all types of cellular membranes is markedly altered [46], alcohol intake may selectively affect high fluid and sensitive zones of postmitotic nerve cells, e.g. synaptic junctional areas.

### 4.3.2. *Intracellular ionic content*

Our present findings support that the permeability of the cellular membrane is markedly affected by ethanol administration in adult rats as documented by the increased intracellular ionic content due to $K^+$ ions (Figure 9). In old animals, ethanol administration appeared to be less harmful in affecting neuronal membrane permeability functions since only a significant decrease in $Na^+$ was demonstrated in treated rats, whereas $K^+$ was unaltered. The rationale for such an interpretation is supported by the fact that the membrane bound $N^+/K^+$-ATPase needs a defined range of membrane fluidity. Conceivably, if the rigidity of the membrane structure is increased, e.g. ageing, a higher number of $N^+/K^+$-ATPase molecules are activated. This assumption is supported by the results of Sun and Sun [41] reporting an increased activity of this enzyme in neurons froms of ethanol intoxicated as well as vitamin E-deficient laboratory animals.

## 5.    CONCLUSIONS

Primary causes of the ageing process, defined as the time-related loss of the cellular capacity to maintain homeostasis, have been claimed to be free radical attacks originating physiologically from cellular oxidative phosphorylation [47]. Living cells, through repeated events in evolution, have improved their defense mechanisms to inactivate free radicals and it has been suggested that the life-span of a given species depends on the balance between free radical production rate and the efficiency of cellular protecting mechanisms [48,49]. In agreement with this rationale, any impairment in free radical control may result in a potential threat to longevity. This is particularly true for postmitotic cells, e.g. neurons, although current data document a lifespan reactive plasticity of the fully differentiated CNS able to counteract time- and/or pathology-related changes. Namely, brain ageing may be thought of as a particular condition in which specific pathological changes are found without clinically evident manifestations: nerve cell alterations are continuously counteracted by compensating reactions. As a consequence, deterioration of function occurs when the number of neurons and of their connections decrease below a critical reserve level and coping with environmental stimulations becomes difficult [50,51]. In this context, our findings lend further support to the concept that the control of free radical rate plays an important role in brain ageing processes. With specific reference to dietary interventions, it appears that nutrients are able to modulate some changes occurring in the senile CNS. Namely, dietary components are critically involved in various stages of the cellular antioxidant defense, thus a proper nutrition may be of help in maintaining cellular homeostasis and in prolonging a healthy lifespan.

## REFERENCES

1.  Masoro EJ. The biological mechanism of aging: is it still an enigma? Age 1996;19: 141–145.
2.  Bertoni-Freddari C, Fattoretti P, Casoli T, Meier-Ruge W, Ulrich J. Morphological adaptive response of the synaptic junctional zones in the human dentate gyrus during aging and Alzheimer's disease. Brain Res 1990; 517: 69–75.
3.  Bertoni-Freddari C, Fattoretti P, Paoloni R, Caselli U, Galeazzi L, Meier-Ruge W. Synaptic structural dynamics and aging. Gerontology 1996;42: 170–180.
4.  Bertoni-Freddari C, Fattoretti P, Ricciuti R, Vecchioni S, Casoli T, Solazzi M, Ducati A. Morphometry of E-PTA stained synapses at the periphery of pathological lesions. Micron 2002; 33:447–451.
5.  Weibel ER. Stereological Methods: Practical Methods for Biological Morphometry. Vol. I. Sampling of Tissue. London, Academic Press, Inc, 1979, pp. 63–100.
6.  Desmond NL, Levy WB. Changes in postsynaptic density with long –term potentiation in the hippocampal dentate gyrus. J Comp Neurol 1986a; 253:476–482.
7.  Desmond NL, Levy WB. Changes in the numerical density of synaptic contacts with long-term potentiation in the hippocampal dentate gyrus. J Comp Neurol 1986; 253: 466–475.
8.  Bertoni-Freddari C, Giuli C, Lustyik G, Zs-Nagy I. *In vivo* effects of vitamin E deficiency on the intracellular monovalent electrolyte concentration in brain and liver of rat. An energy dispersive X-ray microanalytic study. Mech. Age Dev 1981; 16: 169–180.
9.  Bertoni-Freddari, C, Fattoretti, P, Casoli, T, Di Stefano G, Solazzi M, Meier-Ruge, W. Quantitative cytochemical mapping of mitochondrial enzymes in rat cerebella. Micron 2001;32: 405–410.
10. Fattoretti P, Bertoni-Freddari C, Caselli U, Paoloni R, Meier-Ruge W. Impaired succinic dehydrogenase activity of rat Purkinje cell mitochondria during aging. Mech Ageing Dev 1998; 101:175–182.
11. Kerpel-Fronius S, Hajos F. The use of ferricyanide for the light and electron microscopic demonstration of succinic dehydrogenase activity. Histochemie 1968;14: 343–351.
12. Bertoni-Freddari C, Fattoretti P, Casoli T, Spagna C, Meier-Ruge W, Ulrich J. Morphological plasticity of synaptic mitochondria during aging. Brain Res 1993; 628: 193–200, 1993.
13. Zs-Nagy I, Pieri C, Giuli C, Bertoni-Freddari C, Zs-Nagy V. Energy dispersive X-ray microanalysis of the electrolytes in biological bulk specimen. I. Specimen preparation, beam penetration, and quantitative analysis. J Ultrastruct Res 1977;58: 22–33.
14. Zs.-Nagy I, Lustyik Gy, Zs.-Nagy V, Zarandi B, Bertoni-Freddari C. Intracellular N+:K+ ratios in human cancer cells as revealed by energy dispersive X-ray microanalysis. J Cell Biol 1981; 90:769–777.

15. Porta E. A Gomez Dumm C.L.A. A new experimental approach in the study of chronic alcoholism. I. Effects of high alcohol intake in rats fed a commercial laboratory diet. Lab Invest 1968;18: 59, 320–331.

16. Bertoni-Freddari C, Meier-Ruge W, Ulrich J. Age-dependent effect of ethanol on the intranuclear ionic content in hippocampal CA1 pyramidal cells. Adv Biosci 1988;;71, 263–267

17. Hillman D, Chen S. Reciprocal relationship between size of postsynaptic densities and their number: constancy in contact area. Brain Res 1984; 295:325–343.

18. Chen S, Hillman, DE. Robust synaptic plasticity of striatal cells following partial differentiation. Brain Res 1990; 520: 103–114.

19. deToledo-Morrell, L, Geinisman, Y, Morrell, F. Age-dependent alterations in hippocampal synaptic plasticity: relation to memory disorders. Neurobiol Aging 1988;9: 581–590.

20. Geinisman Y, deToledo-Morrell L, Morrell F, Heller RE. Hippocampal markers of age-related memory dysfunction: behavioral, electrophysiological and morphological perspectives. Progr Neurobiol 1995; 45: 223–252.

21. Geinisman Y, de Toledo-Morrell L, Morrell F. Aged rats need a preserved complement of perforated axospinous synapses per hippocampal neuron to mantain good spatial memory. Brain Res 1986; 398: 266–275.

22. Calverley RKS, Jones DG. Contribution of dendritic spines and perforated synapses to synaptic plasticity. Brain Res Rev 1990;15: 215–249.

23. Bailey CH, Kandel ER. Structural changes accompanying memory storage. Ann Rev Physiol 1993; 55:397–426.

24. Carlin RK, Siekevitz P. Plasticity in the central nervous system: do synapses divide? Proc Natl Acad Sci USA 1983; 80: 3517–3521.

25. Yu BP, Masoro EJ, McMahan CA. Nutritional influences on aging of Fischer 344 rats: I. Physical, metabolic and longevity characteristics. J Gerontol 1985;40:657–677.

26. Weindruch R, Walford RL, Fligiel S, Guthrie D. The retardation of aging in mice by dietary restriction: Longevity, cancer, immunity and lifetime energy intake. J Nutr 1986;116: 641–654.

27. Barrows CH, Kokkonen GC. Nutrition in Gerontology. Nutrition and aging: human and animal studies. New York, Raven Press, 1984, pp. 279–322.

28. Algeri S, Biagini L, Manfridi A, Pitsikas N. Age-related ability of rats kept on a life-long hypocaloric diet in a spatial memory test. Longitudinal observations. Neurobiol Aging 1991;12, 277–282.

29. Merry BJ, Holehan AM. Physiological Basis of Aging and Geriatrics. Effects of diet on aging, Boca Raton, CRC Press, 1994, pp. 285–310.

30. Yu BP. Antioxidant action of dietary restriction in the aging process. J Nutr Sci Vitaminol 1993;39: 575–583.

31. Sohal RS, Ku HH, Agarwal S, Forster MJ, Lal H. Oxidative damage, mitochondrial oxidant generation and antioxidant defenses during aging and in response to food restriction in the mouse. Mech Ageing Dev 1994;74:121–133.

32. Choi JH, Yu BP. Brain synaptosomal aging: free radicals and membrane fluidity. Free Rad Biol & Med 1995; 18 (2): 133–139.

33. Ingram DK. The Potential for Nutritional Modulation of Aging Processes. Effects of dietary restriction on brain and behavioural function in aging rodents, Turnbull CT, Food and Nutrition Press, Inc., 1991, pp. 289–310.

34. Tzagoloff A, Myers AM. Genetics of mitochondrial biogenesis. Ann Rev Biochem

1986;55:249–285.

35. Kugler P, Vogel S, Gehm M. Quantitative succinate dehydrogenase histochemistry in the hippocampus of aged rats. Histochem 1988;88: 299–30.

36. Brust R. Viscosity of chromatin solutions increases with increasing ionic strength. Molec Biol Rep 1987;11: 213–217.

37. Higgins CF, Cairney J, Stirling DA, Sutherland L, Booth IR. Osmotic regulation of gene expression: ionic strength as an intracellular signal? Trends Bioch Sci 1987; 12: 339–344.

38. Von Zglinicki T, Bimmler M. Intracellular water and ionic shifts during growth and aging of rats. Mech Ageing Dev 1987; 38:179–187.

39. Damjanovich S, Bot J, Somogyi B, Sumegi J. Protein fluctuation and enzyme activity. J Theor Biol 1983; 105, 25–33.

40. Sun AY, Sun GY. Neurochemical aspects of the membrane hypothesis of aging. In: Interdisciplinary Topics in Gerontology, von Hahn HP (ed), Vol. 15, Karger, Basel, pp.34–53, 1969.

41. Sun AY, Sun GY. Effects of dietary vitamin E and other antioxidant on aging process in rat brain. Adv Exp Med Biol 1978;97: 285–290.

42. Sun YA. Effect of dietary vitamin E and other antioxidants on aging processes in rat brain. Adv Exp Biol 1979; 97: 285–290.

43. Chin JH, Parsons LM, Goldstein DB. Increased cholesterol content of erythrocyte and brain membranes in ethanol-tolerant mice. Biochim Biophys Acta1978; 513: 358–363.

44. Littleton JM, Grieve SJ, Griffiths PJ, John GR. Ethanol-induced alteration of membrane phospholipid composition: possible relationship to development of cellular tolerance to ethanol. In: Begleiter H. (ed) Biological Effects of Alcohol, Advances in Experimental Medicine and Biology, New York, Plenum Press, Vol. 126, pp. 7–19, 1980.

45. Brekenridge WC, Gombos G, Morgan GI. The lipid composition of adult rat brain synaptosomal plasma membranes. Biochim Biophys Acta 1972;226: 659–707.

46. Hegner D. Age-dependence of molecular and functional changes in biological membrane properties. Mech Age Dev 1980;14: 101–118.

47. Harman D. The aging process. Proc Natl Acad Sci USA 1981;78:7124–7128.

48. Ku HH, Brunk UT, Sohal RS. Relationship between mitochondrial superoxide and hydrogen peroxide production and longevity of mammalian species. Free Rad Biol Med 1993;15:621–627.

49. Sohal RS, Ku HH, Agarwal S. Biochemical correlates of longevity in two closely related rodents species. Biochem Biophys Res Commun 1993;196: 7–11, 1993.

50. Meier-Ruge. Senile dementia - a disease of exhaustion of functional brain reserve capacity. Excer. Med Curr Clin Pract Ser 1990; 59: 371–375.

51. Meier-Ruge W, Iwangoff P, Bertoni-Freddari C. What is primary and what secondary for amyloid deposition in Alzheimer's disease. Ann NY Acad Sci 1994; 719:230–237.

*The Neuroendocrine Immune Network in Ageing*
Edited by R.H. Straub and E. Mocchegiani

# Ageing-Related Role of Nitric Oxide in the Brain

SOFIA MARIOTTO[1], MASSIMO MISCUSI[2], TIZIANA PERSICHINI[3], MARCO COLAS-
ANTI[3] and HISANORI SUZUKI[1]

*[1]Department of Neuroscience and Vision, University of Verona, Verona, Italy ;[2]Department of
Neurosurgery, University of Roma, "La Sapienza", Roma, Italy; [3]Department of Biology, University of Roma 3, Roma, Italy*

## ABSTRACT

Production of nitric oxide (NO) in the cell is catalysed by NO synthase (NOS), which using
L-arginine and oxygen as substrates, synthesises NO. Normally cells express neuronal and/or
endothelial NOS (nNOS and eNOS, respectively) producing nanomolar NO. Under inflammatory conditions cells express inducible NOS (iNOS), which, once expressed, produces micromolar NO. Since micromolar NO could elicit damage to cells, iNOS expression is time-spatially
finely regulated.

Recent evidence points out the possible interaction between n/eNOS and iNOS. Under inflammatory conditions interferon-$\gamma$, tumour necrosis factor-$\alpha$ and interleukin 1-$\beta$ rapidly trigger inhibition of e/nNOS activity with successive drop in the amounts of NO. Since NO normally exerts
suppressive action on NF-$\kappa$B activation, drop in NO may create a favourable conditions for
NF-$\kappa$B activation and successive iNOS exptression.

Evidence in the literature indicates that during ageing nNOS activity gradually drops with
concomitant increase in iNOS expression in the brain. However, how nNOS activity may influence iNOS expression remains to be elucidated.

This review points out the possible functional link between age-dependent decrease in nNOS
activity and increase in iNOS expression in the brain. Evidence for the possible scenario that
often underestimated continuous infections to the brain, one of the hallmarks of ageing, may
trigger apparently spontaneous iNOS expression is presented. Future treatment of aged people
based on this scenario is also described.

## 1.    INTRODUCTION

It was only two decades ago that a honour to be a part of biologically active molecule was des-
ignated to nitric oxide (NO), a small free radical molecule, which, until that time, was only considered to be one of the principal components of polluted air. Thereafter, however, an increasing
body of literature evidence has made NO as one of the most popular stars in the world of living
organisms. The first discovery that NO was either a new retrograde neurotransmitter or the
endothelium-derived relaxing factor (EDRF), the nature of which was remained for a long time

a mystery, was *per se* of a tremendous impact. Successive evidence that NO was implicated in a number of important physiopathological events, not only in animals but also in plants, points out that the role played by this small molecule is so fundamental that it is believed that its shadow may be visible in any biochemical pathway one studies.

One of the most critical aspect of the current health condition in the highly industrialised nations is the presence of different pathologies connected strictly to elevated age. Since the brain is one of the organs severely damaged during ageing, there have been numerous reports describing the possible implications of NO in the age-related disorders in the brain. However, either due to the different experimental models examined (different animal species, different regions in the same animals, and so on) or due to the complex biochemical pathways in which NO is involved, a clear-cut role of NO in age-related disorders in the brain has not been so far elucidated.

Recently, we reported a new vision on the physiopathological action of NO, pointing out the complicated network that involves NO molecule coming from different origins inside cells. Accordingly, in the present review, we revisit a number of previous apparently controversial results in order to try to draw a reconciling (re)view on the role of NO played in the course of brain ageing.

## 2.    NITRIC OXIDE

NO, a gas in the body, is principally generated by NO synthase [1]. It freely passes through cell membranes, thus acting not only in the cell where it is synthesised but also in neighbouring cells. Half-life of this free radical in the living cells is estimated to be less than few seconds, due principally to its reaction with compounds such as superoxide, metal ions and heme- or free sulfhydril-containing compounds. Its reactivity with these compounds, together with its time/spatial-dependent concentrations, determine the physiopathological action of NO. At nanomolar concentrations, one of the most important targets is the soluble guanylate cyclase (sGC). Activation of sGC rapidly induces the increase in the intracellular amounts of cGMP which, in virtue of the site of its synthesis, mediates either vasodilation or neurotransmission. At micromolar concentrations, one of the main compounds reacting with NO is superoxide ($O_2^-$). Although, under normal conditions, a small amounts of $O_2^-$ is always present, its amount increases drastically under conditions in which NO is also massively synthesised. Peroxynitrites ($ONOO^-$), a product of the reaction between NO and $O_2^-$, are believed to be responsible in almost all NO-elicited deleterious events during inflammatory processes.

## 3.    NO SYNTHASE

NO synthase (NOS) catalyses the oxygen depending reaction during which L-arginine is converted into L-citrulline and NO [1]. Molecular cloning and sequencing analyses revealed the existence of at least three major types of NOS isoforms, which require FAD, NADPH, and tetrahydrobiopterine as cofactors. Both neuronal NOS (nNOS or NOS-I) and endothelial NOS (eNOS or NOS-III) are $Ca^{2+}$/calmodulin-dependent enzymes and are costitutively expressed in restricted cell types [2]. The third enzyme is an inducible $Ca^{2+}$-independent isoform (iNOS or NOS-II), which is expressed in a variety of cell types in response to lipopolysaccharides (LPS), interferon-$\gamma$ (IFN-$\gamma$) and pro-inflammatory cytokines, such as tumour necrosis factor-$\alpha$ (TNF-$\alpha$) and interleukin 1-$\beta$ (IL1-$\beta$) [3]. Human brain expresses all three isoforms of NOS. NOS-I is widely expressed in the brain with highest levels present in the substantia innominata, cerebellar cortex and nucleus accumbens [4], being generally present only in the distinct group of neurones as well as in astrocytes. It seems to be involved in intracellular signal transduction, in the maintenance of cellular homeostasis and in particular events including regulation of cell metabolism, growth and differentiation, modulation of the reactivity of cells to external noxious stimuli, phasic activity-associated hormone release in neurones located in hypotalamus and regulation of thermic adjustment of body, and finally, memory process in diencephalic neurones. NOS-III, quantitatively compared to NOS-I, seems to be poorly represented in central nervous system, especially in spinal cord, where a minimal vascularisation exists. It is mainly expressed in endothelial cells as well as astrocytes and some neuronal cells. Under inflammatory processes, microglia and astrocytes express NOS-II. Infiltrated macrophages and other inflammatory cells such as neutrophils also express NOS-II in the brain.

## 4.    HOMEOSTASIS OF NO SYSTEM

NO plays major roles in regulating vascular tone, neurotransmission, killing of microorganisms and tumour cells [5]. As already underlined, physiological low amounts of NO (subnanomolar concentrations) produced by NOS-I and NOS-III are mainly involved in the first two events. Whereas, high concentrations of NO (micromolar concentrations) produced transiently and locally by NOS-II exert its toxic action against invading cells (microbes, virus, parasites and tumour cells). A lot of evidence also indicates that elevate NO levels play an important role in the pathogenesis of several chronic disorders and inflammatory processes. High NO levels can also be achieved by hyper-activation of NOS-I, being involved in disorders such as ischemia/reperfusion injury and glutamatergic neurodegenerative processes in the central nervous system.

The body carefully controls the expression of NOS-II through a number of endogenous and exogenous molecules in order to efficiently eliminate invading cells without or with minimal damages to the host tissues. These compounds act either as inducers of the expression of NOS-II mRNA (e.g., LPS, IFN-$\gamma$, TNF-$\alpha$ and IL-1$\beta$), or as suppressors like corticosteroids, estrogens, transforming growth factor-$\beta$ (TGF$\beta$), interleukin-4 (IL-4), interleukin-8 (IL-8), interleukin-10 (IL-10), calcium ionophores and glutamate. Due to the temporally and spatially co-ordinated action among these molecules, the correct NOS-II expression is normally achieved during inflammation [6].

That physiological NO may be a regulator of inducible NOS expression (Figure 1) has been recently indicated also in *in vivo* models: NOS-II superinduction has been observed in the heart of NOS-III-KO mice in an ischemia/reperfusion model [10] and a spontaneous activation of NF-$\kappa$B occurred in the heart of rats treated for a long time with L-NAME, a strong inhibitor of NOS activity [11].

## 5.    ROLE OF NO IN THE BRAIN

As elsewhere, the multiple role of NO in the brain depends on cell types and on the functional district involved. The central nervous system can grossly be divided in a philogenetically new part, the cortex and the basal ganglia and a philogenetically old part, the diencephalon and the forebrain. All these parts are composed by the same cell types: neurones, glial cells (i.e. astro-

294

Figure 1. Putative involvement of NO homeostasis in brain ageing.

(A) In the brain of young and adult persons, cNOS normally produces tonic amounts of NO, enough in keeping suppressed an accidental activation of NF-κB and successive iNOS expression (6). Only when high amounts of LPS/ cytokines are circulating in the brain, especially at the early phase of inflammation, iNOS expression is temporarily induced. Entire system of NO (6) is depicted by a sword, counteracting efficiently insults by microbe/ virus or other stimuli.

(B) In the brain of aged persons, down-regulation of constitutive NOSs may take place, accounting for the gradual loss of NO. This situation may favour triggering the apparently spontaneous iNOS expression, leading to massive NO production. Toxic effect of NO may begin to prevail. Unsuccessful tentative to counteract microbe/virus insult is depicted by a rudimental tool in the hand of aged person.

(C) Successive ageing may reconcile with avoiding or overcoming the potentially harmful situation. A hopeful fan in the hand of a person with sucessful ageing depicts this situation.

Small obejcts flying around persons represent various type of insult.

This vision of the role played by NO homeostasis in the brain ageing leads to the hypothesis on the presence of putative check point (indicated with X) at which down-regulation of constitutive NOSs might begin to take place. Any treatment aimed to prevent it should be promising in counteracting continuous stimuli triggering iNOS expression, a hallmark of ageing. Further elucidation is needed to elucidate the brain NO homeostasis in successful ageing.

cytes, oligodendrocytes and microglia) and endothelial cells. The different connections between neurones represent the key element of the function of the different parts of central nervous system. Neurones require high oxygen supply and, therefore, their function is dependent on functional state of cerebral blood supply. The more functional activation is needed by a group of neurones, the more oxygen is required by a massive and selective loco-regional vasodilatation.

NO produced by NOS-I in the distinct type of cells (e.g., glutamatergic neurones) may exert concentration-dependent effect. Normally, it reacts with heme of soluble guanylate cyclase present in both NO-producing neurones and neighbouring ones. Upon activation of this enzyme, a rapid and transient increase in the intracellular amounts of cGMP occurs. This triggers the formation of potential action in the neurones due to cGMP-dependent depolarisation of the membrane. Thus, NO, being a gas, may act as non-directional or retrograde neurotransmitter, making NO an unique neurotransmitter different from all known neurotransmitters, the action of which is strictly directional (from post-synapsis to pre-synapsis). Neurotransmitters such as acetylcho-line and glutamate, mediated by the rapid and transient increase in intracellular $Ca^{2+}$ levels, may exert their effect also acting as up-regulator of NO. The role played by NO produced by NOS-I in astrocytes has poorly been elucidated. One of its putative role should be the regulation of the transcriptional induction of NOS-II expression in these cells. At micromolar concentrations, it may exert, as elsewhere, deleterious effect to the restricted regions of brain tissues. Neurotoxic effects exerted by elevated concentrations of compounds such as glutamate and kainate are cur-rently believed to be mainly mediated by massive amounts of NO, produced by $Ca^{2+}$-induced hyper-activation of NOS-I.

Release of NO from cerebral vascular endothelial cells appears to play an important role in controlling cerebral vascular tone and circulation in health and disease. [12]. Indeed, NOS-III-derived NO diffuses towards the lumen of blood vessels, thus inhibiting platelet and leukocyte adhesion to the endothelium, a process that may down-regulate pro-inflammatory events. In humans, NO has been shown also to maintain blood fluidity and reduce blood viscosity, thus improving flow [13]. Literature evidence indicates that NOS-III is responsible for the production of a low, basal level of NO, which may exert functionally significant constraints on cell shape. The result of this phenomenon is that NO keeps endothelia stationary and maintains their shape and activation state in the microvasculature and that an absence of basal NO results in cell shape distortions [13].

Inflammatory processes are generally characterised by transient and local production of massive amounts of NO, following the expression of NOS-II at the early phase of inflamma-tion. Here again, as in other tissues, principal goal of NO in the brain may be the elimination of invaded cells, although less-coordinated or badly regulated expression of NOS-II may elicit damages to the brain tissues.

## 6.    AGEING-RELATED CHANGES IN THE NO PRODUCTION

The possible implications of NOS in the brain during ageing was first described in 1993 [14], showing the decreased NO production in aged-rat brain. Thereafter, there have been contro-versial observations regarding the ageing-related changes in the production of NO, making it difficult to draw a clear vision on the role played by NO in the brain ageing. In this context, difficulty of direct estimation of endogenous NO levels has been critical. Furthermore, meas-urements either of the catalytic activity or expression level of NO synthases do not necessarily reflect endogenous production of NO. Difference in animal models and in regions of the same animal species studied by each group add further confusions to understanding the comprehen-sive role played by NO. We attempt to resolve this problem below, focusing the attention to the putative age-correlated change in the cross-talk between NOS-I/III and NOS-II, mediated by the amounts of NO.

In spite of the technical difficulty in assessing the low amounts of NO in animals, there are considerable reports indicating the change in the constitutive NO synthase-derived NO produc-tion during brain ageing. Accordingly, independent of animal species, brain ageing seems to be associated with the general decrease in the amounts of NO produced by constitutive NOS. Indeed, not only the first description of the possible implication of NO in aged-brain [14], but also many other works reported the data indicating the age-dependent decrease in the concentra-tions of constitutive NOS-derived NO [15]. Evidence indicating the increase in NO production was also reported both in the cerebrospinal fluid of aged individuals [16].

Though apparently controversial, many data in the literature show also an ageing-dependent enhancement in the production of NO in the brain mainly due to spontaneous expression of NOS-II. Accordingly, brain ageing seems to be associated, on the one hand, to the decrease in NO production by NOS-I/NOS-III and, on the other, to the increase in the amounts of NO due to spontaneously induced NOS-II. In the following chapters, the possible role played by different isoform of NO synthase in regulating age-dependent changes in the brain NO production will be discussed.

## 7.    ENDOTHELIAL NOS IN THE BRAIN DURING AGEING

Current view on the putative role played by NOS-III in brain ageing is that this enzyme may crucially be involved in impaired endothelium-dependent vasodilation, which is strictly correlated to ageing. Accordingly, down-regulation of the activity and/or expression of NOS-III seems to play deleterious role during ageing.

A number of authors have associated, at least in part, the vascular disease with a decrease in NOS-III-derived NO production. In this respect, vasculopathic evolution may manifest very slowly, due either to a limited short-term presence of NO or to NO from NOS-I, which may partly compensate the limited basal NO [13].

Thus, many measures providing protection from ischemic stroke such as arginine, estrogens replacement, statin drugs, green tea, polyphenols may probably reflect increased activity and/or expression of NOS-III [17]. Although only few notions are available on the role played by NOS-III in astrocytes, its involvement in regulating NOS-II expression seems to be highly reasonable. Although molecular mechanism of age-dependent down-regulation of NOS-III needs further elucidation, phosphorylation of the enzyme is believed to be profoundly involved in this process.

## 8.    NEURONAL NO SYNTHASE IN THE BRAIN DURING AGEING

Previous reports described both down- and up-regulation of this enzyme during the ageing process in the brain. In human brain, decreased neuronal NO synthase was observed during ageing. Consistently, a number of groups reported that in different regions of rat or mice brain, down-regulation of either catalytic activity or expression of NOS-I occurred [18–25], although, in some regions, the decline of NOS-I amounts was not always observed. A significant decline in NADPH-diaphorase-positive neuronal density, size or activity observed in the striatum between 12 to 26 month, or in the axons and axon terminals in old rats, respectively, may be in line with these notions [26, 27]. Furthermore, loss of neuronal NOS in neurones was observed in the aged rat cerebral cortex [28]. All these notions lead to the assumption that the brain ageing may be characterised by the decline in the amounts of NO produced by NOS-I. Astrocyte NOS-I is also believed to be directly involved, at the early phase of inflammation, in the induction of NOS-II gene [6].

However, some groups observed the opposite trend. A remarkable up-regulation of NOS-I expression or activity was reported in the hippocampus, cortex, forebrain and other regions of the brain of aged rats [29–31], respectively. In line with this, the kinetics of NOS-I activity was supposed to be altered during maturation toward the increase in the capacity for NO synthesis [32].

In summary, although literature evidence seems to favour the general notion that NOS-I may be down-regulated in the brain during ageing, further studies are needed to elucidate the timely and regional change in the NOS-I-derived production of NO. The molecular mechanism of the age-dependent modulation of NOS-I, including phosphorylation of the enzyme, awaits further elucidation.

## 9.    INDUCIBLE NO SYNTHASE IN THE BRAIN DURING AGEING

Findings indicate almost unanimously the increase in the expression of NOS-II mRNA in the different regions of the brain of different animal species during the process of ageing. In human, age-dependent increase in the production of NOS-II-derived NO was supposed by measuring 3-nitrotyrsine/tyrosine in the cerebrospinal fluid [33]. In other animal models, age-dependent increase in NO production, activity and expression of the enzyme was largely observed in different regions of the aged-brain [24,26,34–41].

Briefly, an increase of NOS-II expression has been demonstrated in diencephalum, in cerebral cortex, and in the cerebellum hemispheres. In hypotalamus, pituitary gland and hyppocampus (being parts of diencephalum) NOS-II hyper-expression would produce respectively an alteration of physiological process regulating the thermic homeostatic adjustments, hormonal secretion [42] and memory [21, 34]. In the cerebral and cerebellar cortex the above cited metabolic changes would produce a progressive degeneration of neurones and a consequent reduction of neural network responsible of superior neurological functions. Degenerative death would more evident in NOS-I expressing neurones [28,43,44]. Interestingly, in all above-mentioned regions NOS-II expression is inversely associated with NOS-I/III activity.

In clear contrast to the expression of NOS-II mRNA in the young and adult animals including human, which exclusively occurs under acute inflammatory conditions, in aged animals and human it seems to spontaneously be induced in the course of ageing. Since, infections may frequently but silently attack aged individuals, the spontaneous expression of NOS-II could only be circumstantial, since it could be triggered by continuously circulating immune cytokines, such as IFN-$\gamma$ and TNF-$\alpha$. In this scenario, frequently observed down-regulation of both NOS-I and NOS-III (see chapter 4) may play, especially at the early phase of ageing, a pivotal role in the induction of NOS-II gene. Since enzymatic activity of NOS-I/NOS-III could be down-regulated by cytokines such as IFN-$\gamma$, TNF-$\alpha$ and IL1-$\beta$, age-correlated gradual decline in these enzyme could be sustained by gradually increasing amounts of these compounds. A drop in the amounts of NO, at least partly, may create favourable conditions for the apparently spontaneous expression of NOS-II in the brain. Increased synthesis of reactive nitrogen species due to mitochondrial oxidative stress could widely be implicated in this process [45].

As already mentioned, NOS-II, once expressed, usually produces locally and temporally a massive amounts of NO. Although NO, in this case, plays as a double-edged sword acting both beneficially and detrimentally to the body, in the young and adult individuals mechanism to minimise the damages usually prevails. This mainly is due to the intervention of compounds acting as suppressors of inducible NO synthase expression (see chapter 4). However, in the brain of aged-individuals, this mechanism may not work at all or work with less efficiency. Spontaneous production of NO in the brain, on the one hand, may be the tentative to defend against infections but, on the other hand, may become deleterious to the tissue integrity. In this respect, it is interesting to note that there are a number of observations indicating the possible effort of these regions of aged individuals in counteracting the decline in the NO production by neuronal

NOS-I [16,29,30,43].

## 10. CONCLUDING REMARK

Current view on the role of NO in brain ageing is that age-correlated spontaneous production of NO mainly synthesised by inducible NO synthase not only in the cortex and hippocampus but also in all other regions of the brain plays a deleterious role (Figure 1). Many experimental evidences point out, furthermore, the possible, crucial importance of NOS-I/III especially during the early phase of ageing process. Down-regulation of the spontaneous induction of inducible NO synthase expression seems to be a promising intervention against NO-elicited aged-brain damages. In this context, timely and spatially controlled up-modulation of NOS-I/III may provide a new strategy in counteracting the spontaneous production of massive amounts of NO.

## REFERENCES

1.   Moncada S, Palmer RM, Higgs EA. Nitric oxide: physiology, pathophysiology, and pharmacology. Pharmacol Rev 1991; 43: 109–142.
2.   Bredt DS and Snyder SH. Isolation of nitric oxide synthetase, a calmodulin-requiring enzyme. Proc. Natl. Acad.Sci. USA 1990; 87:682–685.
3.   Lowenstein CJ, Glatt CS, Bredt DS and Snyder SH. Cloned and expressed macrophage nitric oxide synthase contrasts with the brain enzyme. Proc. Natl. Acad.Sci. USA 1992; 89: 6711–6715.
4.   Blum-Degen D, Heinemann T, Lan J, Pedersen V, Leblhuber F, Paulus W, Riederer P, Gerlach M. Characterization and regional distribution of nitric oxide synthase in the human brain during normal ageing. Brain Research 1999; 834:128–135.
5.   Mayer B. and Hemmens B. Biosynthesis and action of nitric oxide in mammalian cells. Trends Biochem Sci. 1997; 22: 477–481.
6.   Colasanti M, Suzuki H. The dual personality of NO. TIPS 2000; 21: 249–252.
7.   Colasanti M, Persichini T, Cavalieri E, Fabrizi C, Mariotto S, Menegazzi M, Lauro GM, Suzuki H. Inactivation of NOS-I by lipopolysaccharide plus interferon-gamma-induced tyrosine phosphorylation. J Biol Chem. 1999;274:9915–7.
8.   Komeima K, Hayashi Y, Naito Y, Watanabe Y. Inhibition of neuronal nitric-oxide synthase by calcium/ calmodulin-dependent protein kinase IIalpha through Ser847 phosphorylation in NG108–15. neuronal cells. J Biol Chem. 2000;275:28139–43.
9.   Fleming I, Bauersachs J, Busse R. Calcium-dependent and calcium-independent activation of the endothelial NO synthase. J Vasc Res 1997;34:165–74.
10.  Kanno S, Lee PC, Zhang Y, Ho C, Griffith BP, Shears LL 2nd, Billiar TR. Attenuation of myocardial ischemia/reperfusion injury by superinduction of inducible nitric oxide synthase. Circulation. 2000;101:2742–8.
11.  Kitamoto S, Egashira K, Kataoka C, Koyanagi M, Katoh M, Shimokawa H, Morishita R, Kaneda Y, Sueishi K, Takeshita A. Increased activity of nuclear factor-kappaB participates in cardiovascular remodeling induced by chronic inhibition of nitric oxide synthesis in rats. Circulation. 2000;102:806–12.
12.  Lee TJ. Nitric oxide and the cerebral vascular function. J Biomed Sci 2000;7:16–26.
13.  De la Torre JC, Stefano GB. Evidence that Alzheimer's disease is a microvascular disorder:

the role of constitutive nitric oxide. Brain Research Reviews 2000;34:119–136

14.  Strolin Benedetti M, Dostert P, Marrari P, Cini M. Effect of ageing on tissue levels of amino acids involved in the nitric oxide pathway in rat brain. J Neural Transm Gen Sect 1993;94:21–30.

15.  DiCiero M, de Bruin VMS, Vale MR, Viana GSB. Lipid Peroxidation and Nitrite plus Nitrate Levels in Brain Tissue from Patients with Alzheimer's Disease. Gerontology 2000;46:179–184.

16.  Nagatomo I, Akasaki Y, Uchida M, Tominaga M, Hashiguchi W, Kuchiiwa S, Nakagawa S, Takigawa M. Age-related alterations of nitric oxide production in the brains of seizure-susceptible EL mice. Brain Res Bull 2000;53:301–306.

17.  McCarty MF. Up-regulation of endothelial nitric oxide activity as a central strategy for prevention of ischemic stroke - just say NO to stroke! Med Hypotheses 2000;55:386–403..

18.  Vallebuona F, Raiteri M. Age-related changes in the NMDA receptor/nitric oxide/cGMP pathway in the hippocampus and cerebellum of freely moving rats subjected to transcerebral microdialysis. Eur J Neurosci 1995; 7:694–701.

19.  Mollace V, Rodino P, Massoud R, Rotiroti D, Nistico G. Age-dependent changes of NO synthase activity in the rat brain. Biochem Biophys Res Commun. 1995;215:822–827.

20.  Yamada K, Nabeshima T Changes in NMDA receptor/nitric oxide signaling pathway in the brain with aging Microsc Res Tech 1998;43:68–74.

21.  Yamada K, Nabeshima T. Changes in NMDA receptor/nitric oxide signaling pathway in the brain with aging. Microsc Res Tech. 1998;43:68–74.

22.  Necchi D, Virgili M, Monti B, Contestabile A, Scherini E. Regional alterations of the NO/NOS system in the aging brain: a biochemical, histochemical and immunochemical study in the rat. Brain Res 2002;933:31–41.

23.  Yu W, Juang S, Lee J, Liu T, Cheng J. Decrease of neuronal nitric oxide synthase in the cerebellum of aged rats. Neurosci Lett 2000;291:37–40.

24.  Law A; O'Donnell J; Gauthier S; Quirion R. Neuronal and inducible nitric oxide synthase expressions and activities in the hippocampi and cortices of young adult, aged cognitively unimpaired, and impaired Long–Evans rats. Neuroscience 2002;112:267–75.

25.  Hilbig H, Holler J, Dinse HR, Bidmon HJ. In contrast to neuronal NOS-I, the inducible NOS-II expression in aging brains is modifled by enriched environmental conditions. Exp Toxicol Pathol 2002;53:427–31.

26.  Kuo H, Hengemihle J, Ingram DK. Nitric oxide synthase in rat brain: age comparisons quantitated with NADPH-diaphorase histochemistry. J Gerontol A Biol Sci Med Sci 1997;52:B146–151.

27.  Ma SX, Holley AT, Sandra A, Cassell MD, Abboud FM Increased expression of nitric oxide synthase in the gracile nucleus of aged rats. Neuroscience 1997 76, 659–663.

28.  Cha CI, Uhm MR, Shin DH, Chung YH, Baik SH. Immunocytochemical study on the distribution of NOS-immunoreactive neurons in the cerebral cortex of aged rats. Neuroreport 1998;9:2171–2174.

29.  Law A, Dore S, Blackshaw S, Gauthier S, Quirion R. Alteration of expression levels of neuronal nitric oxide synthase and haem oxygenase-2 messenger RNA in the hippocampi and cortices of young adult and aged cognitively unimpaired and impaired Long-Evans rats. Neuroscience 2000;100:769–75.

30.  La Porta CA, Comolli R. Age-dependent modulation of PKC isoforms and NOS activity and expression in rat cortex, striatum, and hippocampus. Exp Gerontol 1999;34:863–74.

300

31. Vernet D, Bonavera JJ, Swerdloff RS, Gonzalez-Cadavid NF, Wang C. Spontaneous Expression of inducible nitric oxide synthase in the hypothalamus and Other Brain Regions of aging rats. Endocrinology 1998;139:3254.

32. Pearce WJ, Tone B, Ashwal S. Maturation alters cerebral NOS kinetics in the spontaneously hypertensive rat. Am J Physiol 1997, 273(4 Pt 2):R1367–73.

33. Tohgi H, Abe T, Yamazaki K, Murata T, Ishizaki E, Isobe C. Alterations of 3-nitrotyrosine concentration in the cerebrospinal fluid during aging and in patients with Alzheimer's disease. Neurosci Lett 1999;269:52–54.

34. Ferrini M, Wang C, Swerdloff RS, Sinha Hikim AP, Rajfer J, Gonzalez-Cadavid NF. Aging-related increased expression of inducible nitric oxide synthase and cytotoxicity markers in rat hypothalamic regions associated with male reproductive function. Neuroendocrinology 2001;74:1–11.

35. Kyrkanides S, O'Banion MK, Whiteley PE, Daeschner JC, Olschowka JA. Enhanced glial activation and expression of specific CNS inflammation-related molecules in aged versus young rats following cortical stab injury. J Neuroimmunol. 2001;119:269–77.

36. Sloane JA, Hollander W, Moss MB, Rosene DL, Abraham CR. Increased microglial activation and protein nitration in white matter of the aging monkey. Neurobiol Aging. 1999;20:395–405.

37. Abe T, Tohgi H, Murata T, Isobe C, Sato C. Reduction in asymmetrical dimethylarginine, an endogenous nitric oxide synthase inhibitor, in the cerebrospinal fluid during aging and in patients with Alzheimer's disease. Neurosci Lett 2001;312:177–9.

38. Wong BS, Liu T, Paisley D, Li R, Pan T, Chen SG, Perry G, Petersen RB, Smith MA, Melton DW, Gambetti P, Brown DR, Sy MS. Induction of HO-1 and NOS in doppel-expressing mice devoid of PrP: implications for doppel function Mol Cell Neurosci 2001;17:768–75.

39. Heininger K. A unifying hypothesis of Alzheimer's disease. IV. Causation and sequence of events. Rev Neurosci 2000;11:213–328.

40. Jacob JM, Dorheim MA, Grammas P. The effect of age and injury on the expression of inducible nitric oxide synthase in facial motor neurons in F344 rats. Mech Ageing Dev 1999;107:205–18.

41. Wu MD, Kimura M, Inafuku S, Ishigami H, Okajimas S. Effect of aging on the expression of iNOS and cell death in the mouse cochlear spiral ganglion. Folia Anat Jpn 1997;74:155–65.

42. McCann SM. The nitric oxide hypothesis of brain aging. Exp Geront 1997;32:431–440.

43. Uttenthal LO, Alonso D, Fernandez AP, Campbell RO, Moro MA, Leza JC, Lizasoain I, Esteban FJ, Barroso JB, Valderrama R, Pedrosa JA, Peinado MA, Serrano J, Richart A, Bentura ML, Santacana M, Martinez-Murillo R, Rodrigo J. Neuronal and inducible nitric oxide synthase and nitrotyrosine immunoreactivities in the cerebral cortex of the aging rat. Microsc Res Tech 1998;43:75–88.

44. Cheng JT, Lin IM, Yen SI, Juang SW, Lin TP, Chan P. Decrease of neuronal nitric oxide synthase in the cerebellum of aged rats. Neurosc Lett 2000;291:37–40.

45. Sastre J, Pallardo FV, Vina J. Mitochondrial oxidative stress plays a key role in aging and apoptosis. IUMB life 2000, 49: 427–35.

# V. LINKS BETWEEN ONE GLOBAL SYSTEM AND ANOTHER GLOBAL SYSTEM DURING THE AGEING PROCESS

*The Neuroendocrine Immune Network in Ageing*
Edited by R.H. Straub and E. Mocchegiani

# Introduction: Links Between One Global System and Another Global System During the Ageing Process

RAINER H. STRAUB

*Laboratory of Neuroendocrinoimmunology, Dept. of Internal Medicine I, University Hospital, 93042 Regensburg, Germany*

## ABSTRACT

Chapter II to IV of this book demonstrated specific aspects of ageing within the different super-systems, the immune system, the endocrine system, and the nervous system. In these preceding chapters, experts demonstrated how a certain part of a supersystem is altered during ageing. However the link between age-dependent changes of one supersystem and age-dependent changes of another supersystem has not been demonstrated yet. In this respect, chapter V gives four different examples as to how ageing of one supersystem affects ageing of another super-system. I demonstrate how loss of the adrenal hormone dehydroepiandrosterone may affect the age-related increase of the proinflammatory cytokine IL-6. E. Mocchegiani and colleagues demonstrate how loss of zinc during the ageing process modulates function of important thymus–protecting hormones such as melatonin, growth hormone, and thyroid hormones. K. Dinkel and R.M. Sapolsky demonstrate how age-dependent increase of endogenous glucocorticoids can affect brain function. M. Peterlik outlines how loss of growth hormone, sex steroids, and Vitamin D influence osteogenesis, thus, contributing to age-related osteoporosis. Many more examples of such interrelations between different supersystems exist, however, demonstration of which would go beyond the scope of this book.

## 1. AGEING OF ONE PART OF A GLOBAL SYSTEM INVOLVES ANOTHER PART OF ANOTHER GLOBAL SYSTEM

### 1.1. DHEA and IL-6

Serum concentration of IL-6 significantly increases with age [1–6]. It may be that the decline of testosterone [7], estrogens [8] and DHEA [5] will lead to an increase of serum IL-6 during ageing. Indeed, there is some experimental and epidemiological evidence that the decline of DHEA is directly associated with the increase of IL-6: A) Daynes et al. demonstrated in mice that administration of DHEA sulfate, which is converted into the active DHEA by a sulfatase, inhibits spontaneous IL-6 secretion in ageing mice [9]; B) In human subjects, IL-6 secretion increases during ageing and serum levels of IL-6 are inversely correlated to serum levels of DHEA [5]; C) DHEA inhibits IL-6 secretion from human and mouse mononuclear cells [5, 10].

Whether this is a direct effect of DHEA due to activation of the peroxisome proliferator activator receptor alpha and inhibition of NF-kappaB [11] or an indirect effect via downstream androgens or estrogens [12] remains to be investigated.

In another investigation of healthy subjects [13], we found significantly lower serum levels of cortisol in relation to plasma adrenocorticotropic hormone (ACTH) with age only in women [13]. In a multiple linear regression analysis with this particular ratio as the dependent variable, the independent variables 17β-estradiol and IL-6 serum levels were the significant factors in the model [13] Serum levels of TNF did not influence the model. This may indicate that increasing serum levels of IL-6 together with decreasing serum concentrations of 17β-estradiol are responsible for the decrease of serum cortisol in relation to plasma ACTH in women. Thus, a situation may appear where aged women are more prone to inflammatory diseases. Since IL-6 plays an important role in many age-related diseases, its increase is of outstanding importance for the ageing body. This is an example for the interaction of two global systems during the ageing process – the adrenal glands and the immune system.

## 1.2. Zinc, hormones and thymic function

E. Mocchegiani demonstrates how age-dependent loss of zinc may well influence the function of important hormones. He uses the examples of growth hormone and IGF-I, thyroxine and triiodothyronine, and melatonin which all have thymus – protecting effects. He presents evidence that loss of zinc leads to deterioration of hormone function which finally leads to thymic changes typical for aged subjects.

## 1.3. Glucocorticoids and brain function

K. Dinkel and R.M. Sapolsky demonstrate that the relative increase of glucocorticoids in relation to other adrenal hormones leads to deterioration of brain function. Their specific focus is on memory problems, changes of neurogenesis, and loss of hippocampal neurons. They provide evidence that increasing glucocorticoid serum levels and exaggerated glucocorticoid responses during stressful situations may well lead to an age-dependent increase of brain dysfunction.

## 1.4 Hormone loss and osteoporosis

In the final contribution within chapter V, M. Peterlik links the decrease of important hormones to a typical age-related problem, osteoporosis. He demonstrates that an age-dependent increase of central leptin and NPY may lead to an unfavourable influence on bone formation. In addition, loss of growth hormone together with IGF-I and decrease of sex hormones are clearly linked to increased risk of osteoporosis. This is accompanied by a loss of vitamin D in the serum which is an additional unfavourable factor for osteoporosis.

## 2. CONCLUSIONS

All these examples demonstrate that age – dependent deterioration of one supersystem can affect another supersystem far away from the primary site of age-related loss of function. Thus, understanding the many interactions between different neuroendocrine immune pathways will also lead to a better understanding of the ageing process.

REFERENCES

1.  Cheleuitte D, Mizuno S, Glowacki J. In vitro secretion of cytokines by human bone marrow: effects of age and estrogen status. J Clin Endocrinol Metab 1998;83:2043–2051.
2.  Ershler WB, Sun WH, Binkley N, Gravenstein S, Volk MJ, Kamoske G, Klopp RG, Roecker EB, Daynes RA, Weindruch R. Interleukin-6 and aging: blood levels and mononuclear cell production increase with advancing age and in vitro production is modifiable by dietary restriction. Lymphokine Cytokine Res 1993;12:225–230.
3.  Hager K, Machein U, Krieger S, Platt D, Seefried G, Bauer J. Interleukin-6 and selected plasma proteins in healthy persons of different ages. Neurobiol Aging 1994;15:771–772.
4.  Mendall MA, Patel P, Asante M, Ballam L, Morris J, Strachan DP, Camm AJ, Northfield TC. Relation of serum cytokine concentrations to cardiovascular risk factors and coronary heart disease. Heart 1997;78:273–277.
5.  Straub RH, Konecna L, Hrach S, Rothe G, Kreutz M, Schölmerich J, Falk W, Lang B. Serum dehydroepiandrosterone (DHEA) and DHEA sulfate are negatively correlated with serum interleukin-6 (IL-6), and DHEA inhibits IL-6 secretion from mononuclear cells in man in vitro: possible link between endocrinosenescence and immunosenescence. J Clin Endocrinol Metab 1998;83:2012–2017.
6.  Bruunsgaard H, Pedersen BK. Age-related inflammatory cytokines and disease. Immunol Allergy Clin North Am 2003;23:15–39.
7.  Kanda N, Tsuchida T, Tamaki K. Testosterone inhibits immunoglobulin production by human peripheral blood mononuclear cells. Clin Exp Immunol 1996;106:410–415.
8.  Ralston SH, Russell RG, Gowen M. Estrogen inhibits release of tumor necrosis factor from peripheral blood mononuclear cells in postmenopausal women. J Bone Miner Res 1990;5:983–988.
9.  Daynes RA, Araneo BA, Ershler WB, Maloney C, Li GZ, Ryu SY. Altered regulation of IL-6 production with normal aging. Possible linkage to the age-associated decline in dehydroepiandrosterone and its sulfated derivative. J Immunol 1993;150:5219–5230.
10. Padgett DA, Loria RM. Endocrine regulation of murine macrophage function: effects of dehydroepiandrosterone, androstenediol, and androstenetriol. J Neuroimmunol 1998;84: 61–68.
11. Poynter ME, Daynes RA. Peroxisome proliferator-activated receptor alpha activation modulates cellular redox status, represses nuclear factor-kappaB signaling, and reduces inflammatory cytokine production in aging. J Biol Chem 1998;273:32833–32841.
12. Schmidt M, Kreutz M, Löffler G, Schölmerich J, Straub RH. Conversion of dehydroepiandrosterone to downstream steroid hormones in macrophages. J Endocrinol 2000;164:161–169.
13. Zietz B, Hrach S, Schölmerich J, Straub RH. Differential age-related changes of hypothalamus – pituitary – adrenal axis hormones in healthy women and men – Role of interleukin 6. Exp Clin Endocrinol Diabetes 2001;109:1–9.

*The Neuroendocrine Immune Network in Ageing*
Edited by R.H. Straub and E. Mocchegiani

# Plasticity of Neuroendocrine-thymus Interactions During Ontogeny and Ageing: Role of Zinc

EUGENIO MOCCHEGIANI, ROBERTINA GIACCONI, ELISA MUTI, MARIO MUZZIOLI
and CATIA CIPRIANO

*Immunology Ctr. (Section Nutrition, Immunity and Ageing) Res. Dept. INRCA, Ancona, Italy*

ABSTRACT

Thymic re-growth and reactivation of thymic functions may be achieved in old animals by different endocrinological or nutritional manipulations such as, (a) treatment with melatonin, (b) implantation of a growth hormone (GH) secreting tumour cell line (GH3 cells) or treatment with exogenous GH, (c) castration or treatment with exogenous luteinizing hormone-releasing hormone [LHRH], (d) treatment with exogenous thyroxin or triiodothyronine, and (e) nutritional interventions such as arginine or zinc supplementation. These data strongly suggest that thymic involution is a phenomenon secondary to age-related alterations in neuroendocrine-thymus interactions and that it is the disruption of these interactions in old age that is responsible for age-associated immune-neuroendocrine dysfunctions. The mechanisms involved in hormone-induced thymic reconstitution may be direct or indirect involving hormone receptors, cytokines and a trace element such as zinc, which is pivotal for the efficiency of neuroendocrine-immune network during the life-span of an organism. The effect of GH, thyroid hormones, and LHRH are due to specific hormone receptors on thymocytes and on thymic epithelial cells (TECs), which synthesize thymic peptides. Melatonin may also act through specific receptors on T-cells. In this context, the role of zinc, whose turnover is reduced in old age, is fundamental because of its involvement in zinc finger proteins that regulate gene expression for hormone receptors. However, the effects of zinc are multifaceted: it spans from the reactivation of zinc-dependent enzymes, to cell proliferation and apoptosis, to cytokines expression and to the reactivation of thymulin, which is a zinc-dependent thymic hormone required for intrathymic T-cell differentiation and maturation as well as for the homing of stem cells into the thymus. Therefore, the role of zinc is crucial in neuroendocrine-thymus interactions. According to current experimental data in animals and humans, the endocrinological manipulations for the maintenance of thymus function [e.g. by GH, thyroid hormones or melatonin] act via the zinc pool in restoring thymic activity and improving adaptive immunocompetence in ageing.

## 1.  INTRODUCTION

A large body of experimental evidence supports the existence of numerous interactions among the nervous, endocrine, and immune systems. These systems use similar ligands and receptors to establish a physiologic intra- and intersystem communication circuitry that plays an important role in homeostasis. The communications among these networks are mediated by hormones, neurotransmitters, and immune-derived cytokines, which are to a large extent shared by different homeostatic systems. Hormones and neurotransmitters, in addition to regulating various target tissues in the body, also reach lymphoid organs and cells through the circulation or directly through the autonomic nervous system (ANS) connections between the nervous tissue and the organs of the lymphoid system itself [1]. The neuroendocrine-immune interactions supported by circulating humoral mediators are mainly due to, and mediated by, the hypothalamus-pituitary- (HP) axis, which influences the immune system either by releasing various hormones and neuropeptides into the blood with direct modulator action on the immune effectors or by regulating the hormonal secretion of peripheral endocrine glands, which also exert immunomodulating actions [2]. In addition, neuroendocrine-immune interactions are based upon direct neuroimmune connections [3]. Anatomical studies have shown that the nerve endings of the sympathetic and the parasympathetic systems innervate various organs of the immune system such as thymus, spleen, bone marrow, and lymph nodes. Furthermore, ANS-related neurosubstances such as substance P, substance Y, vasoactive-intestinal peptide (VIP)], somatostatin, neurotensin, oxytocin, and vasopressin have been immunocytochemically identified in lymphoid organs [4].

The existence of signals generated within the immune system, capable of modulating various nervous-neuroendocrine functions, has been originally suggested by the alterations that can be induced in the neuroendocrine balance either by removal of relevant lymphoid organs, such as the thymus [5] or by the functioning of the immune system itself, such as reactions to immunogenic or tolerogenic doses of antigen [6]. The discovery that the majority of such effects could be mimicked by various immune-derived factors (e.g., thymic peptides and cytokines) has given support to those findings [7].

It has been found that lymphoid and accessory cells may, in given circumstances, and particularly following antigenic stimulation, synthesize and secrete neurohormonal factors, such as adrenocorticotropin (ACTH), growth hormone (GH), thyrotropin (TSH), prolactin (PRL), gonadotropins and beta-endorphin, which are likely to have an autocrine effect, as well as contribute to the neuroendocrine balance. This discovery has expanded on the number of humoral signals shared by the immune and the neuroendocrine systems [2,8].

Hormones, neurotransmitters, and cytokines may exert developmental actions related to the structural and functional organization of target organs or cells and play roles in the actual performance of mature cells, such as those required to counteract stressful conditions and antigenic insults. The factors involved in neuroendocrine-immune interactions are also responsible for developmental steps. We should, therefore, clearly distinguish them from central and peripheral effects taking also into account the complex autocrine and paracrine influences among various hormones, neurotransmitters and cytokines in the maintenance of neuroendocrine-immune pathway (Figure 1). The stimuli required may obviously differ, both quantitatively and qualitatively, according to the functional demands placed on the organism. In any case, the response to antigenic insults or stressful agents is closely dependent on the age of the organism, because the complex neuroendocrine-immune pathway must be "plastic" for its efficiency. In other words, the variations of the neuroendocrine-immune performances during the circadian cycle are pivotal in conferring the response to antigenic stimuli [9]. Such variations occur in young-adult age, but not in old age. As such, old individuals are "low responders" to stressor agents with the subsequent appearance of age-related degenerative diseases [9].

These considerations suggest two levels of neuroendocrine-immune interrelationships. The first is at central level, based on interactions between the neuroendocrine system and the thymus

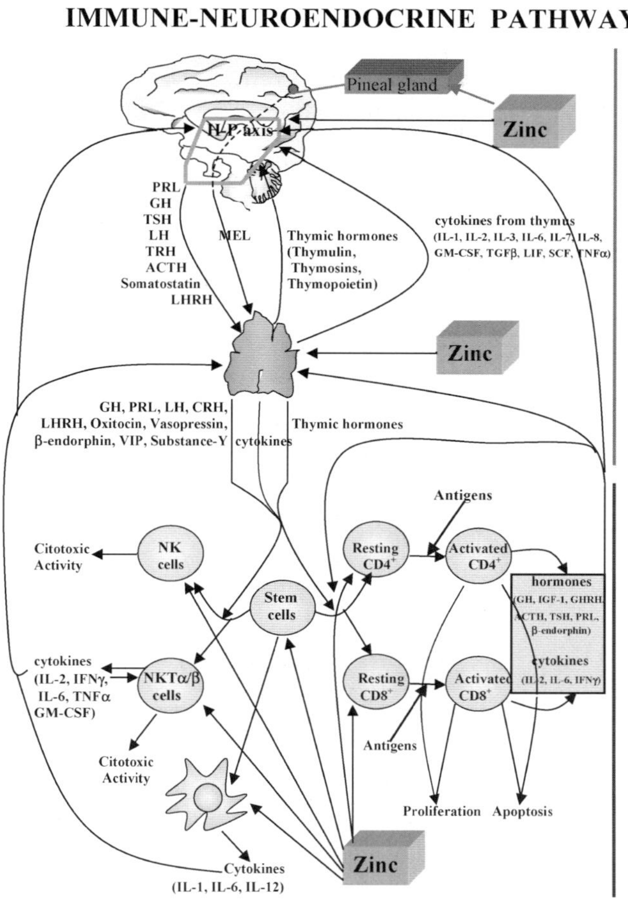

# IMMUNE-NEUROENDOCRINE PATHWAY

Figure 1. Schematic representation of immune-neuroendocrine pathway.

The first level is the *central level*, which consists of the interactions of the neuroendocrine system and the thymus. The second level of interaction is at the *periphery* via neuroendocrine signals, hormones and cytokines, which are secreted by immune cells during specific reactions to various antigens. Cytokines assume the role of feedback towards the central level (e.g. the thymus and the hypothalamus-pituitary axis). Zinc is involved both at central and peripheral levels of immune-neuroendocrine interactions. Please see the text for more details.

(Figure 1), a gland that induces proliferation and differentiation of stem cells into mature T-lymphocytes. Such interactions should take into account the fact that the thymus synthesizes and secretes various hormone-like peptides as well as cytokines, which regulate T cell growth and differentiation in this organ [10]. The second level of interaction is at the periphery (Figure 1), via the neuroendocrine signals, humoral products and cytokines, which are secreted by immune cells during specific reactions to various antigens. These in turn, provide feedback signals towards the central organs (e.g. the thymus and the HPA axis) (ThHP axis) (Figure 1) [2,6,8].

In any case, it is necessary to distinguish two sublevels both at central and peripheral levels because the periphery is closely associated with the presence of antigens and the CNS is affected by psychochemical stressful agents.

310

IMMUNE-NEUROENDOCRINE PATHWAY IN PERMANENT STRESS

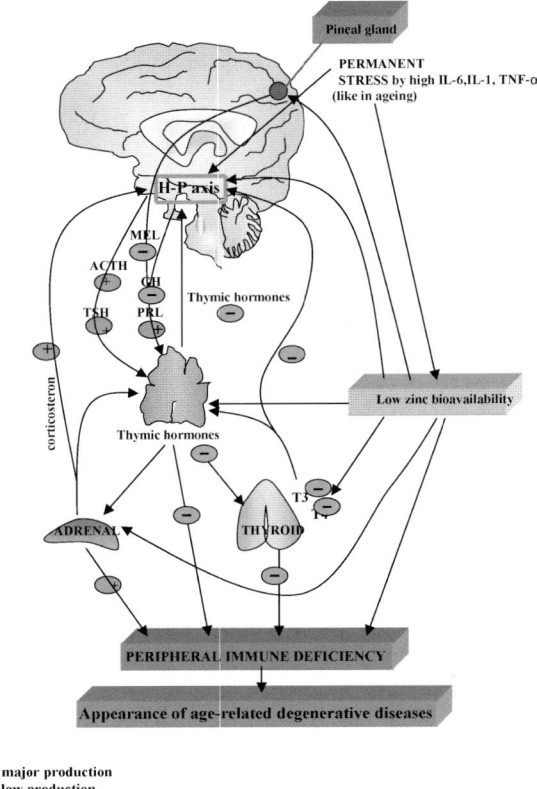

Figure 2. Schematic representation of immune neuroendocrine interactions in permanent stress-like conditions, such as in ageing.

Abnormal production of pro-inflammatory cytokines (IL-1, IL-6 and TNF-alpha) leads to an altered hormonal cascade from the hypothalamus to the peripheral glands. Concomitantly, permanent stress provokes low zinc ion bioavailability for thymic and peripheral immune function This altered immune neuroendocrine interaction in permanent stress leads to immunodeficiency and to the subsequent appearance of age-related diseases.

The rationale for discriminating these two sublevels is based on various parameters. The first sublevel of interactions primarily involves maturation steps of both the immune and neuroendocrine systems, which occur in the absence of pathological events and independently of antigenic stimulation (i.e., under germ-free-conditions). This sublevel of interaction can be considered as *"strategic" circuit*. The second sublevel of interaction requires the presence of fully differentiated immune cells and the occurrence of a specific antigenic or stress-mediated hormonal stimulus. The main role played by these interactions appears to be a return to the normal, homeostatic condition of the neuroendocrine and immune systems following a sudden alteration caused by a stressful cognitive or non-cognitive event [8]. This is, a *"tactical" or "emergency" circuit*. This latter occurs in the presence of transient stimuli, but it is peculiar to ageing where the stress-like condition is permanent. Indeed, during ageing the *"tactical" circuit* is under continuous stimulation by antigens or stressor agents leading to an imbalance at both levels of interaction with final results of altered or diminished functions of the immune-neuroendocrine pathway and subsequent risk of the appearance of age-related pathological conditions (Figure 2) [9]. However, the role of these two sublevels of interactions may be different and closely related to the target and to the kinetics of immune and hormonal mediators. In other words, a *"long wave" action* for the first level may be suggested due by long-acting fluctuations of neurohormonal or thymic peptide turnover; while *"short wave" action* for the second level due to humoral mediators with short-term effects. [1].

In this context, nutritional factors play a key role because they are involved in various steps of neuroendocrine-immune interactions both at central and peripheral levels. They also play a role in transient stress or in permanent stress like-conditions. One of the most relevant nutritional factors is zinc, a trace element, i.e., which has an influence either directly or indirectly via zinc-finger motifs on the biological activity to neurohormones, cytokines, thymic hormones and on T-cell proliferation and maturation [11,12] (Figure 1).

The present article aims to summarize the data outlining the hormonal and nutritional factors that modulate the *strategic circuit* of the adaptive immune system under physiological conditions, during ontogeny and during ageing. In addition, we describe the role of low zinc ion bioavailability during ageing and its influence on the neuroendocrine and immune systems. (Figure 2).

## 2. NEUROENDOCRINE-THYMUS INTERACTIONS: THE STRATEGIC CIRCUIT IN ONTOGENY

The *strategic circuit* of the neuroendocrine-immune interactions includes central lymphoid organs, in which the thymus gland is prominent, and the production and release of various cytokines produced by the thymus itself and peripheral mature T-cells with mechanisms of feedback to the HP axis and to T-cells themselves. In turn, HP axis affects thymus function through the production and release of hormones, releasing hormonal factors and neuropeptides. This complex regulatory cycle between the neuroendocrine and immune systems appears to be indispensable for many homeostatic mechanisms during ontogeny [2].

Such a cycle is multifaceted and includes various aspects: from intrathymic T-cell differentiation and maturation, to cytokine production and, finally, the neuroendocrine influence on thymic functions.

### 2.1. Intrathymic T-cell differentiation and maturation

In the thymus bone marrow-derived T cell precursors undergo a complex process of maturation, eventually leading to migration of positively selected thymus-derived (T) lymphocyte dependent areas of peripheral lymphoid organs like spleen, lymph nodes, Payer's patches and tonsils [13]. Therefore, the thymus is involved in the differentiation and maturation of T-cells with effector (cytotoxic) and regulatory (helper/suppressor) activities. Amongst other effects, this differentiation involves regulation of membrane protein expression. A key cell membrane protein is the T cell antigen receptor (TCR) that is physiologically coupled with a membrane bound molecular complex, termed CD3. Additional accessory molecules, including CD4 and CD8, as well as CD25 and the proteoglycan, CD44, all of which are pivotal to defining various stages of intrathymic T cell differentiation and maturation. The TCR is a heterodimer formed by an $\alpha/\beta$- or a $\gamma/\delta$- chains. $\gamma\delta^+$ thymocytes appear in the early stage of differentiation. By contrast, in adult thymus, , 99% of TCR+ thymocytes express TCRαβ and only 1% express TCRγδT [14]. With regard to accessory molecules, in the first steps of differentiation immature thymocytes are CD3-CD4-CD8-CD25-CD44+, which become subsequently CD25+CD44+. During maturation, the thymocytes lose CD25 and CD44 and become CD4+CD8+, the so-called *double-positive thymocytes*. These cells comprise 80% of total thymocytes and express the rearranged TCRαβ, which in turn reacts with peptides presented by MHC molecules on the surface of thymic antigen presenting (APC) cells. This interaction is fundamental to positive and negative selection events that are crucial for thymocytes differentiation and maturation. The positive selection leads to mature CD4+ or CD8+ *single positive thymocytes*, which are long-lived. These represent about 15% of thymocytes, express TCR/CD3 complex and they leave the thymus for the periphery. Negative selection involves the deletion of T cells that might potentially be autoreactive to self-proteins and self-tolerance is established this way in the T-cell repertoire. Negatively selected cells die by apoptosis [14,15].

The interaction between the TCR/CD3 complex and of class I or class II MHC products on the membranes of thymic APC in the context of CD8 or CD4 accessory molecules respectively, is fundamental for positive *vs.* negative selection. Thymocytes with high avidity are negatively selected and are deleted by apoptosis. This leads to the death of a large number of potentially harmful autoreactive T-cells. By contrast, a small percent of thymocytes with intermediate avidity for recognition of MHC-self peptides are rescued from death and the are positively selected [15].

Thymic epithelial cells (TECs) are responsible for intrathymic T-cell differentiation and maturation and for the subsequent positive selection. The negative selection is executed primarily by dendritic cells (DCs), which derive from the haematopoietic-system, but also by TECs [13]. TECs communicate with each other and with thymocytes by gap junctions (formed by proteins of connexin family) [16] and by ECM ligands, such as fibrinectin and lamin, and their corresponding integrin receptors VLA-4/VLA-5 and VLA-6 [10]. TECs provide the stimulus for the migration of mature thymocytes within the thymus as well as into the circulation through the production of adhesion molecules (LFA-1, ICAM-1, LFA-3, CD2) [17]. In addition, both TECs and DCs produce cytokines that affect thymocyte differentiation and proliferation which also serve as immunotransmitters towards the HP axis [18] (Figure 1) (see below). This fact leads to the notion that several paracrine/endocrine circuits involving TEC-derived factors are likely to have differentiating thymocytes and the HP axis as targets. In addition to cytokines production, TECs synthesize and secrete defined thymic hormones, including thymosin alpha-1, thymulin, and thymopoitein that also act upon the general process of thymocytes maturation [19]. Immunohistochemical studies using polyclonal and monoclonal antibodies have demonstrated the presence of thymulin, thymosin alpha-1 and thymopoietin both in murine and human thymic epithelial cells [2]. The role of the thymic peptides on intrathymic or extrathymic maturation of T-cells is well defined as well as on the chemotactic attraction (homing) of stem cells from the bone marrow into the thymus [20].

## 2.2. Paracrine cycle of cytokines on thymic microenviromental cells

As cited above, thymic micro-environmental cells can influence thymocytes differentiation and proliferation through soluble polypeptides (e.g. cytokines and chemokines). Both TECs and DCs produce IL-1, which stimulates thymocyte proliferation. Actually, various cytokines are produced by thymic epithelium, including IL-2, IL-3, IL-6, IL-7, IL-8, GM-CSF, TGF-beta, leukaemia inhibitory factor (LIF), stem cell factor (SCF, also termed as c-kit ligand) and TNF-alpha

[21]. The production and the release of all these cytokines into the thymic microenvironment are of strong relevance because the task of some cytokines (e,g IL-1, IL-2, IL-6) is to induce proliferation, while others provoke inhibition (TFG-beta) or apoptosis (TNF-alpha) or homing (GM-CSF). Therefore, the large number of cytokines produced by the thymus maintains a balance between thymocyte proliferation and inhibition. As such, the number of thymocytes does not expand during ontogeny and thus assure a correct functionality of the thymus in T-cell differentiation, maturation and proliferation. Such an assumption is supported, for instance, by the fact that an over-expression of some cytokines (IL-6) with respect to others (TGF-beta) in young IL-6 transgenic mice leads to abnormal thymic enlargement and thymic dysfunction [21].

IL-7 plays a peculiar role for thymocytes differentiation. IL-7 promotes the rearrangement of TCR genes by enhancing both the production and activity of recombinases [22]. In conjunction, IL-7-/- as well as IL-7 receptor-deficient mice display a severe reduction in lymphoid development, whereas IL-7 transgene incorporation in nude mice induces T-cell development [23]. A relevant discovery is the influence of IL-7 in growth and differentiation of thymic $\gamma\delta$T cells, as shown in IL-7-/- mice. This fact is peculiar during ontogeny when $\gamma\delta$T cells are abundant in the thymus because of their requirement in TCR rearrangement and, subsequently, in thymocyte positive selection [24].

In addition to classical cytokines, chemokines are also present in the thymic microenvironment and are important in thymus physiology. Chemokines correspond to a family of small polypeptide molecules that control directional migration of leukocytes [25]. In this context, one chemokine should be noted, stromal cell-derived factor (SDF),which is produced by thymic stromal cells that attract preferentially immature CD4-CD8- and CD4+CD8+ thymocytes. Conversely, another chemokine, MIP3beta, exerts chemoattraction for mature single positive thymocytes [25]. This is in keeping with the different expression of corresponding chemokine receptors in distinct CD4/CD8-defined stages of thymocyte differentiation.

All these findings clearly demonstrate the existence of paracrine cycles among cytokines and chemokines that are produced by TECs and thymocytes. In this context, a relevant question arises: do autocrine cycles also exist? In other words, do cytokines produced by TECs affect TECs? Cytokines receptors do not seem to exist on TECs, except for IL-7 and IL-2 (CD25) [26]. However, an intriguing point in this respect is TEC-thymocytes connection by connexin 43 (gap junction), which might substitute for autocrine regulation [10]. Therefore, cytokines may affect TEC proliferation and efficiency through gap junctions, which amounts to the presence of autocrine cycles. These points remain to be clarified by further investigations.

## 2.3. Cytokines as immunotransmitters towards the HP axis

The most likely messengers from the immune system to the nervous system are the cytokines, which, in addition to affecting intrathymic T-cell development and peripheral immune function, serve also as feedback regulators of the HP axis [18]. Therefore, the cytokines produced within the thymus may affect the central and peripheral nervous systems [10]. In this case, cytokines produced by TECs are called immunotransmitters.

A large number of cytokines has been identified. Those currently known to have the most relevance for the nervous system are IL-1, IL-2, IL-6, TNF-$\alpha$ and IFN-$\gamma$ with different role of stimulus or inhibition on hormones and neuropeptides derived from HP axis (Table I). IL-1 acts at hypothalamic, pituitary and adrenal levels, and affect, respectively, corticotropin-releasing factor (CRF), adrenocorticotropin (ACTH) and cortisol production [6]. At the pituitary level, the old literature reports no effects of IL-1 on pituitary hormones releases [27]. There is however

Table I     The effect of cytokines upon hypothalamus-pituitary hormones.

| Cytokines of TECs | ACTH | GH | PRL | MEL | FSH | LH | LHRH | TSH | CRH |
|---|---|---|---|---|---|---|---|---|---|
| Il-1 | + | + | − | + | + | + | + | + | + |
| IL-2 | + | − | + | − | − | − | − | + | + |
| IL-6 | + | + | + | + | + | + | NE | ND | + |
| TNF-α | + | + | + | + | ND | ND | NE | + | + |

| Cytokines of peripheral T-cells | | | | | | | | | |
|---|---|---|---|---|---|---|---|---|---|
| IFN-γ | − | − | − | + | ND | ND | ND | ND | ND |

+ = increase
− = decrease
NE = No Effect; ND = Not Done.

a general consensus that IL-1 stimulates GH, MEL, LH, LHRH, FSH, and TSH secretion and inhibits PRL release [28,29].

IL-1 also increases ACTH secretion, as shown in AtT20 cells [30] and human corticotrophic pituitary adenomas (Table I). Other inflammatory cytokines, such as IL-6 and TNF-α, share the effects of IL-1. They act in stimulating CRH production by the hypothalamus [31]. At pituitary level, IL-6 and TNF-α stimulate the release of ACTH, PRL, GH as well as LH, FSH, TSH release, respectively [see reviews 2,10]. (Table I). Such stimulating effects can be exerted through a direct action on pituitary cells via specific hormone receptors or mediated by hypotalamic factors (e.g. CRF, LHRH or GHRH) [2]. For example, this latter case occurs for the release of pituitary hormones by TNF-α [32]. With regard to melatonin, IL-6 and TNF-α are able to stimulate directly MEL production because specific receptors for both cytokines are present into the pineal gland [33].

IL-2 also acts on the HP axis. When administered to human cancer patients, IL-2 increased beta-endorphin, ACTH, and cortisol levels [34]. In rats, IL-2 enhanced ACTH secretion and corticosterone production through a direct action on rat adrenocortical cells or via CRH release from hypothalamic neurons [35]. Interesting investigations have shown that IL-2 stimulated PRL and TSH secretion and inhibited LH, FSH and GH release, thus mimicking the alterations of pituitary hormone pattern in response to stress [36]. IL-2 enhanced ACTH release in AtT20 cells and PRL secretion in GH3 cells [37]. Conversely, IL-2 and IL-12 injections in cancer patients inhibited melatonin release as a response to inflammation [33].

Although IFN-γ does not seem to be produced by TECs in the thymus, IFN-γ produced by peripheral T-cells, and it also affects the HP axis with an inhibition on GH, PRL and ACTH release, and with a possible involvement of the nitric oxide in this inhibition [38]. By contrast, IFN-γ enhanced melatonin production in the pineal gland in response to stressor stimuli, i.e. isoproterenol [39]. All these findings clearly demonstrate that the cytokines produced either within the thymus or at the periphery are fundamental to maintaining the complex neuroendocrine-immune network. The role of this network is crucial in host response to stressor agents where the level of cytokines increase. As a consequence, counterbalancing must occur from other cytokines. Otherwise, the organism may become a *"low responder" to stress*. Thus, the appearance of degenerative diseases (e.g. cancer, infections, dementia, autoimmune phenomena) may eventually correlate with dysfunctions of the neuroendocrine immune network.

## 2.4. Neuroendocrine influence on thymic functions and vice versa

### 2.4.1. Neuroendocrine influence on thymic functions

Initial evidence for the existence of thymus-neuroendocrine interactions was based on the discovery that congenital mutations affecting pituitary dwarf mice caused concomitant under-development of the thymus and of the thymus dependent systems. These findings have been confirmed in a strain of dwarf dogs (Weimaraner dogs), which display retarded growth, small thymus, absence of the thymus cortex, and deficiency in the lymphocyte mitogenic response. All these immunological defects can be corrected by growth hormone treatment (see review 1).

A number of experimental protocols have further supported these observations. The major-ity of them have been based on the removal of endocrine glands and on the observation of the consequent modification of thymic function, as measured by thymic size, histological evaluation or, indirectly, by the peripheral efficiency of the thymus-dependent lymphoid system [5]. Some thymic factors are secreted into the blood stream and the circulating level of at least one of them, the facteur thymique serique (FTS), now called thymulin in its zinc-bound form [40], strictly reflects the functional activity of the thymus. This has offered a new technical approach to evaluate neuroendocrine-thymus interactions both in animals and in humans. Furthermore, the detection of hormone and neuropeptide receptors both on thymocytes and on thymic epithelial cells has added support to the existence of neuroendocrine-thymus interactions. From all these studies, summarized in various reviews [2,10], the outline of Figure 1 may be presented.

In general, hormones/neurotransmitters, such as melatonin, LHRH, GH, thyroid hormones, beta-adrenergic agonists, and endorphins/enkephalins act positively on thymic function, whereas others, such as corticosteroids and sex hormones, have a negative effect. The role of PRL is still controversial. The effect of hormones/neurotransmitters on thymic cell maturation may be medi-ated by a direct action on thymocytes or through the action exerted on thymic epithelial cells. Evidence supports the presence of receptors for hormones, such as GH, estrogens, testosterone and adrenergic agonists, on thymocytes. In addition, it has been proposed that other pituitary/CNS products may regulate the differentiation of mature T-cells [1].

With regard to the endocrine component of the thymus, it is now apparent that many hor-mones and neuropeptides are capable of modulating thymulin secretion by TECs. Among neuropeptides, Leu-enkephalin and beta-endorphin are able to increase thymulin production by TECs, whereas Met-enkephalin, alpha-endorphin, and gamma-endorphin are inactive [10]. Hormones like GH, PRL, adrenal and sex steroids, and thyroid hormones can modulate thymu-lin production *in vitro*. This *in vitro* effect clearly supports the idea that TEC possesses specific receptors for these hormones and neuropeptides. At present, experimental demonstration of receptors on TEC has been reached for glucocorticoids, progesterone, GH, PRL, and $T_3$ [2]. In addition, apoptosis of lymphoid cells may be influenced by various agents (e.g. glucocorticoids). Glucocorticoids are potent apoptosis-inducing agents at pharmacological doses on immature thymocytes, natural killer (NK) cells, and on cytotoxic T-lymphocytes [41]. Glucocorticoids also induce apoptosis at physiological concentrations *in vitro* [42] or after *in vivo* activation of the pituitary-adrenal axis [43]. Both *in vivo* and *in vitro* data support the notion that thymic func-tion is regulated by the complex neuroendocrine network, which may exert both stimulatory and inhibitory actions (Figure 1).

## 2.4.2. Thymic influence on the neuroendocine sytem

As shown in Figure 1, thymic hormones also modulate the production of hypothalamus pituitary hormones and neuropeptides. Initial experiments revealed that neonatal thymectomy promotes a decrease in the number of secretory granules in acidophic cells of the adenopituitary [44]. In the same vein, athymic nude mice display low levels of various pituitary hormones, such as PRL, GH, LH and FSH [45]. With regard to thymic peptides, thymosin beta-4, when perfused intraventricularly, stimulates LH and LHRH secretion [46]. Similar results were obtained with another thymic peptide, thymulin, in perfused or fragmented pituitary preparations [47]. The administration of thymopoietin (another chemically-defined thymic hormone) in children with Hodgkin's disease increased GH and cortisol serum levels [48]. Moreover, thymopentin (the synthetic biologically active peptide of thymopoietin) enhances *in vitro* the production of ACTH and beta-endorphin [49]. In addition, thymulin exhibits an *in vitro* stimulatory effect on perfused rat pituitaries, enhancing PRL, GH, TSH and LH release [50]. Using short-term cultures of pituitary fragments, an increase in ACTH secretion occurs after *in vitro* thymulin addition, with no changes in GH levels and significant reductions in PRL release [47]. A further thymosin peptide was recently isolated with the task in stimulating IL-6 release from rat glioma cells [51]. By contrast, thymosin alpha-1 is apparently able to down regulate TSH, ACTH and PRL secretion *in vivo* with no modifications on GH levels [52]. These inhibitory effects seem to occur through hypothalamic pathways. Indeed, the production of the corresponding releasing hormones by hypothalamic neurones decreased after *in vitro* addition of thymosin alpha-1 in medial basal hypothalamic fragments [52].

Altogether, these findings point to the complexity of the hypothalamus-pituitary axis as it affects thymic efficiency and vice versa, and reinforce the existence of close relationships between the thymus and the neuroendocrine system [1].

## 3.    THE ROLE OF ZINC

Zinc is an important nutritional factor affecting thymic physiology (Figure 1). Zinc is required as a catalytic component for more than 200 enzymes and it is a structural constituent of many proteins, hormones, neuropeptides, and hormone receptors. Its role in cell division and differentiation, apoptosis, gene transcription, biomembrane function, and many enzymatic activities has led to the consideration that zinc is a crucial element that ensures the correct functioning of body homeostatic mechanisms throughout ontogeny. The role played by zinc in the neuroendocrine-thymus interactions has been elucidated [11,12].

Studies performed in zinc-deprived animals indicated that at neuroendocrine level zinc influences the blood level of various hormones and neurotransmitters. In dietary zinc-deprived animals, insulin, GH, TSH, thyroid hormones, gonadotropins, testosterone, and VIP, are reduced, whereas PRL, cortisol, corticosterone, and catecholamines increase [53,54]. In young-adult humans affected by more or less severe zinc deficiency (e.g. in alcoholism, anorexia nervosa, acrodermatitis enteropathica, HIV, burns, cystic fibrosis, mental depression and dwarfism), testosterone and estrogens are reduced, whereas cortisol and PRL serum levels increase [55]. According to findings obtained in Down's syndrome subjects, zinc is also involved in the pituitary-thyroid axis. Zinc deficiency increased TSH production and reduced reverse triiodothyronine (rT3) plasma levels [55].

With regard to immunocompetence, the data obtained in experimental animals support a crucial role played of zinc. Depending on the severity and duration, zinc deficiency produces hypoplasia of the thymus, spleen, lymph nodes, and Payer's patches. Alterations in lymphocyte populations, including a decreased number of total T-lymphocytes with increased B-lymphocytes, impaired T-helper and T suppressor function and reduced NK cell activity, are common aspects in zinc deficiency [see review 11]. An imbalance of the Th1/Th2/Th3 paradigm also occurs in zinc deficiency with major production and release of Th2 cytokines (e.g. IL-6) and subsequent low resistance to infections [12,56]. A peculiar role of zinc is also exerted on thymic functions. Zinc in an equimolar ratio is required to confer biological activity to a thymic peptide, thymulin [40]. The zinc-bound thymulin is active, whereas the zinc-free form is inactive and prevents the active form in exerting its action [57]. Zinc also influences apoptosis of T-cells induced by high levels of glucocorticoids [58] . The thymic atrophy observed in some pathological conditions is characterized by severe zinc deficiency (e.g. in AIDS patients and in young mice affected by cancer) and is the result of excessive thymocyte apoptosis [59]. Indeed, the chelation of intracellular zinc triggers apoptosis in mature thymocytes [60]. *In vitro* findings have shown that pharmacological zinc concentrations cause an inhibition of the apoptosis induced by dexamethasone, whereas physiological zinc concentrations may induce thymocytes apoptosis. However, this latter point has been well established to be closely dependent on the length of culture (16 hrs), as physiological doses of zinc prevents thymocyte apoptosis at 6h of culture [61]. On the other hand, thymocytes cultivated for a long-time are destined to die due to faulty rearrangement of T-cell receptor genes and due to the induction of apoptosis by endogenous glucocorticoids [61]. The abrogation of murine thymocyte apoptosis in deferiprone (a chelator of iron) treated mice after physiological zinc supplementation [62] strongly supports a role of physiological zinc in preventing apoptosis, via endonuclease enzyme activation [11,63].

Alteration of the zinc pool may occur as a consequence of modifications in intake, absorption, or loss by urine and faeces, as observable in many human diseases [55]. In addition, neurohormonal factors and immune-derived cytokines may influence zinc turnover. Stress-like conditions alter zinc metabolism. Hypozincemia, along with hyperzincuria, is associated with trauma, stress, inflammation, tumours, and burn injury. The hormonal basis of these changes is not yet well defined. It has been shown that high circulating levels of glucocorticoids, epinephrine, and glucagon occur during inflammation coupled with high IL-1, and it is well known that IL-1 provokes consistent zinc loss by urine and faeces [56]. This zinc deficiency may be largely due to abnormal increments at intestinal level of a protein that binds zinc with high affinity. This protein is the cystein-rich intestinal protein (CRIP), which increases in the presence of high Th2 cytokines expression (IL-1, IL-6). As a consequence, a loss of zinc-bound CRIP occurs from the body [64].

Other hormones can modulate zinc turnover. In particular, GH, IGF-I, melatonin and thyroid hormones affect zinc pool with a better redistribution in circulating zinc [11,12]. The mechanisms of action of these hormones may be various. The presence in the intestinal epithelium of hormone specific receptors that require zinc-finger motifs and the influence of some hormones in the induction of zinc-bound metallothioneins, which sequester, dispense and transport zinc ions within the cells (see review 10), may be the mechanisms of a better redistribution of zinc by hormones. In any cases, zinc plays a peculiar role in neuroendocrine-immune interactions, because it is involved as *"zinc fingers" proteins* that regulate gene expression of hormones, neurotransmitters, and transcriptional factors. Moreover, zinc is involved directly or indirectly via metallothionein turnover in the proliferation and maturation of cells of the immune system [11]. Such a role is crucial in ageing because the low zinc ion bioavailability is a common event during ageing as is the impairment of the neuroendocrine-immune interactions [1].

## 4.     NEUROENDOCRINE-THYMUS INTERACTIONS DURING AGEING

The existence of extremely complex micro-environmental factors modulating thymic functions has raised the question, whether age-related thymic involution is the cause or the result of immune ageing [1]. In aged humans and animals, the thymus gland undergoes progressive involution with accompanying loss of function. The thymus attains its maximum size at puberty, after which it involutes, which is characterized by a decrease in weight ([loss of lymphocytes) and infiltration with fat. There is a decrease in the number of thymocytes and of those that remain many are pyknotic. As ageing progresses, epithelial cells decrease in number, and show cystic changes and reduced intracellular granules [65]. The circulating level of thymic factors (i.e., plasma levels of thymulin) declines progressively from birth to old age and is virtually undetectable over 60 years of age in humans [11,46,57].

There are two lines of evidence suggesting that the age-related thymic alterations may not be considered as intrinsic and irreversible phenomena. First of all, according to a more precise determination of thymulin bioassay, which assessed the interference due to the marginal zinc deficiency present with advancing age [57], the age-related decline of thymulin was less pronounced than that previously reported, showing a still significant production in old age both in mice and men [57,66]. Therefore, thymulin molecules are still produced in old age, but they are under the control of zinc ion bioavailability for their biological activity. The *in vitro* addition of zinc to plasma samples from old donors unmasks all thymulin molecules produced, and it is consistently at normal levels [66]. Therefore biologically active thymulin is Produced. Indeed, when thymic explants from old mice are cultured *in vitro* in a zinc-enriched medium, thymulin secretion in the supernatant is similar to that observed in the supernatant of thymic explants from young-adult mice. This phenomenon is associated with increased numbers of TECs and thymocyte subsets [67]. Thus, although the thymus is involuted in old age, these findings suggest that the thymic endocrine activity is not completely lost in ageing, but it is in a quiescent phase or in a possible dormant condition due to low zinc ion bio-availability within the old thymus [11].

The second line of evidence is that the thymus re-growth can be induced in old age by various endocrinological and nutritional manipulations. In particular, the partial reacquisition of the thymic functions can be induced by some procedures shown in Table II. These are: (a) intrathymic transplantation of pineal glands or treatment with melatonin [69-70], (b) implantation of growth hormone secreting tumour cell lines (GH3 cells) [71] or treatment with exogenous growth hormone [72,73] or with IGF-I [74], (c) castration [75], (d) treatment with exogenous LH-RH [76], (e) treatment with exogenous thyroxin or triiodothyronine [77], (f) nutritional interventions with arginine [78], and (g) zinc supplementation [54,66,79].

In old animals, treatments with GH [80], arginine [78,81], and zinc [66,79] are effective in enhancing thymulin plasma concentrations and the number of TECs. Recently, it has been also shown that IL-7 is able to reconstitute the output of the thymus in old animals with a direct mechanism through the activation of the receptor CCR9 (a receptor of the chemokine CCL25) highly expressed in thymic tissue [R. Aspinall, personal communication].

The potential capacity of thyroid hormones in restoring thymic functions in old age is supported by experiments in young propyl-thiouracil (PTU) treated mice (experimental hypothyroidism), which display thymic atrophy, reduced thymulin and decreased number of TECs and thymocyte subsets [82]. Thyroid hormone treatment restores thymic functions [77]. Indirect evidences in the elderly show that hyperthyroidism is associated with thymic enlargement [83] and high circulating levels of thymulin as compared to young normal individuals [84]. These findings suggest an influence of endocrine and nutritional factors on thymic functions. However,

Table II     The reversibility of age-related thymus involution by endocrinological or nutritional interventions.

| Treatment | Species | Thymus weight | Thymic hormone[a] | References |
|-----------|---------|---------------|-------------------|------------|
| Melatonin | mice | 70% | 75% | 68, 70 |
| TRH | mice | 40% | – | 69 |
| TSH | mice | 32% | – | 69 |
| Thyroxine | mice | 65% | 72% | 77 |
| Growth horm. | rat | 50% | – | 72 |
| Growth horm. | dog | 50% | 52% | 73 |
| Growth horm. | mice | 72% | 75% | 80 |
| IGF-I | mice | 50% | 50% | Mocchegiani et al[b] |
| Arginine | mice | 79% | 70% | 78 |
| Arginine | man | – | 50% | 94 |
| Castration | mice | 70% | 75% | 75 |
| Anal. LHRH | rat | 62% | – | 76 |
| Zinc salts | mice | 75% | 82% | 66 |
| Zinc salts | mice | 50% | 50% | 79 |
| Zinc salts | man | – | 45% | 87 |

[a]Percent of recovery when compared with "young" adult values.
[b]Manuscript in preparation.

it is possible that the thymic re-growth, obtained through various manipulations, might not represent a complete reacquisition of the function that normally occurs during ontogeny. In order to answer to this question, zinc, GH and melatonin models have been investigated.

## 5.     THYMIC REJUVENATION BY NUTRITIONAL APPROACH: THE EXAMPLE OF ZINC

Zinc plays a relevant role during the ageing process. With advancing age, marginal zinc deficiencies are frequently encountered in humans [54-56]. Altered intake or metabolism is among the most common causes. Deficiencies in the zinc pool are also observed in aged laboratory animals in spite of unmodified zinc intake [66]. This has raised the question whether zinc deficiency in old age may be the cause of age-related immune-neuroendocrine derangement. Dietary zinc supplementation during the whole life-span of rodents has demonstrated that many of the known age-related immune modifications (e.g., decreased thymic hormone production, reduced T-helper and T-suppressor activity, and depressed NK cytotoxicity) can be prevented. Moreover, oral zinc supplementation in aged mice induces thymus regrowth coupled with both increased thymic hormone production and percentage of TECs. Furthermore, zinc supplementation completely restores the reduced number of T-cell subsets (CD4+ and CD8+) both in the thymus and the spleen, reconstitutes the PHA response and restores NK splenocyte cytotoxic activity. Concomitantly, zinc supplementation increases DNA-repair within the cells. In addition, *in vitro* zinc prevents thymocytes apoptosis in old age [see review 11]. Zinc supplementation in old mice also reverses the low serum levels of thyroxine and triiodothyronine [66]. This beneficial effect of zinc supplementation on thyroid functions also occurs in young PTU mice that display thymic reconstitution and restoration of peripheral immune efficiency. Zinc supplementation in PTU mice also restores ACTH and down-regulates corticosterone plasma levels [82]. All these beneficial effect of zinc in neuroendocrine-immune interactions are not sterile events because significant increments in survival rates were also observed both in old and PTU mice after zinc supplementation with respect to respective untreated mice [11,82]. In humans, few zinc trials have been also performed in showing the effects on thymic endocrine activity and neuroendo-crine-immune interactions, however, the findings are contradictory. Some authors reported the beneficial effect of zinc supplementation on thymic endocrine activity in the elderly [54,85], whereas no effects were observed in other trials [86]. This discrepancy is related to the length of zinc supplementation. In our experience, one month of physiological zinc supplementation restores thymic and neuroendocrine-immune interactions in elderly and in Down's syndrome subjects ( a disease of accelerated ageing) allowing a lack of the immune risk factors (i.e. the loss of CD4+ cells) in the infection relapses [87]. Restored neuroendocrine-immune interactions (down-regulation in corticosterone and enhanced thymic endocrine activity) coupled with increased survival rate have been recently observed in zinc-treated AIDS patients in comparison with untreated ones [59].

## 6.    THYMIC REJUVENATION BY NEUROENDOCRINE APPROACH

### 6.1.    The example of growth hormone

One of the first experimental evidences of the existence of thymus-pituitary interactions was based on the discovery that, on one hand, congenital mutation affecting the pituitary dwarf mice caused concomitant alteration in the thymus and in the thymus-dependent system and, on the other hand, that removal of the thymus in neonatal age was followed by a wasting syndrome, one of its features being a pituitary dysfunction involving growth GH producing cells [88]. With regard to the influence of GH on thymic function, the findings obtained in dwarf mice have gained further support from similar observations in hypopituitary Weimaraner dogs [89] and in hypophysectomized rats [90], both these conditions being characterized by atrophy of the thymus gland and by peripheral immune deficiency. Recent evidences in growth hormone deficient children have shown thymic hypoplasia and low plasma levels of thymulin with respect to age-matched controls [91]. Appropriate GH treatment in dwarfism (animals and humans) corrects these defects [80,91]. With advancing age, GH secretion declines both in animals and in men coupled with impaired thymic and peripheral immune efficiency [94]. The implant of GH3 pituitary adenoma cells, which secrete GH and PRL, in aged rats induced a re-growth of the thymus with a young-like histological picture. At the same time, GH3 implanted cells increased peripheral T-cell functionality and IL-2 production, which were low in old rats [71]. GH treatment recovered thymulin activity and, in some extent, peripheral immune efficiency in old mice, rats [46] and dogs [73]. Such a recovery of thymulin activity by GH is also observed in *in vitro* models with increased number of TECs in old thymus cultures [44]. That GH affects thymus efficiency is indirectly supported in Buffalo rats, which display high GH secretion and an enlargement of the thymus in old age [93]. Indirect evidences also exist in humans. Old people treated with arginine, a compound with secretagogue action on GH production, display increased thymulin plasma levels and restored peripheral immune efficiency [94].

The mechanism of action of GH on the thymus can be direct or indirect. GH receptors have been found on TECs [95] and thymocytes. However, on the functional basis, the action of GH seems to be mediated though IGF-I not only on thymic functions but also on other immu-nomodulatory actions of GH (T-lymphocytes proliferation) due to a close relationship between GH and IFG-I [91,96], and IGF-I receptors are present on TECs [97]. An interesting aspect of the mechanism of GH on the thymus is that GH effects may be mediated by PRL. Indeed, GH may utilize PRL receptors, which in turn exist on TECs and PRL at the same time increases *in vitro* thymulin secretion [98]. However, in this context, a very intriguing possibility is the use of PRL receptors when GH is bound to zinc ions [99]. This discovery suggests that zinc ions are involved in the action of GH on thymic function via PRL receptors. On the other hand, both old mice and growth hormone deficient children, when treated with GH, display restored zinc pool [80]. This occurs because GH receptors are also present in the intestine [91,96]. These findings give clear-cut evidence indicating that GH exerts its immunomodulatory action via the zinc pool. IGF-I is also zinc dependent [100] and zinc has a role in thymic functions (as described above). Zinc turnover is pivotal for the action of GH on thymic function in ageing, independently by direct (receptors) or indirect mechanisms (IGF-I or PRL). The restoration of thymus-neuroendocrine interactions by GH treatment, that involves the zinc pool, is clinically useful in old age. This is indirectly shown in old cancer patients treated with arginine. Arginine treatment improves zinc ion bio-availability and immune efficiency. It blocks the development of metastasis in the long run with a subsequent better efficiency of chemotherapy in comparison with untreated cancer patients, which exhibit a limited efficiency of chemotherapy and succumb precocious death [94].

## 6.2. The example of melatonin

Melatonin affects the efficiency of the immune system at thymic and peripheral levels [69]. Melatonin secretion and production from the pineal gland decrease in ageing with no nocturnal peak during the normal light/dark circadian cycle with respect to young-adult age [101]. Melatonin treatment in old mice provokes a significant recovery of the thymus weight, thymocyte (CD4+ and CD8+) and TEC numbers and thymulin plasma levels [68]. Other parameters, i.e. the phases of the cell cycle (G0/G1, S, G2/M), are not affected by melatonin treatment, whereas melatonin prevents both *in vitro* and *in vivo* the apoptosis of old thymocytes [102]. However, an interesting aspect is related to apoptosis preservation by *in vitro* melatonin. Thymocytes from melatonin treated old mice are more resistant to apoptosis caused by serum deprivation, whereas they have an increased apoptotic response after incubation with dexamethasone (DEX) with a similarity to young mice [102]. These findings suggest that melatonin recovers the reduced sensitivity to apoptosis induced by DEX in old mice to the values of young mice. This means that melatonin may preserve old thymocytes that are useful for positive intrathymic selection and subsequent maturation through the modulation of pituitary-adrenal axis. Indeed, recent findings on the circadian cycle showed that the nocturnal peaks of melatonin in old melatonin treated mice are correlated with nocturnal down-regulation of corticosterone plasma levels and with nocturnal increments of thymulin and TECs number [70]. Therefore, melatonin has a double role: (i) melatonin acts as an anti-stress hormone reducing corticosterone; (ii) melatonin sensitises thymocytes for negative selection and for apoptosis and, at the same time, preserves apoptosis in thymocytes for positive selection and maturation. The re-growth of the thymus, the increments of thymocytes, the enhanced thymulin production and the recovery of peripheral immunocompetence and the increased survival rate in old melatonin treated mice [70], are clear indications that melatonin is pivotal for intrathymic selection and for peripheral immunocompetence.

The mechanisms by which *in vivo* melatonin treatment brings about its immunologic effect might be due to either its direct or indirect action on immune cells. The existence of melatonin binding sites in rat thymocytes [104] would suggest a direct effect of melatonin on immune cells, such as the antioxidant properties of this hormone [105]. It has been demonstrated, for example, that some compounds with an antioxidant activity regulate apoptosis, and melatonin has been

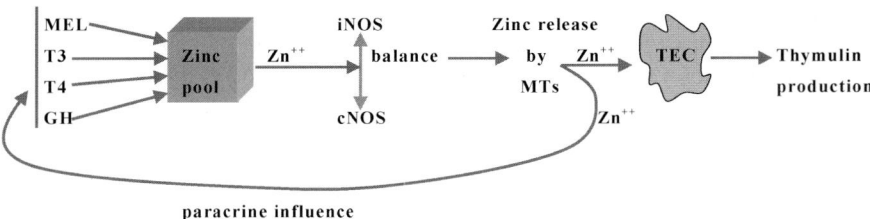

paracrine influence

Figure 3. One mechanism of action of zinc though some zinc-dependent hormones that affect thymic endocrine activity (thymulin production).
Melatonin, thyroid hormones, and growth hormone affect the zinc pool, which induce a balance between NO synthases with subsequent zinc release by metallothionein and, finally, thymulin production and activation by TECs within the thymus. At the same time, free zinc ions coming from the release of zinc by metallothionein affect melatonin, thyroid hormones and growth hormone with the establishment of paracrine circles between zinc and hormones, via metallothionein homeostasis (E. Mocchegiani, unpublished results).

reported to be a free radical scavenger with an antioxidant action more potent than vitamin E [106].

However, indirect mechanisms of action of melatonin on the immune parameters may be suggested. In this context, zinc turnover may be involved because melatonin treatment in old mice restores the negative crude zinc balance to young-like positive values [68]. In addition, the high nocturnal peaks of melatonin are correlated with the high nocturnal peaks of zinc during the circadian cycle in old melatonin treated mice [70]. Moreover, the effect of melatonin on zinc crude balance is comparable to that observed in old zinc-treated mice [66]. This restoration of zinc turnover is associated with a down regulation of corticosterone and increments in thymic and peripheral immune efficiency [70]. Taking into account that a down-regulation of corticosterone increases the zinc ion bioavailability in old zinc treated mice [82], it is evident that the zinc pool, via glucocorticoids, mediates the action of melatonin on thymic function. Indeed, ACTH and glucocorticoids modulate zinc turnover [53], and an inverse relationship has been observed between glucocorticoids and plasma zinc levels [70]. On the other hand, zinc deficiency, like in ageing, provokes an augmented production and release of glucocorticoids [53]. Taking into account these considerations, one mechanism by which melatonin can affect the immune system is thus through the modulation of zinc/glucocorticoid turnover. In other words, melatonin treatment in old age reduces adrenal function that in turn increases zinc ion bioavailability with an influence on the degree of thymocyte differentiation, proliferation, or death. As a final result, thymic morphology and function are restored.

## 7.    IS THERE A UNIQUE MEDIATOR FOR HORMONE-INDUCED THYMIC REJUVENATION?

In our models, the re-growth of the thymus in old age was obtained by different hormonal manipulations. Thymus re-growth was always associated with recovery of active thymulin plasma level, regardless the kind of endocrinological manipulation. Because this implies bioavailability of zinc ions, the question is whether zinc turnover might also be involved in the other models of hormone-induced thymic re-growth. It has been also reported in this review that treatments with thyroid hormones (T3, T4), IGF-I and with an analogue of LHRH (Table II) induces thymic re-growth and recovery of active thymulin. The restoration of plasma zinc levels also occurs after T3 and T4 treatment [82]. These last findings suggest that thyroid hormones may normalize zinc turnover in old age. In humans, hyper-secretion of thyroid hormones is associated with increased blood levels of zinc [84]. Therefore, the involvement of zinc in hormone-induced thymic re-growth is real, and it opens new insights to the relevance of zinc as a mediator of hormonal effects. It does not exclude the possibility that hormones may directly (through specific receptors) modulate the function of target cells. However, for thymic function, the restoration of zinc turnover is a prerequisite in order to observe the desired hormonal effects. These considerations, together with the relevance of zinc for the functional efficiency of the nervous, neuroendocrine, and immune systems [1,11,12], signify that zinc plays a pivotal role for the ageing of the organism and, as such, contributes to successful ageing. Such an assumption is supported by a recent discovery showing a good zinc ion bioavailability, satisfactory thymic function, and decreased corticosterone in very old mice and in nonagenarian/centenarian subjects [9]. The action of zinc-dependent hormones (GH, T3, T4, MEL) on thymic endocrine activity might be regulated by the interrelationship between nitric oxide [NO] and metallothioneins [MTs] as a proper ratio of NO synthases (iNOS and cNOS) provokes the release of zinc from MTs [12]. An altered ratio of NO-synthases during ageing prohibits the release of zinc from its MT reservoirs (see specific chapter in this volume by the same author). In other words, zinc-dependent hormones affect the zinc pool, which in turn sets the proper balance between iNOS and cNOS [Mocchegiani et al., 2003 unpublished results] with subsequent zinc release by zinc-bound MTs. This leads to more thymulin production by TECs and to subsequent paracrine influences by free zinc ions on hormone release itself (Figure 3). Therefore, zinc ion bioavailability is fundamental to thymic reconstitution in ageing by neuroendocrine manipulation.

ACKNOWLEDGMENTS

This paper was supported by INRCA and by the Italian Health Ministry (RF 216/02 to E.M).

REFERENCES

1.  Fabris N, Mocchegiani E, Provinciali M. Plasticity of neuroendocrine-thymus interactions during aging. Exp Gerontol 1997;32:415–429.
2.  Savino W, Dardenne M. Neuroendocrine control of thymus physiology. Endocr Rev 2000;21:412–413.
3.  Madden, KS, Felten DL. Experimental basis for neural-immune interactions. Physiol Rev 1995; 75: 77–106.
4.  Geenen V, Kecha O, Brilot F, Charlet-Renard C, Martens H. The thymic repertoire of neuroendocrine-related self antigens: biological role in T-cell selection and pharmacological implications. Neuroimmunomodulation 1999;6:115–125.
5.  Fabris N, Mocchegiani E. Endocrine control of thymic serum factor production in young-adult and old mice. Cell Immunol 1985;91:325–35.
6.  Besedovsky HO, del Rey A. Immune-neuro-endocrine interactions: facts and hypotheses. Endocr Rev 1996;17:64–102.
7.  McCann SM, Milenkovic L, Gonzales MC, Lyson K, Karanth S, Rettori V. Endocrine aspects of neuroimmunomodulation: Methods and Overview. In: De Souza EB, editor.

Neurobiology of Cytokines-A Volume of Methods in Neurosciences. Boca Raton: Academic Press, 1993;187–201.

8. Weigent DA, Blalock JE. Association between the neuroendocrine and immune systems. Leukoc. Biol 1995; 58:137–150.

9. Mocchegiani E, Giacconi R, Cipriano C, Muzzioli M, Gasparini N, Moresi R, Stecconi R, Suzuki H, Cavalieri E, Mariani E. MtmRNA gene expression, via IL-6 and glucocorticoids, as potential genetic marker of immunosenescence: lessons from very old mice and humans. Exp Gerontol 2002;37:349–357.

10. Savino W, Arzt E, Dardenne M. Immunoneuroendocrine connectivity: the paradigm of the thymus-hypothalamus/pituitary axis. Neuroimmunomodulation 1999;6:126–136.

11. Mocchegiani E, Muzzioli M, Cipriano C, Giacconi R. Zinc, T-cell pathways, aging: role of metallothioneins. Mech Ageing Dev 1998;106:183–204.

12. Mocchegiani E, Muzzioli M, Giacconi R. Zinc and immunoresistance to infection in aging: new biological tools. Trends Pharmacol Sci 2000;21:205–208.

13. Anderson G, Jenkinson EJ. Lymphostromal interactions in thymic development and function. Nat Rev Immunol 2001;1:31–40.

14. Res P, Spits H. Developmental stages in the human thymus. Semin Immunol 1999;11:39–46.

15. Ritter MA, Palmer DB. The human thymic microenvironment: new approaches to functional analysis. Semin Immunol 1999;11:13–21.

16. Alves LA, de Carvalho AC, Savino W. Gap junctions: a novel route for direct cell-cell communication in the immune system? Immunol Today 1998;19:269–275.

17. Patel DD, Haynes BF. Cell adhesion molecules involved in intrathymic T cell development. Semin Immunol 1993;5:282–292.

18. Dunn AJ. Cytokine activation of the HPA axis. Ann N Y Acad Sci 2000;917:608–617.

19. Hadden JW. Thymic endocrinology. Ann N Y Acad Sci 1998;840:352–358.

20. Champion S, Imhof BA, Savagner P, Thiery JP. The embryonic thymus produces chemotactic peptides involved in the homing of hemopoietic precursors. Cell 1986;44:781–790.

21. Moll J, Eibel H, Schmid P, Sansig G, Botteri F, Palacios R, Van der Putten H. Thymic hyperplasia in transgenic mice caused by immortal epithelial cells expressing c-kit ligand. Eur J Immunol 1992;22:1587–1594.

22. Muegge K, Vila MP, Durum SK. Interleukin-7: a cofactor for V(D)J rearrangement of the T cell receptor beta gene. Science 1993;261:93–95.

23. Rich BE, Leder P. Transgenic expression of interleukin 7 restores T cell populations in nude mice. J Exp Med 1995;181:1223–1228.

24. Haks MC, Oosterwegel MA, Blom B, Spits HM, Kruisbeek AM. Cell-fate decisions in early T cell development: regulation by cytokine receptors and the pre-TCR. Semin Immunol 1999;11:23–37.

25. Kim CH, Broxmeyer HE. Chemokines: signal lamps for trafficking of T and B cells for development and effector function. J Leukoc Biol 1999;65:6–15.

26. Klug DB, Carter C, Gimenez-Conti IB, Richie ER. Cutting edge: thymocyte-independent and thymocyte-dependent phases of epithelial patterning in the fetal thymus. J Immunol 2002;169:2842–2845.

27. Uehara A, Gillis S, Arimura A. Effects of interleukin-1 on hormone release from normal rat pituitary cells in primary culture. Neuroendocrinology 1987;45:343–347.

28. Bernton EW, Beach JE, Holaday JW, Smallridge RC, Fein HG. Release of multiple

hormones by a direct action of interleukin-1 on pituitary cells. Science 1987 Oct 23;238(4826):519–21.

29. Arzt E, Buric R, Stelzer G, Stalla J, Sauer J, Renner U, Stalla GK. Interleukin involvement in anterior pituitary cell growth regulation: effects of IL-2 and IL-6. Endocrinology 1993;132:459–467.

30. Fukata J, Usui T, Naitoh Y, Nakai Y, Imura H. Effects of recombinant human interleukin-1 alpha, -1 beta, 2 and 6 on ACTH synthesis and release in the mouse pituitary tumour cell line AtT-20. J Endocrinol 1989;122:33–39.

31. Bateman A, Singh A, Kral T, Solomon S. The immune-hypothalamic-pituitary-adrenal axis. Endocr Rev 1989;10:92–112.

32. Gaillard RC, Turnill D, Sappino P, Muller AF. Tumor necrosis factor alpha inhibits the hormonal response of the pituitary gland to hypothalamic releasing factors. Endocrinology 1990;127:101–106.

33. Lissoni P. The pineal gland as a central regulator of cytokine network. Neuroendocrinol Lett 1999;20:343–349.

34. Denicoff KD, Durkin TM, Lotze MT, Quinlan PE, Davis CL, Listwak SJ, Rosenberg SA, Rubinow DR. The neuroendocrine effects of interleukin-2 treatment. J Clin Endocrinol Metab 1989;69:402–410.

35. Spinedi E, Gaillard RC. Stimulation of the hypothalamo-pituitary-adrenocortical axis by the central serotonergic pathway: involvement of endogenous corticotropin-releasing hormone but not vasopressin. J Endocrinol Invest 1991;14:551–557.

36. Karanth S, McCann SM. Anterior pituitary hormone control by interleukin 2. Proc Natl Acad Sci 1991;88:2961–2965.

37. Smith LR, Brown SL, Blalock JE. Interleukin-2 induction of ACTH secretion: presence of an interleukin-2 receptor alpha-chain-like molecule on pituitary cells. J Neuroimmunol 1989;21:249–254.

38. Vankelecom H, Matthys P, Denef C. Involvement of nitric oxide in the interferon-gamma-induced inhibition of growth hormone and prolactin secretion in anterior pituitary cell cultures. Mol Cell Endocrinol 1997;129:157–167.

39. Withyachumnarnkul B, Nonaka KO, Santana C, Attia AM, Reiter RJ. Interferon-gamma modulates melatonin production in rat pineal glands in organ culture. J Interf Res 1990;10:403–411.

40. Dardenne M, Pleau JM, Nabama B, Lefancier P, Denien M, Choay J, Bach JF. Contribution of zinc and other metals to the biological activity of the serum thymic factor. Proc Natl Acad Sci USA 1982; 79: 5370–5375.

41. Migliorati G, Nicoletti F, D'Adamio F, Spreca A, Pagliacci C, Riccardi C. Dexamethasone induces apoptosis in mouse natural killer cells and cytotoxic T. lymphocytes. Immunology 1994;81: 21–26.

42. Iwata M, Hanaoka S, Sato K. Rescue of thymocytes and T-cell hybridomas from glucocorticoidsinduced apoptosis by stimulation via the T cell receptors/CD3 complex: A possible in vitro model for positive selection of the T cell repertoire. Eur J Immunol 1991; 21: 643–649.

43. Sei Y, Yoshimoto K, Mcintyre A, Skolnick P, Arora PK Morphine-induced thymic hypoplasia is glucocorticoid dependent. J Immunol 1991; 146: 194–198.

44. Goya RG, Sosa YE, Console GM, Dardenne M. Altered thyrotropic and somatotropic responses to environmental challenges in congenitally athymic mice. Brain Behav Immun 1995;9:79–86.

45. Daneva T, Spinedi E, Hadid R, Gaillard RC. Impaired hypothalamo-pituitary-adrenal axis function in Swiss nude athymic mice. Neuroendocrinology 1995;62:79–86.

46. Goya RG, Brown OA, Bolognani F. The thymus-pituitary axis and its changes during aging. Neuroimmunomodulation 1999;6:137–142.

47. Hadley AJ, Rantle CM, Buckingham JC. Thymulin stimulates corticotrophin release and cyclic nucleotide formation in the rat anterior pituitary gland. Neuroimmunomodulation 1997;4:62–69.

48. Angioni S, Iori G, Cellini M, Sardelli S, Massolo F, Petraglia F, Genazzani AR. Acute beta-interferon or thymopentin administration increases plasma growth hormone and cortisol levels in children. Acta Endocrinol (Copenh) 1992;127:237–241.

49. Malaise MG, Hazee-Hagelstein MT, Reuter AM, Vrinds-Gevaert Y, Goldstein G, Franchimont P. Thymopoietin and thymopentin enhance the levels of ACTH, beta-endorphin and beta-lipotropin from rat pituitary cells in vitro. Acta Endocrinol (Copenh) 1987 Aug;115(4):455–60.

50. Goya RG, Sosa YE, Brown OA, Dardenne M. In vitro studies on the thymus-pituitary axis in young and old rats. Ann N Y Acad Sci 1994;741:108–114.

51. Tijerina M, Gorospe WC, Bowman KL, Badamchian M, Goldstein AL, Spangelo BL. A novel thymosin peptide stimulates interleukin-6 release from rat C6 glioma cells in vitro. Neuroimmunomodulation 1997;4:163–170.

52. Milenkovic L, Lyson K, Aguila MC, McCann SM. Effect of thymosin alpha 1 on hypothalamic hormone release. Neuroendocrinology 1992;56:674–679.

53. Prasad AS. Clinical, endocrinological and biochemical effects of zinc deficiency Clin Endocrinol Metab 1985 Aug;14(3):567–89.

54. Prasad AS, Fitzgerald JT, Hess JW, Kaplan J, Pelen F, Dardenne M. Zinc deficiency in elderly patients. Nutrition 1993; 9: 218–224.

55. Fabris N, Mocchegiani E. Zinc, human diseases and aging. Review article. Aging Clin Exp Res 1995; 7: 77–93.

56. Shankar AH, Prasad AS. Zinc and immune function: the biological basis of altered resistance to infection. Am J Clin Nutr 1998;68:447S-463S

57. Fabris N, Mocchegiani E, Amadio L, Zannotti M, Licastro F, Franceschi C. Thymic hormone deficiency in normal aging and Down's syndrome: Is there a primary failure of the thymus? Lancet 1984; 1: 983–986.

58. Garvy BA, Telford WG, King LE, Fraker PJ. Glucocorticoids and irradiation-induced apoptosis in normal murine bone marrow B-lineage lymphocytes as determined by flow cytometry. Immunology 1993;79:270–277.

59. Mocchegiani E, Muzzioli M. Therapeutic application of zinc in human immunodeficiency virus against opportunistic infections. J Nutr 2000;130:1424S-1431S

60. McCabe MJ Jr, Aijiang S, Orrenius S. Chelation of intracellular zinc triggers apoptosis in mature thymocytes. Lab. Invest. 1993;69:101–110.

61. Fraker PJ, Telford WG. A reappraisal of the role of zinc in life and death decisions of cells. Proc Soc Exp Biol Med 1997;215:229–236.

62. MacLean KH, Cleveland JL, Porter JB. Cellular zinc content is a major determinant of iron chelator-induced apoptosis of thymocytes. Blood 2001;98:3831–3839.

63. Barbieri D, Troiano L, Grassilli E, Agnesini C, Cristofalo EA, Monti D, Capri M, Cossarizza A, Franceschi C. Inhibition of apoptosis by zinc: a reappraisal. Biochem Biophys Res Commun 1992;187:1256–1261.

64. Lanningham-Foster L, Green CL, Langkamp-Henken B, Davis BA, Nguyen KT, Bender

BS, Cousins RJ. Overexpression of CRIP in transgenic mice alters cytokine patterns and the immune response. Am J Physiol Endocrinol Metab 2002;282:E1197-E1203

65. Aspinall R, Andrew D. Thymic involution in aging. J Clin Immunol 2000;20:250–256.

66. Mocchegiani E, Santarelli L, Muzzioli M, Fabris N. Reversibility of the thymic involution and of age-related immune dysfunctions by zinc supplementation in old mice. Int J Immunopharmacol 1995; 17: 703–7. 18

67. Mocchegiani E, Fabris N. Age-related thymus involution zinc reverses in vitro the thymulin secretion defect. Int J Immunopharmacol 1995; 17: 745–749.

68. Mocchegiani E, Builian D, Santarelli L, Tibaldi A, Muzzioli M, Pierpaoli W, Fabris N. The immuno-reconstituting effect of melatonin of pineal grafting and its relation to zinc pool in aging mice. J Neuroimmunol 1994; 53: 189–201.

69. Pierpaoli W, Regelson W. Pineal control of aging: Effect of melatonin and pineal grafting on aging mice. Proc Natl Acad Sci USA 1994; 91: 787–791.

70. Mocchegiani E, Santarelli L, Tibaldi A, Muzzioli M, Bulian D, Cipriano K, Olivieri F, Fabris N. Presence of links between zinc and melatonin during the circadian cycle in old mice: effects on thymic endocrine activity and on the survival. J Neuroimmunol 1998;86: 111–122.

71. Kelley KW, Brief S, Westly HJ, Novakofsky J, Bechitel P J, Simon J, Walker EB. GH3 pituitary adenoma cells can reverse thymic aging in rats. Proc Natl Acad Sci USA 1986; 83: 5663–5667.

72. Davila DR, Brief S, Simon J, Hammer RE, Brinster RL, Kelly KW. Role of growth hormone in regulating T-dependent immune events in aged, nude and transgenic rodents. J Neurosci Res 1987; 18:108–116.

73. Goff BL, Roth JA, Arp LH, Incefy G.S. Growth hormone treatment stimulates thymulin production in aged dogs. Clin Exp Immunol 1987; 65: 580–587.

74. Binz K, Joller P, Froesch P, Binz H, Zapf J, Froesch ER. Repopulation of the atrophied thymus in diabetic rats by insulin-like growth factor I. Proc Natl Acad Sci U S A 1990;87: 3690–3694.

75. Greenstein BD, Fitzpatrick FT, Kendall MD, Wheeler MJ. Regeneration of the thymus in old male rats treated with a stable analogue of LHRH. J Endocrinol 1987; 112: 345–350.

76. Marchetti B, Morale MC, Batticane N, Gallo F, Farinelli Z, Cioni M. Aging of the reproductive-neuroimmune axis. Ann NY Acad Sci 1991; 621: 159–173.

77. Fabris N, Muzzioli M, Mocchegiani E. Recovery of age-dependent immunological deterioration in Balb/c mice by short-term treatment with L-thyroxine. Mech Ageing Dev 1982;18:327–338.

78. Fabris N, Mocchegiani E, Muzzioli M, Piloni S. Recovery of age related decline of thymic endocrine activity and PHA response by lysine-arginine combination. Int J Jmmunopharmacol 1986 ; 8 : 677–685.

79. Dardenne M, Bukaiba N, Gagnerault MC, Homo-Delarche F, Chappuis P, Lemmonnier D, Savino W. Restoration of the thymus in aging mice by in vivo zinc supplementation. Clin Immunol Immunopathol 1993; 66: 127–135.

80. Fabris N, Mocchegiani E. Immunomodulating role of growth hormone. In: Bercu BB, Walker RF, editors. Growth Hormone II. New York: Springer Verlag, 1994; 116–130.

81. Fabris N, Mocchegiani E. Arginine-containing compounds and thymic endocrine activity. Thymus 1992;19:S21-S30

82. Mocchegiani E, Giacconi R, Cipriano C, Gasparini N, Orlando F, Stecconi R, Muzzioli M, Isani G, Carpenè E. Metallothioneins (I+II) and thyroid-thymus axis efficiency in old

mice: role of corticosterone and zinc supply. Mech Ageing Dev 2002;123:675–694.

83. Simpson JC, Gray ES, Michie W, Beck JS. The influence of preoperative drug treatment on the extent of hyperplasia of the thymus in primary thyrotoxicosis. Clin Exp Immunol 1975; 22: 249–253.

84. Fabris N, Mocchegiani E, Mariotti S, Pacini F, Pinchera A. Thyroid function modulates thymus endocrine activity. J. Clin Endocrinol Metab 1986; 62: 474–478.

85. Boukaiba N, Flament C, Acher S, Chappuis P, Piau A, Fusselier M, Dardenne M, Lemonnier D. A physiological amount of zinc supplementation: effects on nutritional, lipid, and thymic status in an elderly population. Am J Clin Nutr 1993;57:566–572.

86. Bogden JD, Oleske JM, Lavenhar MA, Munves EM, Kemp FW, Bruening KS, Holding KJ, Denny TN, Guarino MA, Holland BK. Effects of one year of supplementation with zinc and other micronutrients on cellular immunity in the elderly. J Am Coll Nutr 1990;9: 214–225.

87. Mocchegiani E, Muzzioli M, Gaetti R, Veccia S, Viticchi C, Scalise G. Contribution of zinc to reduce CD4+ risk factor for 'severe' infection relapse in aging: parallelism with HIV. Int J Immunopharmacol 1999;21:271–281.

88. Fabris N, Pierpaoli W, Sorkin E. Hormones and the immunological capacity. 3. The immunodeficiency disease of the hypopituitary Snell-Bagg dwarf mouse. Clin Exp Immunol 1971;9:209–225.

89. Roth JA, Laeberle ML, Grier DL, Hopper JG, Spiegel HE, Macallister HA. Improvement in clinical condition and thymic morphological features associated with growth treatment of immuno-deficient dwarf-dogs. Am J Vet Res 1984; 45: 1151–1155.

90. Wise T, MacDonald GJ, Klindt J, Ford JJ. Characterization of thymic weight and thymic peptide thymosin-beta 4: effects of hypophysectomy, sex, and neonatal sexual differentiation. Thymus 1992;19:235–244.

91. Mocchegiani E, Sartorio A, Santarelli L, Ferrero S, Fabris N. Thymulin, zinc and insulin-like growth factor-I (IGF-I) activity before and during recombinant growth hormone (rec-GH) therapy in children and adults with GH deficiency. J Endocrinol Invest 1996;19: 630–637.

92. Muller EE, Rigamonti AE, Colonna Vde G, Locatelli V, Berti F, Cella SG. GH-related and extra-endocrine actions of GH secretagogues in aging. Neurobiol Aging 2002;23: 907–919.

93. Hirokawa K, Utsuyama M, Kasai M, Konno A, Kurashima C, Moriizumi E. Age-related hyperplasia of the thymus and T-cell system in the Buffalo rat. Immunological and immunohistological studies. Virchows Arch B Cell Pathol Incl Mol Pathol 1990;59: 38–47.

94. Mocchegiani E, Cacciatore L, Talarico M, Lingetti M, Fabris N. Recovery of low thymic hormone levels in cancer patients by lysine-arginine combination. Int J Immunopharmacol 1990;12:365–371.

95. Ban E, Gagnerault MC, Jammes H, Postel-Vinay MC, Haour F, Dardenne M. Specific binding sites for growth hormone in cultured mouse thymic epithelial cells. Life Sci 1991;48:2141–2148.

96. Mocchegiani E, Paolucci P, Balsamo A, Cacciari E, Fabris N. Influence of growth hormone on thymic endocrine activity in humans. Horm Res 1990;33:248–255.

97. Timsit J, Savino W, Safieh B, Chanson P, Gagnerault MC, Bach JF, Dardenne M. Growth hormone and insulin-like growth factor-I stimulate hormonal function and proliferation of thymic epithelial cells. J Clin Endocrinol Metab 1992;75:183–188.

98.  Dardenne M, Kelly PA, Bach JF, Savino W Identification and functional activity of prolactin receptors in thymic epithelial cells. Proc Natl Acad Sci U S A 1991;88:9700–9704.

99.  Cunningham BC, Bass S, Fuh G, Wells JA. Zinc mediation of the binding of human growth hormone to the human prolactin receptor. Science 1990;250:1709–1712.

100. MacDonald RS. The role of zinc in growth and cell proliferation. J Nutr 2000;130:1500S-1508S

101. Maestroni GJ. The immunoneuroendocrine role of melatonin. J Pineal Res 1993;14:1–10.

102. Sainz RM, Mayo JC, Uria H, Kotler M, Antolin I, Rodriguez C, Menendez-Pelaez A. The pineal neurohormone melatonin prevents in vivo and in vitro apoptosis in thymocytes. J Pineal Res 1995;19:178–188.

103. Provinciali M, Di Stefano G, Bulian D, Tibaldi A, Fabris N. Effect of melatonin and pineal grafting on thymocyte apoptosis in aging mice.Mech Ageing Develop 1996; 90: 1–19.

104. Martin-Cacao A, Lopez-Gonzales MA, Calvo JR, Seguraand JJ, Guerrero JM Melatonin binding sites in rat thymus during development. In: Touitou Y, Arendt J, Pevet P, editors. Melatonin and the Pineal Gland. Amsterdam: Elsevier, 1993; 379–382.

105. Reiter RJ, Tan DX, Poeggeler B, Chen LD, Menendez-Pelaez A. Melatonin, free radicals and cancer initiation. In: Maestroni GJM, Conti A, Reiter R, editors. Advances in Pineal Research. London: John Libbey, 1994: 211–228.

106. Pieri C, Marra M, Moroni F, Recchioni R, Marcheselli F. Melatonin: A peroxil radical scavenger more effective than vitamin E. Life Sci 1994;55:271–276.

*The Neuroendocrine Immune Network in Ageing*
Edited by R.H. Straub and E. Mocchegiani

# Adverse Glucocorticoid Actions and their Relevance to Brain Ageing

KLAUS DINKEL[1] and ROBERT M. SAPOLSKY[2]

*[1]Leibniz Institute for Neurobiology, Dept of Neuropharmacology, 39118 Magdeburg, Germany;*
*[2]Dept. Biol. Sc. Stanford University. Stanford, CA 94305 5020, USA*

## ABSTRACT

Ageing is typically associated with hormonal unbalances such as an age-related increase of glucocorticoids (GC). GC are the adrenal steroids that are normally secreted during stress. Although this physiological GC response is critical for successful adaptation to acute physical and/or psychological stressors GC can have a variety of adverse effects if secreted in excess or chronically. These harmful GC actions also include damage to and impairment of the funtion of the central nervous system such as disruption of synaptic plasticity, atrophy of dendritic processes and compromising the ability of neurons to survive a neurological insult. This review considers some of these adverse GC actions and their relevance to brain ageing. This also includes a look on the effects of GC on immunosenescence and nervous system inflammation since brain injury such as stroke is accompanied by a marked inflammatory reaction.

## 1.     INTRODUCTION

The central concept of physiology and health, in many ways, is the concept of homeostasis, the notion that there are optimal levels of function for a range of physiological endpoints. The fact that what is optimal can shift under differing circumstances, that homeostatic setpoints are not static, has given rise to recent concepts such as "homeodynamics" or "allostasis," which can be viewed as more sophisticated versions of the homeostasis concept. In either the more classical or more modern formulations of physiological systems having their points of optimal function, it is obvious that a central challenge to physiological systems is how to reestablish homeostatic balance when it has been perturbed. A "stressor" can be viewed as anything in the outside environment which disrupts the homeostatic balance of an organism. When considering cognitively sophisticated species such as primates, including ourselves, the definition needs to be expanded to include the fact that a stressor can also be the anticipation that one is about to be thrown out of homeostatic balance. When this perception is accurate, such anticipatory stress can be highly adaptive. In contrast, when such a perception is incorrect and frequent, this is the realm of anxiety.

Since the time of Walter Cannon and Hans Selye, it has come to be recognized that the body activates an array of adaptations meant to re-establish homeostasis. Known in older terminology as the "fight or flight" response, or the "General Adaptation Syndrome," this is now generally known as the "stress-response." One major branch of the stress-response involves the activation of the sympathetic nervous system, resulting in the secretion of epinephrine and norepinephrine. The other main branch is the release during stress, by the adrenal glands, of glucocorticoids (GCs). These steroid hormones include naturally occurring versions such as cortisol (also known as hydrocortisone, predominately found in primates) and corticosterone (predominately found in rodents), and synthetic versions, such as prednisone and dexamethasone.

The secretion of GCs is essential for adapting to acute physical stressors; they divert energy to exercising muscle, enhance cardiovascular tone, suppress unessentials such as digestion, growth and reproduction, suppress immune responses that might otherwise spiral into autoimmunity, and sharpen cognition. In contrast, excessive GCs (as seen with chronic stress, exogenous administration, or with the GC hypersecretion of Cushing's disease) can cause or worsen various disorders, including myopathy, adult-onset diabetes, hypertension, amenorrhea and impotency, as well as be highly immunosuppressive. Because many of those disorders are the diseases of slow accumulation of damage of Western ageing, the links between stress and ageing have often been framed in the context of the pathogenic potential of GCs.

The adverse effects of an excess of GCs also includes damage to, and impairment of the function of the central nervous system. This finding is made more pertinent by the fact that in the rodent, there is a fairly linear increase in basal GC concentrations with age, and a non-linear increase that emerges with extreme old age in humans. In this review, we consider some of these adverse effects that may be relevant to understanding brain ageing. This includes some surprising and recent findings regarding the interactions between GCs and inflammation.

## 2.    GCS AND AGEING

Ageing is typically associated with hormonal unbalances such as hyperglucocorticoidemia and it has been proposed, that these hormonal changes may be responsible for the ageing-associated increased vulnerability of neurons to degeneration [1,2]. Brain ageing as well as brain degenerative processes with accompanying cognitive impairments are generally associated with hyperactivity of the HPA-axis which leads to an age-related increase of glucocorticoids (GCs).

### 2.1.    An overview of the adverse CNS actions of glucocorticoids

An excess of GCs has been shown to have disruptive effects in a number of areas of the brain. Nonetheless, the most dramatic ones occur in the hippocampus, a primary GC target. The hippocampus contains high concentrations of both the high affinity mineralocorticoid receptor (MR), which is heavily occupied by basal GC levels, and a lower affinity glucocorticoid receptor (GR), which is only heavily occupied by stress levels of GCs [3]. While MR occupancy mediates salutary effects in the hippocampus (e.g., enhancement of cognition and synaptic plasticity), GR occupancy mediates adverse effects. These opposing MR and GR effects result in an "inverse-U" for many GC actions – pathological hyposecretion of GCs has adverse consequences; basal to low stress concentrations of GCs (i.e., what would result from what would typically be called stimulation) have beneficial effects, and levels in the range of moderate stress or higher again have adverse effects. Whether this is an "inverse-U" or "U"-shaped effect depends, of course, on how the dependent variable is framed (e.g., neuron survival versus neuron death). For simplicity, this non-monolithic effect will be referred to as "U"-shaped throughout.

The adverse effects of GCs in the hippocampus take a variety of forms. There are a number of subtypes of memory, and the hippocampus plays a central role in what is termed "declarative" or "explicit" memory, which is knowledge about facts and conscious awareness of such knowledge. This would include knowing where one is, what month it is, what one had for breakfast that morning, the name of the head of state, the name of one's children. As a first example of the effects of GCs, the hormone disrupts hippocampal-dependent declarative cognition. Moreover, GCs also disrupt the sort of synaptic plasticity that is thought to underlie such cognition (termed "long term potentiation" [LTP]). Within minutes of the onset of stress and GC secretion, declarative memory formation is enhanced, as is hippocampal LTP. This is a case of short-term effects of stress and GCs having salutary, stimulatory actions, via occupancy of MR. However, if the GC exposure is severe (even if brief), or prolonged (even if only moderately elevated), there is suppression of LTP and disruption of declarative memory consolidation and retrieval. This is a GR-mediated phenomenon, and has been demonstrated in rodents, non-human primates and humans as well. Mechanistically, GCs causes increases in calcium currents in hippocampal neurons. This leads to prolonged opening of calcium-gated potassium channels, leading to neuronal hyperpolarization. This causes decreased neuronal excitability, specifically in the form of a prolonged afterhyperpolarization following an action potential. Such decreased excitation decreases the likelihood of LTP occurring.

As a next example, exposure to elevated levels of GCs or stress (over the course of days to weeks) can cause retraction of dendritic processes in hippocampal neurons; this is likely to have adverse consequences for the functioning of networks within the hippocampus. The atrophy has been demonstrated in both rodents and non-human primates, and indirect evidence suggests that this may occur in humans with Cushing's Syndrome (any of a number of tumors that produce highly elevated GC concentrations). The retraction is dependent upon the release of the neurotransmitter glutamate, and its interaction with the NMDA subtype of its receptors; as evidence, the atrophy can be reversed with drugs that block either glutamate release or NMDA receptors. Most interestingly, with the cessation of stress or GC exposure, processes gradually regrow.

A next example of an adverse GC effect reflects a recent revolution in neurobiology. A long-standing dogma of the field has been that neurogenesis ceases in the perinatal brain, with all extant neurons at that point being post-mitotic. Beginning with pioneering and generally ignored studies in the 1960's, evidence has accumulated to demonstrate that there is, in fact, adult neurogenesis from neuronal stem cells, and that one of the two sites in the brain where this occurs is the hippocampus. Moreover, there is increasing evidence that such neurons form functional synapses with other neurons, and that they may even subserve aspects of learning and memory.

With the general acceptance of these findings has come an intense interest in factors that modulate the rate of such neurogenesis. Factors that promote neurogenesis include stimulation and environmental enrichment, exercise, exposure to estrogen, and an array of mechanistically unrelated antidepressant drugs. Of great interest, a highly reliable inhibitor of neurogenesis is exposure to stress or GCs. This is a fairly new observation, and the mechanisms underlying it are virtually unknown at this point.

As another example, truly prolonged exposure to very high levels of stress or GCs can cause the overt loss of hippocampal neurons (i.e., cell death). The first evidence for this was reported in the late 1970's with the observation that decreasing life-long exposure to GCs in the rat decreased age-related loss of hippocampal neurons. Such GC-induced neurotoxicity has been replicated since then, including in non-human primates. However, it is a fairly controversial finding for a number of reasons: a) first, it is unclear whether this is a physiological phenomenon, as opposed to a purely pharmacological one; b) more modern stereological techniques for counting of neurons have generally failed to show that GCs can a decrease in the numbers of hippocampal neurons; c) the relevance to ageing has also been called to question by these stereological techniques, with their demonstration that there is far less neuron loss in the ageing hippocampus than once thought. Collectively, these findings suggest that overt neurotoxicity is the weakest of the adverse CNS effects demonstrated to be caused by GCs.

These findings are quite pertinent in the context of brain ageing and the rising basal GC levels frequently reported. This is because hippocampal ageing involves an attenuated capacity of LTP, impairment of declarative memory, retraction of dendritic arbor, and decreased rates of neurogenesis. Moreover, a few studies have demonstrated that these age-related impairments are reversible when GCs levels are minimized.

Stress and GCs have an additional adverse effect in the hippocampus, which has been termed "endangerment." Specifically, the hormone compromises the ability of neurons to survive co-incident insults, many of which are more common with age. The mechanistic underpinnings of this phenomenon are the best understood of these instances of adverse GC actions, and will be briefly reviewed.

## 2.2. Necrotic neurological insults and "endangerment" by stress and/or glucocorticoids

The hippocampus is particularly vulnerable to necrotic insults such as seizure, the hypoxia-ischemia, following stroke or cardiac arrest and hypoglycemia [4]. Neuron death following such insults arises from excessive synaptic levels of excitatory amino acid (EAA) neurotransmitters such as glutamate and aspartate, resulting in excessive post-synaptic mobilization of free cytosolic calcium. Calcium excess causes promiscuous overactivation of calcium-dependent processes, leading to the generation of reactive oxygen species (ROS), cytoskeletal damage and DNA fragmentation [4–7]. Apoptotic cell death is activated in a subset of insulted neurons; the severity and duration of the necrotic insult, the age of the cells, and the extent of energy availability all influence the likelihood of death by apoptosis [8,9].

GCs and stress can impair the capacity of neurons to survive necrotic insults. Insults whose toxicities are worsened by GCs in the rodent hippocampus include EAAs [10–15], global ischemia [16–19], hypoglycemia, antimetabolites [9,10], and cholinergic and serotonergic toxins [20–22]. These effects occur in all hippocampal cell fields, reflecting the high levels of GR expression throughout the rat hippocampus [3]. The exacerbation of damage by GCs is apparent with just a few days of GC exposure, can involve a 10-fold increase in toxicity, and is steroid-specific and mediated by the GR.

Insults such as seizure or cardiac arrest stimulate substantial GC secretion. Preventing such secretion (by adrenalectomy), or blocking GR decreases neurotoxicity [15,16,18,23]. Thus, the "typical" extent of damage following an insult reflects, in part, GC-induced endangerment.

As noted, GCs have numerous peripheral actions. Studying endangerment in vitro, absent these confounding actions, has aided in the identification of mediating mechanisms. GCs increase the toxicity of EAAs both in hippocampal cultures and cell lines [24–27], hypoglycemia [28], cyanide [29], reactive oxygen species [24,26,30], the β-amyloid peptide [26], and constituents of HIV such as gp120 [31–36]. This endangerment also occurs in other brain regions, including the striatum, cortex and nucleus basalis, albeit less consistently or to a lesser extent than in the hippocampus [14,16,17,37–39]. Of note, GCs do not worsen all insults examined. Exceptions include zinc neurotoxicity in hippocampal cultures [40], neonatal hypoxia-ischemia in rats [19,41–44] and spinal cord trauma [45–47].

2.3.    Mechanisms mediating the GC endangerment

### 2.3.1. *Disruption of neuronal energetics* [48]

For most animals, "stress" is a physical challenge such as escaping a predator. At such times, GC adaptively divert energy to exercising muscle. This involves GCs inhibiting glucose uptake throughout the body. GCs do the same in the brain, inhibiting glucose uptake 15–25%, particularly in the hippocampus (or in hippocampal cultures) [49–54].

These GC actions appear to have consequences during an insult. For example, GCs accelerate the decline in ATP levels in hippocampal neurons and glia during insults [55,56], and the endangering effects of GCs are lessened if neurons are supplemented with excess energy [57].

Necrotic insults such as stroke, global ischemia, hypoglycemia and seizure constitute energy crises (disrupting energy production or excessively consuming energy). In such crises, hippocampal neurons fail in the costly tasks of regulating EAA and calcium levels. GCs, by disrupting energetics, exacerbate the EAA and calcium accumulation during these insults; as evidence, energy supplementation can reverse these effects [13,26,58–61] (this capacity of GCs to increase synaptic concentration so of glutamate and of cytosolic calcium is probably relevant to understanding GC-induced disruption of LTP via opening of calcium-dependent potassium channels, and of GC-induced atrophy of dendritic processes). While the source of the excess calcium influences the extent of damage at least as much as the absolute amount of calcium mobilized [62], a GC-induced increase in calcium load worsens the resulting oxygen adical accumulation and cytoskeletal damage [13,30,63–66]. GCs also decrease levels of reduced glutathione (the substrate for the antioxidant glutathione peroxidase), an effect secondary to the disruptive effects of GCs on energetics [67]. These deleterious effects of GCs on various steps of this necrotic cascade are lessened when neurons are supplemented with excess energy substrates.

### 2.3.2. *Endangering GC actions that are independent of energetic effects*

There are also GC effects on glutamate, calcium and reactivce oxygen species that are likely to be non-energetic, in that they occur in the absence of shifts in energetic profiles, or occur too quickly to be secondary to an energy deficit. For example, some GC actions on EAA or calcium trafficking are rapid [68,69]. Moreover, in the absence of an insult, GCs rapidly increase voltage-dependent calcium conductance, calcium spike duration, and calcium-dependent afterhyperpolarizations in CA1 neurons [70,71]. Within hours, GCs also decrease the expression of hippocampal PMCA-1 (plasma membrane calcium ATPase-1), which extrudes cytosolic calcium [72]. Finally, GCs decrease the activity of catalase and gluathione peroxidase in the hippocampus in a manner that is at least partially energy-independent [66,67].

### 2.3.3. *Disruption of cellular defenses*

Neurons mobilize numerous defenses in response to insult [73,74]. GCs impair some of these defenses. For example, the inhibition of synaptic glutamate removal or calcium extrusion by GCs can be viewed as impairments of defenses [72,75]. Moreover, GCs block the defensive upregulation of glutathione peroxidase activity post-insult. Thus, GCs have varied effects which, collectively, impair the capacity of hippocampal neurons to survive necrotic insults.

### 2.3.4. Triggering/worsening apoptosis

As noted, a subset of neurons die with at least elements of apoptosis following necrotic insults. GCs have long been recognized for their capacity to induce apoptosis in immune cells. This has suggested to some that the endangering effects of GCs in the hippocampus are likely to involve increasing the number of neurons dying apoptotically [76]. Insofar as GCs, under circumstances where they increase the neurotoxicity of excitotoxic insults, do not increase indices of apoptosis, including caspase activation, DNA condensation, or internucleosomal DNA fragmentation [66], this appears not to be the case.

### 2.4. Glucocorticoid endangerment and a seeming paradox

The earliest report of the endangering effects of GCs in 1985 [10] brought up the issue of inflammation. It has been known for decades that GCs can be profoundly anti-inflammatory; these actions account for most of the approximately 16 million prescriptions for synthetic GCs written annually in the United States. Critically, it has become apparent that necrotic insults cause considerable inflammation in the brain, and that such inflammation can be extremely damaging (leading to "secondary" toxicity to neurons otherwise destined to survive the primary insult).

This seemingly generates a clear problem with the notion of GC endangerment. If GCs decrease inflammation, and if such inflammation adds to the neurotoxicity following necrotic insults, then GCs should be <u>protective</u>, rather than endangering. And thus if a quite consistent literature demonstrates that GCs, nonetheless, increase the neurotoxicity of necrotic insults in the hippocampus, then the adverse endangering effects (summarized above) must be, collectively, more powerful than the opposing protective anti-inflammatory actions.

The resolution is one that a number of investigators have suggested [69]. However, the basic premise of this resolution has been questioned in recent years. Specifically, it appears that following necrotic neurological insults, GCs are far less anti-inflammatory than previously assumed and, in fact, have some pronounced pro-inflammatory effects.

## 3.    GCS AND THE IMMUNE SYSTEM

GCs have been used widely for the treatment of diseases associated with activation of the immune system since their original application in the late 1940s [77]. This work led to the winning of the Nobel Prize for medicine and provided the foundation for the dogma that GCs are uniformly immunosuppressive. The use of GCs in the treatment of various clinical disorders such as autoimmune diseases, chronic inflammation or transplant rejection has been proven successful. The therapeutic value of GCs is attributed to their potent anti-inflammatory and immunomodulatory effects (Table I) on T-cell activation, adhesion molecule expression, cell migration and cytokine production [70]. It is furthermore well-known that GCs induce apoptosis in certain immune cells like T-lymphocytes and macrophages (see Table I).

In aged populations with elevated GC levels, infectious diseases are generally more serious and cause higher mortality. This increased susceptibility to infections is attributed to a peripheral suppression/decline of function of the immune response [78]. In general this and other age-related effects are not caused by elevated GC levels alone; other factors such as shortening of telomeres and impairment of DNA repair are most likely also contributing to the observed phenomenon of immunosenescence. GC mediated reduction of leukocyte infiltrate for example

Table I    Immunosuppressive/Anti-inflammatory Glucocorticoid effects.

**Lymphocytes (adaptive)**

Decreased cytokine-induced proliferation
Decreased cytotoxicity
Decreased cytokine production (IL-1, IL-2)
Induction of apoptosis (` Lymphopenia)

**Neutrophil granulocytes (innate)**

Decreased extravasation
Decreased adhesion molecule expression
Decreased phagocytosis
Decreased free radical generation
Decreased chemotaxis

**Macrophages/monocytes (innate, adaptive)**

Decreased extravasation
Inhibition of differentiation
Decreased phagocytosis
Decreased MHC I and II expression
Decreased antigen presentation

**Miscellanous**

Decreased adhesion molecule expression (e.g. ICAM-1, VCAM-1)
Decreased pro-inflammatory cytokine production (e.g. IL-1$\alpha$/$\beta$, TNF-$\alpha$, IL-6, IL-2)

occurs via the downregulation of adhesion molecules such as ICAM-1, ELAM-1 and VCAM-1 [79]. In addition monocyte and neutrophil recruitment during acute inflammation has been found to be under the negative modulatory control of the GC-induced lipocortin-1 [80]. This clearly shows that GCs have anti-inflammatory effects in the periphery. Infections in the elderly are not only more frequent and more severe, but they also have distinct features with respect to clinical presentation, laboratory results, microbial epidemiology, treatment, and infection control [81].

## 3.1.    GCs and immunosenescence

Age related alteration of T cell function represents one of the most consistent and dramatic effects observed. It has been postulated that these changes may be due to alterations of the cytokine profile. Data from rodent studies support an age-related and probably GC-mediated shift from a Th1-like to a Th2-like cytokine pattern in the periphery. However an extensive review of over 60 studies by Gardner et al. [82] suggested that age-associated cytokine changes in humans are inconsistent and do not always show a shift towards a Th2-like pattern. It has been observed that elderly caregivers of dementia patients showed blunted mitogen-induced proliferation [83], NK-cell activity [84] and impaired antibody responses following influenza vaccination [85] compared to age-matched non-caregivers. Another study showed that chronic stress in elderly caregivers is associated with impaired lymphocyte immunity [86].

Interestingly in elderly men a decrease of GC-sensitivity in the periphery immediately after stress has been found compared to young controls [87]. The increase in GC sensitivity after stress in younger individuals might serve as a protective mechanism from a harmful increase of pro-inflammatory cytokines.

## 3.2.    GC effects on CNS vs. peripheral inflammation

Amid the textbook picture of GCs being anti-inflammatory in the periphery, recent work suggests some instances of pro-inflammatory GC effects in the CNS.

Several investigators found a GC-induced selective suppression of cellular adaptive immunity (T cells) and enhancing of humoral (B cells, antibodies) immunity. In a recent study, dexamethasone treatment failed to downregulate the cytokines IL-1$\alpha$ and TNF-$\alpha$ after chemically-induced hippocampal injury in mice [88]. GCs are of no therapeutic benefits in stroke patients [89], nonsteroidal anti-inflammatory drugs exert a stronger protective effect in Alzheimer's disease [90,91] and several studies argue even against their use in the treatment of post-stroke edema [92,93]. Furthermore, GC treatment failed to have beneficial effects in a recent clinical trial with Alzheimers patients [94].

Even more surprisingly, some studies even demonstrated pro-inflammatory GC effects in the CNS: 5-Lipoxygenase (5-LO), the enzyme crucial for the biosynthesis of inflammatory leukotrienes is present in neurons and 5-LO expression has been found to be increased in the rat brain after GC treatment [95] and during ageing [96]. Despite a general immunosenescence (decrease of immunological parameters like phagocytosis, cell trafficking, etc.) some increased immune functions like IL-4, IL-6 and TNF$\alpha$ production could be observed [2]. It is well accepted that infectious diseases are generally more serious and cause higher mortality in ageing populations [81]. For example influenza-infected, aged mice had a blunted immune response and a lower survival rate than age-matched controls [97]. However the situation seems to be quite different regarding viral infections of the nervous system. It was found that newborn guinea pigs were significantly more susceptible than adult animals to the development of a herpes simplex virus infection of the trigeminal ganglion [98]. In a very recent study on Sindbis virus infected mice, Labrada et al. confirmed that several mammalian neurotropic viruses produce an age-dependent encephalitis characterized by more severe disease in younger hosts [99]. Furthermore 24 month old rats showed higher IL-6 levels and equal TNF-$\alpha$ levels after intracerebrovenrticular lipopolysaccharide treatment compared to 3 month old rats [100]. Since aged animals seem to deal better with CNS infections, these studies suggest that within the CNS elevated GC levels do not have an immunosuppressive effect. In fact a considerable number of studies demonstrate an age-related increase in immunological activity in the brain despite elevated GC levels: The age related increase of glial fibrillary acidic protein (GFAP) expression demonstrates an increased activation of astrocytes [101–103]. There is also evidence that microglial cells show an increased activation with age [104–108]. On a functional level these microglia show increased phagocytosis [109], increased major histocompatibility class II (MHC II) expression, and an increased secretion of proinflammatory cytokines [110,111]. In a very recent study using microarray and real time RT-PCR technology on 3, 12, 18, and 24 month old mice, Terao and colleagues found that in the ageing hippocampus immune response gene expression was significantly increased. Among those were T cell activation genes, proinflammatory cytokines and chemokines [112].

All these examples indicate that GCs not only suppress immune function as known in the periphery but have also the potential to enhance certain parts of the immune system especially in the CNS. We could also confirm this in a recent study from our group [113]:

We investigated the GC effect on lesion size, cellular inflammatory infiltrate and mRNA cytokine pattern after excitotoxic brain injury in the rat hippocampus. Compared to basal GC levels acute high GC levels reduced these inflammatory cells at early timepoints but cell numbers were increased later on, suggesting a delaying and pro-inflammatory effect of GCs. Even more surprisingly, chronically elevated GC levels resulted in rapid infiltration and a further increase of cellular infiltrate compared to the other two groups. In contrast to their immunosuppressive effects in the periphery (reduction of cellular inflammatory infiltrate) GCs seemed to have a pro-inflammatory effect (increase of total inflammatory cell numbers) in this model of CNS inflammation. We found elevated mRNA levels of the pro-inflammatory cytokines IL-$1\alpha$, IL-$1\beta$, IL-6 and TNF-$\alpha$ in all three groups after kainate injection. Consistent with known peripheral anti-inflammatory GC effects (Table I), acute high GC levels inhibited IL-$1\alpha$, IL-$1\beta$, TNF-$\alpha$ but not IL-6 mRNA synthesis compared to basal GC levels. In contrast, chronically high GC levels caused an increase of these four pro-inflammatory cytokine messages compared to basal GC levels, thus revealing yet another unexpected pro-inflammatory effect of GCs in CNS inflammation. These findings suggest that the physiological role of GCs, especially in the CNS, is not merely immunosuppressive, but can also be pro-inflammatory.

Although immunological function in the periphery has been demonstrated to decline with age [114], the immune response in the brain seems on the contrary seems to become more activated with age. This is especially interesting since these age-related increases in the inflammatory potential of the brain could be an important factor in the pathogenesis of neurodegenerative diseases like Alzheimer´s disease that are associated with a inflammatory reaction [115,116]. For example the pro-inflammatory cytokines IL-$1\beta$, IL-6 and TNF-$\alpha$ have all been reported to play a role in the formation of neuritic plaques in AD [117–119].

## 4.    STRESS AND AGEING

The hippocampus, a brain part vital for learning and memory, has one of the highest GC receptor concentrations in the brain [120,121], and is therefore particularly sensitive to GC effects. In a striking correlation, immune response genes such as cytokines and their respective receptors are also highly expressed in the hippocampus [122–124]. Regarding (neuro-)inflammation it is important to note that besides psychological and physiological stimuli, proinflammatory cytokines such as IL-1 proved to be potent stimulators of the HPA axis [125–127].

A part of the well-known pharmacological effects of glucocorticoid hormones (GC) as immunodepressive agents, the direct effects of many HPA axis hormones on immune functions is actually documented also in physiologic conditions. Conversely, soluble mediators of the immune system are reported to affect the HPA axis functions at various steps of HPA axis regulation, in both physiologic and pathologic conditions. Stress and ageing may represent two paradigmatic conditions to show the relevance of the bidirectional network between the HPA axis and the immune system.

## ACKNOWLEDGEMENTS

Work in our lab was supported by grants from the DFG to KD. Funding was also provided to RMS by NIH grant RO1 MH53814.

## REFERENCES

1.    Lupien SJ, de Leon M, de Santi S, Convit A, Tarshish C, Nair NP, et al. Cortisol levels during human aging predict hippocampal atrophy and memory deficits. Nat Neurosci

1998;1(1):69–73.

2.  Straub RH, Miller LE, Scholmerich J, Zietz B. Cytokines and hormones as possible links between endocrinosenescence and immunosenescence. J Neuroimmunol 2000;109(1): 10–5.

3.  Reul JM, de Kloet ER. Two receptor systems for corticosterone in rat brain: microdistribution and differential occupation. Endocrinology 1985;117(6):2505–11.

4.  Siesjo B. Cellular and Molecular Mechanisms of Ischemic Brain Damage. NY: Lippincott-Raven; 1996.

5.  Lee JM, Zipfel GJ, Choi DW. The changing landscape of ischaemic brain injury mechanisms. Nature 1999;399(6738 Suppl):A7–14.

6.  Perez-Pinzon MA, Sick TJ, Rosenthal M. Mechanism(s) of mitochondrial hyperoxidation after global cerebral ischemia. Adv Exp Med Biol 1999;471:175–80.

7.  Sattler R, Tymianski M. Molecular mechanisms of calcium-dependent excitotoxicity. J Mol Med 2000;78(1):3–13.

8.  Choi DW. Ischemia-induced neuronal apoptosis. Curr Opin Neurobiol 1996;6(5):667–72.

9.  Roy M, Sapolsky R. Neuronal apoptosis in acute necrotic insults: why is this subject such a mess? Trends Neurosci 1999;22(10):419–22.

10. Sapolsky RM. A mechanism for glucocorticoid toxicity in the hippocampus: increased neuronal vulnerability to metabolic insults. J Neurosci 1985;5(5):1228–32.

11. Sapolsky RM. Glucocorticoid toxicity in the hippocampus: reversal by supplementation with brain fuels. J Neurosci 1986;6(8):2240–4.

12. Sapolsky RM. Glucocorticoid toxicity in the hippocampus. Temporal aspects of synergy with kainic acid. Neuroendocrinology 1986;43(3):440–4.

13. Stein-Behrens B, Mattson MP, Chang I, Yeh M, Sapolsky R. Stress exacerbates neuron loss and cytoskeletal pathology in the hippocampus. J Neurosci 1994;14(9):5373–80.

14. Supko DE, Johnston MV. Dexamethasone potentiates NMDA receptor-mediated neuronal injury in the postnatal rat. Eur J Pharmacol 1994;270(1):105–13.

15. Smith-Swintosky VL, Pettigrew LC, Sapolsky RM, Phares C, Craddock SD, Brooke SM, et al. Metyrapone, an inhibitor of glucocorticoid production, reduces brain injury induced by focal and global ischemia and seizures. J Cereb Blood Flow Metab 1996;16(4):585–98.

16. Sapolsky RM, Pulsinelli WA. Glucocorticoids potentiate ischemic injury to neurons: therapeutic implications. Science 1985;229(4720):1397–400.

17. Koide T, Wieloch TW, Siesjo BK. Chronic dexamethasone pretreatment aggravates ischemic neuronal necrosis. J Cereb Blood Flow Metab 1986;6(4):395–404.

18. Morse JK, Davis JN. Regulation of ischemic hippocampal damage in the gerbil: adrenalectomy alters the rate of CA1 cell disappearance. Exp Neurol 1990;110(1):86–92.

19. Tuor UI, Chumas PD, Del Bigio MR. Prevention of hypoxic-ischemic damage with dexamethasone is dependent on age and not influenced by fasting. Exp Neurol 1995;132(1): 116–22.

20. Amoroso D, Kindel G, Wulfert E, Hanin I. Long-term exposure to high levels of corticosterone aggravates AF64A-induced cholinergic hypofunction in rat hippocampus in vivo. Brain Res 1994;661(1–2):9–18.

21. Hortnagl H, Berger ML, Havelec L, Hornykiewicz O. Role of glucocorticoids in the cholinergic degeneration in rat hippocampus induced by ethylcholine aziridinium (AF64A). J Neurosci 1993;13(7):2939–45.

22. Johnson M, Stone DM, Bush LG, Hanson GR, Gibb JW. Glucocorticoids and 3,4-

methylenedioxymethamphetamine (MDMA)-induced neurotoxicity. Eur J Pharmacol 1989;161(2–3):181–8.

23. Stein BA, Sapolsky RM. Chemical adrenalectomy reduces hippocampal damage induced by kainic acid. Brain Res 1988;473(1):175–80.

24. Sapolsky RM, Packan DR, Vale WW. Glucocorticoid toxicity in the hippocampus: in vitro demonstration. Brain Res 1988;453(1–2):367–71.

25. Packan DR, Sapolsky RM. Glucocorticoid endangerment of the hippocampus: tissue, steroid and receptor specificity. Neuroendocrinology 1990;51(6):613–8.

26. Goodman Y, Bruce AJ, Cheng B, Mattson MP. Estrogens attenuate and corticosterone exacerbates excitotoxicity, oxidative injury, and amyloid beta-peptide toxicity in hippocampal neurons. J Neurochem 1996;66(5):1836–44.

27. Rajan V, Edwards CR, Seckl JR. 11 beta-Hydroxysteroid dehydrogenase in cultured hippocampal cells reactivates inert 11-dehydrocorticosterone, potentiating neurotoxicity. J Neurosci 1996;16(1):65–70.

28. Tombaugh GC, Yang SH, Swanson RA, Sapolsky RM. Glucocorticoids exacerbate hypoxic and hypoglycemic hippocampal injury in vitro: biochemical correlates and a role for astrocytes. J Neurochem 1992;59(1):137–46.

29. Chou YC. Corticosterone exacerbates cyanide-induced cell death in hippocampal cultures: role of astrocytes. Neurochem Int 1998;32(3):219–26.

30. McIntosh LJ, Sapolsky RM. Glucocorticoids increase the accumulation of reactive oxygen species and enhance adriamycin-induced toxicity in neuronal culture. Exp Neurol 1996;141(2):201–6.

31. Limoges J, Persidsky Y, Bock P, Gendelman HE. Dexamethasone therapy worsens the neuropathology of human immunodeficiency virus type 1 encephalitis in SCID mice. J Infect Dis 1997;175(6):1368–81.

32. Brooke S, Chan R, Howard S, Sapolsky R. Endocrine modulation of the neurotoxicity of gp120: implications for AIDS-related dementia complex. Proc Natl Acad Sci U S A 1997;94(17):9457–62.

33. Brooke SM, Howard SA, Sapolsky RM. Energy dependency of glucocorticoid exacerbation of gp120 neurotoxicity. J Neurochem 1998;71(3):1187–93.

34. Howard SA, Nakayama AY, Brooke SM, Sapolsky RM. Glucocorticoid modulation of gp120-induced effects on calcium-dependent degenerative events in primary hippocampal and cortical cultures. Exp Neurol 1999;158(1):164–70.

35. Yusim A, Franklin L, Brooke S, Ajilore O, Sapolsky R. Glucocorticoids exacerbate the deleterious effects of gp120 in hippocampal and cortical explants. J Neurochem 2000;74(3):1000–7.

36. Iyer AM, Brooke SM, Sapolsky RM. Glucocorticoids interact with gp120 in causing neurotoxicity in striatal cultures. Brain Res 1998;808(2):305–9.

37. Uhler TA, Frim DM, Pakzaban P, Isacson O. The effects of megadose methylprednisolone and U-78517F on toxicity mediated by glutamate receptors in the rat neostriatum. Neurosurgery 1994;34(1):122–7; discussion 127–8.

38. Abraham I, Veenema AH, Nyakas C, Harkany T, Bohus BG, Luiten PG. Effect of corticosterone and adrenalectomy on NMDA-induced cholinergic cell death in rat magnocellular nucleus basalis. J Neuroendocrinol 1997;9(9):713–20.

39. Abraham I, Harkany T, Horvath KM, Veenema AH, Penke B, Nyakas C, et al. Chronic corticosterone administration dose-dependently modulates Abeta(1–42)- and NMDA-induced neurodegeneration in rat magnocellular nucleus basalis. J Neuroendocrinol

2000;12(6):486–94.

40. Sunanda, Rao BS, Raju TR. Corticosterone attenuates zinc-induced neurotoxicity in primary hippocampal cultures. Brain Res 1998;791(1–2):295–8.

41. Barks JD, Post M, Tuor UI. Dexamethasone prevents hypoxic-ischemic brain damage in the neonatal rat. Pediatr Res 1991;29(6):558–63.

42. Tuor UI, Simone CS, Arellano R, Tanswell K, Post M. Glucocorticoid prevention of neonatal hypoxic-ischemic damage: role of hyperglycemia and antioxidant enzymes. Brain Res 1993;604(1–2):165–72.

43. Tuor UI, Simone CS, Barks JD, Post M. Dexamethasone prevents cerebral infarction without affecting cerebral blood flow in neonatal rats. Stroke 1993;24(3):452–7.

44. Chumas PD, Del Bigio MR, Drake JM, Tuor UI. A comparison of the protective effect of dexamethasone to other potential prophylactic agents in a neonatal rat model of cerebral hypoxia-ischemia. J Neurosurg 1993;79(3):414–20.

45. Bracken MB, Shepard MJ, Collins WF, Holford TR, Young W, Baskin DS, et al. A randomized, controlled trial of methylprednisolone or naloxone in the treatment of acute spinal-cord injury. Results of the Second National Acute Spinal Cord Injury Study. N Engl J Med 1990;322(20):1405–11.

46. Braughler JM, Hall ED. Current application of "high-dose" steroid therapy for CNS injury. A pharmacological perspective. J Neurosurg 1985;62(6):806–10.

47. Young W, Flamm ES. Effect of high-dose corticosteroid therapy on blood flow, evoked potentials, and extracellular calcium in experimental spinal injury. J Neurosurg 1982;57(5): 667–73.

48. Sapolsky RM. Glucocorticoids, stress, and their adverse neurological effects: relevance to aging. Exp Gerontol 1999;34(6):721–32.

49. Kadekaro M, Ito M, Gross PM. Local cerebral glucose utilization is increased in acutely adrenalectomized rats. Neuroendocrinology 1988;47(4):329–34.

50. Doyle P, Rohner-Jeanrenaud F, Jeanrenaud B. Local cerebral glucose utilization in brains of lean and genetically obese (fa/fa) rats. Am J Physiol 1993;264(1 Pt 1):E29–36.

51. Doyle P, Rohner-Jeanrenaud F, Jeanrenaud B. Alterations of local cerebral glucose utilization in lean and obese fa/fa rats after acute adrenalectomy. Brain Res 1994;655(1–2): 115–20.

52. Freo U, Holloway HW, Kalogeras K, Rapoport SI, Soncrant TT. Adrenalectomy or metyrapone-pretreatment abolishes cerebral metabolic responses to the serotonin agonist 1-(2,5-dimethoxy-4-iodophenyl)-2-aminopropane (DOI) in the hippocampus. Brain Res 1992;586(2):256–64.

53. Horner HC, Packan DR, Sapolsky RM. Glucocorticoids inhibit glucose transport in cultured hippocampal neurons and glia. Neuroendocrinology 1990;52(1):57–64.

54. Virgin CE, Jr., Ha TP, Packan DR, Tombaugh GC, Yang SH, Horner HC, et al. Glucocorticoids inhibit glucose transport and glutamate uptake in hippocampal astrocytes: implications for glucocorticoid neurotoxicity. J Neurochem 1991;57(4):1422–8.

55. Tombaugh GC, Sapolsky RM. Corticosterone accelerates hypoxia- and cyanide-induced ATP loss in cultured hippocampal astrocytes. Brain Res 1992;588(1):154–8.

56. Lawrence MS, Sapolsky RM. Glucocorticoids accelerate ATP loss following metabolic insults in cultured hippocampal neurons. Brain Res 1994;646(2):303–6.

57. Chou YC, Lin WJ, Sapolsky RM. Glucocorticoids increase extracellular [3H]D-aspartate overflow in hippocampal cultures during cyanide-induced ischemia. Brain Res 1994;654(1):8–14.

58. Stein-Behrens BA, Elliott EM, Miller CA, Schilling JW, Newcombe R, Sapolsky RM. Glucocorticoids exacerbate kainic acid-induced extracellular accumulation of excitatory amino acids in the rat hippocampus. J Neurochem 1992;58(5):1730–5.

59. Lowy MT, Wittenberg L, Yamamoto BK. Effect of acute stress on hippocampal glutamate levels and spectrin proteolysis in young and aged rats. J Neurochem 1995;65(1):268–74.

60. Elliott EM, Sapolsky RM. Corticosterone enhances kainic acid-induced calcium elevation in cultured hippocampal neurons. J Neurochem 1992;59(3):1033–40.

61. Elliott EM, Mattson MP, Vanderklish P, Lynch G, Chang I, Sapolsky RM. Corticosterone exacerbates kainate-induced alterations in hippocampal tau immunoreactivity and spectrin proteolysis in vivo. J Neurochem 1993;61(1):57–67.

62. Sattler R, Charlton MP, Hafner M, Tymianski M. Distinct influx pathways, not calcium load, determine neuronal vulnerability to calcium neurotoxicity. J Neurochem 1998;71(6): 2349–64.

63. Elliott EM, Sapolsky RM. Corticosterone impairs hippocampal neuronal calcium regulation--possible mediating mechanisms. Brain Res 1993;602(1):84–90.

64. McIntosh LJ, Hong KE, Sapolsky RM. Glucocorticoids may alter antioxidant enzyme capacity in the brain: baseline studies. Brain Res 1998;791(1–2):209–14.

65. Brooke SM, Sapolsky RM. The effects of steroid hormones in HIV-related neurotoxicity: a mini review. Biol Psychiatry 2000;48(9):881–93.

66. Brooke SM, McLaughlin JR, Cortopassi KM, Sapolsky RM. Effect of GP120 on glutathione peroxidase activity in cortical cultures and the interaction with steroid hormones. J Neurochem 2002;81(2):277–84.

67. Patel R, McIntosh L, McLaughlin J, Brooke S, Nimon V, Sapolsky R. Disruptive effects of glucocorticoids on glutathione peroxidase biochemistry in hippocampal cultures. J Neurochem 2002;82(1):118–25.

68. Abraham I, Juhasz G, Kekesi KA, Kovacs KJ. Corticosterone peak is responsible for stress-induced elevation of glutamate in the hippocampus. Stress 1998;2(3):171–81.

69. Venero C, Borrell J. Rapid glucocorticoid effects on excitatory amino acid levels in the hippocampus: a microdialysis study in freely moving rats. Eur J Neurosci 1999;11(7): 2465–73.

70. Cato AC, Wade E. Molecular mechanisms of anti-inflammatory action of glucocorticoids. Bioessays 1996;18(5):371–8.

71. Marx J. How the glucocorticoids suppress immunity. Science 1995;270(5234):232–3.

72. Kern JA, Lamb RJ, Reed JC, Daniele RP, Nowell PC. Dexamethasone inhibition of interleukin 1 beta production by human monocytes. Posttranscriptional mechanisms. J Clin Invest 1988;81(1):237–44.

73. Goulding NJ, Euzger HS, Butt SK, Perretti M. Novel pathways for glucocorticoid effects on neutrophils in chronic inflammation. Inflamm Res 1998;47 Suppl 3:S158–65.

74. Perretti M, Flower RJ. Cytokines, glucocorticoids and lipocortins in the control of neutrophil migration. Pharmacol Res 1994;30(1):53–9.

75. Chrousos GP. The hypothalamic-pituitary-adrenal axis and immune-mediated inflammation. N Engl J Med 1995;332(20):1351–62.

76. Reagan LP, McEwen BS. Controversies surrounding glucocorticoid-mediated cell death in the hippocampus. J Chem Neuroanat 1997;13(3):149–67.

77. Hench P KE, Slocumb CH, Polley HF. T. he effects of a hormone of the adrenal cortex and of pituitary adrenocorticotropic hormone on rheumatoid arthritis. Proc Mayo Clin 1949;24:181–197.

78.    Lloberas J, Celada A. Effect of aging on macrophage function. Exp Gerontol 2002;37(12): 1325–31.

79.    Cronstein BN, Kimmel SC, Levin RI, Martiniuk F, Weissmann G. A mechanism for the antiinflammatory effects of corticosteroids: the glucocorticoid receptor regulates leukocyte adhesion to endothelial cells and expression of endothelial-leukocyte adhesion molecule 1 and intercellular adhesion molecule 1. Proc Natl Acad Sci U S A 1992;89(21):9991–5.

80.    Getting SJ, Flower RJ, Perretti M. Inhibition of neutrophil and monocyte recruitment by endogenous and exogenous lipocortin 1. Br J Pharmacol 1997;120(6):1075–82.

81.    Gavazzi G, Krause KH. Ageing and infection. Lancet Infect Dis 2002;2(11):659–66.

82.    Gardner EM, Murasko DM. Age-related changes in Type 1 and Type 2 cytokine production in humans. Biogerontology 2002;3(5):271–90.

83.    Kiecolt-Glaser JK, Dura JR, Speicher CE, Trask OJ, Glaser R. Spousal caregivers of dementia victims: longitudinal changes in immunity and health. Psychosom Med 1991;53(4):345–62.

84.    Castle S, Wilkins S, Heck E, Tanzy K, Fahey J. Depression in caregivers of demented patients is associated with altered immunity: impaired proliferative capacity, increased CD8+, and a decline in lymphocytes with surface signal transduction molecules (CD38+) and a cytotoxicity marker (CD56+ CD8+). Clin Exp Immunol 1995;101(3):487–93.

85.    Vedhara K, Cox NK, Wilcock GK, Perks P, Hunt M, Anderson S, et al. Chronic stress in elderly carers of dementia patients and antibody response to influenza vaccination. Lancet 1999;353(9153):627–31.

86.    Bauer ME, Vedhara K, Perks P, Wilcock GK, Lightman SL, Shanks N. Chronic stress in caregivers of dementia patients is associated with reduced lymphocyte sensitivity to glucocorticoids. J Neuroimmunol 2000;103(1):84–92.

87.    Rohleder N, Kudielka BM, Hellhammer DH, Wolf JM, Kirschbaum C. Age and sex steroid-related changes in glucocorticoid sensitivity of pro-inflammatory cytokine production after psychosocial stress. J Neuroimmunol 2002;126(1–2):69–77.

88.    Bruccoleri A, Pennypacker KR, Harry GJ. Effect of dexamethasone on elevated cytokine mRNA levels in chemical-induced hippocampal injury. J Neurosci Res 1999;57(6):916–26.

89.    Millikan CH MF, Easton JD. Progressing stroke. Philadelphia, PA: Lea&Febiger; 1987.

90.    Asanuma M, Nishibayashi-Asanuma S, Miyazaki I, Kohno M, Ogawa N. Neuroprotective effects of non-steroidal anti-inflammatory drugs by direct scavenging of nitric oxide radicals. J Neurochem 2001;76(6):1895–904.

91.    Breitner JC. The role of anti-inflammatory drugs in the prevention and treatment of Alzheimer's disease. Annu Rev Med 1996;47:401–11.

92.    Fishman RA. Steroids in the treatment of brain edema. N Engl J Med 1982;306(6):359–60.

93.    Tominaga T, Katagi H, Ohnishi ST. Is Ca2+ -activated potassium efflux involved in the formation of ischemic brain edema? Brain Res 1988;460(2):376–8.

94.    Aisen PS, Davis KL, Berg JD, Schafer K, Campbell K, Thomas RG, et al. A randomized controlled trial of prednisone in Alzheimer's disease. Alzheimer's Disease Cooperative Study. Neurology 2000;54(3):588–93.

95.    Uz T, Dwivedi Y, Savani PD, Impagnatiello F, Pandey G, Manev H. Glucocorticoids stimulate inflammatory 5-lipoxygenase gene expression and protein translocation in the brain. J Neurochem 1999;73(2):693–9.

96.    Uz T, Pesold C, Longone P, Manev H. Aging-associated up-regulation of neuronal 5-

lipoxygenase expression: putative role in neuronal vulnerability. Faseb J 1998;12(6): 439–49.

97. Padgett DA, MacCallum RC, Sheridan JF. Stress exacerbates age-related decrements in the immune response to an experimental influenza viral infection. J Gerontol A Biol Sci Med Sci 1998;53(5):B347–53.

98. Tenser RB, Hsiung GD. Pathogenesis of latent herpes simplex virus infection of the trigeminal ganglion in guinea pigs: effects of age, passive immunization, and hydrocortisone. Infect Immun 1977;16(1):69–74.

99. Labrada L, Liang XH, Zheng W, Johnston C, Levine B. Age-dependent resistance to lethal alphavirus encephalitis in mice: analysis of gene expression in the central nervous system and identification of a novel interferon-inducible protective gene, mouse ISG12. J Virol 2002;76(22):11688–703.

100. Terrazzino S, Perego C, De Luigi A, De Simoni MG. Interleukin-6, tumor necrosis factor and corticosterone induction by central lipopolysaccharide in aged rats. Life Sci 1997;61(7):695–701.

101. Goss JR, Finch CE, Morgan DG. Age-related changes in glial fibrillary acidic protein mRNA in the mouse brain. Neurobiol Aging 1991;12(2):165–70.

102. Morgan TE, Rozovsky I, Goldsmith SK, Stone DJ, Yoshida T, Finch CE. Increased transcription of the astrocyte gene GFAP during middle-age is attenuated by food restriction: implications for the role of oxidative stress. Free Radic Biol Med 1997;23(3): 524–8.

103. Nichols NR, Day JR, Laping NJ, Johnson SA, Finch CE. GFAP mRNA increases with age in rat and human brain. Neurobiol Aging 1993;14(5):421–9.

104. Mattiace LA, Davies P, Dickson DW. Detection of HLA-DR on microglia in the human brain is a function of both clinical and technical factors. Am J Pathol 1990;136(5):1101–14.

105. Sheng JG, Mrak RE, Griffin WS. Enlarged and phagocytic, but not primed, interleukin-1 alpha-immunoreactive microglia increase with age in normal human brain. Acta Neuropathol (Berl) 1998;95(3):229–34.

106. Peters A, Josephson K, Vincent SL. Effects of aging on the neuroglial cells and pericytes within area 17 of the rhesus monkey cerebral cortex. Anat Rec 1991;229(3):384–98.

107. Sheffield LG, Berman NE. Microglial expression of MHC class II increases in normal aging of nonhuman primates. Neurobiol Aging 1998;19(1):47–55.

108. Rozovsky I, Finch CE, Morgan TE. Age-related activation of microglia and astrocytes: in vitro studies show persistent phenotypes of aging, increased proliferation, and resistance to down-regulation. Neurobiol Aging 1998;19(1):97–103.

109. Dickson DW, Wertkin A, Kress Y, Ksiezak-Reding H, Yen SH. Ubiquitin immunoreactive structures in normal human brains. Distribution and developmental aspects. Lab Invest 1990;63(1):87–99.

110. Nanamiya W, Takao T, Asaba K, De Souza EB, Hashimoto K. Effect of orchidectomy on the age-related modulation of IL-1beta and IL-1 receptors following lipopolysaccharide treatment in the mouse. Neuroimmunomodulation 2000;8(1):13–9.

111. Takao T, Nagano I, Tojo C, Takemura T, Makino S, Hashimoto K, et al. Age-related reciprocal modulation of interleukin-1beta and interleukin-1 receptors in the mouse brain-endocrine-immune axis. Neuroimmunomodulation 1996;3(4):205–12.

112. Terao A, Apte-Deshpande A, Dousman L, Morairty S, Eynon BP, Kilduff TS, et al. Immune response gene expression increases in the aging murine hippocampus. J Neuroimmunol

2002;132(1–2):99–112.

113. Dinkel K, MacPherson A, Sapolsky RM. Novel glucocorticoid effects on acute inflammation in the CNS. J Neurochem 2003;84(4):705–16.

114. Miller RA. The aging immune system: primer and prospectus. Science 1996;273(5271): 70–4.

115. McGeer PL, McGeer EG. The inflammatory response system of brain: implications for therapy of Alzheimer and other neurodegenerative diseases. Brain Res Brain Res Rev 1995;21(2):195–218.

116. McGeer EG, McGeer, PL. Neurodegeneration and the immune system. Philadelphia, PA: Saunders; 1994.

117. Griffin WS, Sheng JG, Roberts GW, Mrak RE. Interleukin-1 expression in different plaque types in Alzheimer's disease: significance in plaque evolution. J Neuropathol Exp Neurol 1995;54(2):276–81.

118. Patterson PH. Cytokines in Alzheimer's disease and multiple sclerosis. Curr Opin Neurobiol 1995;5(5):642–6.

119. Strauss S, Bauer J, Ganter U, Jonas U, Berger M, Volk B. Detection of interleukin-6 and alpha 2-macroglobulin immunoreactivity in cortex and hippocampus of Alzheimer's disease patients. Lab Invest 1992;66(2):223–30.

120. McEwen BS, De Kloet ER, Rostene W. Adrenal steroid receptors and actions in the nervous system. Physiol Rev 1986;66(4):1121–88.

121. Sapolsky RM. The physiological relevance of glucocorticoid endangerment of the hippocampus. Ann N Y Acad Sci 1994;746:294–304; discussion 304–7.

122. Besedovsky HO, del Rey A. Immune-neuro-endocrine interactions: facts and hypotheses. Endocr Rev 1996;17(1):64–102.

123. Haas HS, Schauenstein K. Neuroimmunomodulation via limbic structures--the neuroanatomy of psychoimmunology. Prog Neurobiol 1997;51(2):195–222.

124. Murray CA, Clements MP, Lynch MA. Interleukin-1 induces lipid peroxidation and membrane changes in rat hippocampus: An age-related study. Gerontology 1999;45(3): 136–42.

125. Besedovsky H, del Rey A, Sorkin E, Dinarello CA. Immunoregulatory feedback between interleukin-1 and glucocorticoid hormones. Science 1986;233(4764):652–4.

126. Berkenbosch F, van Oers J, del Rey A, Tilders F, Besedovsky H. Corticotropin-releasing factor-producing neurons in the rat activated by interleukin-1. Science 1987;238(4826): 524–6.

127. Sapolsky R, Rivier C, Yamamoto G, Plotsky P, Vale W. Interleukin-1 stimulates the secretion of hypothalamic corticotropin-releasing factor. Science 1987;238(4826):522–4.

The Neuroendocrine Immune Network in Ageing
Edited by R.H. Straub and E. Mocchegiani

# Neuroendocrine Immune Aspects of Osteoporosis During the Ageing Process

MEINRAD PETERLIK

*Department of Pathophysiology, University of Vienna Medical School, A-1090 Vienna, Waehringer Guertel 18-20, Austria*

## ABSTRACT

Osteoporosis (Greek for "porous bone") describes a pathological condition of the skeleton as it results from accelerated bone loss. The disease is multifactorial in origin and afflicts both women and men with increasing frequency at advanced age. Reduction of bone mass and mineral density can be caused by any perturbation of the bone remodeling process that leads to an increase of bone resorption at the expense of bone formation. This review describes (i) the role of a complicated network of neuroendocrine and immune factors, which involves steroid hormones (1,25-dihydroxyvitamin $D_3$, estrogens, androgens, corticosteroids), neuropeptides and neurotransmitters (NPY, $\alpha$-MSH, leptin, calcitonin, CGRP, VIP) as well as cytokines (IL-1, IL-4, IL-6, IFN-$\gamma$), in systemic and local control of bone turnover, (ii) focuses on age-related changes in the cross-talk between the neuro-endocrine and the immune system, and (iii) evaluates their relevance for the development of osteoporosis.

## 1. INTRODUCTION

After cessation of longitudinal skeletal growth, bone can still acquire additional mass and mineral density until the so-called peak bone mass is reached. This is the case between 25-30 years of age. From then on, bone mass is invariably declining. Therefore, both the absolute amount of the peak bone mass as well as the rate of involutional bone loss must be considered the two critical factors that determine the extent of bone mineral density in later life. Physiologically, involution of mineralized bone leads to osteopenia, a status of reduced bone mass and bone mineral density. Osteoporosis is the manifestation of a systemic skeletal disorder which afflicts both genders with increasing incidence in elderly people. The disease is characterized by severe loss of bone mass in combination with typical microarchitectural deterioration of bone tissue (for review, see [1]). Compromised bone strength predisposes a person to an increased risk of fracture at typical sites of the skeleton [2]. The disease is certainly of multifactorial origin, since apart from ageing (for recent review, [3]) and genetic disposition [4], mechanical, nutritional as well as hormonal factors [5-7] have been identified as causes of accelerated bone loss. It must be noted, that low bone mineral density is the most important though not the only contributor to the impairment of bone strength [2].

## 2.   THE BONE REMODELING PROCESS

During the entire lifetime of a vertebral organism, skeletal tissue is undergoing cellular and metabolic turnover in a cyclic process that is known as "bone remodeling" [8]. This allows the adult skeleton to regenerate continuously and, at the same time, to maintain the mineral density of individual bones at fairly constant levels. Under physiologic conditions, bone loss normally does not exceed the amount of 1% per year. This can be the case only if the rate of bone resorption by osteoclasts is matched by the rate of bone formation by osteoblasts. The tight coupling of bone formation and resorption is achieved through the coordinate action of systemic osteotropic hormones and local factors on proliferation, differentiation and function of osteoblasts as well as of osteoclasts. Hence, any defect in control of bone remodeling leading to an excess of resorption over formation of bone, must be seen as a potential risk factor for the development of osteoporosis [9,10].

### 2.1.   Bone formation/resorption coupling

The bone forming osteoblasts are mesenchymal cells, which originate from bone marrow stromal cells, whereas bone-resorbing osteoclasts derive from undifferentiated hematopoetic cells. Systemic hormones and locally acting growth factors and cytokines initiate commitment and expansion of undifferentiated precursor cells along either the osteoblastic or the monocytic lineage, and promote their further differentiation into mature osteoblasts [11] or osteoclasts, respectively [12].

### 2.1.1.   Regulation of osteoblast differentiation and activity

Stromal bone marrow cells are pluripotent mesenchymal stem cells [13] which can differentiate into osteoblasts and chondrocytes, but also into other cell types such as fibroblasts, adipocytes or myoblasts [14]. Members of the family of bone morphogenetic proteins induce the transition from pluripotent stem cells into commited osteoprogenitor cells [15]. Further growth expansion is achieved by growth factors such as TGF-ß, PDGF, FGF, and IGF [16]. It has been confirmed recently that also parathyroid hormone (PTH), under certain conditions, causes proliferation of osteoblastic cells [17]. The four phases of the bone formation process, e.g. proliferation, matrix deposition, maturation and mineralization, are the result of stepwise further differentiation of osteoprogentior cells and pre-osteoblasts into mature osteoblasts. The process is driven by phase-specific genetic programming of the respective cellular activities [16]. Apart from growth factors, estrogens and androgens stimulate proliferation and matrix deposition, whereas these processes are negatively affected by glucocorticoids. 1,25-dihydroxyvitamin $D_3$ ($1,25-(OH)_2D_3$) [18] and also triiodothyronine [19] are indispensable for final transition of osteoblasts into a fully differentiated state (for detailed review on osteoblast functions in health and disease, see [20]).

### 2.1.2.   Regulation of osteoclast differentiation and activity

It must be noted that formation of multinucleated mature osteoclasts requires the close contact of osteoclast progenitors with cells of the osteoblast lineage within a unique spatial and temporal structure known as the basic multicellular unit. The osteoblast is the only bone cell, which exhibits the complete repertoire of plasma membrane and nuclear receptors for all systemic osteotropic hormones, except for calcitonin, as well as for local bone-acting factors including cytokines and prostaglandins [21]. Therefore, the osteoblast is the prime target cell for factors that stimulate bone resorption, viz., PTH, TNF-$\alpha$, interleukin (IL)-1, prostaglandins (PG), or glucocorticoids, but also for factors that inhibit osteoclast differentiation and activation such as estrogens [22], TGF-$\beta$ [23], and IFN-$\gamma$ [24].

Formation of osteoclasts is initiated by the induction of a membrane-associated "osteoclast differentiation factor" (ODF) in stromal/osteoblast cells. ODF has been shown to be identical with the receptor activator of NF-$\kappa$B ligand (RANK-L), a member of the TNF ligand and receptor family, which is also expressed on T lymphocytes and dendritic cells [25]. Its cognate receptor, RANK, is expressed on mononuclear osteoclast precursors. Signaling from RANK/RANK-L interaction initiates fusion and differentiation into multinucleated mature osteoclasts. Osteoclast differentiation is inhibited by osteoprotegerin (OPG), a protein released from stromal/osteoblastic cells [26]. OPG serves as a soluble "decoy" receptor that competes with RANK for RANK-L, and thereby acts as osteoclastogenesis inhibitory factor (OCIF) [27].

According to a model proposed by Hofbauer et al. [27], the actions of all bone acting hormones, growth factors and cytokines converge on the level of RANK-L and OPG expression on stromal/osteoblastic cells and thereby modulate the pool size of active osteoclasts. The direction in which the number of active osteoclasts changes depends on whether bone-resorbing factors up-regulate the expression of RANK-L over that of OPG, or, conversely, whether osteogenic factors shift the balance towards increased OPG expression.

Not included in this model is the action of IL-6 and other gp130-signaling cytokines on bone turnover. Although considered a potent stimulator of osteoclastic bone resorption, IL-6 in fact does not act on RANK-L expression on osteoblastic cells [28]. This is consistent with earlier reports that IL-6 alone has no effect on osteoclast generation in bone marrow cultures, although signaling from the IL-6/IL-6 receptor/gp130 complex augments the osteoclast generating action of 1,25-$(OH)_2D_3$, triiodothyronine, PTH or $PGE_2$, respectively [29-31].

3.    AGEING OF BONE: ALTERATIONS IN BONE CELL PRECURSOR POOL SIZES AND DIFFERENTIATION

Ageing is the result of multiple genetically programmed mechanisms as well as stochastic events and leads to accumulation of damaging alterations in genes that code for vital cellular functions. With respect to bone formation, a number of studies demonstrate an age-related deficit in osteoblast maturation and function. Bergman and colleagues [32] found that primary marrow stromal mesenchymal stem cell cultures from older BALB/c mice yielded clearly fewer osteogenic progenitor cell colonies than those from younger animals. Similar data were reported by Kahn and coworkers [33] who also found a considerable decrease in the number of osteogenic stromal cells in bone marrow from 24-month old compared to 4-6 month-old mice. Quarto et al. [34] found a lower number of osteoprogenitor cells in bone marrow from adult when compared to aged rats. In addition, a defect in maturation of pre-osteoblasts into osteoblasts as well as in their ability to form bone was demonstrated [35]. Battmann et al. [36] demonstrated an impaired growth of human endosteal bone cells from men aged over 50 years. Pfeilschifter and coworkers [37] studying cellular outgrowth from human bone explants, observed a significant negative correlation between donor age and responsiveness of osteoblast-like cells to mitogenic cytokines and growth factors.

The senescence-accelerated mouse-P6 (SAMP6) is an interesting model of involutional osteopenia. SAMP6 mice exhibit decreased osteoblastogenesis in the bone marrow that is associated with a low rate of bone formation and reduced bone mineral density [38]. Secondary to impaired osteoblast formation, also osteoclastogenesis is decreased in these mice.

There is strong evidence that age-related osteopenia results from inversely related changes in the pool size of hematopoetic osteoclast precursor cells and osteogenic stromal cells. For example, Perkins et al. [39] found a significant age-related increase in the pool size of hematopoetic osteoclast precursor cells in bone marrow from mouse long bones. In addition, co-culture of bone marrow cells from aged mice with a stromal cell line gave rise to twice as many osteoclast-like cells than marrow cells from young animals. Recently, Makhluf et al. [40] reported that OPG mRNA expression in human bone marrow cells declines with age. Consequently, this may increase the ability of stromal cells/osteoblastic cells to support formation of osteoclasts and hence active bone resorption.

## 4.      NEURO/ENDOCRINE/IMMUNE CIRCUITS IN CONTROL OF BONE TURNOVER

Neuropeptides play an important role in bone remodelling. The hypothalamic-pituitary endocrine system regulates the production of osteotropic steroid/thyroid/vitamin D hormones. Secondly, neuropeptides can regulate bone turnover by activation of presynaptic receptors in the hypothalamic region. Alternatively, when released from nerve terminals of neurons that are widely distributed in mineralized tissue, neuropeptides may exert direct effects on bone cell functions, and thus participate in the local control of bone remodeling. Furthermore, neuropeptides affect the release of cytokines from immune cells, which, on the one hand, are potent effectors in systemic and local bone turnover, but, on the other hand, participate in feed-back regulation of the neuroendocrine system [41,42].

### 4.1.    Peripheral effects of neuropeptides on bone turnover

An intense network of nerve fibers can be demonstrated in skeletal tissues, not only in the periosteum but also within cortical bone and bone marrow. This neuro-osteogenic network expresses neuropeptides and neurotransmitters with specific effector functions on bone cell activities [43,44]. For example, calcitonin gene-related peptide (CGRP), vasoactive intestinal peptide (VIP) and neuropeptide Y (NPY) can modulate intracellular cAMP levels in osteoblastic cells in a receptor-mediated fashion [45], whereas substance P can induce resorption of cultured mouse calvarial bone [46]. These observations underscore the notion that nerves containing these neuropeptides may play a role in focal bone remodeling at various sites of the skeleton, e.g. in alveolar bone remodeling [47] or pathogenesis of arthritis [48,49].

#### 4.1.1.  Calcitonin, CGRP, and VIP

Calcitonin and calcitonin gene-related peptide (CGRP) are derived by alternative splicing from the same calcitonin gene, CALC 1. Calcitonin must be considered a neuropeptide, because it is mainly produced in the so-called C-cells of the thyroid gland, which are ultimately derived from the neural crest. It has been known since long that calcitonin by interaction with high-affinity receptors on the osteoclast plasma membrane is a potent inhibitor of osteoclastic bone resorption (for review, [50]) Nevertheless, it is still a matter of debate whether calcitonin has any physiological significance for the regulation of calcium homeostasis [51].

Calcitonin gene-related peptide (CGRP) is the major processed calcitonin gene product in neurons. Although CGRP, as a weak agonist for the calcitonin receptor, can act as an inhibitor of osteoclastic bone resorption, its main effect on skeletal metabolism seems to stem from its effect on osteoblastic cell proliferation [52]. CGRP has been detected in nerve fibers in rat periosteum, bone, bone marrow and in soft tissue adjacent to bone. The wide distribution of CGRP-containing nerve fibers in the vicinity of bone cells suggests a role for CGRP in local regulation of bone remodelling [45,53]. CGRP plasma membrane receptors are present on osteoblasts of many species including humans [54]. Activation of CGRP receptors is transduced into intracellular proliferative signaling via the PKC pathway [55].

Periosteum and bone are also innervated by sympathetic nerve fibers with immunoreactivity to VIP [56,57]. VIP has been shown to stimulate mRNA and protein expression of the osteoblastic differentiation marker, alkaline phosphatase, in mouse calvarial osteoblasts. In conjunction with the observation that VIP promotes bone nodule formation in osteoblast cultures, it is suggested that the VIP possibly controls anabolic processes in bone [58]. Another facet of VIP action on bone in vitro was described by Lundberg et al. [59], who observed a transient, probably receptor-mediated inhibition of activity of rat osteoclasts cultured on dentin slices.

### 4.1.2. NPY and α-melanocyte stimulating hormone (α-MSH)

Sympathetic nerve fibers containing NPY have been identified within and around mineralized skeletal tissue [57]. NPY may have a direct effect on bone cell function, since in vitro it modulates intracellular cAMP levels in osteoblastic cells in a receptor-mediated fashion [45,54]. α-MSH stimulates proliferation of osteoblasts in vitro through activation of the melanocortin receptor IV, which is expressed also on osteoblasts [60]. α-MSH seems to have a dual peripheral effect on bone remodeling inasmuch as it stimulates also formation of osteoclasts in bone marrow cultures. The consequent bone-resorbing effect of α-MSH apparently prevails in vivo, since subcutaneous injection of α-MSH into mice causes extensive trabecular bone loss [60].

### 4.2. Central nervous control of bone turnover

NPY and α-MSH are both downstream effectors of leptin and play a central role in the hormone's action on food intake. Signals from hypothalamic leptin receptors are transduced in a dual mode: activity of NPY-containing neurones is depressed, leading to reduced food intake, whereas release of the anorexigenic POMC gene product, α-MSH, is stimulated. Ducy et al. [61] put forward the notion that leptin is not only a regulator of body fat but also modulates central nervous control of bone remodeling. This notion has been substantiated by a wealth of experimental evidence although the exact mechanisms by which leptin exerts control on bone remodeling are not yet fully understood.

At present it is not clear to which extent α-MSH is involved in central nervous control of bone turnover [60]. There is, however, substantial evidence from the work of Herzog and his group, that NPY mediates hypothalamic control of bone remodeling [62]. It had been reported that intra-ventricular infusion of NPY leads to bone loss in mice. However, it was not clear whether this effect of NPY is mediated by secondary changes in the activity of systemic osteotropic hormones and related factors including leptin, or whether the observed reduction in bone density reflected primarily a central regulatory effect of NPY on bone remodeling. Baldock et al. [62] studied parameters of bone remodeling in neuropeptide receptor Y2-deficient mouse models. They presented evidence, which allowed to identify the Y2 receptor, which is abundant in the arcuate nucleus of the hypothalamus, as the most likely candidate for mediating the effect of NPY on bone formation. It was suggested that the effect of NPY on bone mass is selectively transduced through activation of the Y2 receptor, whereas signaling from receptors Y1 and Y3-5 mediates the effects of NPY on food intake.

Similarly, a dissociation of the effects of leptin on body adipose tissue and bone mass was reported by Takeda et al. [63] These authors concluded from their study on leptin action in appropriate knock-out and transgenic mouse models, that leptin exerts its effects on bone density solely by activation of the peripheral sympathetic nerve system. In leptin-deficient animals, pharmacological activation of β-adrenergic receptors on osteoblasts mimics the anti-osteogenic effect of leptin.

## 4.3.  Leptin, NPY and ageing

The apparent negative effects, which leptin and NPY have on bone formation by modulating central nervous control of bone remodeling, are contrasted by opposite peripheral effects on bone cell function. NPY, similar to other neuropeptides, promotes osteoblast differentiation through activation of a specific plasma membrane receptor, which is expressed together with various other neuropeptide receptors on human osteoblasts and osteogenic sarcoma cells [45,54]. Similarly, leptin regardless whether derived from adipose tissue or produced by osteoblastic cells could exert an osteogenic effect (for review, [64]), for example, through its ability to promote differentiation of osteoblasts [65] and to inhibit osteoclastogenesis through induction of OPG and inhibition of RANKL expression in bone marrow stromal cells [66,67].

Gordeladze and Reseland [68] as well as Thomas [64] have proposed a unified model for the dual mode of action of leptin on bone turnover. Accordingly, the net result of the antagonism between central and peripheral effects may depend on specific conditions, which modify local production of leptin in either the skeletal tissue, such as mechanical strain, or in the adipose tissue, such as fat deposition. An important determining factor for the efficiency of leptin is the development of central resistance to leptin with ageing [69] and with increasing body fat mass [70]. In this case, bone formation induced by peripheral leptin action prevails over the extent of bone resorption resulting from the central nervous action of the hormone. Development of central resistance to leptin can therefore be considered a physiological mechanism by which otherwise accelerated bone loss can be counteracted and, hence, the negative effects of ageing on bone turnover can be mitigated. A strong indication that an osteoprotective mechanism is increasingly active with advancing age comes from a recent report by Kudlacek et al. [71] showing that serum levels of OPG increase with age in a healthy adult population.

## 4.4.  Neuroendocrine regulation of cytokine expression: relevance for bone turnover

Immune cytokines such as the three members of the TNF-α receptor/ligand family, RANK, RANK-L and OPG, are the most distal effectors of bone turnover inasmuch as they integrate osteoblast functions with generation and activation of osteoclasts (see paragraph 2.1.2). According to the convergence hypothesis of Hofbauer et al. [27], a variety of upstream cytokines (IL-1, TNF-α) and hormones (PTH, 1,25-$(OH)_2D_3$, sex hormones, glucocorticoids) act on the expression of RANK-L and OPG and eventually control the pool size of active osteoclasts.

Periosteum, bone and bone marrow are innervated by a dense network of sympathetic nerve fibers exhibiting immunoreactivity to NPY and other neuropeptides, e.g. VIP, CGRP or substance P [57]. Upon release from nerve terminals, these neuropeptides can affect bone cell function (see paragraph 3.1) but also modulate cytokine production by immune cells [42,72], which are present in bone marrow and in the basic multicellular bone forming unit. Particularly NPY induces the release of bone-resorbing cytokines, such as IL-1, TNF-$\alpha$ and IL-6, from human peripheral blood mononuclear cells

The presence of IL-1 in autonomous nerve fibers in bone [73] indicates another feature of the neuroimmune crosstalk, since IL-1 not only acts on bone cells but also affects neurotransmitter release from nerve terminals [42]. That bidirectional communication between the neuroendocrine and the immune system is of relevance for the control of bone turnover, is illustrated best by the fact, that the most powerful bone-resorbing cytokines, IL-1, TNF-$\alpha$, IL-6, particularly when released from monocytes and T lymphocytes during an acute or chronic inflammatory response, increase the activity of the hypothalamic/pituitary/adrenal axis [41,42].

4.5.    The hypothalamic/pituitary/endocrine gland axis in control of bone remodeling: influence of ageing

Most of the hormonal changes associated with ageing are secondary to age-related changes in nutritional status, body composition or increased prevalence of certain illnesses. However, some may well have a primary role in the emergence of various phenotypic changes of ageing such as involutional bone loss. Age-dependent changes in activity of the hypothalamic/pituitary endocrine system, as reported by Mooradian and Wong [74], Kappeler et al. [75] or Goncharova and Lapin [76], are therefore of particular relevance for the pathogenesis of osteoporosis.

The mechanisms, by which age acts as a negative regulator of GH secretion, encompass increased somatostatin release, decreased GH-releasing hormone secretion, or, most likely, reduced pituitary responsiveness to GH-RH or other endogenous secretagogues [77]. Since post-pubertal falloff in GH circulating levels is twice as marked in men as in age-matched premenopausal women, whereas the rate of involutional bone in this time period is the same in both sexes, it is difficult to assess how much the physiological decrease in GH secretion contributes to osteoporotic bone loss.

Of all pituitary hormones, only GH may have a direct effect on bone remodeling inasmuch as it causes stimulation of IGF-1 production by bone cells [16]. IGF-1 in an autocrine/paracrine fashion enhances proliferation of preosteoblasts and augments bone matrix synthesis in more differentiated osteoblasts. Its expression can be upregulated, for example, by $PGE_2$ [78] and thyroid hormone [79].

Hypogonadism is an important risk factor for osteoporosis in both men and women [80]. A severe negative impact on the maintenance of an adequate bone mass and mineral density can therefore be expected to result from age- or developmental-related disturbances of neuroendocrine regulation of gonadotropin secretion causing retardation of puberty in both sexes or prolonged periods of amenorrhoe in women, respectively.

Apart from symptomatic hypogonadism, low serum levels of sex steroids in otherwise healthy individuals must be considered risk factors for osteoporosis in both women and men. There is good correlation between the decline of testosterone levels with decreased bone mineral density in elderly men (see among others, [81]). Interestingly, there is increasing evidence that androgens also contribute to normal bone mineral density in females [82], and that, consequently, low bioavailable serum testosterone is an independent risk factor for osteoporotic fractures in post-menopausal women [83]. The importance of estrogens for the maintenance of bone mass and mineral density in women is well documented (for review, see [22]). In addition, there is increasing evidence that low bioavailable estrogen is a risk factor also for male osteoporosis (see among others, [7,84]). This implies that particularly with fading of gonadal functions, production of osteoprotective sex hormones by the adrenal gland becomes increasingly important for the maintenance of bone mass. However, there is evidence for an age-related impairment in the activity of the hypothalamic/pituitary/adrenal gland axis, which specifically reduces production of adrenal androgens and consequently also of estrogens. At the same time, synthesis of anti-osteogenic glucocorticoids is apparently unaffected by the ageing process [76].

According to Mooradian and Wong [74] also the decline in activity of the hypothalamus/pituitary/thyroid gland axis is primarily an age-related phenomenon, which, in combination with end-organ resistance to thyroid hormone action may explain why clinical features of hypothyroidism are frequently observed in elderly people. It has been known for a long time that hypothyroidism is associated with reduced rates of bone formation [8].

With respect to the pathogenesis of osteoporosis one has also to take into consideration that of all pituitary cells particularly the thyreotropes exhibit high levels of the vitamin D receptor (VDR) [85], and are thus prime targets for genomic actions of $1,25\text{-}(OH)_2D_3$. Interestingly, Zofkova and Bednar [86] reported a significant stimulatory effect of orally administered $1,25\text{-}(OH)_2D_3$ on TRH-induced TSH secretion in healthy human volunteers. Thus, vitamin D deficiency, which is observed more frequently with increasing age [87], could have a negative effect on TSH secretion and consequently on thyroid gland function. The fact, that thyroid hormones have a profound effect on osteoblast maturation and function [19] and that thyroid hormones often act in a synergistic manner with $1,25\text{-}(OH)_2D_3$ in homeostatic control of mineral metabolism [88], certainly warrants the conclusion, that reduced thyroid hormone activity in the elderly could contribute to the development of involutional osteoporosis.

REFERENCES

1.    Kanis JA. Osteoporosis. Oxford: Blackwell Science, 1994.
2.    Anonymous. Osteoporosis prevention, diagnosis, and therapy. Jama 2001;285: 785–795.
3.    Pietschmann P, Gruber R, Peterlik M. Pathophysiology and Aging of Bone. In: Grampp S, editor. Radiology of Osteoporosis. Heidelberg: Springer-Verlag, 2003; 3–24.
4.    Seeman E, Hopper JL, Bach LA, Cooper ME, Parkinson E, McKay J, Jerums G. Reduced bone mass in daughters of women with osteoporosis. N Engl J Med 1989;320: 554–558.
5.    Frost HM. Vital biomechanics: proposed general concepts for skeletal adaptations to mechanical usage. Calcif Tissue Int 1988;42: 145–156.
6.    Bronner F. Calcium and osteoporosis. Am J Clin Nutr 1994;60: 831–836.
7.    Riggs BL, Khosla S, Melton LJ, 3rd. A unitary model for involutional osteoporosis: estrogen deficiency causes both type I and type II osteoporosis in postmenopausal women and contributes to bone loss in aging men. J Bone Miner Res 1998;13: 763–773.
8.    Eriksen EF, Axelrod DW, Melsen F. Bone Histomorphometry. New York: Raven Press, 1994.
9.    Manolagas SC. Birth and death of bone cells: basic regulatory mechanisms and implications for the pathogenesis and treatment of osteoporosis. Endocr Rev 2000;21: 115–137.
10.   Raisz LG. Local and systemic factors in the pathogenesis of osteoporosis. N Engl J Med 1988;318: 818–828.
11.   Karsenty G. The genetic transformation of bone biology. Genes Dev 1999;13: 3037–3051.
12.   Zaidi M, Blair HC, Moonga BS, Abe E, Huang CL. Osteoclastogenesis, bone resorption,

and osteoclast-based therapeutics. J Bone Miner Res 2003;18: 599–609.

13. Caplan AI, Elyaderani M, Mochizuki Y, Wakitani S, Goldberg VM. Principles of cartilage repair and regeneration. Clin Orthop 1997 (342): 254–269.

14. Pittenger MF, Mackay AM, Beck SC, Jaiswal RK, Douglas R, Mosca JD, Moorman MA, Simonetti DW, Craig S, Marshak DR. Multilineage potential of adult human mesenchymal stem cells. Science 1999;284: 143–147.

15. Canalis E, Economides AN, Gazzerro E. Bone morphogenetic proteins, their antagonists and the skeleton. Endocrine Rev. 2003;in press.

16. Lian JB, Stein GS, Canalis E, Robey PG, Boskey AL. Bone Formation: osteoblast lineage cells, growth factors, matrix proteins, and the mineralization process. In: Favus M, editor. Primer on the metabolic bone diseases and disorders of mineral metabolism. Philadelphia: Lippincott Williams & Wilkins, 1999; 14–29.

17. Schmidt IU, Dobnig H, Turner RT. Intermittent parathyroid hormone treatment increases osteoblast number, steady state messenger ribonucleic acid levels for osteocalcin, and bone formation in tibial metaphysis of hypophysectomized female rats. Endocrinology 1995;136: 5127–5134.

18. Owen TA, Aronow MS, Barone LM, Bettencourt B, Stein GS, Lian JB. Pleiotropic effects of vitamin D on osteoblast gene expression are related to the proliferative and differentiated state of the bone cell phenotype: dependency upon basal levels of gene expression, duration of exposure, and bone matrix competency in normal rat osteoblast cultures. Endocrinology 1991;128: 1496–1504.

19. Varga F, Rumpler M, Luegmayr E, Fratzl-Zelman N, Glantschnig H, Klaushofer K. Triiodothyronine, a regulator of osteoblastic differentiation: depression of histone H4, attenuation of c-fos/c-jun, and induction of osteocalcin expression. Calcif Tissue Int 1997;61: 404–411.

20. Peterlik M. Targeting the osteoblast for prevention and treatment of bone diseases. In: Bronner F, Farach-Carson MC, editors. Bone Formation. London: Springer-Verlag, 2003; 138–153.

21. Vaes G. Cellular biology and biochemical mechanism of bone resorption. A review of recent developments on the formation, activation, and mode of action of osteoclasts. Clin Orthop 1988 (231): 239–271.

22. Harris SA, Tau KR., Turner RT, Spelsberg TC. Estrogens and Progestins. In: Bilezikian JP, Raisz LG, Rodan GA, editors. Principles of Bone Biology. San Diego: Academic Press, 1996; 507–520.

23. Chenu C, Pfeilschifter J, Mundy GR, Roodman GD. Transforming growth factor beta inhibits formation of osteoclast-like cells in long-term human marrow cultures. Proc Natl Acad Sci U S A 1988;85: 5683–5687.

24. Peterlik M, Hoffmann O, Swetly P, Klaushofer K, Koller K. Recombinant gamma-interferon inhibits prostaglandin-mediated and parathyroid hormone-induced bone resorption in cultured neonatal mouse calvaria. FEBS Lett 1985;185: 287–290.

25. Yasuda H, Shima N, Nakagawa N, Yamaguchi K, Kinosaki M, Mochizuki S, Tomoyasu A, Yano K, Goto M, Murakami A, Tsuda E, Morinaga T, Higashio K, Udagawa N, Takahashi N, Suda T. Osteoclast differentiation factor is a ligand for osteoprotegerin/osteoclastogenesis-inhibitory factor and is identical to TRANCE/RANKL. Proc Natl Acad Sci U S A 1998;95: 3597–3602.

26. Simonet WS, Lacey DL, Dunstan CR, Kelley M, Chang MS, Luthy R, Nguyen HQ, Wooden S, Bennett L, Boone T, Shimamoto G, DeRose M, Elliott R, Colombero A, Tan

356

HL, Trail G, Sullivan J, Davy E, Bucay N, Renshaw-Gegg L, Hughes TM, Hill D, Pattison W, Campbell P, Boyle WJ, et al. Osteoprotegerin: a novel secreted protein involved in the regulation of bone density. Cell 1997;89: 309–319.

27. Hofbauer LC, Khosla S, Dunstan CR, Lacey DL, Boyle WJ, Riggs BL. The roles of osteoprotegerin and osteoprotegerin ligand in the paracrine regulation of bone resorption. J Bone Miner Res 2000;15: 2–12.

28. Hofbauer LC, Lacey DL, Dunstan CR, Spelsberg TC, Riggs BL, Khosla S. Interleukin-1beta and tumor necrosis factor-alpha, but not interleukin- 6, stimulate osteoprotegerin ligand gene expression in human osteoblastic cells. Bone 1999;25: 255–259.

29. Schiller C, Gruber R, Redlich K, Ho GM, Katzgraber F, Willheim M, Pietschmann P, Peterlik M. 17Beta-estradiol antagonizes effects of 1alpha,25-dihydroxyvitamin D3 on interleukin-6 production and osteoclast-like cell formation in mouse bone marrow primary cultures. Endocrinology 1997;138: 4567–4571.

30. Schiller C, Gruber R, Ho GM, Redlich K, Gober HJ, Katzgraber F, Willheim M, Hoffmann O, Pietschmann P, Peterlik M. Interaction of triiodothyronine with 1alpha,25-dihydroxyvitamin D3 on interleukin-6-dependent osteoclast-like cell formation in mouse bone marrow cell cultures. Bone 1998;22: 341–346.

31. Gruber R, Nothegger G, Ho GM, Willheim M, Peterlik M. Differential stimulation by PGE(2) and calcemic hormones of IL-6 in stromal/osteoblastic cells. Biochem Biophys Res Commun 2000;270: 1080–1085.

32. Bergman RJ, Gazit D, Kahn AJ, Gruber H, McDougall S, Hahn TJ. Age-related changes in osteogenic stem cells in mice. J Bone Miner Res 1996;11: 568–577.

33. Kahn A, Gibbons R, Perkins S, Gazit D. Age-related bone loss. A hypothesis and initial assessment in mice. Clin Orthop 1995 (313): 69–75.

34. Quarto R, Thomas D, Liang CT. Bone progenitor cell deficits and the age-associated decline in bone repair capacity. Calcif Tissue Int 1995;56: 123–129.

35. Roholl PJ, Blauw E, Zurcher C, Dormans JA, Theuns HM. Evidence for a diminished maturation of preosteoblasts into osteoblasts during aging in rats: an ultrastructural analysis. J Bone Miner Res 1994;9: 355–366.

36. Battmann A, Jundt G, Schulz A. Endosteal human bone cells (EBC) show age-related activity in vitro. Exp Clin Endocrinol Diabetes 1997;105: 98–102.

37. Pfeilschifter J, Diel I, Pilz U, Brunotte K, Naumann A, Ziegler R. Mitogenic responsiveness of human bone cells in vitro to hormones and growth factors decreases with age. J Bone Miner Res 1993;8: 707–717.

38. Jilka RL, Weinstein RS, Takahashi K, Parfitt AM, Manolagas SC. Linkage of decreased bone mass with impaired osteoblastogenesis in a murine model of accelerated senescence. J Clin Invest 1996;97: 1732–1740.

39. Perkins SL, Gibbons R, Kling S, Kahn AJ. Age-related bone loss in mice is associated with an increased osteoclast progenitor pool. Bone 1994;15: 65–72.

40. Makhluf HA, Mueller SM, Mizuno S, Glowacki J. Age-related decline in osteoprotegerin expression by human bone marrow cells cultured in three-dimensional collagen sponges. Biochem Biophys Res Commun 2000;268: 669–672.

41. Chesnokova V, Melmed S. Minireview: Neuro-immuno-endocrine modulation of the hypothalamic- pituitary-adrenal (HPA) axis by gp130 signaling molecules. Endocrinology 2002;143: 1571–1574.

42. Straub RH, Westermann J, Scholmerich J, Falk W. Dialogue between the CNS and the immune system in lymphoid organs. Immunol Today 1998;19: 409–413.

43.  Lundberg P, Lerner UH. Expression and regulatory role of receptors for vasoactive intestinal peptide in bone cells. Microsc Res Tech 2002;58: 98–103.

44.  Sisask G, Bjurholm A, Ahmed M, Kreicbergs A. The development of autonomic innervation in bone and joints of the rat. J Auton Nerv Syst 1996;59: 27–33.

45.  Bjurholm A, Kreicbergs A, Schultzberg M, Lerner UH. Neuroendocrine regulation of cyclic AMP formation in osteoblastic cell lines (UMR- 106–01, ROS 17/2.8, MC3T3-E1, and Saos-2) and primary bone cells. J Bone Miner Res 1992;7: 1011–1019.

46.  Sherman BE, Chole RA. A mechanism for sympathectomy-induced bone resorption in the middle ear. Otolaryngol Head Neck Surg 1995;113: 569–581.

47.  Hill EL, Turner R, Elde R. Effects of neonatal sympathectomy and capsaicin treatment on bone remodeling in rats. Neuroscience 1991;44: 747–755.

48.  Hukkanen M, Konttinen YT, Rees RG, Santavirta S, Terenghi G, Polak JM. Distribution of nerve endings and sensory neuropeptides in rat synovium, meniscus and bone. Int J Tissue React 1992;14 : 1–10.

49.  Schwab W, Bilgicyildirim A, Funk RH. Microtopography of the autonomic nerves in the rat knee: a fluorescence microscopic study. Anat Rec 1997;247: 109–118.

50.  Copp DH. Calcitonin comparative endocrinology. In: DeGroot L, editor. Endocrinology. New York: Grune & Stratton, 1979; 637–645.

51.  Deftos LJ, Roos, BA., Oates EL. Calcitonin. In: Favus MJ, editor. Primer on the Metabolic Bone Diseases and Disorders of Mineral Metabolism. Philadelphia: Lippincott Williams & Wilkins, 1999; 99–104.

52.  Zaidi M, Chambers TJ, Bevis PJ, Beacham JL, Gaines Das RE, MacIntyre I. Effects of peptides from the calcitonin genes on bone and bone cells. Q J Exp Physiol 1988;73: 471–485.

53.  Irie K, Hara-Irie F, Ozawa H, Yajima T. Calcitonin gene-related peptide (CGRP)-containing nerve fibers in bone tissue and their involvement in bone remodeling. Microsc Res Tech 2002;58: 85–90.

54.  Togari A, Arai M, Mizutani S, Koshihara Y, Nagatsu T. Expression of mRNAs for neuropeptide receptors and beta-adrenergic receptors in human osteoblasts and human osteogenic sarcoma cells. Neurosci Lett 1997;233: 125–128.

55.  Villa I, Dal Fiume C, Maestroni A, Rubinacci A, Ravasi F, Guidobono F. Human osteoblast-like cell proliferation induced by calcitonin-related peptides involves PKC activity. Am J Physiol Endocrinol Metab 2003;284:E627-E633.

56.  Hohmann EL, Elde RP, Rysavy JA, Einzig S, Gebhard RL. Innervation of periosteum and bone by sympathetic vasoactive intestinal peptide-containing nerve fibers. Science 1986;232: 868–871.

57.  Hill EL, Elde R. Distribution of CGRP-, VIP-, D beta H-, SP-, and NPY-immunoreactive nerves in the periosteum of the rat. Cell Tissue Res 1991;264: 469–480.

58.  Lundberg P, Bostrom I, Mukohyama H, Bjurholm A, Smans K, Lerner UH. Neuro-hormonal control of bone metabolism: vasoactive intestinal peptide stimulates alkaline phosphatase activity and mRNA expression in mouse calvarial osteoblasts as well as calcium accumulation mineralized bone nodules. Regul Pept 1999;85: 47–58.

59.  Lundberg P, Lie A, Bjurholm A, Lehenkari PP, Horton MA, Lerner UH, Ransjo M. Vasoactive intestinal peptide regulates osteoclast activity via specific binding sites on both osteoclasts and osteoblasts. Bone 2000;27: 803–810.

60.  Cornish J, Callon KE, Mountjoy KG, Bava U, Lin JM, Myers DE, Naot D, Reid IR. α-Melanocyte stimulating hormone (α-MSH) is a novel regulator of bone. Am J Physiol

Endocrinol Metab 2003;4:4.

61. Ducy P, Amling M, Takeda S, Priemel M, Schilling AF, Beil FT, Shen J, Vinson C, Rueger JM, Karsenty G. Leptin inhibits bone formation through a hypothalamic relay: a central control of bone mass. Cell 2000;100: 197–207.

62. Baldock PA, Sainsbury A, Couzens M, Enriquez RF, Thomas GP, Gardiner EM, Herzog H. Hypothalamic Y2 receptors regulate bone formation. J Clin Invest 2002;109: 915–921.

63. Takeda S, Elefteriou F, Levasseur R, Liu X, Zhao L, Parker KL, Armstrong D, Ducy P, Karsenty G. Leptin regulates bone formation via the sympathetic nervous system. Cell 2002;111: 305–317.

64. Thomas T, Burguera B. Is leptin the link between fat and bone mass? J Bone Miner Res 2002;17: 1563–1569.

65. Thomas T, Gori F, Khosla S, Jensen MD, Burguera B, Riggs BL. Leptin acts on human marrow stromal cells to enhance differentiation to osteoblasts and to inhibit differentiation to adipocytes. Endocrinology 1999;140: 1630–1638.

66. Burguera B, Hofbauer LC, Thomas T, Gori F, Evans GL, Khosla S, Riggs BL, Turner RT. Leptin reduces ovariectomy-induced bone loss in rats. Endocrinology 2001;142: 3546–3553.

67. Holloway WR, Collier FM, Aitken CJ, Myers DE, Hodge JM, Malakellis M, Gough TJ, Collier GR, Nicholson GC. Leptin inhibits osteoclast generation. J Bone Miner Res 2002;17: 200–209.

68. Gordeladze JO, Reseland JE. A unified model for the action of leptin on bone turnover. J Cell Biochem 2003;88: 706–712.

69. Li H, Matheny M, Nicolson M, Tumer N, Scarpace PJ. Leptin gene expression increases with age independent of increasing adiposity in rats. Diabetes 1997;46: 2035–2039.

70. Fernandez-Galaz C, Fernandez-Agullo T, Perez C, Peralta S, Arribas C, Andres A, Carrascosa JM, Ros M. Long-term food restriction prevents ageing-associated central leptin resistance in wistar rats. Diabetologia 2002;45: 997–1003.

71. Kudlacek S, Schneider B, Woloszczuk W., Pietschmann P, Willvonseder, R. Serum levels of osteoprotegerin increase with age in a healthy adult population. Bone 2003:in print.

72. Bedoui S, Kawamura N, Straub RH, Pabst R, Yamamura T, von Horsten S. Relevance of neuropeptide Y for the neuroimmune crosstalk. J Neuroimmunol 2003;134: 1–11.

73. Bjurholm A, Ahmed M, Svenson SB, Kreicbergs A, Schultzberg M. Interleukin-1 immunoreactive nerve fibres in rat joint synovium. Clin Exp Rheumatol 1994;12: 583–587.

74. Mooradian AD, Wong NC. Age-related changes in thyroid hormone action. Eur J Endocrinol 1994;131: 451–461.

75. Kappeler L, Gourdji D, Zizzari P, Bluet-Pajot MT, Epelbaum J. Age-associated changes in hypothalamic and pituitary neuroendocrine gene expression in the rat. J Neuroendocrinol 2003;15: 592–601.

76. Goncharova ND, Lapin BA. Effects of aging on hypothalamic-pituitary-adrenal system function in non-human primates. Mech Ageing Dev 2002;123 : 1191–1201.

77. Iranmanesh A, Veldhuis JD. Clinical pathophysiology of the somatotropic (GH) axis in adults. Endocrinol Metab Clin North Am 1992;21: 783–816.

78. McCarthy TL, Centrella M, Raisz LG, Canalis E. Prostaglandin E2 stimulates insulin-like growth factor I synthesis in osteoblast-enriched cultures from fetal rat bone. Endocrinology 1991;128: 2895–2900.

79. Varga F, Rumpler M, Klaushofer K. Thyroid hormones increase insulin-like growth

factor mRNA levels in the clonal osteoblastic cell line MC3T3-E1. FEBS Lett 1994;345: 67–70.

80. Stepan JJ, Lachman M, Zverina J, Pacovsky V, Baylink DJ. Castrated men exhibit bone loss: effect of calcitonin treatment on biochemical indices of bone remodeling. J Clin Endocrinol Metab 1989;69: 523–527.

81. Murphy S, Khaw KT, Cassidy A, Compston JE. Sex hormones and bone mineral density in elderly men. Bone Miner 1993;20: 133–140.

82. Daniel M, Martin AD, Drinkwater DT. Cigarette smoking, steroid hormones, and bone mineral density in young women. Calcif Tissue Int 1992;50: 300–305.

83. Jassal SK, Barrett-Connor E, Edelstein SL. Low bioavailable testosterone levels predict future height loss in postmenopausal women. J Bone Miner Res 1995;10: 650–654.

84. Pietschmann P, Kudlacek S, Grisar J, Spitzauer S, Woloszczuk W, Willvonseder R, Peterlik M. Bone turnover markers and sex hormones in men with idiopathic osteoporosis. Eur J Clin Invest 2001;31: 444–451.

85. Stumpf WE. Vitamin Dn-soltriol the heliogenic steroid hormone: somatotrophic activator and modulator. Discoveries from histochemical studies lead to new concepts. Histochemistry 1988;89: 209–219.

86. Zofkova I, Bednar J. Effect of 1.25 (OH)2 vitamin D3 (calcitriol) on TRH-induced thyrotropin secretion in man. Exp Clin Endocrinol 1988;91: 7–12.

87. Kudlacek S, Schneider B, Peterlik M, Leb G, Klaushofer K, Weber K, Woloszczuk W, Willvonseder R. Assessment of vitamin D and calcium status in healthy adult Austrians. Eur J Clin Invest 2003;33: 323–331.

88. Cross HS, Klaushofer K, Peterlik M. Triiodothyronine and 1,25-dihydroxyvitamin D3: Actions and interactions at the gene, cell and organ level. In: Bronner F, Peterlik M, editors. Extra- and Intracellular Calcium and Phosphate Regulation: From Basic Research to Clinical Medicine. Boca Raton: CRC Press, 1992; 93–109.

# VI. THE AGEING PROCESS AND CHRONIC INFLAMMATORY DISEASES

# Neuroendocrine Immune Mechanisms of Accelerated Ageing in Patients with Chronic Inflammatory Diseases

RAINER H. STRAUB[1], JÜRGEN SCHÖLMERICH[1] and MAURIZIO CUTOLO[2]

[1]*Laboratory of Neuroendocrinoimmunology, Dept. of Internal Medicine I, University Hospital, 3042 Regensburg, Germany;* [2]*Division of Rheumatology, Department of Internal Medicine and Medical Specialties, University of Genova, Italy*

## ABSTRACT

Among the processes that characterize ageing in healthy individuals the best recognized include atherosclerosis and osteoporosis. Similar as in healthy subjects, premature atherosclerosis has been recognized as an important factor for the morbidity and mortality in patients with chronic inflammatory diseases [1–8]. In chronic inflammatory diseases, independent of glucocorticoid therapy, we observe other phenomena of premature ageing such as osteoporosis [9–12], risk of thrombosis and hypercoagulation [13–15], actinically degenerate elastic tissue [16], muscle wasting [17,18], and sleep disorders [19–22]. At the moment, the general rule behind accelerated ageing is not known. This review tries to link phenomena of premature ageing in chronic inflammatory diseases with systemic inflammation.

## 1. SYSTEMIC INFLAMMATION IN CHRONIC INFLAMMATORY DISEASES

Elderly healthy subjects demonstrate continuous immune activation which is evident as a mild increase of serum TNF, serum interleukin (IL)-6, serum monocyte chemoattractant protein-1, serum $\beta$2-microglobulin, serum neopterin and many others [23–31]. This smoldering proinflammatory stimulation may lead to subsequent deterioration of other global systems [32]. In chronic inflammatory diseases, this situation is much more pronounced because serum levels of TNF and IL-6 are much higher as compared to the situation in elderly healthy people: For example, patients with rheumatoid arthritis have much higher serum levels of TNF and IL-6 as compared to age-matched healthy controls [33]. In contrast, elderly healthy people demonstrate approximately 2 to 3 times elevated serum levels of these cytokines in relation to young people [24,25]. It may be that a slow increase of cytokines steadily influence global systems in healthy individuals during 3 to 5 decades of ageing (slow-motion) whereas the situation in a chronic inflammatory disease is accelerated due to inflammation (time-lapse within months). How might this happen ?

## 2. CHANGES OF THE ENDOCRINE SYSTEM

### 2.1. ACTH, cortisol, and adrenal androgens

Plasma ACTH and serum cortisol remain relatively stable during ageing and in chronic inflammatory diseases, but these hormones increase in relation to other steroid hormones. This hormonal shift may lead to uncoupling of regulation because we know that synchronized release or action of hormones and neurotransmitters are necessary for their additive or synergistic effects [34]. This hormone shift is particularly obvious in relation to androgens because their production is modulated by proinflammatory cytokines: In vitro, TNF, interleukin (IL)-1, and interferon-γ (IFN-γ) inhibit androgen secretion of adrenocortical and gonadal cells, particularly, due to inhibition of the P450c17 which may lead to the cortisol pathway predominance [35–38]. I.v. injection of endotoxin to healthy subjects increased serum levels of IL-6 and TNF which was related to an increase of cortisol relative to adrenal androgens [39]. A similar increase of cortisol relative to adrenal androgens was also observed in short-term inflammatory cholestasis in humans [40] which confirmed earlier studies in rats [41]. Furthermore, in patients with rheumatoid arthritis, we were able to demonstrate that anti-TNF therapy normalized function of the adrenal glands, particularly secretion of adrenal androgens relative to cortisol [42]. All these factors indicate that cytokines can inhibit adrenal androgen secretion and lead to the cortisol pathway predominance. We believe that the dissociation of cortisol relative to adrenal androgens is common in early chronic inflammatory diseases (similar during healthy ageing). Loss of adrenal androgens is important because, in the periphery, these hormones are precursors of downstream testosterone and estrogens, which is particularly evident when gonadal hormone secretion declines.

### 2.2. Gonadal hormones

During healthy ageing, we observe a gradual decrease of gonadal hormone production. The consequence of peripheral sex hormone loss is a compensatory increase of hypothalamic and pituitary hormones such as gonadotropin releasing hormone, follicle stimulating hormone, and luteinizing hormone [43]. In chronic inflammatory diseases, we observe an inflammation-dependent decrease of serum testosterone [44–46]. On the other hand, serum 17β-estradiol remains relatively stable and only tends to decrease [47–50]. The decline of serum testosterone is most probably due to the cytokine – induced inhibition of the P450c17 in gonadal glands (see above) [36–38]. Stable 17β-estradiol serum levels are due to increased conversion of remaining androgens to 17β-estradiol due to a cytokine – induced activation of the aromatase and an activation of the reducing 17β-hydroxysteroid dehydrogenase in inflamed tissue [50]. Due to relatively low serum levels of testosterone and 17β-estradiol, increased serum levels of follicle stimulating hormone and luteinizing hormone are measurable in patients with chronic inflammatory diseases [46,51]. Thus, a state of increased inflammation leads to down-regulation of serum testosterone and other androgens whereas proinflammatory estrogens remain stable.

## 3. CHANGES OF THE METABOLIC SYSTEM

### 3.1. Homocysteine and zinc

A moderately increased serum level of homocysteine is an independent risk factor of vascular diseases such as coronary artery disease, stroke, and thrombosis [52]. During normal ageing, serum levels of homocysteine increase [53,54], which was attributed to a deficiency of folate, vitamin B6, and vitamin B12. Interestingly, serum levels of homocysteine are also increased in patients with chronic inflammatory diseases [55], which may contribute to thrombosis in these patients and other vascular sequelae [56]. At present, it remains unclear whether systemic inflammation directly leads to hyperhomocysteinemia since the direct influence of inflammatory mediators on homocysteine metabolism has not been studied. Malnutrition was suggested to be an important factor in patients with rheumatoid arthritis which leads to deficiency of folate, vitamins, pyridoxine, zinc, and magnesium [57]. These nutritional deficits would be strong factors for hyperhomocysteinemia. Another element may be malabsorption of these important nutrients which is influenced by concomitant DMARD therapy in patients with rheumatoid arthritis [58,59]. In this respect, low serum levels of zinc were found in these patients [60], which is also a phenomenon of the normal ageing process [61]. These changes may largely depend on deterioration of gastrointestinal function such as intestinal motility and exocrine secretion [62–64], which may be disturbed by proinflammatory cytokines [65]. Thus, inflammation may induce a sequence of events which can lead to hyperhomocysteinemia.

### 3.2. Lipids and lipoproteins

Another typical feature of the normal ageing process is an increase of triglycerides, oxidized LDL cholesterol and a decrease of HDL cholesterol serum levels [25,66]. Decreased serum levels of HDL cholesterol were also observed in patients with rheumatoid arthritis independent of glucocorticoid therapy [67–69]. The decreased HDL cholesterol and increased LDL cholesterol serum levels were suggested to play a significant role in premature atherosclerosis in these patients. It has been repeatedly demonstrated that proinflammatory cytokines such as TNF increase serum levels of triglycerides and cholesterol and decrease HDL cholesterol serum levels [70,71]. Thus, in patients with chronic inflammatory diseases altered lipoprotein serum levels are to be expected on the basis of inflammation.

### 3.3. Advanced glycosylation products

Another interesting ageing phenomenon is tissue accumulation of products of advanced protein glycosylation, which has been first described in diabetic patients [reviewed in 72]. Products of advanced protein glycosylation induce permanent abnormalities in extracellular matrix components, stimulate production of cytokines and reactive oxygen species, and alter intracellular proteins [72]. It is also a characteristic feature of normal ageing and has recently been described in patients with rheumatoid arthritis [73]. In this study non-enzymatic glycosylation of IgG was found, and autoantibodies against this altered IgG were described in half of the patients. These autoantibodies may prolong the half-life of altered IgG due to a diminished clearance of the immune complex. Furthermore, the diabetes–typical advanced protein glycosylation product pentosidine was found to be significantly higher in plasma of rheumatoid arthritis patients as compared to plasma of osteoarthritis patients, diabetic patients, and normal subjects [74]. Pentosidine synovial fluid levels highly significantly correlate with pentosidine plasma levels which indicates that the inflamed synovium may be the source. These glycated proteins have receptors on monocytes and generate reactive oxygen species (ROS) which leads to stimulation of TNF and IL-1 secretion from macrophages [75, 76]. ROS were found to be increased during the normal ageing process and in the synovial microenvironment in rheumatoid arthritis [77–79]. Neutrophil – derived oxygen metabolites contribute to the development of the tissue injury and inflammation. Since inflammatory mediators such as TNF stimulate production of ROS and ROS stimulate TNF a vicious cycle may appear which leads to progression of the inflammatory disease. In rheumatoid arthritis as compared to healthy ageing, inflammation might induce a much earlier entry into the vicious cycle.

## 4.  CHANGES OF THE NERVOUS SYSTEM

### 4.1.  Neuronal innervation

Data are available from normal old rats and mice which show that, in comparison to young animals, sympathetic innervation is reduced in various tissues which depends on location and species [rev. in 80]. Up to now, this phenomenon was not investigated in humans due to the difficult access to tissue. Interestingly, sympathetic nerve fibers are dramatically reduced in the inflamed synovial microenvironment of patients with rheumatoid arthritis [81, 82]. In chronic inflammatory diseases, dysregulation of autonomic nervous reflexes, increased sympathetic nervous tone, and increased plasma levels of sympathetic neurotransmitters most likely are a consequence of inflammation.

### 4.2.  Autonomic nervous reflexes

Flattening of autonomic nervous reflex amplitudes of the cardiovascular, pupillary, gastrointestinal and other systems is typical during normal ageing [83, 84]. The interaction of these autonomic nervous pathways become altered which leads to attenuation of autonomic reflex amplitudes. It is now relatively clear that similar changes of the autonomic nervous system appear in patients with different chronic inflammatory diseases [85–87]. The observed changes of the autonomic nervous reflexes partly depends on the state of inflammation [87–89]. Proinflammatory cytokines stimulate the sympathetic nervous system [90] so that a imbalance between the two portions of the autonomic nervous system may appear. Chronic inflammation may lead to uncoupling of the sympathetic and parasympathetic pathways which is detectable as autonomic nervous dysfunction. Furthermore, levels of epinephrine, norepinephrine, and neuropeptide Y were also increased in patients with chronic inflammatory diseases [88, 91–93]. This is paralleled by reduced responses to sympathetic stimuli [88, 94]. During chronic inflammation, autonomic nervous dysfunction and increased levels of sympathetic neurotransmitters are most probably a consequence of chronic inflammation.

### 4.3.  Brain functions

In elderly people, several reports describe memory and cognitive deficits and sleep disturbances. Young individuals exposed to i.v. endotoxin, which increases serum levels of TNF and IL-6, show similar changes and sometimes even overt signs of depression [95, 96]. In chronic inflam-

Table I    Some therapeutical concepts as additions to immunosuppressive therapy. Abbreviations: ACE, angiotensin converting enzyme; DHEA, dehydroepiandrosterone; GF, growth hormone; HRT, hormone replacement therapy; IGF-1, insulin-like growth factor 1.

| aging phenomenon | therapy |
| --- | --- |
| atherosclerosis | lipid lowering drugs, acetylsalicylic acid |
| osteoporosis | vitamin D, calcium, bisphosphonates, HRT |
| risk of thrombosis and hypercoagulation (von Willebrand factor, fibrinogen, platelet hyperactivity) | anticoagulants |
| actinically degenerate elastic tissue | adequate cream formulation |
| muscle wasting | training, adequate nutrition double-blind, placebo-controlled trials using GF or IGF-1 are needed |
| sleep disorders | decrease of proinflammatory load (anti-TNF therapy has good effect) |
| inadequately low cortisol | cortisol substitution (not more than substitution) |
| loss of adrenal androgens (DHEAS, DHEA, androstenedione) | DHEA substitution (the increased conversion to estrogens may be unfavorable which would need additional therapy with an aromatase inhibitor – trials are needed) |
| loss of testosterone | testosterone in male subjects, double-blind, placebo-controlled trials are needed |
| loss of progesterone | trials are needed |
| loss of IGF-1 | double-blind, placebo-controlled trials are needed |
| loss of vitamin D | vitamin D substitution (together with calcium) |
| elevated homocysteine serum levels (loss of folate, B12, B6) | folate, vitamin B12, vitamin B6 substitution |
| dyslipoproteinemia (elevated triglycerides, oxLDL, loss of HDL) | lipid lowering drugs |
| elevated advanced non-enzymatic glycosylation products activate macrophages | double-blind, placebo-controlled trials with antioxidants are needed |
| elevated reactive oxygen species | vitamin E, ascorbic acid, trials are needed |
| low zinc levels | zinc substitution |
| alteration of autonomic nervous reflexes | decrease of proinflammatory load |
| increased sympathetic nervous tone (blood pressure, heart rate) | decrease of proinflammatory load ACE inhibitor, centrally acting sympatholytic drugs |
| elevated plasma norepinephrine, epinephrine, or neuropeptide Y (relative to cortisol and other steroid hormones) | decrease of proinflammatory load ACE inhibitor, centrally acting sympatholytic drugs |
| loss of peripheral sympathetic nervous innervation | decrease of proinflammatory load, antagonizing local substance P effects (in arthritis, trials are needed), locally applied corticosteroids |
| memory deficits, cognitive deficits | decrease of proinflammatory load |

matory diseases, some studies demonstrated similar changes [97–99]. In experimental animal models, IL-1 seems to be an extremely important cytokine for memory and cognitive function [100, 101]. Most probably, the observed cognitive and memory changes in patients with chronic inflammatory diseases are due to systemic inflammation with increased systemic cytokines [102].

## 5.    THE SEQUENCE OF EVENTS OF ACCELERATED AGEING

As mentioned, during normal ageing a slow increase of serum and tissue levels of circulating pro-inflammatory cytokines is observed. This is accompanied by typical cellular ageing phenomena such as telomeric loss, oxidative damage, DNA defects, accumulation of advanced glycosylation products, cellular loss and others. These defects may be stimulators of cytokine secretion, and cytokines are released into the systemic circulation. This results in a slowly progressive endocrinosenescence and neurosenescence (slow-motion). The different factors influence each other in form of a vicious spiral [32]. For example, the loss of DHEA may cause increased secretion of IL-6 which itself may alter adrenal steroidogenesis [24].

In chronic inflammatory diseases, the proinflammatory situation is an important trigger of the downstream premature ageing process with functional deterioration of the CNS, the autonomic nervous system, the hypothalamus, the pituitary gland, peripheral endocrine glands, the gastrointestinal tract, the liver, bone, muscles, vessels, skin, kidneys and others. These changes may happen within a few months (time-lapse), depending on the degree of inflammation. In contrast to the situation during normal ageing, acute inflammation rapidly leads to deterioration of all systems in a parallel fashion.

## 6.    ARE THERE THERAPEUTIC CONSEQUENCES?

In chronic inflammatory disease, prevention of premature ageing would be the goal and this can only be reached by inhibition of inflammation. Since the entire body is involved, the inflammatory process needs to be suppressed on the systemic and local tissue level. At the moment, we use immunosuppressive or antiproliferative drugs which inhibit the immune system in a very general way (e.g. methotrexate, leflunomide, cyclophosphamide). If the available immunosuppressive drugs were to affect disease-inducing pathways in the immune system, we should see longer-lasting therapeutic effects. However, we do not see these long-lasting effects due to an incomplete consideration of many other homeostatic factors. Therapeutic approaches must be also directed towards other organ systems exterior of the immune system (Table I). The deterioration of other organ systems besides the immune system induces unfavorable feedback on local inflammatory processes, which contribute to maintenance of inflammation.

## REFERENCES

1.    Meller J, Conde CA, Deppisch LM, Donoso E, Dack S. Myocardial infarction due to coronary atherosclerosis in three young adults with systemic lupus erythematosus. Am J Cardiol 1975;35:309–314.
2.    Wolfe F, Mitchell DM, Sibley JT, Fries JF, Bloch DA, Williams CA, Spitz PW, Haga

M, Kleinheksel SM, Cathey MA. The mortality of rheumatoid arthritis. Arthritis Rheum 1994;37:481–494.

3. Geissler A, Andus T, Roth M, Kullmann F, Caesar I, Held P, Gross V, Feuerbach S, Schölmerich J. Focal white-matter lesions in brain of patients with inflammatory bowel disease. Lancet 1995;345:897–898.

4. Levy PJ, Tabares AH, Olin JW. Lower extremity arterial occlusions in young patients with Crohn's colitis and premature atherosclerosis: report of six cases. Am J Gastroenterol 1997;92:494–497.

5. Bruce IN, Gladman DD, Urowitz MB. Premature atherosclerosis in systemic lupus erythematosus. Rheum Dis Clin North Am 2000;26:257–278.

6. George J, Afek A, Gilburd B, Harats D, Shoenfeld Y. Autoimmunity in atherosclerosis: lessons from experimental models. Lupus 2000;9:223–227.

7. Park YB, Ahn CW, Choi HK, Lee SH, In BH, Lee HC, Nam CM, Lee SK. Atherosclerosis in rheumatoid arthritis: morphologic evidence obtained by carotid ultrasound. Arthritis Rheum 2002;46:1714–1719.

8. Kumeda Y, Inaba M, Goto H, Nagata M, Henmi Y, Furumitsu Y, Ishimura E, Inui K, Yutani Y, Miki T, Shoji T, Nishizawa Y. Increased thickness of the arterial intima-media detected by ultrasonography in patients with rheumatoid arthritis. Arthritis Rheum 2002;46:1489–1497.

9. Henderson NK Sambrook PN. Relationship between osteoporosis and arthritis and effect of corticosteroids and other drugs on bone. Curr Opin Rheumatol 1996;8:365–369.

10. Bjarnason I, Macpherson A, Mackintosh C, Buxton-Thomas M, Forgacs I, Moniz C. Reduced bone density in patients with inflammatory bowel disease. Gut 1997;40:228–233.

11. El Maghraoui A, Borderie D, Cherruau B, Edouard R, Dougados M, Roux C. Osteoporosis, body composition, and bone turnover in ankylosing spondylitis. J Rheumatol 1999;26:2205–2209.

12. Teichmann J, Lange U, Stracke H, Federlin K, Bretzel RG. Bone metabolism and bone mineral density of systemic lupus erythematosus at the time of diagnosis. Rheumatol Int 1999;18:137–140.

13. Federici AB, Fox RI, Espinoza LR, Zimmerman TS. Elevation of von Willebrand factor is independent of erythrocyte sedimentation rate and persists after glucocorticoid treatment in giant cell arteritis. Arthritis Rheum 1984;27:1046–1049.

14. Devine ST, Pettitt D, Arons RR. Plasma fibrinogen levels in US arthritis poupulation. Arthritis Rheum 2002;46(Suppl):S459.

15. Ames PR, Alves J, Pap AF, Ramos P, Khamashta MA, Hughes GR. Fibrinogen in systemic lupus erythematosus: more than an acute phase reactant? J Rheumatol 2000;27:1190–1195.

16. O'Brien JP Regan W. Actinically degenerate elastic tissue is the likely antigenic basis of actinic granuloma of the skin and of temporal arteritis. J Am Acad Dermatol 1999;40:214–222.

17. Walsmith J Roubenoff R. Cachexia in rheumatoid arthritis. Int J Cardiol 2002;85:89–99.

18. Hakkinen A, Sokka T, Kotaniemi A, Paananen ML, Malkia E, Kautiainen H, Hannonen P. Muscle strength characteristics and central bone mineral density in women with recent onset rheumatoid arthritis compared with healthy controls. Scand J Rheumatol 1999;28:145–151.

19. Bloom BJ, Owens JA, McGuinn M, Nobile C, Schaeffer L, Alario AJ. Sleep and its

relationship to pain, dysfunction, and disease activity in juvenile rheumatoid arthritis. J Rheumatol 2002;29:169–173.

20. Valencia-Flores M, Resendiz M, Castano VA, Santiago V, Campos RM, Sandino S, Valencia X, Alcocer J, Ramos GG, Bliwise DL. Objective and subjective sleep disturbances in patients with systemic lupus erythematosus. Arthritis Rheum 1999;42:2189–2193.

21. Zamir G, Press J, Tal A, Tarasiuk A. Sleep fragmentation in children with juvenile rheumatoid arthritis. J Rheumatol 1998;25:1191–1197.

22. Drewes AM, Svendsen L, Taagholt SJ, Bjerregard K, Nielsen KD, Hansen B. Sleep in rheumatoid arthritis: a comparison with healthy subjects and studies of sleep/wake interactions. Br J Rheumatol 1998;37:71–81.

23. Saxne T, Palladino MA, Jr., Heinegard D, Talal N, Wollheim FA. Detection of tumor necrosis factor alpha but not tumor necrosis factor beta in rheumatoid arthritis synovial fluid and serum. Arthritis Rheum 1988;31:1041–1045.

24. Straub RH, Konecna L, Hrach S, Rothe G, Kreutz M, Schölmerich J, Falk W, Lang B. Serum dehydroepiandrosterone (DHEA) and DHEA sulfate are negatively correlated with serum interleukin-6 (IL-6), and DHEA inhibits IL-6 secretion from mononuclear cells in man in vitro: possible link between endocrinosenescence and immunosenescence. J Clin Endocrinol Metab 1998;83:2012–2017.

25. Bruunsgaard H, Skinhoj P, Pedersen AN, Schroll M, Pedersen BK. Ageing, tumour necrosis factor-alpha (TNF-alpha) and atherosclerosis. Clin Exp Immunol 2000;121: 255–260.

26. Swaak AJ, van Rooyen A, Nieuwenhuis E, Aarden LA. Interleukin-6 (IL-6) in synovial fluid and serum of patients with rheumatic diseases. Scand J Rheumatol 1988;17:469–474.

27. Koch AE, Kunkel SL, Harlow LA, Johnson B, Evanoff HL, Haines GK, Burdick MD, Pope RM, Strieter RM. Enhanced production of monocyte chemoattractant protein-1 in rheumatoid arthritis. J Clin Invest 1992;90:772–779.

28. Inadera H, Egashira K, Takemoto M, Ouchi Y, Matsushima K. Increase in circulating levels of monocyte chemoattractant protein-1 with aging. J Interferon Cytokine Res 1999;19:1179–1182.

29. Manicourt D, Brauman H, Orloff S. Plasma and urinary levels of beta2 microglobulin in rheumatoid arthritis. Ann Rheum Dis 1978;37:328–332.

30. Satoh T, Brown LM, Blattner WA, Maloney EM, Kurman CC, Nelson DL, Fuchs D, Wachter H, Tollerud DJ. Serum neopterin, beta2-microglobulin, soluble interleukin-2 receptors, and immunoglobulin levels in healthy adolescents. Clin Immunol Immunopathol 1998;88:176–182.

31. Hannonen P, Tikanoja S, Hakola M, Mottonen T, Viinikka L, Oka M. Urinary neopterin index as a measure of rheumatoid activity. Scand J Rheumatol 1986;15:148–152.

32. Straub RH, Cutolo M, Zietz B, Schölmerich J. The process of aging changes the interplay of the immune, endocrine and nervous systems. Mech Ageing Dev 2001;122:1591–1611.

33. Straub RH, Müller-Ladner U, Lichtinger T, Schölmerich J, Menninger H, Lang B. Decrease of interleukin 6 during the first 12 months is a prognostic marker for clinical outcome during 36 months treatment with disease-modifying anti-rheumatic drugs. Br J Rheumatol 1997;36:1298–1303.

34. Buijs RM, Van Eden CG, Goncharuk VD, Kalsbeek A. The biological clock tunes the organs of the body: timing by hormones and the autonomic nervous system. J Endocrinol 2003;177:17–26.

35. Jäättelä M, Ilvesmaki V, Voutilainen R, Stenman UH, Saksela E. Tumor necrosis factor as a potent inhibitor of adrenocorticotropin- induced cortisol production and steroidogenic P450 enzyme gene expression in cultured human fetal adrenal cells. Endocrinology 1991;128:623–629.

36. Xiong Y Hales DB. Differential effects of tumor necrosis factor-alpha and interleukin-1 on 3 beta-hydroxysteroid dehydrogenase/delta 5-->delta 4 isomerase expression in mouse Leydig cells. Endocrine 1997;7:295–301.

37. Orava M, Voutilainen R, Vihko R. Interferon-gamma inhibits steroidogenesis and accumulation of mRNA of the steroidogenic enzymes P450scc and P450c17 in cultured porcine Leydig cells. Mol Endocrinol 1989;3:887–894.

38. Hales DB. Interleukin-1 inhibits Leydig cell steroidogenesis primarily by decreasing 17 alpha-hydroxylase/C17–20 lyase cytochrome P450 expression. Endocrinology 1992;131: 2165–2172.

39. Straub RH, Schuld A, Mullington J, Haack M, Schölmerich J, Pollmächer T. The endotoxin-induced increase of cytokines is followed by an increase of cortisol relative to dehydroepiandrosterone (DHEA) in healthy male subjects. J Endocrinol 2002;175: 467–474.

40. Zietz B, Wengler I, Messmann H, Lock G, Schölmerich J, Straub RH. Early shifts of adrenal steroid synthesis before and after relief of short-term cholestasis. J Hepatol 2001;35:329–337.

41. Swain MG, Patchev V, Vergalla J, Chrousos G, Jones EA. Suppression of hypothalamic-pituitary-adrenal axis responsiveness to stress in a rat model of acute cholestasis. J Clin Invest 1993;91:1903–1908.

42. Straub RH, Pongratz G, Schölmerich J, Kees F, Schaible TF, Antoni C, Kalden JR, Lorenz HM. Long-term anti-TNF antibody therapy in rheumatoid arthritis patients sensitizes the pituitary gland and favors adrenal androgen secretion. Arthritis Rheum 2003;48:1504–1512.

43. Greenspan FS. Basic and clinical endocrinology. East Norwalk: Appleton and Lange, 1991.

44. Gordon D, Beastall GH, Thomson JA, Sturrock RD. Androgenic status and sexual function in males with rheumatoid arthritis and ankylosing spondylitis. Q J Med 1986;60:671–679.

45. Cutolo M, Balleari E, Giusti M, Monachesi M, Accardo S. Sex hormone status of male patients with rheumatoid arthritis: evidence of low serum concentrations of testosterone at baseline and after human chorionic gonadotropin stimulation. Arthritis Rheum 1988;31: 1314–1317.

46. Martens HF, Sheets PK, Tenover JS, Dugowson CE, Bremner WJ, Starkebaum G. Decreased testosterone levels in men with rheumatoid arthritis: effect of low dose prednisone therapy. J Rheumatol 1994;21:1427–1431.

47. Deighton CM, Watson MJ, Walker DJ. Sex hormones in postmenopausal HLA-identical rheumatoid arthritis discordant sibling pairs. J Rheumatol 1992;19:1663–1667.

48. Hall GM, Perry LA, Spector TD. Depressed levels of dehydroepiandrosterone sulphate in postmenopausal women with rheumatoid arthritis but no relation with axial bone density. Ann Rheum Dis 1993;52:211–214.

49. Foppiani L, Cutolo M, Sessarego P, Sulli A, Prete C, Seriolo B, Giusti M. Desmopressin and low-dose ACTH test in rheumatoid arthritis. Eur J Endocrinol 1998;138:294–301.

50. Castagnetta LA, Cutolo M, Granata OM, Di Falco M, Bellavia V, Carruba G. Endocrine

end-points in rheumatoid arthritis. Ann N Y Acad Sci 1999;876:180–191.

51. Athreya BH, Rafferty JH, Sehgal GS, Lahita RG. Adenohypophyseal and sex hormones in pediatric rheumatic diseases. J Rheumatol 1993;20:725–730.

52. McCully KS. Homocysteine and vascular disease. Nat Med 1996;2:386–389.

53. Herrmann W, Quast S, Ullrich M, Schultze H, Bodis M, Geisel J. Hyperhomocysteinemia in high-aged subjects: relation of B-vitamins, folic acid, renal function and the methylene tetrahydrofolate reductase mutation. Atherosclerosis 1999;144:91–101.

54. Selhub J, Jacques PF, Wilson PW, Rush D, Rosenberg IH. Vitamin status and intake as primary determinants of homocysteinemia in an elderly population. JAMA 1993;270: 2693–2698.

55. Roubenoff R, Dellaripa P, Nadeau MR, Abad LW, Muldoon BA, Selhub J, Rosenberg IH. Abnormal homocysteine metabolism in rheumatoid arthritis. Arthritis Rheum 1997;40: 718–722.

56. Seriolo B, Fasciolo D, Sulli A, Cutolo M. Homocysteine and antiphospholipid antibodies in rheumatoid arthritis patients: relationships with thrombotic events. Clin Exp Rheumatol 2001;19:561–564.

57. Kremer JM Bigaouette J. Nutrient intake of patients with rheumatoid arthritis is deficient in pyridoxine, zinc, copper, and magnesium. J Rheumatol 1996;23:990–994.

58. Weber J, Werre JM, Julius HW, Marx JJ. Decreased iron absorption in patients with active rheumatoid arthritis, with and without iron deficiency. Ann Rheum Dis 1988;47: 404–409.

59. Phelan MJ, Taylor W, van Heyningen C, Williams E, Thompson RN. Intestinal absorption in patients with rheumatoid arthritis treated with methotrexate. Clin Rheumatol 1993;12: 223–225.

60. Honkanen V, Konttinen YT, Sorsa T, Hukkanen M, Kemppinen P, Santavirta S, Saari H, Westermarck T. Serum zinc, copper and selenium in rheumatoid arthritis. J Trace Elem Electrolytes Health Dis 1991;5:261–263.

61. Uza G Vlaicu R. Serum zinc and copper in patients with atherosclerosis and thromboangiitis obliterans. Biol Trace Elem Res 1989;20:197–206.

62. Henriksson K, Uvnas-Moberg K, Nord CE, Johansson C, Gullberg R. Gastrin, gastric acid secretion, and gastric microflora in patients with rheumatoid arthritis. Ann Rheum Dis 1986;45:475–483.

63. Kanerud L, Hafstrom I, Berg A. Effects of antirheumatic treatment on gastric secretory function and salivary flow in patients with rheumatoid arthritis. Clin Exp Rheumatol 1991;9:595–601.

64. Kulkarni SG, Parikh SS, Shankhpal PD, Desai SA, Borges NE, Desai SB, Vora IM, Kalro RH. Gastric emptying of solids in long-term NSAID users: correlation with endoscopic findings and Helicobacter pylori status. Am J Gastroenterol 1999;94:382–386.

65. Nishibayashi H, Kanayama S, Shinomura Y, Kawata S, Matsuzawa Y. Delayed gastric emptying during interferon-alpha therapy in patients with chronic hepatitis C: relief by cisapride. Scand J Gastroenterol 1997;32:547–551.

66. Tronel H, Antebi H, Felden F, Guerci B, Fremont S, Drouin P, Nicolas JP, Alcindor LG. Low-density lipoprotein ability to generate lipoperoxides in healthy subjects: variations according to age. Ann Nutr Metab 1997;41:160–165.

67. Heldenberg D, Caspi D, Levtov O, Werbin B, Fishel B, Yaron M. Serum lipids and lipoprotein concentrations in women with rheumatoid arthritis. Clin Rheumatol 1983;2: 387–391.

68. Lakatos J Harsanyi A. Serum total, HDL, LDL cholesterol, and triglyceride levels in patients with rheumatoid arthritis. Clin Biochem 1988;21:93–96.

69. Magaro M, Altomonte L, Zoli A, Mirone L, Ruffini MP. Serum lipid pattern and apolipoproteins (A1 and B100) in active rheumatoid arthritis. Z Rheumatol 1991;50: 168–170.

70. Feingold KR, Hardardottir I, Grunfeld C. Beneficial effects of cytokine induced hyperlipidemia. Z Ernahrungswiss 1998;37 Suppl 1:66–74:66–74.

71. Song H, Saito K, Fujigaki S, Noma A, Ishiguro H, Nagatsu T, Seishima M. IL-1 beta and TNF-alpha suppress apolipoprotein (apo) E secretion and apo A-I expression in HepG2 cells. Cytokine 1998;10:275–280.

72. Brownlee M. Advanced protein glycosylation in diabetes and aging. Annu Rev Med 1995;46:223–34.:223–234.

73. Newkirk MM, LePage K, Niwa T, Rubin L. Advanced glycation endproducts (AGE) on IgG, a target for circulating antibodies in North American Indians with rheumatoid arthritis (RA). Cell Mol Biol (Noisy -le-grand) 1998;44:1129–1138.

74. Miyata T, Ishiguro N, Yasuda Y, Ito T, Nangaku M, Iwata H, Kurokawa K. Increased pentosidine, an advanced glycation end product, in plasma and synovial fluid from patients with rheumatoid arthritis and its relation with inflammatory markers. Biochem Biophys Res Commun 1998;244:45–49.

75. Vlassara H, Brownlee M, Manogue KR, Dinarello CA, Pasagian A. Cachectin/TNF and IL-1 induced by glucose-modified proteins: role in normal tissue remodeling. Science 1988;240:1546–1548.

76. Mullarkey CJ, Edelstein D, Brownlee M. Free radical generation by early glycation products: a mechanism for accelerated atherogenesis in diabetes. Biochem Biophys Res Commun 1990;173:932–939.

77. Droge W. Free radicals in the physiological control of cell function. Physiol Rev 2002;82: 47–95.

78. Rister M. Superoxide anion and superoxide dismutase activity in arthritic conditions. Agents Actions Suppl 1981;8:137–43.:137–143.

79. Chiu PL, Davis P, Wong K, Dasgupta M. Superoxide production in neutrophils of patients with rheumatoid arthritis and Felty's syndrome. J Rheumatol 1983;10:694–700.

80. Bellinger DL, Madden KS, Lorton D, Thyagarajan S, Felten DL. Age-related alterations in neural-immune interactions and neural strategies in immunosenescence. In: Ader R, Felten DL, Cohen N, editors. Psychoneuroimmunology. San Diego: Academic Press, 2001;241–286.

81. Pereira SJ, Carmo-Fonseca M. Peptide containing nerves in human synovium: immunohistochemical evidence for decreased innervation in rheumatoid arthritis. J Rheumatol 1990;17:1592–1599.

82. Miller LE, Jüsten HP, Schölmerich J, Straub RH. The loss of sympathetic nerve fibers in the synovial tissue of patients with rheumatoid arthritis is accompanied by increased norepinephrine release from synovial macrophages. FASEB J 2000;14:2097–2107.

83. Wieling W, van Brederode JF, de Rijk LG, Borst C, Dunning AJ. Reflex control of heart rate in normal subjects in relation to age: a data base for cardiac vagal neuropathy. Diabetologia 1982;22:163–166.

84. Straub RH, Thies U, Jeron A, Palitzsch KD, Schölmerich J. Valid parameters for investigation of the pupillary light reflex in normal and diabetic subjects shown by factor analysis and partial correlation. Diabetologia 1994;37:414–419.

85.  Edmonds ME, Jones TC, Saunders WA, Sturrock RD. Autonomic neuropathy in rheumatoid arthritis. Br Med J 1979;2:173–175.

86.  Geenen R, Godaert GL, Jacobs JW, Peters ML, Bijlsma JW. Diminished autonomic nervous system responsiveness in rheumatoid arthritis of recent onset. J Rheumatol 1996;23:258–264.

87.  Straub RH, Antoniou E, Zeuner M, Gross V, Schölmerich J, Andus T. Association of autonomic nervous hyperreflexia and systemic inflammation in patients with Crohn's disease and ulcerative colitis. J Neuroimmunol 1997;80:149–157.

88.  Kuis W, Jong-de Vos v, Sinnema G, Kavelaars A, Prakken B, Helders PM, Heijnen CJ. The autonomic nervous system and the immune system in juvenile rheumatoid arthritis. Brain Behav Immun 1996;10:387–398.

89.  Straub RH, Glück T, Zeuner M, Schölmerich J, Lang B. Association of pupillary parasympathetic hyperreflexia and systemic inflammation in patients with systemic lupus erythematosus. Br J Rheumatol 1998;37:665–670.

90.  Niijima A. The afferent discharges from sensors for interleukin 1 beta in the hepatoportal system in the anesthetized rat. J Auton Nerv Syst 1996;61:287–291.

91.  Straub RH, Herfarth H, Falk W, Andus T, Schölmerich J. Uncoupling of the sympathetic nervous system and the hypothalamic-pituitary-adrenal axis in inflammatory bowel disease? J Neuroimmunol 2002;126:116–125.

92.  Alstergren P, Ernberg M, Kopp S, Lundeberg T, Theodorsson E. TMJ pain in relation to circulating neuropeptide Y, serotonin, and interleukin-1 beta in rheumatoid arthritis. J Orofac Pain 1999;13:49–55.

93.  Hirano D, Nagashima M, Ogawa R, Yoshino S. Serum levels of interleukin 6 and stress related substances indicate mental stress condition in patients with rheumatoid arthritis. J Rheumatol 2001;28:490–495.

94.  Glück T, Oertel M, Reber T, Zietz B, Schölmerich J, Straub RH. Altered function of the hypothalamic stress axes in patients with moderately active systemic lupus erythematosus. I. The hypothalamus-autonomic nervous system axis. J Rheumatol 2000;27:903–910.

95.  Pollmächer T, Schreiber W, Gudewill S, Vedder H, Fassbender K, Wiedemann K, Trachsel L, Galanos C, Holsboer F. Influence of endotoxin on nocturnal sleep in humans. Am J Physiol 1993;264:R1077-R1083.

96.  Reichenberg A, Yirmiya R, Schuld A, Kraus T, Haack M, Morag A, Pollmächer T. Cytokine-associated emotional and cognitive disturbances in humans. Arch Gen Psychiatry 2001;58:445–452.

97.  Ferstl R, Niemann T, Biehl G, Hinrichsen H, Kirch W. Neuropsychological impairment in auto-immune disease. Eur J Clin Invest 1992;22 Suppl 1:16–20.:16–20.

98.  Carbotte RM, Denburg SD, Denburg JA. Cognitive deficit associated with rheumatic diseases: neuropsychological perspectives. Arthritis Rheum 1995;38:1363–1374.

99.  Wright GE, Parker JC, Smarr KL, Johnson JC, Hewett JE, Walker SE. Age, depressive symptoms, and rheumatoid arthritis. Arthritis Rheum 1998;41:298–305.

100. Schneider H, Pitossi F, Balschun D, Wagner A, Del Rey A, Besedovsky HO. A neuromodulatory role of interleukin-1beta in the hippocampus. Proc Natl Acad Sci U S A 1998;95:7778–7783.

101. Yirmiya R, Winocur G, Goshen I. Brain interleukin-1 is involved in spatial memory and passive avoidance conditioning. Neurobiol Learn Mem 2002;78:379–389.

102. Dantzer R, Wollman EE, Yirmiya R. Brain Behavior and Immunity: Special issue on cytokines and depression. San Diego: Academic Press, 2002.

# Thyroid Autoimmunity and Ageing

STEFANO MARIOTTI[1], GIOVANNI PINNA[1] and ALDO PINCHERA[2]

*[1]Endocrinology, Department of Medical Sciences, University of Cagliari, Italy; [2]Department of Endocrinology – University of Pisa, Italy*

ABSTRACT

Ageing is associated with an increased prevalence of thyroid autoantibodies and subclinical hypothyroidism. Thyroid autoantibodies are rare in centenarians and in other highly selected aged populations, while they are frequently observed in hospitalized or unselected elderly subjects. These data suggest that thyroid autoimmune phenomena are not the consequence of the ageing process itself, but rather an expression of age-associated disease. Thyroid autoimmunity and/or thyroid dysfunction may contribute to the development of several age-associated diseases, although the precise mechanisms involved remain to be elucidated.

## 1.    INTRODUCTION

A significant proportion of the population in developed countries is growing progressively older as a consequence of increased life expectancy and reduced birth rate. The ageing thyroid shows a wide range of morphological and functional changes, especially represented by decreased serum triiodothyronine ($T_3$) and thyrotropin (TSH) concentrations, partly related to non-thyroid illnesses and/or drugs. The above alterations can also be associated with several autoimmune manifestations against the thyroid, which have been described to develop during the physiological ageing process as well as other immunological abnormalities [1]. General immune functions decrease as a consequence of senescence, and their decline is responsible for the predisposition to infections and cancer, probably related to the physiological decrease in overall functions, such as protein synthesis and gene expression. Conversely, the increase in autoimmunity could be related with an elevation in other parameters, such as mutation of genes and apoptosis. It has been proposed that the degree of apoptosis causes an increase in the number of cell components to be eliminated, which in turn could lead to an increase in autoimmune reactivity [2]. In fact, one of the age-associated immune abnormalities is the general trend toward development of autoimmune phenomena and production of both organ- and nonorgan-specific autoantibodies, including thyroid autoantibodies.

## 2. THE RELEVANCE OF THYROID AND THYROID AUTOIMMUNE DISEASE IN THE AGEING PROCESS

The age-dependent increase in the prevalence of positive serum anti-thyroglobulin (anti-TG) and anti-microsomal/thyroperoxidase (anti-TPO) autoantibodies has been recognized since the early sixties. This finding is particularly prevalent in females and over 60 years of age [3], but its precise biological and clinical significance is still only partially understood. In particular, it is not clear whether it represents the lower end of the clinical spectrum of thyroid autoimmune disease or it is one of the consequences of the ageing process on the immune system, without direct and specific involvement of the target gland.

In contrast with circulating anti-thyroid autoantibodies, the prevalence of clinically overt thyroid autoimmune diseases is generally not increased in the elderly, with the remarkable exception of those associated to hypothyroidism. The peak incidence of Graves' disease and goitrous Hashimoto's thyroiditis is reported between the third and the fourth decade, while only primary myxedema peaks over 60 years [3,4]. Furthermore, a trend to a reduction of both prevalence and titers of circulating anti-thyroid autoantibodies has been reported in older patients with clinically established thyroid autoimmune disorders [5].

It has been also suggested that anti-TG autoantibodies detected in sera of euthyroid elderly might recognize different thyroglobulin (TG) epitopes when compared to those detected in overt thyroid diseases [6]. This finding, however, has not been confirmed for anti-TPO autoantibodies [7].

Taken together, the above data suggest that age-associated thyroid autoimmune phenomena and fully expressed thyroid autoimmune disease are often separate entities. In addition, an increased incidence of both clinical and subclinical primary autoimmune hypothyroidism is commonly observed in the elderly, with values ranging from 0.5% to 6% for overt and from 4% to 10–15% for subclinical thyroid failure [3,8]. These findings seem to be sufficient to underscore that the preferential hypothyroid expression of thyroid autoimmunity in the elderly suggests an age-dependent modulation of the pathogenetic mechanisms involved.

### 2.1. Thyroid autoimmunity and ageing: animal models

The preferential hypothyroid expression of thyroid autoimmunity in the elderly could be due either to increased destructive effector mechanisms and/or to increased target organ susceptibility, and the available experimental data mainly support the second hypothesis. In fact, lymphocytes from young mice with experimental autoimmune thyroiditis transferred to aged animals cause a thyroiditis which is more severe than that observed in younger recipients, suggesting that old thyroid glands are more susceptible to the autoimmune attack [9]. On the other hand, lymphocytes from old mice with thyroiditis are less active in transferring the disease when compared to mononuclear cells obtained from young donors, suggesting an age-dependent decrease of effector autoimmune activation [9,10]. In keeping with these results are also previous studies showing a lower anti-TG autoantibody response in old mice immunized with xenogeneic TG [11]. A similar behavior has been more recently reported in the experimental autoimmune myasthenia gravis model, since aged rats immunized with acetylcholine receptor (AChR) display lower anti-AChR antibody titers and more disturbed neuromuscular transmission as compared to young animals [12]. Whether and to what extent these findings can be extrapolated to human subjects remains matter of speculation.

## 2.2. Thyroid autoimmunity and ageing: the lesson of centenarians

Taking advantage of selected healthy elderly populations, it has been possible to approach the problem of thyroid autoimmunity and ageing from a different point of view. In a first study, the presence of circulating thyroid autoantibodies has been assessed in a large group of healthy centenarians [13]. The results showed that both anti-TG and anti-TPO prevalence in centenarians was not significantly different from that observed in a large series of unselected subjects aged <50 years, and significantly lower than that found in unselected elderly subjects. A low prevalence of other organ-specific autoantibodies such as gastric parietal cell antibody has been also documented in healthy centenarians, while several nonorgan-specific autoantibodies are increased as in younger elderly groups [13,14]. This finding suggests a different behavior of age-dependent organ-specific and nonorgan-specific autoimmune phenomena in humans, although the underlying mechanism remains to be elucidated.

The low prevalence of thyroid and other autoantibodies in centenarians could be accounted for disappearance of thyroid autoantibodies present at "younger" age. An alternative explanation is selection bias. Since approximately only 1 out of 10,000 human beings becomes a centenarian [15], it can be assumed that survival might be associated with an unusually efficient immune system activity, devoid of the age-related abnormalities frequently observed in "younger" elderly. In keeping with this concept are studies on immune function in healthy subjects over 100 years [15]. An inverse relationship between "successful ageing" and appearance of autoantibodies is also supported by a brief report that the increased prevalence of serum anti-TG antibodies occurred almost exclusively in hospitalized elderly persons [16].

A study was then carried out in three selected groups of older individuals [17], including healthy centenarians, healthy elderly subjects (aged 65–80 years) selected according to the strict EURAGE-SENIEUR protocol and elderly hospitalized patients (aged 61–95 years) with various acute or chronic diseases. An increased prevalence of thyroid autoantibodies was found in unselected or hospitalized elderly groups, while the prevalence of circulating thyroid autoantibodies in centenarians and in "younger" healthy elderly was similar to that of young and middle aged adults. These results strongly support the hypothesis that the appearance of thyroid autoantibodies might be related to age-associated disease, rather than to the ageing process itself. It is conceivable that, as hypothesized for centenarians, healthy older subjects selected with strict criteria represent a population with an unusually efficient immune system. Increased prevalence of circulating anti-thyroid autoantibodies has been confirmed in recent surveys carried out in hospitalized patients, although no clear correlation was found between antibody frequency and/ or titer and the severity or the nature of the underlying non-thyroidal illnesses [18].

The prevalence of antithyroid autoantibodies has been also recently investigated in a large cohort of octo-nonagerians. In keeping with the hypothesis previously suggested, thyroid autoantibodies prevalence was increased only in disabled but not in free-living subjects [19].

## 3. RELEVANCE OF THYROID ABNORMALITIES IN THE AGEING PROCESS AND IN AGEING-ASSOCIATED DISEASES

### 3.1. Abnormalities of thyroid function

Thyroid dysfunction develops as the consequence of the autoimmune attack which has been triggered in affected glands. The clinical effect in patients and the decision to treat them is influenced by several situations, such as the entity of the dysfunction and the presence of other chronic diseases (usually present in elderly).

Autoimmune thyroiditis, both goitrous (Hashimoto's thyroiditis) or atrophic, represents the main cause of hypothyroidism in the elderly [3,20,21] and, as stated before, the prevalence of hypothyroidism in the elderly is increased.

The effect of thyroid insufficiency on the central nervous system and cognitive function in the elderly is well recognized [3,8]. Dementia due to hypothyroidism occurring in the elderly is rare, but treatable [3,8]. Several neuropsychological disturbances, including bipolar depression, slight cognitive dysfunction and impaired memory, have been described in association with different degrees of clinical and subclinical hypothyroidism. Most, but not all, of these disturbances are reversed by thyroid hormone therapy and, if they are not, other causes for the mental abnormalities are likely [3,8].

Overt hypothyroidism is associated with hyperlipidaemia, which may represent a risk factor for coronary artery disease (CAD). The decision to treat elderly patients with subclinical hypothyroidism requires careful consideration of potential adverse reactions such as aggravation of myocardial ischemia.

Although early investigations indicated a slight increase of CAD in subjects with positive thyroid antibodies, with or without subclinical hypothyroidism [3,8], a recent longitudinal study carried out at the English Community of Whickham reported no association between autoimmune thyroiditis (as defined by hypothyroidism, raised serum TSH, or positive thyroid antibodies) and CAD [22]. The effect of subclinical hypothyroidism and its treatment upon circulating lipid is also controversial [3,8]. Hypothyroidism is 2–3 times more frequent in people with elevated plasma total cholesterol (TC), and TC is slightly elevated in patients with subclinical hypothyroidism. Restoration of normal TSH levels with L-thyroxine (L-$T_4$) decreases TC by about 0.4 mmol/L, producing an average 6% reduction in TC levels that in hypercholesterolaemic patients might contribute in reducing ischemic heart disease [3,8]. There is also some, but not univocal evidence for a decrease of low-density lipoprotein cholesterol levels, mainly if pretreatment TSH values are >10 mU/L [3,8]. In addition, a more recent meta-analysis has reported that L-$T_4$ in individuals with mild thyroid failure lowers mean serum TC and LDL cholesterol concentrations, without effects in serum HDL and triglyceride concentrations. The reduction in serum TC appears to be mostly evident in those individuals with higher pre-treatment cholesterol levels and in hypothyroid individuals taking suboptimal L-$T_4$ doses [23].

The question whether aged patients with subclinical hypothyroidism should or should not be treated with thyroid hormone has been debated for more than two decades [24]. The potential advantages of treatment include: a) prevention of the progression to overt hypothyroidism, b) improvement of serum lipid profile, and c) reversal of symptoms including psychiatric and cognitive abnormalities [24]. The arguments against treatment are its expense, the uncertainty of benefits and the danger of overtreatment [25,26]. Although the controversy is still open [26,27], due to the lack of sufficient prospective randomized studies, there is a general agreement that the majority of patients with mild thyroid failure, particularly those with symptoms, hyperlipemia, goiter or high titers of anti-thyroid antibodies should be treated [24,27].

The prevalence of autoimmune hyperthyroidism (Graves' disease) in the elderly is lower than autoimmune hypothyroidism and ranges 0.5–2.3% [3,21]. It is important to remember here only the reported high incidence of embolism (10–40%) in old hyperthyroid patients with atrial fibrillation, and the increased activation of blood coagulation and fibrinolysis which has been described in hyperthyroid Graves' disease [28].

## 3.2.    Autoimmunity

Independently of accompanying changes in thyroid function, thyroid autoimmune phenomena in the elderly may be related to the development of some age-associated diseases, such as hypertension and atherosclerosis, in which autoimmune mechanisms are believed to play a relevant role [29]. Limiting our discussion to atherosclerosis, the involvement of the immune system in this condition is mainly suggested by the presence of activated T cells within the atherosclerotic lesions and of circulating autoantibodies to plaque components. Recent studies focused on the involvement of two major autoantigenic determinants in the atherogenic process, namely the heat shock protein (hsp) 60/65 [29] and oxidized low density lipoprotein [30]. According to this view, after early activation of the immune system the progression of the atherosclerotic lesion would be enhanced by focal expression of adhesion molecules and local secretion of cytokine and growth factor networks [29]. In the framework of the above *scenario*, the increased prevalence of circulating thyroid (and/or other) autoantibodies often observed in the elderly could reflect an ongoing autoimmune activation involved in the atherosclerotic process and leading to increased CAD risk. However, no direct support to this hypothesis has been provided so far. In one of the few studies addressing more directly this question, a significant correlation was found in humans between atherosclerotic lesions and circulating anti-hsp 65 antibodies, while in the same patients the arterial damage was unrelated to serum thyroid and other autoantibodies [31].

Age-associated endocrine and/or immune dysfunctions may be involved in the pathogenesis of Alzheimer Dementia (Alzheimer Dementia Type, DAT) and in other forms leading to age-associated cognitive decline. In particular, autoimmune processes, including autoimmunity against neural components, have been reported as possible pathogenic mechanisms [32]. According to early reports, thyroid disease appeared to represent a significant risk factor for DAT [33], but these data were not confirmed by several subsequent case-control studies [34,35]. The potential relevance of thyroid autoimmunity in the pathogenesis of DAT is suggested by experiments showing that cerebral vascular amyloid deposition may be induced in rabbits immunized with TG [36] and by the significant association found between thyroid autoimmunity and a familial variety of DAT [37]. The latter finding, however, could be the consequence rather the cause of the biochemical abnormalities typical of this rare neurodegenerative disease. Familial forms of DAT are indeed characterized by mutations of the Alzheimer β-amyloid precursor protein (APP) favoring the generation of soluble amyloidogenic peptides (sAPP) [38]. Since thyroid epithelial cells are able to actively synthesize APP and to secrete sAPP [39,40], increased/altered thyroid APP expression (as in familiar DAT) may induce thyroid epithelial cell alterations and damage, which could be responsible of the frequent occurrence of thyroid autoimmunity.

## 4.    CONCLUSIONS

In conclusion, ageing is associated with increased prevalence of thyroid autoantibodies and subclinical hypothyroidism. Thyroid autoantibodies are rare in centenarians and in other highly selected aged populations, while are frequently observed in hospitalized elderly. Taken together, these data suggest that thyroid autoimmune phenomena are not the consequence of the ageing process itself, but rather expression of age-associated disease. Thyroid autoimmunity and/or thyroid dysfunction may contribute to the development of several age-associated diseases, although the precise mechanisms involved remain to be elucidated.

REFERENCES

1.  Wick G, Grubeck-Loebenstein B. Primary and secondary alterations of immune reactivity in the elderly: impact of dietary factors and disease. Immunol Rev 1997;160:171–184.
2.  Higami Y, Shimokawa I, Tomita M, Okimoto T, Koji T, Kobayashi N, et al. Aging accelerates but life-long dietary restriction suppresses apoptosis-related Fas expression on hepatocytes. Am J Pathol 1997;151[3]:659–663.
3.  Mariotti S, Franceschi C, Cossarizza A, Pinchera A. The aging thyroid. Endocr Rev 1995;16[6]:686–715.
4.  Vanderpump MP, Tunbridge WM, French JM, Appleton D, Bates D, Clark F, et al. The incidence of thyroid disorders in the community: a twenty-year follow-up of the Whickham Survey. Clin Endocrinol [Oxf] 1995;43[1]:55–68.
5.  Aizawa T, Ishihara M, Hashizume K, Takasu N, Yamada T. Age-related changes of thyroid function and immunologic abnormalities in patients with hyperthyroidism due to Graves' disease. J Am Geriatr Soc 1989;37[10]:944–948.
6.  Bouanani M, Dietrich G, Hurez V, Kaveri SV, Del Rio M, Pau B, et al. Age-related changes in specificity of human natural autoantibodies to thyroglobulin. J Autoimmun 1993;6[5]:639–648.
7.  Jaume JC, Costante G, Nishikawa T, Phillips DI, Rapoport B, McLachlan SM. Thyroid peroxidase autoantibody fingerprints in hypothyroid and euthyroid individuals. I. Cross-sectional study in elderly women. J Clin Endocrinol Metab 1995;80:994–999.
8.  Chiovato L, Mariotti S, Pinchera A. Thyroid diseases in the elderly. Baillières Clin Endocrinol Metab 1997;11[2]:251–270.
9.  Romball CG, Weigle WO. The effect of aging on the induction of experimental autoimmune thyroiditis. J Immunol 1987;139[5]:1490–5.
10. Okayasu I, Hatakeyama S, Kong YC. Long-term observation and effect of age on induction of experimental autoimmune thyroiditis in susceptible and resistant mice. Clin Immunol Immunopathol 1989;53[2 Pt 1]:254–267.
11. Goidl EA, Michelis MA, Siskind GW, Weksler ME. Effect of age on the induction of autoantibodies. Clin Exp Immunol 1981;44[1]:24–30.
12. Hoedemaekers AC, Verschuuren JJ, Spaans F, Graus YF, Riemersma S, van Breda Vriesman PJ, et al. Age-related susceptibility to experimental autoimmune myasthenia gravis: immunological and electrophysiological aspects. Muscle Nerve 1997;20[9]:1091–1101.
13. Mariotti S, Sansoni P, Barbesino G, Caturegli P, Monti D, Cossarizza A, et al. Thyroid and other organ-specific autoantibodies in healthy centenarians. Lancet 1992;339[8808]:1506–1508.
14. Mariotti S, Barbesino G, Caturegli P, Bartalena L, Sansoni P, Fagnoni F, et al. Complex alteration of thyroid function in healthy centenarians. J Clin Endocrinol Metab 1993;77[5]:1130–1134.
15. Franceschi C, Monti D, Sansoni P, Cossarizza A. The immunology of exceptional individuals: the lesson of centenarians. Immunol Today 1995;16[1]:12–16.
16. Moulias R, Proust J, Wang A, Congy F, Marescot MR, Deville Chabrolle A, et al. Age-related increase in autoantibodies. Lancet 1984;1[8386]:1128–1129.
17. Pinchera A, Mariotti S, Barbesino G, Bechi R, Sansoni P, Fagiolo U, et al. Thyroid autoimmunity and ageing. Horm Res 1995;43[1–3]:64–68.
18. Szabolcs I, Bernard W, Horster FA. Thyroid autoantibodies in hospitalized chronic geriatric

patients: prevalence, effects of age, nonthyroidal clinical state, and thyroid function. J Am Geriatr Soc 1995;43[6]:670–673.

19. Mariotti S, Barbesino G, Chiovato L, Marino M, Pinchera A, Zuliani G, et al. Circulating thyroid autoantibodies in a sample of Italian octo-nonagenarians: relationship to age, sex, disability, and lipid profile. Aging [Milano] 1999;11[6]:362–366.

20. Felicetta JV. The thyroid and aging. In: Sowers JR, Felicetta JV, editors. The Endocrinology of Aging. New York: Raven; 1988. p. 15–39.

21. Mokshagundam S, Barzel US. Thyroid disease in the elderly. J Am Geriatr Soc 1993;41[12]:1361–9.

22. Vanderpump MP, Tunbridge WM, French JM, Appleton D, Bates D, Clark F, et al. The development of ischemic heart disease in relation to autoimmune thyroid disease in a 20-year follow-up study of an English community. Thyroid 1996;6[3]:155–160.

23. Danese MD, Ladenson PW, Meinert CL, Powe NR. Clinical review 115: effect of thyroxine therapy on serum lipoproteins in patients with mild thyroid failure: a quantitative review of the literature. J Clin Endocrinol Metab 2000;85[9]:2993–3001.

24. Cooper DS. Clinical practice. Subclinical hypothyroidism. N Engl J Med 2001;345[4]: 260–265.

25. Marqusee E, Haden ST, Utiger RD. Subclinical thyrotoxicosis. Endocrinol Metab Clin North Am 1998;27[1]:37–49.

26. Chu JW, Crapo LM. The treatment of subclinical hypothyroidism is seldom necessary. J Clin Endocrinol Metab 2001;86[10]:4591–4599.

27. McDermott MT, Ridgway EC. Subclinical hypothyroidism is mild thyroid failure and should be treated. J Clin Endocrinol Metab 2001;86[10]:4585–4590.

28. Marongiu F, Conti M, Murtas ML, Mameli G, Sorano GG, Martino E. Activation of blood coagulation and fibrinolysis in Graves' disease. Horm Metab Res 1991;23[12]:609–611.

29. Wick G, Schett G, Amberger A, Kleindienst R, Xu Q. Is atherosclerosis an immunologically mediated disease? Immunol Today 1995;16[1]:27–33.

30. Palinski W, Tangirala RK, Miller E, Young SG, Witztum JL. Increased autoantibody titers against epitopes of oxidized LDL in LDL receptor-deficient mice with increased atherosclerosis. Arterioscler Thromb Vasc Biol 1995;15[10]:1569–1576.

31. Xu Q, Willeit J, Marosi M, Kleindienst R, Oberhollenzer F, Kiechl S, et al. Association of serum antibodies to heat-shock protein 65 with carotid atherosclerosis. Lancet 1993;341[8840]:255–259.

32. Singh VK. Neuroautoimmunity: pathogenic implications for Alzheimer's disease. Gerontology 1997;43[1–2]:79–94.

33. Heyman A, Wilkinson WE, Hurwitz BJ, Schmechel D, Sigmon AH, Weinberg T, et al. Alzheimer's disease: genetic aspects and associated clinical disorders. Ann Neurol 1983;14[5]:507–515.

34. Yoshimasu F, Kokmen E, Hay ID, Beard CM, Offord KP, Kurland LT. The association between Alzheimer's disease and thyroid disease in Rochester, Minnesota. Neurology 1991;41[11]:1745–1747.

35. Kokmen E, Beard CM, O'Brien PC, Kurland LT. Epidemiology of dementia in Rochester, Minnesota. Mayo Clin Proc 1996;71[3]:275–282.

36. Premachandra BN, Naidu RG, Williams IK, Blumenthal HT. Cerebral vascular amyloid deposition in rabbits with induced thyroglobulin immunity. Neurosci Lett 1995;188[1]: 65–69.

37. Ewins DL, Rossor MN, Butler J, Roques PK, Mullan MJ, McGregor AM. Association

between autoimmune thyroid disease and familial Alzheimer's disease. Clin Endocrinol [Oxf] 1991;35[1]:93–96.

38. Nussbaum RL, Ellis CE. Alzheimer's disease and Parkinson's disease. N Engl J Med 2003;348[14]:1356–1364.

39. Graebert KS, Popp GM, Kehle T, Herzog V. Regulated O-glycosylation of the Alzheimer beta-A4 amyloid precursor protein in thyrocytes. Eur J Cell Biol 1995;66[1]:39–46.

40. Schmitt TL, Steiner E, Klingler P, Lassmann H, Grubeck-Loebenstein B. Thyroid epithelial cells produce large amounts of the Alzheimer beta- amyloid precursor protein [APP] and generate potentially amyloidogenic APP fragments. J Clin Endocrinol Metab 1995;80[12]: 3513–3519.

*The Neuroendocrine Immune Network in Ageing*
Edited by R.H. Straub and E. Mocchegiani
383

# The Clinical Importance of Proinflammatory Cytokines in Elderly Populations

HELLE BRUUNSGAARD and KAREN KRABBE

*Department of Infectious Diseases, Rigshospitalet, University of Copenhagen, Denmark*

ABSTRACT

Low-grade increases in the plasma level of inflammatory mediators are related to age-associated pathology including cardiovascular diseases, dementia and sarcopenia. Moreover, inflammatory cytokines such as tumour necrosis factor and interleukin-6 act as independent predictors of high mortality risk. In this review, we discuss how the biological activities of inflammatory cytokines may cause morbidity in elderly populations and we suggest that systemic low-grade inflammation provide a common link between ageing, the body composition, and life style factors on one hand and age-related diseases and mortality on the other.

## 1. INTRODUCTION

Old-age survival has improved substantially since 1950 in developed countries and in the period 1950–1990 the number of octogenarians has increased four-fold, the number of nonagenarians eight-fold and the number of centenarians twenty-fold [1]. Whether the added years at the end of the life cycle are healthy, enjoyable, and productive depends upon preventing and controlling a number of chronic diseases and conditions.

Ageing is associated with a complex reshaping of the immune system and immunosenescence is defined as a deterioration of the immune system [2]. The clinical relevance is reflected by reports of increased incidence of several infections as well as a higher mortality rate from infections in elderly populations [3]. Moreover, pneumonia together with influenza are among the top 10 causes of death in persons aged 65 years or older [4]. Even more importantly, it has also become increasingly clear during the last decade that immune mechanisms play a central part in the pathology of the major age-related, chronic diseases. Thus, atherosclerosis has been defined as an age-related inflammatory disease, representing the principal contributor to ischaemic cardiovascular diseases (CVD), which constitute the leading cause of death in elderly populations [5]. Furthermore, the prevalence of senile dementia increases exponentially until the age of 95 years [6] and dysregulation of the local as well as the peripheral cytokine production is believed to contribute importantly to cognitive dysfunction in elderly populations and the pathogenesis of Alzheimer's disease (AD) [7]. Accordingly, inflammation in the senile immune system is an integrated part of age-related pathological processes.

Circulating levels of inflammatory cytokines and their clinical importance have not been explored in population-based samples of elderly people until recently because immunogerontological studies have mainly focused on small groups of healthy elderly subjects selected by "The SENIEUR protocol" in an attempt to separate the influence of disease from physiological ageing. In our opinion, it is equally important to perform longitudinal studies of epidemiological well-described cohorts, representing approximations to normal populations in order to assess the clinical relevance of the senile immune system.

The purpose of the present review is to discuss the clinical importance of proinflammatory cytokines in elderly populations.

## 2.    THE INFLAMMATORY RESPONSE

In response to tissue damage elicited by trauma or infection the acute phase response (APR) set in, constituting a complex network of molecular and cellular interactions with the purpose to aid tissue repair and to facilitate a return to physiological homeostasis [8]. APR is composed of immediate local events as well as systemic activation.

Tumour necrosis factor (TNF)-$\alpha$ and interleukin (IL)-1$\beta$ are classical proinflammatory cytokines that are produced at the local inflammatory site by activated macrophages. They induce a second wave of cytokines including IL-6 and chemokines (eg., IL-8 and macrophage inflammatory proteins), which regulate the influx of leukocytes to the site of infection [8]. IL-6 represents a key point in the regulation of the APR. This important cytokine is often classified as a proinflammatory cytokine but it has in deed also many anti-inflammatory and immunosuppressive effects, $e.g.$, IL-6 stimulates the pituitary-adrenal axis, inhibits the synthesis of TNF-$\alpha$, stimulates the production of anti-inflammatory cytokines such as IL-1 receptor antagonist (Ra) and IL-10, and induces the shedding of TNF receptors that secondly bind and suppress the function of circulating TNF-$\alpha$ [9]. Furthermore, IL-6 induces the synthesis of acute phase proteins in the liver such as C reactive protein (CRP) [8].

The balance between proinflammatory and anti-inflammatory cytokines and the magnitude of the inflammatory response are crucial. It has been suggested that insufficient responses results in immunodeficiency, which can lead to infection and cancer whereas excessive responses cause morbidity and mortality in diseases such as atherosclerosis, diabetes, Alzheimer's disease, and autoimmune diseases or shock during acute infections [10].

## 3.    CHRONIC ELEVATIONS OF CIRCULATING CYTOKINES IN ELDERLY POPU-LATIONS

Circulating levels of TNF-$\alpha$, IL-6, IL-1Ra, and sTNFR-I and II are increased in elderly populations compared to younger controls [11,12] whereas IL-1$\beta$ is difficult to detect in the circulation. Furthermore, old people have elevated concentrations of acute phase proteins such as CRP and Serum Amyloid A [13]. Age-related increases in circulating IL-6 and mediators downstream in the APR can already be detected in middle-aged people whereas similar increases are not found before the age of 80 years with regard to TNF-$\alpha$ [11], (Figure 1). This likely explains why most immunogerontological studies have focused on IL-6.

Elevations in inflammatory mediators are, however, only 2 fold in octogenarians and 4 fold in centenarians compared to young healthy volunteers and such increases are far from levels observed during acute infections. It has naturally been questioned if this systemic low-

# Cytokines and aging

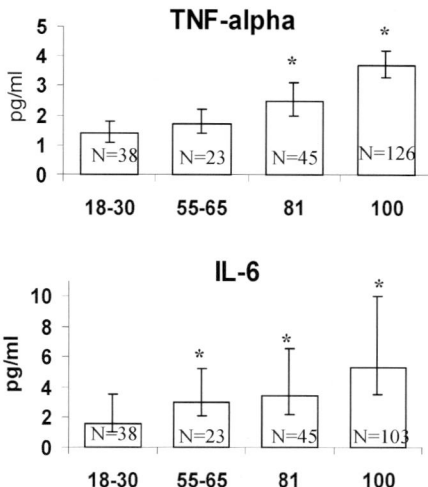

Figure 1. Plasma levels of TNF-α and IL-6 in different age groups.
Medians and interquartile range are shown. *, p<0.05 compared to the youngest group (from Ref. [11]).

grade inflammation is of any clinical importance in elderly populations. The highest levels of cytokines are found among centenarians who thus represent a good basis for studies of associations between low-grade inflammation and chronic diseases at the end stage of the ageing process whereas longitudinal studies of younger elderly such as relatively healthy octogenarians make it possible to evaluate if inflammation induces versus reflects morbidity.

## 4. CYTOKINES, MORBIDITY AND MORTALITY IN THE OLDEST OLD

### 4.1. TNF-α is associated with dementia and atherosclerosis in centenarians

A study was done on 75% (N=207) of Danes who celebrated their 100th birthday in the period April 1995 to May 1996. This Longitudinal Study of Danish Centenarians has demonstrated that the extreme life span is accompanied by a high prevalence of several common diseases and chronic conditions [14]. More than half of this population suffers from dementia and at least 70% has a history of cardiovascular diseases (CVD), representing the major disease categories in the cohort [14]. Plasma levels of TNF-α, IL-6, sTNFR-II and CRP were measured in a subset of 126 centenarians and the hypothesis was tested that high circulating levels of inflammatory mediators were related to dementia and atherosclerosis [11]. High plasma levels of TNF-α were associated with both dementia and an ankle-brachial blood pressure index below 0.9, which is a good indicator of peripheral atherosclerosis and CVD, (Figure 2). These associations were independent of each other as well as of other severe medical disorders.

The association between circulating levels of TNF-α and atheroscerlosis was confirmed in a population of 130 relatively healthy 80-year-old people who were divided into three groups with low, medium, and high plasma levels of TNF-α [15]. The prevalence of CVD was 45%

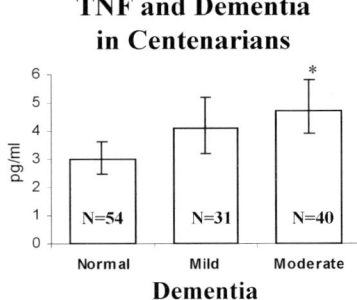

Figure 2. Plasma levels of TNF-α and age-associated diseases in centenarians.
Medians and interquartile range are shown. *, p<0.05 (from Ref. [11].

in the group with high TNF-α levels versus only 20% in the two other groups. Consistent with this, TNF-α affects several classical risk factors in CVD, *e.g.*, TNF-α causes insulin resistance, dyslipidaemia, endothelial dysfunction, and the endothelial upregulation of cellular adhesion molecules in experimental studies (reviewed in [16]). TNF-α has thus the potential to be a causative factor in atherosclerosis although it is likely that activated endothelial cells also contribute to increases in circulating levels of cytokines including TNF-α, providing a positive feedback mechanism.

The association between high plasma levels of TNF-α and dementia in centenarians seems more complex. Alzheimer's Disease (AD) and vascular dementia (VD) constitute the major categories of age-related dementia but in centenarians it is difficult to distinguish between the two categories [17]. Both types of dementia are accompanied by increased TNF-α levels in the cerebrospinal fluid and in the circulation [7]. High plasma levels of TNF-α in demented centenarians represent either a spill over from CNS inflammatory processes or an influence of peripheral cytokine dysregulation on cognitive processing. In support of the latter hypothesis, peripheral cytokines are known to penetrate the blood-brain barrier directly via active transport mechanisms or indirectly via vagal stimulation. Moreover, peripheral administration of cytokines are able to produce adverse cognitive effects in animals and humans, making it plausible that peripheral cytokines interact with cognitive ageing [7]. The role of TNF-α in brain degeneration is, however, very controversial. It is thus unresolved if local inflammatory mechanisms cause damage in CNS or if they are present to remove the detritus from other, more primary pathological processes or if endogenous TNF-α is neurotoxic (based largely on acute interventions) versus neuroprotective (based largely on studies on genetically modified animals) during inflammatory processes in CNS [18]. For instance, functional outcomes in TNF-α knock-out mice were improved early after brain injury compared with wild type mice, but TNF-α knock-out mice showed permanent deficits and reduced recovery [19].

If systemic inflammation is a cause of age-related pathology such as dementia, plasma levels of TNF-α is a marker of the magnitude of the inflammatory response and the severity of pathology and genetic polymorphisms that determine the rate of TNF-α production are important risk factors. The TNF −308G>A promoter polymorphism is common and it has been demonstrated that −308A is a much stronger transcriptional activator than the more common G allele but the effect is dependent on the cell line and the stimulus [20]. Preliminary data in our lab has demonstrated that the TNF −308GA genotype is associated with a decreased prevalence of age-related dementia compared to GG carriers among centenarians, suggesting that a balanced

TNF-α response has a protective effect [Bruunsgaard et al, unpublished data]. Consistent with this finding it has been suggested that TNF-α contributes to early neuronal injury, but improves recovery [18].

## 4.2.    TNF-α is a prognostic risk marker in centenarians

In order to further explore the relation between low-grade inflammation and morbidity we postulated that TNF-α was a specific biological risk factor in centenarians and we aimed to assess whether TNF-α was a prognostic marker of the all-cause mortality risk [21]. In support of this hypothesis, TNF-α was prognostic marker of early mortality risk in the following 5 years independently of dementia and a history of cardiovascular diseases in the previously described cohort of 126 centenarians. Accordingly, TNF-α did not simply act as a marker of these disorders although especially dementia was also strongly associated with mortality. IL-6 and IL-8 did not affect the survival function and CRP had an effect that disappeared when TNF-α was included in the analysis. These results demonstrate together that TNF-α is an independent prognostic risk marker in very old people and support the hypothesis that TNF-α has specific biological effects and causes morbidity in very old humans.

## 4.3.    TNF-α is a possible candidate to be a marker of the frailty syndrome

The independent effect of TNF-α in the survival analysis of centenarians may partly be hidden in the synonymous name of TNF-α, which is cachectin. Thus, TNF-α causes increased basal energy expenditure, anorexia, and loss of muscle and bone mass *in vivo* and TNF-α has been associated with wasting/cachexia in chronic inflammatory disorders including HIV infection, rheumatoid arthritis, and cancer [22]. Considering the high systemic levels in centenarians, it is likely that the catabolic activity of TNF-α plays a clinical role in this population. In support of this, high plasma concentrations of TNF-α and IL-6 were associated with lower muscle mass [23] and decreased muscle strength in well-functioning older men and women [24]. Furthermore, in a case-control study of frail nonagenarians who performed 12 weeks of resistance training or occupational activities, high activity in the TNF system at baseline was inversely correlated with the final muscle strength in the exercise group [25]. Sarcopenia in elderly populations has a central role in the syndrome of frailty, which has been defined as a clinical syndrome of decreased reserve and resistance to stressors characterized by weight loss, muscle weakness, poor endurance, slowness and low activity levels [26]. TNF-α has the biological potential to be a central player in such a syndrome. In accordance with this, frailty has been defined as a result from a metabolic imbalance caused by overproduction of catabolic cytokines such as TNF-α and by diminished availability or action of anabolic hormones, resulting from ageing itself and the presence of associated chronic conditions [27].

## 4.4.    The clinical significance of other inflammatory mediators in centenarians

TNF-α, IL-6, and IL-8 were strongly correlated to one another in centenarians but IL-6 and IL-8 were not associated with dementia, atherosclerosis or mortality in this population. TNF-α was also correlated with sTNFR-II and similar results emerged if TNF-α was replaced by soluble TNF receptor-II in the statistical analyses. It is often postulated that soluble TNF receptors reflect an earlier peak in TNF-α because the former is more stable in the circulation. On the other hand, soluble TNF receptors neutralize the biological functions of TNF-α and it is possible that low levels promote TNF-α related pathology. However, statistical analyses of the TNF-α/sTNFR ratio as well as simultaneous inclusions of both parameters in survival analyses have demonstrated that sTNFRs mainly reflect activities in the TNF system.

Based on our studies in centenarians, we suggest that TNF-α is an important link between inflammatory activity, chronic diseases, and mortality in frail and very old populations and TNF-α is a possible marker of the frailty syndrome. Increased levels of inflammatory mediators seem to represent a bystander phenomenon in this population. IL-6 has, however, turned out to be a major risk factor in populations of younger elderly.

## 5. THE CLINICAL SIGNIFICANCE OF CYTOKINES AND THE ROLE OF IL-6 IN YOUNGER ELDERLY

In contrast to the findings in Danish centenarians, low-grade elevations of IL-6 were associated with cognitive decline in high-functioning men and women aged 70–79 years [28], increased risk of myocardial infarcts in healthy middle-aged men [29], and mortality in relatively healthy people aged 65+ years [30] and disabled older women with a history of CVD [31]. Furthermore, IL-6 acted as a marker of subclinical CVD in a case-control study of men and women with a mean age of 73 years [32] and cross-sectional studies of different age groups indicated that the IL6-174G>C promoter gene polymorphism was associated with longevity in men beyond the age of 100 years [33].

It is possible that interleukin-6 acts as a surrogate marker for TNF-α in some epidemiological studies because the production of TNF-α and interleukin-6 is tightly linked. For instance, in a large study of American women IL-6 predicted the development of type 2 diabetes [34] even though, in contrast to TNF-α, IL-6 does not induce insulin resistance in experimental studies and IL-6 knock-out mice develop hyperglycaemia [35]. Furthermore, IL-6 has been related to decreased muscle mass/strength [24] and functional disability associated with sarcopenia [36] although mice with IL-6 producing tumours have a preserved lean body mass in spite of an increased energy expenditure [37]. The role of IL-6 in senile dementia is still controversial [38]. However, in a study of 333 relatively healthy 80-year-old Danes low-grade elevations in IL-6 was strongly associated with mortality in the following 6 years independently of TNF-α, as well as of other traditional risk factors for death, including smoking, blood pressure, physical exercise, total cholesterol, co-morbidity, BMI, and intake of anti-inflammatory drugs [39], (Figure 3). In this population TNF-α was only a predictor of mortality in men, (Figure 3). These findings indicate that low-grade elevations of TNF-alpha and IL-6 have separate biological functions that trigger age-associated pathology and cause mortality in octogenarians.

If high levels of IL-6 cause age-related pathology it is expectable that the IL6 -174G>C promoter gene polymorphism, which regulates the rate of IL-6 production will be a risk factor of morbidity and mortality in elderly populations. In accordance with this hypothesis, preliminary data from our lab indicate that this polymorphism is associated with the prevalence of CVD and mortality in the population of octogenarians but the effect appears to be complex and interacts with life style factors such as smoking status (Bruunsgaard et al, unpublished data). The IL6 -174G>C polymorphism has also been associated with plasma levels of IL-6 in elderly populations [32] and the risk of cardiovascular events and related risk factors in middle-aged populations [40]. Consistent with this, IL-6 induces dyslipidaemia [37] and a prothrombotic state [41] in experimental studies. These studies suggest together that IL-6 is a risk factor in the development of CVD.

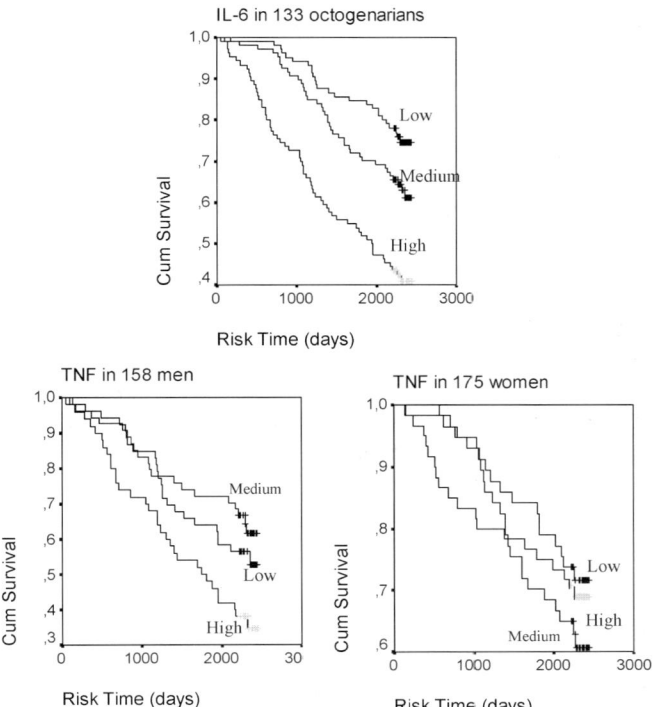

Figure 3. IL-6 in 333 octogenarians.
IL-6 was associated with mortality in both men and women. TNF-α was only associated with mortality in men (significant interaction between TNF-α and gender). The follow-up time was 5–6 years. A total of 133 people died during this period (from Ref. [39]).

Based on the literature and our own studies in relatively healthy octogenarians we suggest that TNF-α and IL-6 have separate biological activities in elderly populations. One may speculate that IL-6 is especially a risk marker of thromboembolic complications and this activity is probably very important in middle-aged populations and in relatively healthy old people but not in centenarians who have already survived to advanced age despite considerable co-morbidity. Consistent with this hypothesis, IL-6 was associated with mortality in disabled old women with a history of CVD but not in women with a high prevalence of co-morbidity but without CVD [31].

6.   TRIGGERING SIGNALS OF LOW-GRADE INFLAMMATION IN ELDERLY POP-
      ULATIONS

Cytokines are associated with mortality in elderly populations independently of co-morbidity, indicating that cytokines cause age-related disorders rather than reflect underlying pathological processes. Thus, cytokines have a wide range of biological activities and are more than just markers of disease in elderly populations. It has been hypothesized that a progressive increase in proinflammatory activity results from a continuous antigenic load and stress concomitant with a global reduction in the capacity to cope with a variety of stressors and this "inflammaging" is a major characteristic of the ageing process [42]. Consistent with this, a wide range of factors may contribute to low-grade inflammatory activity in elderly populations including life style factors, infections, and anatomical/physiological changes. For instance, smoking is associated with increased IL-6 levels among octogenarians [39]. Fatty tissue produces cytokines and ageing is accompanied by an increased percentage of body fat that is directly correlated with plasma levels of TNF-$\alpha$ and IL-6 [23]. Moreover, TNF-$\alpha$ was directly correlated with plasma levels of Leptin, which is a fat cell-derived hormone [43]. The prevalence of chronic asymptomatic infections is also increased in elderly populations and it has been demonstrated that inflammatory markers are associated with high levels of chlamydia specific IgA antibody titres in centenarians [44] and asymptomatic bacteriuria in frail elderly patients [45]. Sex hormones inhibit the production of proinflammatory cytokines and a loss of this inhibition could also contribute to low-grade inflammation [46]. Moreover, there is some evidence, suggesting that ageing is associated with a dysregulated acute phase response, which may be important in the pathogenesis of chronic age-related diseases as well as explain the increased susceptibility and the high mortality from acute infections in elderly people.

6.1.    Is ageing associated with a dysregulated cytokine response?

*In vitro* stimulations of isolated peripheral mononuclear cells (PMNC) and/or whole blood are often used to study age-associated changes in the cytokine production and based on this type of studies it has been suggested that ageing is associated with a shift in the balance between the production of proinflammatory versus anti-inflammatory cytokines, resulting in a relative increase in proinflammatory activity [47]. In a recent review of the literature we conclude that monocytes from elderly subjects are in a preactivated state reflected by a larger initial production of TNF-$\alpha$ but no difference in the peak production, (Figure 4). Controversial reports reflect that most studies include a low number of subjects and only measure cytokine production at one time point, which will largely affect the conclusion [16]. With regard to T lymphocytes most studies report increased production of TNF-$\alpha$ and IL-6 in supernatants as well as on a single cell level [16]. In accordance with this, an increased percentage and number of T lymphocytes from 80-year-old people expressed TNF-$\alpha$ following stimulation with PMA and ionomycin but centenarians did not show this altered TNF-$\alpha$ secretion profile, suggesting that T cells may contribute to the elevated levels of plasma TNF-$\alpha$ in healthy octogenarians whereas other mechanisms are responsible in very old individuals [48].

Only a few studies have evaluated the capacity of cytokine production *in vivo*. The finding of an age-related preactivation of monocytes reflected by a more pronounced initial production of TNF-$\alpha$ but no difference in peak levels was confirmed in a human in vivo sepsis model in which an intravenously bolus of E.coli LPS was given to young and elderly healthy volunteers [49], Figure 4. Furthermore, this model demonstrated that elderly participants had a prolonged inflammatory response reflected by a slower of sTNFRs, CRP, and body temperature [49], (Figure 4). A similar phenomenon was observed in elderly patients with pneumoccocal infections [50]. Thus, in vivo studies indicate that ageing is associated with defect termination of inflammatory activity that may contribute to a preactivation of cytokine producing cells. The clinical significance of this phenomenon during severe infections remains unclear.

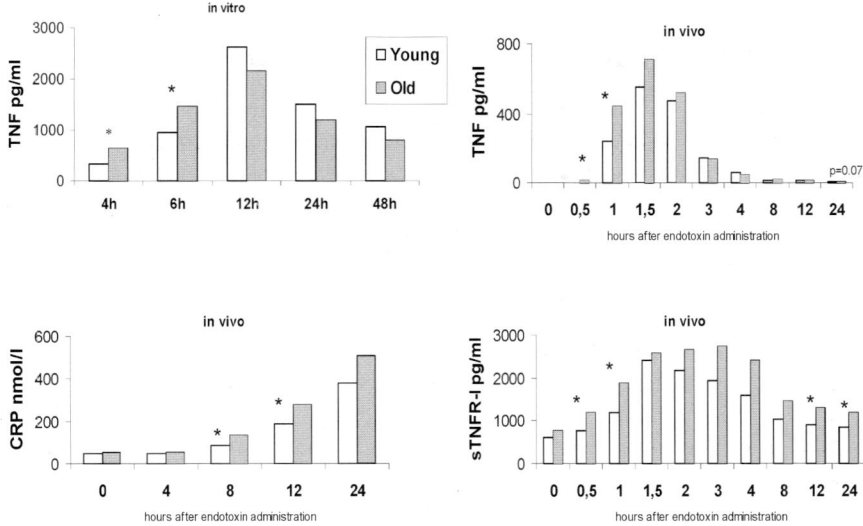

Figure 4. Age-associated differences in levels of inflammatory markers following LPS stimulation in vitro and in vivo. In vitro: Whole blood diluted 1:4 from 8 healthy elderly volunteers aged 77–81 years and 8 young controls aged 20–30 years (from Ref. [16]). In vivo: Plasma levels in 9 elderly volunteers aged 61–69 years and 8 young controls aged 20–40 years. Geometric means are shown. *, p<0.05 (from Ref. [49]).

## 7.    CONCLUDING REMARKS

Ageing is associated with systemic low-grade inflammation that is related to age-associated pathology and high mortality risk. TNF-α and IL-6 appear to have separate biological activities in elderly populations and could provide independent links between life style factors and physiological changes on one hand and risk factors and age-related disorders such as CVD, dementia, and the syndrome of frailty on the other. Furthermore, we suggest that the biological effects of TNF-α and IL-6 are of different clinical importance during ageing. Thus, TNF-α seems to be a strong risk marker in very old and frail populations whereas IL-6 is the best predictor of thromboembolic complications and functional decline in relatively healthy populations of younger elderly. Circulating levels of IL-6 are probably a better marker of the sum of ongoing inflammation in the body than systemic levels of TNF-α in younger elderly because a local production of TNF-α may not escape into the circulation although it induces the systemic production of IL-6 and mediators downstream in the inflammatory cascade. Concurrently with progression of pathological processes increasing levels of TNF-α also appear in the circulation and it gradually becomes a stronger risk marker in the oldest old. Activities related to IL-6 could also be of less importance in very old people who have already survived to advanced age despite considerable morbidity. We consider age-related increases in circulating levels of inflammatory mediators downstream in the inflammatory cascade to be secondary to increased production of TNF-α and/or IL-6. It remains unclear to which extent a dysregulated acute phase response affects low-grade inflammation but ageing is accompanied by prolonged inflammatory activity in response to acute stimulations and PMNC appear to be in a preactivated state. Intervention strategies with the purpose to reduce age-related low-grade inflammation have yet to be assessed but could include prevention of risk factors such as smoking and obesity as well as a more aggressive treatment of infections in order to reduce the burden of inflammatory stimuli. Moreover, physical exercise, supplementation with antioxidants or treatments with statins possess anti-inflammatory effects that may be beneficial in elderly populations. Future studies should address the specific biological effects of TNF-$\alpha$ and IL-6; the extent to which different peripheral tissues contribute to systemic low-grade inflammation; how peripheral cytokines interact with local inflammatory processes in CNS and cognitive function; and if therapy directed specifically against TNF-$\alpha$ and IL-6 is able to reverse age-related inflammatory pathology.


## ACHNOWLEDGEMENTS

The authors thank The Danish Research Council (22-02-0261 and 52-00-0851) for financial support (22-02-0261 and 52-00-0851).


## REFERENCES

1.  Vaupel JW, Carey JR, Christensen K, Johnson TE, Yashin AI, Holm NV et al. Biodemographic trajectories of longevity. Science 1998; 280(5365):855–860.
2.  Pawelec G, Solana R. Immunosenescence. Immunol Today 1997; 18(11):514–516.
3.  Yoshikawa TT. Perspective: aging and infectious diseases: past, present, and future. J Infect Dis 1997; 176(4):1053–1057.
4.  Sahyoun NR, Lentzner H, Hoyert D, Robinson KN. Trends in Causes of Death Among the Elderly. In: National Center for Health Statistics. Aging trends. Hyattsville: 2001.
5.  Ross R. Atherosclerosis - an inflammatory disease. N Engl J Med 1999; 340(2):115–126.
6.  Ritchie K, Kildea D. Is senile dementia »age-related« or »ageing-related«?--evidence from meta-analysis of dementia prevalence in the oldest old. Lancet 1995; 346(8980): 931–934.
7.  Wilson CJ, Finch CE, Cohen HJ. Cytokines and cognition – the case for a head-to-toe inflammatory paradigm. J Am Geriatr Soc 2002; 50(12):2041–2056.
8.  Richards C, Gauldie J. Role of cytokines in the acute-phase response. In: Aggarwal BB, Puri RK, editors. Human Cytokines: Their roles in disease and therapy. Cambridge, Massachussetts, USA: Blackwell Science, 1995: 253–269.
9.  Tilg H, Trehu E, Atkins MB, Dinarello CA, Mier JW. Interleukin-6 (IL-6) as an anti-inflammatory cytokine: induction of circulating IL-1 receptor antagonist and soluble tumor necrosis factor receptor p55. Blood 1994; 83(1):113–118.
10. Tracey KJ. The inflammatory reflex. Nature 2002; 420(6917):853–859.
11. Bruunsgaard H, Andersen-Ranberg K, Jeune B, Pedersen AN, Skinhoj P, Pedersen BK. A high plasma concentration of TNF-alpha is associated with dementia in centenarians. J Gerontol Med Scien 1999; 54A(7):M357-M364.
12. Catania A, Airaghi L, Motta P, Manfredi MG, Annoni G, Pettenati C et al. Cytokine antagonists in aged subjects and their relation with cellular immunity. J Gerontol Med Sci 1997; 52(2):B93-B97.
13. Ballou SP, Lozanski FB, Hodder S, Rzewnicki DL, Mion LC, Sipe JD et al. Quantitative and qualitative alterations of acute-phase proteins in healthy elderly persons. Age Ageing 1996; 25(3):224–230.
14. Andersen-Ranberg K, Schroll M, Jeune B. Healthy centenarians do not exist, but

autonomous centenarians do: a population-based study of morbidity among Danish centenarians. J Am Geriatr Soc 2001; 49(7):900–908.

15. Bruunsgaard H, Skinhøj P, Pedersen AN, Schroll M, Pedersen BK. Ageing, TNF-alpha and atherosclerosis. Clin Exp Immunol 2000; 121:255–260.

16. Bruunsgaard H, Pedersen BK. Age-related inflammatory Cytokines and Disease. Immunol Allergy Clin N Am 2003; 23:15–39.

17. Andersen-Ranberg K, Vasegaard L, Jeune B. Dementia is not inevitable. A population-based study of Danish Centenarians. J Gerontol 2001; 56B(3):P152-P159.

18. Allan SM, Rothwell NJ. Cytokines and acute neurodegeneration. Nat Rev Neurosci 2001; 2(10):734–744.

19. Scherbel U, Raghupathi R, Nakamura M, Saatman KE, Trojanowski JQ, Neugebauer E et al. Differential acute and chronic responses of tumor necrosis factor-deficient mice to experimental brain injury. Proc Natl Acad Sci U S A 1999; 96(15):8721–8726.

20. Kroeger KM, Steer JH, Joyce DA, Abraham LJ. Effects of stimulus and cell type on the expression of the -308 tumour necrosis factor promoter polymorphism. Cytokine 2000; 12(2):110–119.

21. Bruunsgaard H, Andersen-Ranberg K, Hjelmborg JvB, Pedersen B, Jeune B. TNF-α is an independent prognostic marker of mortality risk in centenarians. Am J Med. In press.

22. Tisdale MJ. Wasting in cancer. J Nutr 1999; 129(1S Suppl):243S-246S.

23. Pedersen M, Bruunsgaard H, Weis N, Hendel HW, Andreassen BU, Eldrup E et al. Circulating levels of TNF-alpha and IL-6 - Relation to truncal fat mass and muscle mass in healthy elderly individuals and patients with type 2 diabetes. Mech Ageing Dev 2003. In press.

24. Visser M, Pahor M, Taaffe DR, Goodpaster BH, Simonsick EM, Newman AB et al. Relationship of interleukin-6 and tumor necrosis factor-alpha with muscle mass and muscle strength in elderly men and women: the Health ABC Study. J Gerontol Med Sci 2002; 57(5):M326-M332.

25. Bruunsgaard H, Bjerregaard E, Schroll M, Pedersen BK. High levels of soluble TNF receptors limit exercise-induced increases in muscle strength in the oldest old. J Am Geriatr Soc 2003.

26. Fried LP, Tangen CM, Walston J, Newman AB, Hirsch C, Gottdiener J et al. Frailty in older adults: evidence for a phenotype. J Gerontol A Biol Sci Med Sci 2001; 56(3):M146-M156.

27. Hamerman D. Toward an Understanding of Frailty. Ann Intern Med 1999; 130(11):945–950.

28. Weaver JD, Huang MH, Albert M, Harris T, Rowe JW, Seeman TE. Interleukin-6 and risk of cognitive decline: MacArthur studies of successful aging. Neurology 2002; 59(3):371–378.

29. Ridker PM, Rifai N, Stampfer MJ, Hennekens CH. Plasma concentration of interleukin-6 and the risk of future myocardial infarction among apparently healthy men. Circulation 2000; 101(15):1767–1772.

30. Harris TB, Ferrucci L, Tracy RP, Corti MC, Wacholder S, Ettinger WHJ et al. Associations of elevated interleukin-6 and C-reactive protein levels with mortality in the elderly. Am J Med 1999; 106(5):506–512.

31. Volpato S, Guralnik JM, Ferrucci L, Balfour J, Chaves P, Fried LP et al. Cardiovascular disease, interleukin-6, and risk of mortality in older women: the women's health and aging study. Circulation 2001; 103(7):947–953.

394

32. Jenny NS, Tracy RP, Ogg MS, Luong lA, Kuller LH, Arnold AM et al. In the elderly, interleukin-6 plasma levels and the -174G>C polymorphism are associated with the development of cardiovascular disease. Arterioscler Thromb Vasc Biol 2002; 22(12): 2066–2071.

33. Bonafe, Olivieri F, Cavallone L, Giovagnetti S, Mayegiani F, Cardelli M et al. A gender-dependent genetic predisposition to produce high levels of IL-6 is detrimental for longevity. Eur J Immunol 2001; 31(8):2357–2361.

34. Pradhan AD, Manson JE, Rifai N, Buring JE, Ridker PM. C-reactive protein, interleukin 6, and risk of developing type 2 diabetes mellitus. JAMA 2001; 286(3):327–334.

35. Wallenius V, Wallenius K, Ahren B, Rudling M, Carlsten H, Dickson SL et al. Interleukin-6-deficient mice develop mature-onset obesity. Nat Med 2002; 8(1):75–79.

36. Ferrucci L, Penninx BW, Volpato S, Harris TB, Bandeen-Roche K, Balfour J et al. Change in muscle strength explains accelerated decline of physical function in older women with high interleukin-6 serum levels. J Am Geriatr Soc 2002; 50(12):1947–1954.

37. Metzger S, Hassin T, Barash V, Pappo O, Chajek-Shaul T. Reduced body fat and increased hepatic lipid synthesis in mice bearing interleukin-6-secreting tumor. Am J Physiol Endocrinol Metab 2001; 281(5):E957-E965.

38. Akiyama H, Barger S, Barnum S, Bradt B, Bauer J, Cole GM et al. Inflammation and Alzheimer's disease. Neurobiol Aging 2000; 21(3):383–421.

39. Bruunsgaard H, Ladelund S, Pedersen AN, Schroll M, Jorgensen T, Pedersen BK. Predicting death from TNF-alpha and IL-6 in 80-year-old people. Clin Exp Immunol 2003; 24–31.

40. Humphries SE, Luong LA, Ogg MS, Hawe E, Miller GJ. The interleukin-6 -174 G/C promoter polymorphism is associated with risk of coronary heart disease and systolic blood pressure in healthy men. Eur Heart J 2001; 22(24):2243–2252.

41. Kerr R, Stirling D, Ludlam CA. Interleukin 6 and haemostasis. Br J Haematol 2001; 115(1):3–12.

42. Franceschi C, Bonafé M, Valensin S, Olivieri.F., De Luca M, Ottaviani E et al. Inflamm-aging. An Evolutionary Perspective of Immunosenescence. Annals New York Academy of Sciences 2000;244–254.

43. Bruunsgaard H, Pedersen AN, Schroll M, Skinhoj P, Pedersen BK. TNF-α, Leptin, and lymphocyte function in human aging . Life Sci 2000; 67:2721–2731.

44. Bruunsgaard H, Østergaard L, Andersen-Ranberg K, Jeune B, Pedersen BK. Proinflammatory cytokines, antibodies to Chlamydia pneumoniae and age-associated diseases in Danish centenarians – Is there a link? Scand J Infect Dis 2002; 34:493–499.

45. Priø TK, Bruunsgaard H, Røge B, Skinhoj P, Pedersen BK. Asymtomatic bacteriuria in elderly humans is associated with increased levels of circulating TNF receptors. Experimental Gerontology 2002; 37(5):693–699.

46. Straub RH, Miller LE, Schölmerich J, Zietz B. Cytokines and hormones as possible links between endocrinosenescence and immunosenescence. J Neuroimmunol 2000; 109: 10–15.

47. Saurwein-Teissl M, Blasko I, Zisterer K, Neuman B, Grubeck-Loebensten B. An imbalance between pro- and anti-inflammatory cytokines, a characteristic feature of old age. Cytokine 2000; 12(7):1160–1161.

48. Sandmand M, Bruunsgaard H, Kemp K, Andersen-Ranberg K, Schroll M, Jeune B. TNF-production is not increased in T lymphocytes from centenarians. Gerontol. In press.

49. Krabbe K, Bruunsgaard H, Hansen CM, Møller K, Fonsmark L, Quist J et al. Aging

is associated with a prolonged fever response in human endotoxemia. Clin Diagn Lab Immunol 2001; 8(2):333–338.

50. Bruunsgaard H, Skinhøj P, Quist J, Pedersen BK. Elderly humans show prolonged in vivo inflammatory activity during Pneumococcal infections. J Infect Dis 1999; 180(2): 551–554.

# VII.   CONCLUSIONS

*The Neuroendocrine Immune Network in Ageing*
Edited by R.H. Straub and E. Mocchegiani

# Possible New Anti-Ageing Strategies Related to Neuroendocrine-Immune Interactions

EUGENIO MOCCHEGIANI[1] and RAINER H. STRAUB[2]

[1]*Immunology Ctr. (Section Nutrition, Immunity and Ageing) Res. Dept. INRCA, Ancona, Italy;*
[2]*Lab. of Neuroendocrinoimmunology, Dept. Int. Medicine I, University Medical Centre Regensburg, 93053 Regensburg, Germany*

## ABSTRACT

The ageing process demonstrates gradual and spontaneous changes, resulting in maturation through childhood, puberty, and young adulthood and then decline through middle and late age. However, animals and humans are capable of reaching the extreme limit of life span characteristic for the species with a very efficient network of anti-ageing mechanisms. Among them, neuroendocrine-immune interactions play a pivotal role. The loss of the capacity of the organism in remodelling the neuroendocrine-immune response leads to appearance of age-associated pathologies. We herein report some substances, which have been proposed to be useful in new anti-ageing strategies because of their capacity to remodel some biological functions in old animals and humans. These substances are: L-deprenyl, leptin, ghrelin, carnosine and NO-donors. Their role as possible anti-ageing substances in man in relation to the improvement of neuroimmune responses is reported and discussed.

## 1.    INTRODUCTION

The ageing process shows gradual and spontaneous changes, resulting in maturation through childhood, puberty, and young adulthood and then decline through middle and late age. Ageing refers to the time-dependent deterioration that occurs in most individuals involving weakness, increased susceptibility to diseases and adverse environmental conditions, loss of mobility and agility and age-related physiological changes. However, there are animals and humans capable of reaching the extreme limit of life span characteristic for the species with a very efficient network of anti-ageing mechanisms. Focusing on humans, and on the changes occurring with age, neuroendocrine-immune interactions play a key role in ageing. As reported in this volume, the neuroendocrine system is strictly correlated with immune efficiency. The loss of the capacity of the organism in remodelling the neuroimmune response may lead to appearance of age- associated pathologies, such as cancer, infections, neurodegenerative diseases, diabetes, osteoporosis and osteoarthritis, among others. Thus, as a new paradigm, the *remodelling theory of ageing*, was proposed [1]. Such a remodelling naturally occurs in very old people involving some hormones and trace elements, as reported in specific chapters of this volume. Without reporting again the beneficial effects of some hormones (e.g. GH, melatonin, steroid hormones), trace elements (zinc) and vitamins (Vit.E) (please, see specific chapters in this volume), we present in this short chapter some of the substances that may be considered as new anti-ageing remedies in order to induce a remodelling of neuroendocrine-immune interactions. Although, these substances are known for many years, their role as possible anti-ageing drugs have been recognised only during the last decade as the result of specific experiments in old animals. These substances are: L-deprenyl, ghrelin, leptin, carnosine and NO-donors. Some of them are drugs (L-depreny)], or pro-hormones (ghrelin, leptin), or natural antioxidants (carnosine and NO-donor) with specific properties affecting neuroendocrine and immune functions. It is interesting and, at the same time, intriguing that some of these substances are directly or indirectly linked to the bioavailability of free zinc ions through the activation of zinc-dependent antioxidant enzymes. Therefore, these new anti-ageing strategies may also be pertinent to zinc ion bioavailability and to the capacity of zinc to remodel the neuroimmune interactions during ageing, as reported in Section V of this Volume [Mocchegiani et al.]. It is clear that the bioavailability of this trace element has not to be considered an "axiom" because other trace elements (for example Cu and Mn) are involved in conferring biological activity to antioxidant enzymes, but for some substances that we will report herein, the relevance to zinc has to be considered. This is important for future perspectives and for the development of possible new anti-ageing strategies.

## 2.    L-DEPRENYL

L-deprenyl has been discovered many years ago as an anti-depressive drug [2]. L-deprenyl inhibits monoamine oxidases (MAO), which is an enzyme that functions in the brain to break-down (inactivate) neurotransmitters [2]. MAO exists in two forms A and B. MAO-A is found in most neurons and it is most effective for breaking down neurotransmitters such as serotonin, adrenaline and noradrenaline; MAO-B is found in astrocytes and is more effective in breaking down the neurotransmitter dopamine [3]. With advanced ageing, MAO activity increases in the human brain, often to the point of severely depressed levels of necessary monoamine transmitters [2]. L-deprenyl at low doses [20mg/day] was used as antidepressant during ageing due to its inhibiting effect on MAO-A [4], and it was used for treating the chronic dopamine depletion of Parkinson patients inhibiting MAO-B without elevating blood-pressure [5]. More recent studies have however pointed out that L-deprenyl extends remaining life span of both laboratory animals and Alzheimer patients. Old rats treated 3 times a week with L-deprenyl at the age of 24 months lived for a significantly longer period than saline-treated controls animals [6]. The beneficial effect for prolonged survival was also associated with improved spatial learning and long-term memory [7]. Studies on Alzheimer patients showed a 15% improvement in behavioural symptoms [8] and an increase in median survival of 215 days when comparing placebo [9] with 10 mg/day deprenyl. A further intriguing study provided convincing evidence that deprenyl retards the death of ageing female beagle dogs, which appears to be largely explained by the significant reduction of the incidence of mammary cancer in treated animals [10]. The possible causes of a prolonged survival in aged animals with an antitumorigenic effect by the drug are strictly related to three factors. First, the increased activity of antioxidant enzymes, such as superoxide dismutase (SOD) and catalase (CAT) [11]. Second, the increased activity of the immune system through an enhanced production of some cytokines (e.g. IL-2, IL-6, TNF-alpha, IFN-gamma) [12] and the improved natural killer (NK) cell cytotoxicity [13]. Third, the increased activity of the endocrine system through a recovery from low IGF-I production, suggesting that growth hormone secretion may be stimulated by this drug [14]. In this context, it has also been shown that the drug decreases prolactin concentrations in sera of aged animals, thereby probably decreasing the incidence and progression of breast cancer [15]. All these findings support the idea that L-deprenyl prolongs survival because of its effect on the neuroimmune system. It has a promise for treatment also in humans. However, it is noteworthy that the beneficial effects of the drug are mainly related to zinc ion bioavailability, as it relates to changes in neuroendocrine and immune functions. Indeed, the production of cytokines listed above, the activity of NK cells as well as the action of hormones (e.g. GH, IGF-I and prolactin) are supported by the zinc pool [16]. Therefore, for the treatment of aged humans, the status of the zinc ion bioavailability has to be considered before initiating the treatment. In our experience, zinc bioavailability declines in adult age (e.g. 40–50 years) [16]. The treatment would start in adults in order to delay ageing and thus to prevent cancer, which is the main cause of death in old individuals. Caution should be exercised to avoid the possible effects of this agent on hypertension [4]. The improvement in cognitive functions after treatment with deprenyl [8] is a further supportive factor to use this drug for anti-ageing treatment, taking into account the impairment of cognitive functions in the elderly [17].

## 3.   GHRELIN AND LEPTIN

Ghrelin, a 28-amino-acid peptide predominantly produced by the stomach, displays strong growth hormone (GH)-releasing actions mediated by the activation of GH secretagogue (GHS) receptor (GHS-R) type 1a [18]. Although the presence of substances with a secretagogue role on GH production has been discovered 20 years ago, the presence of ghrelin and its relevance on the efficiency of the endocrine and immune system has only been recently discovered [19]. In particular, the existence of GHS-R within the hypothalamus and pituitary gland [20], and in peripheral organs (e.g. pancreas, testis, ovary, cardiac tissues) [21], and also on B and T cells [22], has lead to suggest that ghrelin affects the HPA axis, gonadal function, energy balance, and finally immunocompetence. These all impaired factors during ageing [16,23]. Ghrelin is mainly produced by the enteroendocrine cells called X/A-like cells representing the major endocrine population in the stomach oxyntic mucosa [24]. Ghrelin increases during fasting and malnutrition as well in patients with anorexia nervosa [25]. However, the ghrelin production is balanced by another peptide, called leptin, which is synthesised by adipose tissue and has been identified as the critical hormone for the gene defect responsible for the obese syndrome in ob/ob mice [26]. High leptin levels signal the presence of sufficient energy stores to leptin receptors (OB-R) in the hypothalamus, which respond by reducing appetite and increasing energy expenditure, and thus prevent severe obesity [27]. Low leptin levels occur during starvation leading to an increased susceptibility to infectious and inflammatory stimuli and this is associated with a dysregulation of cytokine production and with immunosuppression [28]. In experimental obesity, which is a condition of leptin signalling defect [26], leptin increases, as shown in obese db/db mice [29], with subsequent continuous appetite [27]. Similarly ghrelin is increased in experimental obesity with ghrelin signalling defects [30]. Anyway, protein energy malnutrition coupled with an altered metabolism and impaired neuroendocrine and immune functions are observed in obesity [28] as well as in anorexia [31]. During ageing, protein energy malnutrition is present [32] and many old people are obese [33]. Therefore, a dysbalance between leptin and ghrelin production might occur during ageing with subsequent changes of receptor signalling in the hypothalamus. This fact may trigger on the one hand side metabolic alterations and, on the other hand, impairments in neuroendocrine-immune responses with subsequent appearance of age-related diseases. It is not a simple coincidence that a dysbalance in leptin and ghrelin production, as well as the presence of leptin and ghrelin polymorphisms are coupled with a strong dysregulation of neuroimmune responses, which are observed in cancer and in infectious diseases [34]. In turn these conditions are characterised by protein-energy malnutrition [35]. Therefore, new anti-ageing strategies do not have to include treatments of leptin or ghrelin in a separate way, as reported [4,28], because of their possible inefficiency during long-term treatment. But rather, these strategies have to be addressed by restoring the balance between leptin and ghrelin production though the modulation of neuropeptide Y (NPY) in the hypothalamus, which regulates both leptin and ghrelin action [36,37]. The loss of neurons producing NPY [38] coupled with neuroendocrine-immune dysfunctions [39] and protein-energy malnutrition [32] are characteristic of elderly people.

## 4. CARNOSINE

Carnosine (beta-alanyl-L-histidine) is a dipeptide, a non-enzymatic free radical scavenger and a natural antioxidant [40]. It is widely distributed in tissues and exists at particularly high concentrations in muscle and brain [41]. Throughout the brain, the most important carnosine containing cells are astrocytes and oligodendrocytes. Carnosine has been discovered as an antioxidant and a free radical scavenger ten years ago with a role in rejuvenating senescent cells, particularly fibroblasts [42]. At the clinical level, carnosine can delay vision impairment during ageing while simultaneously exerting a remarkable effect on prevention and treatment of senile cataract [43]. However, recently intriguing findings showed that an enhanced release of carnosine by astrocytes is under the control of zinc [44]. These findings are very important, if one takes into account the relevance of zinc for the neuroimmune interactions and, at the same time, for the protective role of carnosine itself. Zinc has a specific role of in the brain functions, such as synaptic transmission, for instance [for details see reference 45]. Recent findings have suggested that carnosine is able to inhibit the aggregation of amyloid peptide and that the toxic effects of amyloid peptide can be prevented or reduced by carnosine [46]. Therefore, carnosine, via zinc ion bioavailability, may be used as a potential anti-ageing drug. Carnosine may also increase the lifespan of CD4-positive T cell clones [47] and may attenuate the deleterious effects of high TNF-alpha concentrations in injured rats, thus protecting immune cells through its antioxidant activity [48]. Good intracellular zinc ion bioavailability, which is required for the maintenance of the immune system in ageing people, may be largely due to metallothionein homeostasis [49], and to the positive antioxidant effect of carnosine, which are exerted via the zinc pool. On the other hand, intriguing findings show the positive effects of zinc complex of L-carnosine (generic name: Polaprezinc), which is capable of increasing spermatogenesis in a mouse experimental model of accelerated ageing (SAMP1) [50] and it also reduces the inflammation during the course of gastric ulcer [44]. This gives further support to the relevance of carnosine and zinc (like a complex) as a potential new anti-ageing factors related to neuroendocrine-immune interactions.

## 5. NO-DONORS

It has been reported that metallothionein (MT) acts as a sensor of the cellular redox status by reacting to a shift to more oxidizing conditions with the release of zinc, whereas a shift to a more reducing environment leads to binding of zinc [51]. Nitric oxide is capable of interacting with MT to form either dinitrosyl iron sulfur complexes [52] or S-nitrosylation to result in the NO-mediated release of zinc [53]. The S-nitrosylation (oxidizing of SH groups of cysteine in the MT molecule) of cysteines in zinc thiolate clusters are very important for zinc-dependent transcription factors [54], antioxidant enzyme activity [55] and probably for the maintenance of the neuroimmune network [39]. The phenomenon of zinc release from MT via NO is under the control of pro-inflammatory cytokines (e.g. IL-1 and IL-6) [56]. These cytokines rise in young-adults under transient stress-like conditions [57]. During ageing, the release of zinc from MT via NO is limited because of a strong increase of iNOS activity that upsets the proper balance between iNOS and cNOS [57]. This leads to a dysregulation of the intracellular redox status. On the other hand, an overexpression of MT, as it occurs during ageing [49], causes an inhibition of NO-induced changes in labile $Zn^{++}$ that can be reversed by zinc supplementation [53]. In this context, NO-donors (e.g. SNAP and NOC 18) may play a relevant role because they are capable to inducing a proper balance between iNOS and cNOS [58] and thus promote zinc release from MT in chronic stress-like conditions and during ageing. Indeed, the treatment of stressed (by $H_2O_2$) endothelial cells with the NO-donor, SNOC, releases zinc (as detected by a fluorescent fluorophore, zinquin) by the overexpressed MT [53]. Such a phenomenon of NO-donors may be relevant for brain functions as well as for the maintenance of the neuroendocrine-immune network during ageing. As reported in the chapter of Suzuki [this Volume] a strong increase of NO within the brain during ageing leads to neuronal cell-death. An abnormal increase of MT was also reported in the old brain coupled with low zinc ion bioavailability [45] Zinc is necessary for optimal cognitive brain functions [60], for synaptic transmission [59], for the production of hormone releasing factors [see Mocchegiani, this Volume] and for the prevention of cell death [16]. It is evident from these observations that NO-donors may be useful therapeutic agents for the enhancement of neuroimmune function during ageing.

## 6.    CONCLUSIONS

As reported in this short chapter, new anti-ageing drugs may possibly be added to the classical strategies, such as growth hormone, melatonin, zinc, vitamin E and caloric restriction. The latter has been demonstrated to be very efficient in relation to increased longevity in experimental animals. We have not reported the beneficial effect of caloric restriction because of its discovery many years ago by Masoro [61]. Therefore, it may not be considered as a possible new anti-ageing strategy. In contrast, we have preferred to report some substances, which have been proposed during this last decade as new anti-ageing strategies in animals and humans. The choice is based on the fact that all these substances affect neuroendocrine-immune interactions with a capacity to remodel this network. The capacity of remodelling is a condition "sine qua non" to support healthy ageing [49]. All these substances positively affect remodelling but, to our opinion, the more intriguing drugs are NO-donors because NO is also involved in zinc release from metallothionein [MT]. This role is very important in ageing as indicated by the fact that the impaired neuroendocrine-immune responses can be restored by zinc supplementation [35]. Zinc acts on many biological functions, including the maintenance of normal neuroendocrine-immune function. Zinc deficiency and high MT levels are characteristic of ageing, with limited zinc release from MT [45]. Consequently, NO-donors may play an important role during ageing, which could provide us with the tools for a new anti-ageing strategy. NO-donors are natural substances with very limited toxic effects. Experiments NO donors in old animals are in progress in our laboratory.

All of the above presented concepts focused attention on more or less healthy ageing people. In contrast, ageing patients with chronic inflammatory diseases may not profit from these therapies because all of the above discussed agents would stimulate the immune system. Such a stimulation would lead to an unwanted aggravation of chronic inflammatory diseases, for instance. Chapter VI gives some therapeutic ideas how to treat patients with chronic inflammatory diseases in harmony with the principles of neuroimmune biology. This important difference has to be taken into account when anti-ageing strategies are discussed.

REFERENCES

1.  Paolisso G, Barbieri M, Bonafe M, Franceschi C. Metabolic age modelling: the lesson from centenarians. Eur J Clin Invest 2000;30:888–894.
2.  Knoll J. Deprenyl (selegiline): the history of its development and pharmacological action. Acta Neurol Scand 1983;95:57–80.
3.  Holschneider DP, Chen K, Seif I, Shih JC. Biochemical, behavioral, physiologic, and neurodevelopmental changes in mice deficient in monoamine oxidase A or B.Brain Res Bull. 2001; 56 :453–462
4.  Knoll J. The striatal dopamine dependency of life span in male rats. Longevity study with (-) deprenyl. Mech Ageing Dev 1988;46:237–62.
5.  Birkmayer W, Knoll J, Riederer P, Youdim MB, Hars V, Marton J. Increased life expectancy resulting from addition of L-deprenyl to Madopar treatment in Parkinson's disease: a long term study. J Neural Transm 1985;64:113–127.
6.  Stoll S, Hafner U, Kranzlin B, Muller WE. Chronic treatment of Syrian hamsters with low-dose selegiline increases life span in females but not males. Neurobiol Aging 1997;18: 205–211.
7.  Bickford PC, Adams CE, Boyson SJ, Curella P, Gerhardt GA, Heron C, Ivy GO, Lin AM, Murphy MP, Poth K, Wallace DR, Young DA, Zahniser NR, Rose GM. Long-term treatment of male F344 rats with deprenyl: assessment of effects on longevity, behavior, and brain function. Neurobiol Aging 1997;18:309–318.
8.  Tolbert SR, Fuller MA. Selegiline in treatment of behavioral and cognitive symptoms of Alzheimer disease. Ann Pharmacother 1996;30:1122–1129.
9.  Sano M, Ernesto C, Thomas RG, Klauber MR, Schafer K, Grundman M, Woodbury P, Growdon J, Cotman CW, Pfeiffer E, Schneider LS, Thal LJ. A controlled trial of selegiline, alpha-tocopherol, or both as treatment for Alzheimer's disease. The Alzheimer's Disease Cooperative Study. N Engl J Med 1997;336:1216–1222.
10. Ruehl WW, Entriken TL, Muggenburg BA, Bruyette DS, Griffith WC, Hahn FF. Treatment with L-deprenyl prolongs life in elderly dogs. Life Sci 1997;61:1037–1044.
11. Kitani K, Minami C, Isobe K, Maehara K, Kanai S, Ivy GO, Carrillo MC. Why (--)deprenyl prolongs survivals of experimental animals: increase of anti-oxidant enzymes in brain and other body tissues as well as mobilization of various humoral factors may lead to systemic anti-aging effects. Mech Ageing Dev 2002;123:1087–1100.
12. Muller T, Kuhn W, Kruger R, Przuntek H. Selegiline as immunostimulant--a novel mechanism of action? J Neural Transm Suppl 1998;52:321–328.
13. ThyagaRajan S, Madden KS, Kalvass JC, Dimitrova SS, Felten SY, Felten DL. L-deprenyl-induced increase in IL-2 and NK cell activity accompanies restoration of noradrenergic

nerve fibers in the spleens of old F344 rats. J Neuroimmunol 1998 ;92:9–21.

14. De la Cruz CP, Revilla E, Rodriguez-Gomez JA, Vizuete ML, Cano J, Machado A. (-)-Deprenyl treatment restores serum insulin-like growth factor-I (IGF-I) levels in aged rats to young rat level. Eur J Pharmacol 1997;327:215–220.

15. ThyagaRajan S, Felten DL. Modulation of neuroendocrine-immune signaling by L-deprenyl and L-desmethyldeprenyl in aging and mammary cancer. Mech Ageing Dev. 2002;123:1065–1079.

16. Mocchegiani E, Muzzioli M, Cipriano C, Giacconi R. Zinc, T-cell pathways, aging: role of metallothioneins. Mech Ageing Dev 1998;106:183–204.

17. Ferro JM, Madureira S. Age-related white matter changes and cognitive impairment. J Neurol Sci 2002;203–204:221–225.

18. Smith RG, Leonard R, Bailey AR, Palyha O, Feighner S, Tan C, Mckee KK, Pong SS, Griffin P, Howard A. Growth hormone secretagogue receptor family members and ligands. Endocrine 2001;14:9–14.

19. Broglio F, Gottero C, Benso A, Prodam F, Destefanis S, Gauna C, Maccario M, Deghenghi R, van der Lely AJ, Ghigo E. Effects of ghrelin on the insulin and glycemic responses to glucose, arginine, or free fatty acids load in humans. J Clin Endocrinol Metab 2003;88: 4268–4272.

20. Kojima M, Hosoda H, Kangawa K. Purification and distribution of ghrelin: the natural endogenous ligand for the growth hormone secretagogue receptor. Horm Res 2001;56: 93–97.

21. Gnanapavan S, Kola B, Bustin SA, Morris DG, McGee P, Fairclough P, Bhattacharya S, Carpenter R, Grossman AB, Korbonits M. The tissue distribution of the mRNA of ghrelin and subtypes of its receptor, GHS-R, in humans. J Clin Endocrinol Metab 2002 ;87:2988.

22. Hattori N, Saito T, Yagyu T, Jiang BH, Kitagawa K, Inagaki C. GH, GH receptor, GH secretagogue receptor, and ghrelin expression in human T cells, B cells, and neutrophils. J Clin Endocrinol Metab 2001;86:4284–4291.

23. Donini LM, Savina C, Cannella C. Eating habits and appetite control in the elderly: the anorexia of aging. . Int Psychogeriatr 2003;15:73–87.

24. Date Y, Kojima M, Hosoda H, Sawaguchi A, Mondal MS, Suganuma T. Matsukura S, Kangawa K, Nakazato M. Ghrelin, a novel growth hormone-releasing acylated peptide, is synthesized in a distinct endocrine cell type in the gastrointestinal tracts of rats and humans. Endocrinology 2000;141:4255–4261.

25. Ariyasu H, Takaya K, Tagami T, Ogawa Y, Hosoda K, Akamizu T, Suda M. Koh T, Natsui K, Toyooka S, Shirakami G, Usui T, Shimatsu A, Doi K, Hosoda H, Kojima M, Kangawa K, Nakao K. Stomach is a major source of circulating ghrelin, and feeding state determines plasma ghrelin-like immunoreactivity levels in humans. J Clin Endocrinol Metab 2001;86: 4753–4758.

26. Zhang Y, Proenca R, Maffei M, Barone M, Leopold L, Friedman JM. Positional cloning of the mouse obese gene and its human homologue. Nature 1994;372:425–432.

27. Maffei M, Halaas J, Ravussin E, Pratley RE, Lee GH, Zhang Y, Fei H, Kim S, Lallone R, Ranganathan S, et al. Leptin levels in human and rodent: measurement of plasma leptin and ob RNA in obese and weight-reduced subjects. Nat Med 1995;1:1155–1161.

28. Faggioni R, Feingold KR, Grunfeld C. Leptin regulation of the immune response and the immunodeficiency of malnutrition. FASEB J 2001;15:2565–2571.

29. Lee GH, Proenca R, Montez JM, Carroll KM, Darvishzadeh JG, Lee JI, Friedman JM. Abnormal splicing of the leptin receptor in diabetic mice. Nature 1996;379:632–635.

30. Rosicka M, Krsek M, Jarkovska Z, Marek J, Schreiber V. Ghrelin, a new endogenous growth hormone secretagogue. Physiol Res 2002;51: 435–441.

31. Brambilla F, Ferrari E, Panerai A, Manfredi B, Petraglia F, Catalano M, Sacerdote P. Psychoimmunoendocrine investigation in anorexia nervosa. Neuropsychobiology 1993;27: 9–16.

32. Constans T. Malnutrition in the elderly. Rev Prat 2003;53:275–279.

33. Rossner S. Obesity in the elderly, a future matter of concern? Obes Rev 2001;2:183–188.

34. Loktionov A. Common gene polymorphisms and nutrition: emerging links with pathogenesis of multifactorial chronic diseases. J Nutr Biochem 2003; 14: 426–451.

35. Groopman JE. Fatigue in cancer and HIV/AIDS. Oncology 1998;12:335–344.

36. Shintani M, Ogawa Y, Ebihara K, Aizawa-Abe M, Miyanaga F, Takaya K, Hayashi T, Inoue G, Hosoda K, Kojima M, Kangawa K, Nakao K. Ghrelin, an endogenous growth hormone secretagogue, is a novel orexigenic peptide that antagonizes leptin action through the activation of hypothalamic neuropeptide Y/Y1 receptor pathway. Diabetes 2001;50: 227–232.

37. Kalra SP, Kalra PS. Neuropeptide Y: A Physiological Orexigen Modulated by the Feedback Action of Ghrelin and Leptin. Endocrine 2003;22:49–56.

38. Schmidt RE. Neuropathology and pathogenesis of diabetic autonomic neuropathy. Int Rev Neurobiol 2002;50:257–292.

39. Fabris N, Mocchegiani E, Provinciali M. Plasticity of neuroendocrine-thymus interactions during aging. Exp Gerontol 1997;32:415–429.

40. Hipkiss AR. Carnosine, a protective, anti-ageing peptide? Int J Biochem Cell Biol 1998;30: 863–868.

41. De Marchis S, Melcangi RC, Modena C, Cavaretta I, Peretto P, Agresti C, Fasolo A. Identification of the glial cell types containing carnosine-related peptides in the rat brain. Neurosci Lett 1997;237:37–40.

42. McFarland GA, Holliday R. Retardation of the senescence of cultured human diploid fibroblasts by carnosine. Exp Cell Res 1994;212:167–175.

43. Wang AM, Ma C, Xie ZH, Shen F.. Use of carnosine as a natural anti-senescence drug for human beings. Biochemistry (Mosc) 2000;65:869–871.

44. Trombley PQ, Horning MS, Blakemore LJ Interactions between carnosine and zinc and copper: implications for neuromodulation and neuroprotection. Biochemistry (Mosc) 2000;65:807–816.

45. Mocchegiani E, Giacconi R, Cipriano C, Muzzioli M, Fattoretti P, Bertoni-Freddari C, Isani G, Zambenedetti P, Zatta P Zinc-bound metallothioneins as potential biological markers of ageing. Brain Res Bull 2001;55:147–153.

46. Preston JE, Hipkiss AR, Himsworth DT, Romero IA, Abbott JN. Toxic effects of beta-amyloid(25–35) on immortalised rat brain endothelial cell: protection by carnosine, homocarnosine and beta-alanine. Neurosci Lett 1998; 242:105–108.

47. Hyland P, Duggan O, Hipkiss A, Barnett C, Barnett Y The effects of carnosine on oxidative DNA damage levels and in vitro lifespan in human peripheral blood derived CD4+T cell clones. Mech Ageing Dev 2000;121:203–215.

48. Naito Y, Yoshikawa T, Yagi N, Matsuyama K, Yoshida N, Seto K, Yoneta T. Effects of polaprezinc on lipid peroxidation, neutrophil accumulation, and TNF-alpha expression in rats with aspirin-induced gastric mucosal injury. Dig Dis Sci 2001;46:845–851.

49. Mocchegiani E, Giacconi R, Cipriano C, Muzzioli M, Gasparini N, Moresi R, Stecconi R, Suzuki H, Cavalieri E, Mariani E. MtmRNA gene expression, via IL-6 and glucocorticoids,

as potential genetic marker of immunosenescence: lessons from very old mice and humans. Exp Gerontol 2002;37:349–357.

50. Zakhidov ST, Gopko AV, Semenova ML, Mikhaleva YY, Makarov AA, Kulibin AY Carnosine modifies the incidence of genetically abnormal sex cells in the testes of senescence-accelerated mice. Bull Exp Biol Med 2002;134:78–80.

51. Maret W, Vallee BL. Thiolate ligands in metallothionein confer redox activity on zinc clusters. Proc Natl Acad Sci U S A 1998;95:3478–3482.

52. Kennedy MC, Gan T, Antholine WE, Petering DH.. Metallothionein reacts with Fe2+ and NO to form products with A g = 2.039 ESR signal. Biochem Biophys Res Commun 1993;196:632–635.

53. St Croix CM, Wasserloos KJ, Dineley KE, Reynolds IJ, Levitan ES, Pitt BR. Nitric oxide-induced changes in intracellular zinc homeostasis are mediated by metallothionein/thionein. Am J Physiol Lung Cell Mol Physiol 2002;282:L185–192.

54. Kroncke KD, Carlberg C. Inactivation of zinc finger transcription factors provides a mechanism for a gene regulatory role of nitric oxide. FASEB J 2000;14:166–173.

55. Gergel D, Cederbaum AI. Inhibition of the catalytic activity of alcohol dehydrogenase by nitric oxide is associated with S nitrosylation and the release of zinc. Biochemistry 1996;35:16186–16194.

56. Zangger K, Oz G, Haslinger E, Kunert O, Armitage IM. Nitric oxide selectively releases metals from the amino-terminal domain of metallothioneins: potential role at inflammatory sites. FASEB J 2001;15:1303–1305.

57. Mocchegiani E, Muzzioli M, Giacconi R Zinc and immunoresistance to infection in aging: new biological tools. Trends Pharmacol Sci 2000;21:205–208.

58. Delledonne M, Polverari A, Murgia I. The functions of nitric oxide-mediated signaling and changes in gene expression during the hypersensitive response. Antioxid Redox Signal 2003;5:33–41.

59. Weiss JH, Sensi SL, Koh JY. Zn(2+): a novel ionic mediator of neural injury in brain disease. Trends Pharmacol Sci. 2000;21:395–401.

60. Black MM. Zinc deficiency and child development. Am J Clin Nutr 1998;6:464S-469S.

61. Masoro EJ. Caloric restriction and aging: an update. Exp Gerontol 2000;35:299–305.

*The Neuroendocrine Immune Network in Ageing*
Edited by R.H. Straub and E. Mocchegiani

# Concluding Remarks and Future Directions

ISTVAN BERCZI[1] and ANDOR SZENTIVANYI[2]

[1]*Department of Immunology, Faculty of Medicine, The University of Manitoba, Winnipeg, MB Canada, R3E 0W3;* [2]*Department of Internal Medicine, Faculty of Medicine, The University of South Florida, Tampa, Florida 33612, USA*

## ABSTRACT

It is indicated by this volume that normal age related changes in the neuroimmune regulatory network are highly relevant to "successful", or healthy ageing. Abnormal regulation is associated with age related diseases. Clinical observations provide compelling evidence for the relevance of Neuroimmune Biology (NIB) to practical Medicine. Thus the Science of NIB provides novel perspectives for the investigation of highly complex biological processes, such as ageing and lead to original observations that allows for better understanding of higher organisms in their entire complexity. The future is very challenging both scientifically and from the moral standpoint.

## 1. INTRODUCTION

Ageing is one of the unresolved problems in Biology. However, the rule is abundantly clear: we all must be born in order to exist and eventually must die. Without this iron-fisted law of Mother Nature the evolution of the species would have been difficult if not impossible. Further, the proper balance of the Plant and Animal Kingdoms would have been endangered. Clearly, immortality could be considered as the cancer of the evolutionary process and of normal biological ecosystems.

By now it is firmly established that programmed cell death (apoptosis) is part of normal embryonic and post-partum development as well as it is fundamental to the regulation of various systems in the body, including the immune system. Immune killer cells use apoptosis as efficient weapons for the elimination of infected and cancerous target cells. Clearly all of our cells possess the genetic machinery for apoptosis and there are multiple receptors that are capable of committing various somatic cells to the suicide pathway [1–3]. One may pose the question: if all of our cells can be induced to die, are we also programmed to die? Common sense would give us a Yes answer. In other words, under normal conditions ageing may be a maturation process, which will proceed smoothly, without major life-threatening problems. This is often termed as "successful" ageing. However, it seems that in most people ageing is associated with major abnormalities, including of those of the Neuroimmune Regulatory System and consequently, diseases prevail during ageing. The chapters in this book lend support to this hypothesis.

## 2.    AGEING OF THE IMMUNE SYSTEM

### 2.1.    Immunosenescence

The currently accepted wisdom is that as we age, we get sicker and sicker, and then we die. Early immunological studies support very well this common belief. Nearly all immune parameters have been shown to decline with ageing, especially when male subjects were examined. However, most clinical experiments are done in hospitals, where only sick people are available for study. Some recent data indicate that healthy old people have healthy immune-, endocrine- and nervous systems. Nevertheless, the Immune System shows age-related changes, but it may be regarded as maturation, rather then a disastrous break down. Thus thymus involution is possible because the thymus has less and less work to do with our immunological adaptation to our external and internal environment. This hypothesis is supported by the fact that antigen presentation does not decline, if anything it is improving with age. So are memory T lymphocytes. Few immunological phenomena carry the same significance than antigen presentation and immunological memory [4–6].

The above parameters are fundamental pre-requisites of maintaining an effective immune defense. So, if older people are better in some aspects of immunity than are younger individuals, should they be examined with the expectation that they are deficient? Should not we consider the possibility that a process of immune maturation takes place during ageing that under normal conditions is capable of maintaining good health and survival? Should one consider the possibility of ripen old age without disease? Studies of healthy old people in order to learn more about health, disease, the immune system and ageing represent a new dimension of research in ageing [7–9]

Apparently now there is evidence for the relationship between the oligoclonal expansion of lymphocytes with age and endogenous viruses (e.g. Herpes, Cytomegalovirus and Epstein-Barr virus) [10–13]. Traditionally these viruses are viewed as potential pathogens, rather than anything else. However, it is possible to envision a different perspective for these persistent organisms. It is clear that they become activated from time-to-time and in immunodeficient individuals they may even cause disease and death. But during temporary deficiency they actually may serve as powerful and relatively harmless immunological adjuvants that activate the cytokine system, which has enormous restorative and stimulatory power for the immune system. For instance, EBV is a polyclonal B cell activator. Therefore, it is relevant to boosting (natural) antibody mediated defense. Herpes and CMV would be relevant to boosting the T cell and NK cell-mediated immunity by the stimulation of the right cytokines, such as interferons. Indeed the maintenance of persistent viruses may be symbiotic, rather than antagonistic to the organism. An analogous situation exists with LPS, which is released by intestinal bacteria and will absorb and boost the immune system during emergency situations, such as trauma and shock [14,15] The boosting of host defense mechanisms with the aid of symbiotic microorganisms is likely to acquire more and more significance during the ageing process. This is a very effective way of self-medication. Clearly, the biological significance of persistent viruses need further clarification .

### 2.2.    Metallothioneins as potential biological markers of immunosenescence

The zinc-zinc-binding protein system regulates the activity of key hormones, cytokines and enzymes in the body. Moreover, this system is affected by stress and the zinc binding proteins are known as acute phase proteins (APP). Apparently Zn is essential for the normal function of the adaptive immune system as it controls the activity of key hormones, such as growth hormone (GH) and of cytokines, as well as enzymes. Indeed, the over-expression of zinc binding proteins during the acute phase response (APR) may play an essential role in the conversion of the immune system from the adaptive mode of reactivity to the boosting of natural immune host defense [16,17] Because there is a drift during ageing from prevalent adaptive immune reactions towards natural immune mechanisms and to frequent APRs, it comes natural that zinc-binding proteins show elevations with ageing [15] However, memory T cells are all right in old age, in spite of Zn deficiency. Moreover, these changes are not inevitable as centenarians were found to have normal Zn and binding protein values [18]. It is also noteworthy that liver regeneration was not affected in old mice with high MT and Zn levels in this organ[19]. This observation indicates that cell proliferation *per se* may not be deleteriously affected by the zinc homeostatic regulatory system.

## 2.3. Neutrophil ageing and immunosenescence

Neutrophil numbers are not lowered with age [20] and there is normal neutophilia to infection [21]. Adhesion molecules are normal or may even be increased [22,23] and bactericid innate recognition is all right [24]. Contradictory observations were made for neutrophil extravasation, chemotaxis and superoxide production. Neutrophil reaction was reduced to GCSF but not to GMCSF. Significant reduction was found in phagocytosis and cytotoxicity was declined towards bacteria and yeast [25]. However, it is not clear, how much of this functional impairment is due to the ageing of neutrophils and to what extent is it the consequence of glucocorticoid excess, which is characteristic of ageing animals and man [15,26,27]. Hypopituitary patients show excess mortality due to respiratory disease [28] which gives further emphasis to the relevance of age related neuroendocrine alterations to immune abnormalities.

## 2.4. Apoptosis and ageing

Apoptotic (APO) phenomena are increased in ageing [29]. Tumor necrosis factor (TNF) is also increased, as well as tumor necrosis factor receptor-I (TNFR-I), which mediates APO on various target cells. In contrast, TNFR-II, which conveys other actions of TNF, is decreased [30,31,]. Effector memory T cells (e.g. CD28+) are increased during ageing and these cells are susceptible to apoptosis [32]. However some subsets of memory T cells may be APO resistant [33]. Replicative senescence may block APO [34] Several functions of polymorfonuclear cells (e.g. killing, free radicals, chemotaxis, O3, lytic enzymes) are decreased with ageing [1,35]. Cholesterol, which is often elevated in the elderly, is important for signal transduction by membrane bound receptors [1].

It is apparent from this chapter that APO occurs more frequently in the elderly. However, it is also apparent that programmed cell death remains an important physiological regulatory mechanism for memory T cells, which show an increase, rather than a decline with ageing.

## 2.5. MHC-unrestricted cytotoxity in ageing

Monocyte and macrophage numbers and cytokine production are increased in the elderly [36], and dendritic cells are unimpaired as well [37]. This assures excellent antigen presentation as well as phagocytic/cytotoxic activity by these cells. TNF alpha was not changed, whereas IFN gamma was increased in centenarians [38]. Mature NK (CD56+) cells were elevated in the elderly [39], and NK functions were not diminished with ageing [40]. High NK numbers were observed in centenarians [41]. Gamma/delta T cell cytotoxicity was preserved in centenarians [42] and higher gamma-delta-T activation was observed in old individuals [43]. Bone and muscle remodeling by the immune system was maintained in the elderly [44,45].

This chapter testifies that innate cytotoxic mechanisms are actually superior in elderly people when compared to young individuals. This is coupled with excellent antigen presentation and immunological memory. Moreover, immune capability of centenarians is truly remarkable. All these facts support the idea, that during successful ageing an immune disaster is not inevitable, on the contrary, it is possible to suggest that the immune system keeps adapting to age related changes in order to optimally fulfill its normal functions.

2.6.    MHC Polymorphism and ageing

Caruso and colleagues conclude that there is no statistically significant relationship between MHC-polymorphism and successful ageing [46] [Candore et al. 2004]. However, one must not forget that MHC polymorphism and immunological diversity play a major role in survival at the time of major epidemics.

3.    AGEING OF THE ENDOCRINE SYSTEM

3.1.    Neuroplasticity in the human hypothalamus during ageing

Hofman and Swaab [47] provides an accout of hypothalamic changes during ageing. They point out that each group of cells (e.g. nuclei) in the hypothalamus have their own sex-specific pattern of ageing. Some nuclei become less active, others become more active with age. Steroid hormones play a key role in neuroplasticity and ageing.

This story is very similar to what was said about the ageing immune system. Both positive and negative changes occur, which implies regulation rather then simple deterioration during ageing. Current evidence indicates that the hypothalamus is the ultimate immunoregulator [48] and it is fundamental for the regulation of all other organs and tissues in higher organisms, as is obvious from this volume. On this basis one may suggest that the process of ageing also is under hypothalamic control. Clearly this part of the central nervous sytem is involved in setting of biological rhythms and adaptive variations. The lifespan of the organism may be suggested as the ultimate rhythm to regulate. Elderly animals and humans are typically characterized by the loss of rhythmicity of physiological parameters and by the inability to adapt to environmental challenges, as repeatedly pointed out in this volume.

3.2.    The role of growth hormone signaling in the control of ageing

Growth hormone plasma levels gradually decline during adult life. It is apparent from the relevant literature that role of GH changes in different stages of life. Excess GH during adulthood in animals and man shortens their lifespan. Tradeoffs must exist between growth, reproduction and longevity. Low GH levels may have a role in protecting the organism from cancer and other age-related diseases. Subnormal GH levels affect body composition, muscle and brain function, and are suspected of contributing to the deterioration of quality of life in the elderly. However, the risks and benefits of anti-ageing therapy with GH are not well understood and the concept of GH replacement during "somatopause" is controversial [49]

It seems very intriguing that GH has an influence on lifespan. Being an anabolic hormone, excess GH must promote metabolic processes that are deleterious for longevity. Growth hormone is the member of the growth and lactogenic hormone (GLH) family, which include multiple iso-forms of GH, prolactin (PRL) and placental lactogenic hormones (PL). These hormones have overlapping functions and are of fundamental importance in governing the growth, development and bodily functions of higher animals for their life-cycle. On the basis of the immunological effects of GLH hormones it was postulated that they function as competence hormones for the immune system, and likely for all other tissues and organs in higher animals [50,51] Because of overlapping functions between GH and PRL, GH or PRL deficiency alone rarely causes severe problems due to compensation by the other hormone, if it is present at normal levels. However, joint and complete deficiency has not been adequately demonstrated to date [52]. Our experiments in hypophysectomized rats indicated that animals lacking both hormones die of bone marrow and immune failure and of cachexia within 6 weeks [53]. These observations indicate that PRL alone is capable of maintaining vital bodily functions in hypophysectomized rats. Redundancy exists within the GLH family and these hormones are indispensable for the growth, development and functional maintenance of the body throughout the entire lifespan of higher animals [50–53]. Consequently, it is reasonable to suggest that GLH, rather than GH alone, should be viewed as functional unit, when it comes to the assessment of biological functions and of the impact on health and longevity.

3.3.    Ageing and the adrenal cortex

Ageing in healthy people is associated with the gradual decrease of adrenal dehydroepian-drosterone (DHEA) secretion. This leads to a relative glucocorticoid excess during ageing. At present, there is no clear indication that DHEA therapy would work in the elderly [54]

DHEA functions as a steroid hormone precursor, and its decline leads to androgen and estro-gen deficiency, whereas glucocorticoid levels are not changed or may even be elevated during ageing. Whether or not these changes represent adaptation to the *changing internal milieu char-acteristic of ageing*, or may function as an important link between immune-, endocrine- and neuronal-*senescence*, remains to be determined. In woman a major trigger of changes in steroid hormone secretion is *menopause*. It is abundantly clear that steroid hormones are major regula-tors in the body that includes the immune system and that altered levels do have far reaching consequences [3,55].

3.4.    Hormonal changes in ageing men

A number of age-related changes in men can ultimately lead to androgen deficiency. Some of the symptoms and disorders frequently seen in elderly men (e.g. loss of body mass, sexual dysfunc-tion, osteoporosis, depression) have been linked to low androgen levels. Up to one third of men beyond the age of 60 have low serum testosterone levels. Androgen supplementation for these men is controversial at this time [56].. Circadian changes of testosterone are lost in man during ageing [57,58].

There is little doubt that testosterone declines with age, even wen levels do not get subnormal in elderly man. A more serious problem may be the loss of the ability for adaptive changes. This is indicated by the loss of the circadian rhythm. Testosterone declines during trauma and inflammatory disease, and exerts a major effect on the immune system, which has a bearing on mortality [3]. Therefore, this hormone has a much wider biological potential than originally anticipated, and this should be taken into consideration in future investigations.

### 3.5.    Ageing of the neuroendocrine circadian system

Changes do occur in the neuroendocrine circadian system. The most frequent observation is the reduction of circadian amplitude during ageing. With the reduction of melatonin the information on time and seasons is decreased. This leads to reduced adaptability, the loss of seasonal adjustments and more susceptibility to climatic change and in general to stress [59]

Adrenal cortisol is maintained, whereas DHEA abruptly declines during ageing. The circadian rhythm of ACTH decreases or is abolished in aged rats. In man melatonin decreases during the sixties and seventies, but later it levels off. Many factors influence the hypothalamus-pituitary-thyroid axis. It is inhibited by sleep deprivation, iodine deficiency. Fasting suppresses TSH levels, which may be restored by leptin. Geographical changes and climatic changes, such as cold increase T4 levels, but not in the elderly [59].

The circadian rhythm of prolactin continues in the elderly. High levels are secreted during REM sleep. Nursing over-rules this regulation. Numerous medications stimulate PRL secretion (e.g. psychoactive drugs). In sexually mature woman, cycling PRL is increased. Stress also increases PRL serum levels [59].

Catecholamines drop with age and beta-adrenergic receptors are impaired. However dopamine (DOP) cells increase in the substancia nigra up to age 60. Sympathetic nervous responses to stress increase with age. Norepinephrine production and release is up and blood pressure is elevated, which may be decreased by bromocriptine. Urinary catecholamine levels are higher in hypertension. Melatonin antagonizes these changes [59].

There is no doubt that the neuroendocrine system plays a major role in adaptation to environmental factors and to stressors. The diurnal rhythm of hormones is related to the rhythm of daytime activity, energy spending and catabolism and night-time regeneration and anabolism. The loss of circadian rhythm of various hormones impairs adaptability as well as proper regeneration and maintenance of good health.

It is remarkable that PRL levels and rhythms are maintained in the elderly. Indeed, PRL is required to maintain vital bodily functions. Animal experiments showed, that if this hormone is lost, it has fatal consequences [53]. Therefore, GH may decline with age because PRL is able to maintain all the vital functions GH would do, without the deleterious effects that elevated GH levels may produce [49].

### 3.6.    Melatonin rhythms, melatonin supplementation and sleep in old age

After 80 years of age, there are no rhythms of melatonin secretion and there is no diurnal periodicity. Rhythms are regulated by noradrenaline, which is increased. However, melatonin did not decline in extremely healthy people. At present there is no clear evidence for the benefit of melatonin treatment [60,61].

As already pointed out, the lack of biological rhythms leads to the loss of adaptation and impair the daily regeneration of the body. These are very significant problems of most elderly people.

## 4. AGEING OF THE NERVOUS SYTEM

### 4.1. Age-related changes of the human autonomic nervous system

Ageing is accompanied by significant structural and functional modifications of the cardiovascular system. Cardiovagal modulation is decreased with increasing age as indicated by a decrease of cardiovagally mediated indices of heart rate variability. Sympathetic outflow to the heart is elevated, however, it is not well transformed into an enhanced end-organ response because of a decrease of α-and ß-adrenergic receptor potency. Orthostatic dysregulation is common in the elderly for several reasons; one major point is that the baroreflexes are impaired with increasing age [62].

### 4.2. Age-related alterations in autonomic nervous innervation

The thymus is innervated, alpha- and beta adrenergic receptors are present and there is evidence for the neural regulation of T cell subsets. Of the neuropeptides substance P, calcitonin gene related peptide, meth-enkephalin, vasoactive intestinal peptide, cholecystokinin, neurotensin, intreleukin–1, C-reactive protein and arginine vasopressin are present in thymic nerves. Noradrenalin (NE) is diminished in spleen of old rats in all compartments. There is peripheral neuropathy suggesting *metabolic insults*. Heightened sympathetic activity is present in the remaining nerves. NE treatment depletes NE nerves in the rat. Bacterial lipopolysaccharide (LPS) has no effect, but IL-2 releases NE. *L-deprenyl*, a monoamine oxidase-B (MAO-B) inhibitor, partially restores NA in 21 month-old rats. Natural killer cell activity, Concanavalin-A and IL-2 responses are increased by such treatment.

Immune changes in aged humans and animals are variable. Sympathetic nerves may stimulate immunity and inhibit B cells and the TH1 response. The effect is variable, depending on the circumstances. Compensatory mechanisms operate in order to achieve "fine tuning" of the immune system. T cell memory is increased with ageing and this requires adjustments [63–71].

This chapter sums up very well the facts and dilemmas of ageing and neural regulation of the immune system. Yes, there are definite signs of deterioration and malfunction, but neither the regulatory influence of sympathetic nerves, nor the age related deterioration of immune function is etched in stone. One reason for this must be that no differentiation is made between ageing with disease and healthy ageing. As for the fine-tuning of the immune system by the autonomic nerves system, this must be viewed as an important mechanism for the adaptive modulation of immune function. However, it seems certain that T cell memory is maintained, no matter what age-related changes are present.

### 4.3. Ageing and the neuroendocrine system of the gut

Everything declines in the gut with ageing, somatostatin is decreased and so is gastrin, more so in males than in females. Endocrine cells are all right in the large intestine. Serotonin is increased in the small intestine, so are peptide YY and enteroglucagon . In BALB/c/nunu mice, which do not have thymuses, endocrine cells are increased in the gastrointestinal tract with age. In general gastrointestinal nerve cells decrease in man and in the guinea pig with ageing, but not in the mouse [72–74].

Here again the general rule is decline, but the mouse is exception and in thymus-less nude mice actually there is an increase of gastrointestinal nerve cells with age. Does this signify the existence of as yet unsuspected influence of the thymus or of T cells on gastrointestinal innervation? Future studies are to decide this question.

## 4.4.  Nutrition and brain ageing

Ageing is characterized by the loss of capacity to maintain homeostasis. During dietary caloric restriction more synapses are present in the brain with larger transmission area. Vitamin E deficiency leads to less and larger synapses. Synapses enlarge in order to compensate for the decreased numbers. Ageing is accelerated under these conditions. Alcohol impairs synaptic potential by disturbing lipid metabolism [75,76]

Brain ageing may be thought of as a particular condition in which specific pathological changes are found without clinically evident manifestations, because nerve cell alterations are continuously counteracted by compensating reactions. As a consequence, deterioration of function occurs when the number of neurons and of their connections decrease below a critical reserve level and coping with environmental stimulations becomes difficult [77,78].

There is no doubt that optimal nutrition is fundamental to good health. Over-nutrition is just as damaging as is under-nutrition. Food does not only pose "oxidative damage", but always contain multiple carcinogenic and other harmful substances. Toxic substances are also produced during the process of digestion. These insults must be adequately handled by the defence systems of the body in order to survive and to maintain good health. Clearly, understanding and practising proper nutrition is one of the key concerns of ageing and of public health today.

## 4.5.  Ageing-related role of nitric oxide in the brain

This review points out the possible functional link between age-dependent decrease in neural nitric oxide synthase (nNOS) activity and increase in inducible NOS (iNOS) expression in the brain. Evidence is presented for the existence of infections in the brain, which is one of the hallmarks of ageing, and apparently trigger "spontaneous" iNOS expression. The future treatment of aged people is also described, which takes into consideration these facts [79].

The problem of persistent infections is not resolved at the present time. Most scientists tend to regard them as harmful, yet they were observed to afford protection to the host on repeated occasions. For instance it is well known about the herpes simplex virus persists within the CNS in most people and that it becomes readily activated in certain individuals in response to various stressors [80]. In the overwhelming majority of the cases there is a harmless self-resolving infectious lesion, which heals spontaneously, and hardly can be classified as dangerous. Actually there is a real possibility that the cytokines generated during this benevolent immune activation are quite beneficial to the individual. Recent observations revealed, that inflammation itself has protective value for the nerves system, possibly via T cell derived nerve growth factor [81]. Moreover, the lipopolysaccharide endotoxin of intestinal gram negative bacteria, which are absorbed from the gut during trauma and shock, functions as a powerful immune activator. Clearly, LPS functions to activate the innate immune system in stressful conditions, and therefore, it may be considered as one of the remedies for self–medication in emergency situations [82]. Persistent neurotrop viruses may in fact fulfill a similar mission. It is very well established that iNOS is an important weapon against infectious agents, and may be used as such in the ageing brain. It is also clear that LPS, iNOS and excessive activation of persistent viruses may have harmful effects. These questions await further clarification.

5.    LINKS BETWEEN ONE GLOBAL SYSTEM AND ANOTHER GLOBAL SYSTEM
      DURING THE AGEING PROCESS

5.1.    Plasticity of neuro-endocrine-thymus interactions during the ontogeny of ageing: the
        role of zinc

Current evidence strongly suggests that thymic involution is a phenomenon secondary to age-related alterations in neuroendocrine-thymus interactions. The disruption of these interactions in old age is responsible for age-associated immune-neuroendocrine dysfunctions. Thymic reconstitution may be achieved by GH, thyroid hormones, and LHRH, which act on specific hormone receptors on thymocytes and on thymic epithelial cells. Melatonin may also act through specific receptors on T-cells. Zinc finger proteins are important because they regulate gene expression for hormone receptors. However, the effect of zinc is multifaceted. Zinc-dependent thymic hormone is required for intrathymic T-cell differentiation and maturation as well as for the homing of stem cells into the thymus. Therefore, the role of zinc is crucial in neuroendocrine-thymus interactions and for the maintenance of thymus function (e.g. by GH, thyroid hormones or melatonin) and for improving adaptive immunocompetence during ageing [83]

The traditional view that thymic involution is responsible for immune deterioration during ageing may now be challenged. As is obvious from this volume, memory T lymphocytes increase with age. This means that memory cells are being accumulated gradually towards most, if not all, of the environmental pathogens. Therefore, there is little if any need for naïve T cells freshly released by the thymus. However, the thymus can be re-activated again, even at advanced age [84,85]. This suggests, that thymic involution may be a physiological phenomenon, at least in part, during successful ageing. Thymic involution may also occur under pathophysiological conditions, during the acute phase response (APR), or febrile illness, which is initiated by the immune system in response to infection or to other forms of traumatic injury. Cells of the monocyte/macrophage lineage release massive amounts of IL-1, IL-6 and TNF-alpha into the circulation, which in turn trigger profound neuroendocrine and metabolic changes. The hypothalamus-pituitary-adrenal axis is activated, the thymus undergoes a profound involution, due to the apoptotic effect of glucocorticoids and of TNF, and to the relative deficiency of growth and lactogenic hormones, which is coupled with zinc deficiency. The biological significance of APR is that it is capable of switching over the immune system from the adaptive mode of reactivity to the amplification of natural immune mechanisms. APR is a highly coordinated emergency defense reaction, which mobilizes all the resources of the body in the interest of survival. Many individuals exhibit signs of low-grade APR during ageing [15,17]. Therefore, thymic involution may not be the primary cause for immunodeficiency in the aged, but age-related diseases, which may be the consequence of immunodeficiency, would accelerate thymic involution. Hence, thymic involution is also associated with pathological conditions. Would the improvement of thymic function lead to health benefit in the elderly? Current indications are that it would indeed do so.

5.2.    Adverse glucocorticoid actions and their relevance to brain ageing

This chapter discusses the concepts of stress and of homeostasis. It is presented that the HPA axis becomes gradually more active with ageing. Glucocorticoids affect the CNS, immunocompetence, and inflammation. Although a fair amount of information is available, the biological significance of the observations is not always clear [86,87].

5.3.    Neuroendocrine immune aspects of osteoporosis during the ageing process

 Age related alterations of gonadal and thyroid hormones and of vitamin D leads to loss of bone mass. Alpha melanocyte stimulating hormone is a novel bone regulator as is leptin. Serum osteo-protegrin is increased with age [88–91].

One may remark here that osteoclasts belong to the monocyte/macrophage family of myeloid cells and they respond to immune-derived cytokines, especially to IL-6. IL-6 has a stimulatory effect on osteoclasts, which leads to accelerated bone resorption. Estrogens suppress IL-6 production. Estrogen deficiency develops abruptly in females after the menopause and to a lesser extent also in males, due to androgen deficiency. This will lead to increased secretion of IL-6 and consequently to accelerated bone loss [55].

## 6.    THE AGEING PROCESS AND CHRONIC INFLAMMATORY DISEASE

6.1.    Neuroendocrine immune mechanisms of accelerated ageing in patients with chronic
         inflammatory diseases

Straub and colleagues consider the deleterious effects of chronic inflammatory disease, which accelerates ageing. Indeed inflammatory disease is characterized by immune activation and cytokine release, which invariably elicit neuroendocrine and metabolic alterations as correctly pointed out by the authors. Changes in glucocorticoids, androgens, estrogens and vitamin D and the altered metabolism affect immune function, the nervous system and bone metabolism. It is concluded that more information is required before a rational approach may be adapted for therapy [92]

Although there is little doubt that chronic inflammation has an enormous thaw on the patient, one should keep in mind that inflammation is a defense reaction, which should protect the host. Is this possible in chronic inflammation? Recent observations in experimental autoimmune encephalitis of rats, a chronic inflammatory disease with relapses, demonstrated that indeed, inflammation has protective value [81].

6.2.    Thyroid autoimmunity and ageing

Thyroid autoantibodies and subclinical hypothyroidism increase with ageing. However, thyroid autoantibodies are rare in centenarians and in other highly selected aged populations. This suggests that thyroid autoimmunity is not the consequence of the ageing process itself, but rather, an expression of age-associated disease. Thyroid autoimmunity and/or thyroid dysfunction may have an etiological role in several age-associated diseases, although the precise mechanisms involved remain to be elucidated [93].

This chapter clearly demonstrates that successful ageing is not characterized by autoimmune reactions, whereas autoimmunity itself may be the cause of age-associated diseases. Indeed, it is highly desirable to keep these findings in mind when trying to investigate the problem of ageing. Even if most people age with disease, the goal should be to achieve successful, healthy ageing, whenever possible.

6.3.    The clinical importance of proinflammatory cytokines in elderly populations

Low-grade increases in plasma level of inflammatory mediators are characteristic of age-asso-ciated pathology, including cardiovascular diseases, dementia and sarcopenia. Inflammatory cytokines, such as TNF and IL-6, act as independent predictors of high mortality risk. It is suggested that systemic low-grade inflammation provide a common link between ageing, body composition, and life style factors on one hand and age-related diseases and mortality on the other [94].

The senile immune system is clinically relevant. It is likely that TNF and IL-6 trigger age associated pathology. Cardiovascular disease and dementia are prevalent in the elderly. TNF is associated with dementia in the very old (>80 years of age) and IL-6 may play a role in throm-boembolic complications and cardiovascular disease in middle-aged people. Monocytes are in a pre-activated state in the elderly. Some of the possible reasons for the development of disease are: continuous antigenic load, stressors, "inflam-aging" and dys-regulated acute phase response [94,95].

TNF and IL-6 promoter polymorphism may play an important role in disease development [96,97]. Body fat directly correlates with TNF [98] and with leptin [99] Endocrinosenescence contributes to immunosenescence. The loss of sex hormones leads to immunosuppression and to a low grade inflammatory milieu in the elderly [100]

The results presented in Figure 4 of this chapter support the idea of a proinflammatory milieu in the elderly. Peripheral blood cells from young individuals produced more TNF after *in vitro* stimulation by LPS than did cells from old people, yet the blood level of TNF in the elderly was significantly higher after LPS treatment. Infections, especially *Chlamydia* and bacteriuria are frequent in the elderly [101,102].

This chapter is an excellent demonstration of the relevance of Neuroimmune Biology to Clini-cal Practice. Physicians have always been concerned with the patient's well being, and for that they had to understand how the body works. This chapter presents information from the area of Molecular Biology, Immunology, Microbiology, Endocrinology, Neurology, Food Science, Pathology, Gerontology, all of which are integrated into a modern version of Clinical Medicine. The pathophysiological mechanisms that are discussed in the context of age-related diseases are compelling and are testable for further confirmation. Indeed, this is what should be the ultimate mission of Neuroimmune Biology, to understand the biology of man and of higher animals in their entire complexity, and to apply the newly found knowledge to the benefit of all.

7.    CONCLUSIONS

7.1.    New anti ageing strategies related to neuroendocrine immune interactions

This chapter demonstrates that the agents/methods used for the prolongation of life span influ-ence growth hormone, prolactin, zinc or apoptosis. The new agents act on one of these basic factors in ageing [103]

This chapter illustrates very well again the relevance of Neuroimmune Biology to practical Medicine. It was postulated a decade ago that members of the growth and lactogenic hormone family function as *competence hormones* that regulate growth and all functions in higher animals [50]. It has also been established that hypohysectomized rats rely on residual PRL for survival. If this PRL is neutralized by antibodies, they will die within 6 weeks [53]. Now several chapters, including this one, lends further support to the hypothesis that *GLH maintain vital bodily functions and are essential for survival and for longevity.*

## 7.2. The future

Much of the global problems of our times come from the fact that man aspires to bypass and overrule the laws of Mother Nature. Ageing is no exception, the quest is on for longer and healthier life, and in actual fact significant increases are seen in the mean life-span globally, with the Western world in the lead. Some countries are much overpopulated, and in others the percentage of elderly people is steadily increasing. This poses ever-increasing problems on Society and on future generations. Global overpopulation is already threatening us with catastrophe, so why push our luck any further? Why study ageing at all?

Well, for most people ageing means diseases, often suffering from lengthy and debilitating conditions. Nobody will argue against studying and preventing age related diseases. Better health and better quality of life for all people are noble causes and is highly desirable to pursue. One may also argue that healthy people would be productive longer, so the burden posed on Society would decrease. The scientific value of understanding the ageing process is immense from the point of view of understanding our biology, but also for gaining better insights to the pathobiology of diseases.

This volume sheds light on many aspect of the problem of ageing. The most important massage is that *the Neuroimmune Regulatory Network plays a fundamental role in the ageing process.* Indeed, this book is living testimony for the validity and significance of *Neuroimmune Biology* to ageing. NIB provides for a comprehensive way of viewing and investigating biological problems, such as ageing. For the first time ageing is looked upon from the special perspectives of NIB, the science of systemic regulation of higher organisms. Already there is compelling evidence that this science is highly relevant to Clinical Medicine.

Is it possible to achieve immortality? It is now clear that most if not all of the biological processes, which decline during ageing, are reversible. This would suggest that it would be feasible to achieve immortality in macro-organisms. Ageing seems to be genetically programmed. However some aspects of ageing may be related to the very fundamental properties of matter and energy that dominate our world. So far we have not achieved a clear understanding of the rules governing the fundamental forces of our universe.

## REFERENCES.

1. Larbi A, Fülöp T. Apoptosis and Aging. In "The Neuroendocrine Immune Network in Ageing", Neuroimmune Biology Volume 4. Straub RH, Mocchegiani E. Eds. Berczi I, Szentivanyi A. Series Eds. Elsevier, Amsterdam, 2004; pp. 57–72.
2. Nagy E, Berczi I, Baral E. Combination immunotherapy of cancer. In "Neuroimmune Biology Volume 1: New foundation of Biology" Berczi I, Gorczynski R, Editors, Elsevier, 2001;pp.417–432. S.C.
3. Berczi I, Nagy E, Baral E, Szentivanyi, A. Sterid Hormones. In "Neuroimmmune Biology, Volume 3: The Immune-Neuroendocrine Circuitry. History and Progress". Berczi I, Szentivanyi A, Editors, Elsevier, Amsterdam, 2003; pp. 221–270.
4. Solana R, Pawelec G. Immunosenescence. . In "The Neuroendocrine Immune Network in Ageing", Neuroimmune Biology Volume 4. Straub RH, Mocchegiani E. Eds., Berczi I,

Szentivanyi A. Series Eds. Elsevier, Amsterdam, 2004; pp. 9–21.

5.  Castle, K. Uyemura, W. Crawford, W. Wong and T. Makinodan, Antigen presenting cell function is enhanced in healthy elderly. Mech Ageing Dev 1999;107:137–145.

6.  Chamberlain WD, Falta MT,. Kotzin BL. Functional subsets within clonally expanded CD8(+) memory T cells in elderly humans. Clin Immunol 2000;94:160–172.

7.  Wikby A, Johansson B, Ferguson F, Olsson J. Age-related changes in immune parameters in a very old population of Swedish people: a longitudinal study. Exp Gerontol 1994;29: 531–541.

8.  Wikby A, Maxson P, Olsson J, Johansson B, Ferguson FG. Changes in CD8 and CD4 lymphocyte subsets, T cell proliferation responses and non-survival in the very old: the Swedish longitudinal OCTO-immune study. Mech Ageing Dev1998;1002:187–198.

9.  Pawelec G, Ouyang Q, Colonna-Romano G, Candore G, Lio D, Caruso C. Is human immunosenescence clinically relevant? Looking for 'immunological risk phenotypes'. Trends Immunol 2002;23:330–332.

10. Khan N, Shariff N, Cobbold M, Bruton R, Ainsworth JA, Sinclair AJ, Nayak L, Moss PA. Cytomegalovirus seropositivity drives the CD8 T cell repertoire toward greater clonality in healthy elderly individuals. J Immunol 2002;169:1984–92.

11. Ouyang Q, Wagner WM, Wikby A, Walter S, Aubert G, Dodi AL, Travers P, Pawelec G. Large numbers of dysfunctional CD8+ T lymphocytes bearing receptors for a single dominant CMV epitope in the very old. J Clin Immunol 2003;23:247–257.

12. Ouyang Q, Wagner WM, Walter S, Muller CA, Wikby A, Aubert G, Klatt T, Stevanovic S, Dodi T, Pawelec G. An age-related increase in the number of CD8(+) T cells carrying receptors for an immunodominant Epstein-Barr virus (EBV) epitope is counteracted by a decreased frequency of their antigen-specific responsiveness. Mech Ageing Dev 2003;124: 477–485.

13. Ouyang Q, Wagner WM, Voehringer D, Wikby A, Klatt T, Walter S, Muller CA, Pircher H, Pawelec G. Age-associated accumulation of CMV-specific CD8(+) T cells expressing the inhibitory killer cell lectin-like receptor G1 (KLRG1). Exp Gerontol 2003;38:911–920.

14. Berczi I. Neuroendocrine defence in endotoxin shock. A review. Acta Microbiol Hung 1993:40:265–302.

15. Berczi I, Szentivanyi A. The acute phase response. In Neuroimmmune Biology, Volume 3: The Immune-Neuroendocrine Circuitry. History and Progress. Berczi I, Szentivanyi A, Editors, Elsevier, Amsterdam, 2003; pp. 463–494.

16. Mocchegiani E, Giacconi R, Muti E, Muzzioli M, Cipriano C. Zinc-binding proteins (metallothionein and α-2 macroglobulin) as potential biological markers of immunosenescence. In "The Neuroendocrine Immune Network in Ageing", Neuroimmune Biology Volume 4. Straub RH, Mocchegiani E. Eds., Berczi I, Szentivanyi A. Series Eds. Elsevier, Amsterdam, 2004; pp. 23–40.

17. Haeryfar SMM, Berczi I. The thymus and the acute phase response. Cell Mol Biol 47: 145–156,2001.

18. Mocchegiani E, Giacconi R, Cipriano C, Muzzioli M, Gasparini N, Moresi R, Stecconi R, Suzuki H, Cavalieri E, Mariani E. MtmRNA gene expression, via IL-6 and glucocorticoids, as potential genetic marker of immunosenescence: lessons from very old mice and humans. Exp Gerontol 2002;37:349–357.

19. Mocchegiani E, Verbanac D, Santarelli L, Tibaldi A, Muzzioli M, Radosevic-Stasic B, Milin C. Zinc and metallothioneins on cellular immune effectiveness during liver regeneration in young and old mice. Life Sci 1997;61:1125–1145.

422

20.  Chatta GS, Andrews RG, Rodger E, Schrag M, Hammond WP, Dale DC. Hematopoietic progenitors and aging: alterations in granulocytic precursors and responsiveness to recombinant human G-CSF, GM-CSF, and IL-3. J Gerontol 1993;M207–M212.

21.  Butcher S, Chahal H, Lord JM. The effect of age on neutrophil function: loss of CD16 associated with reduced phagocytic capacity. Mech Ageing Dev 2001;122:20.

22.  Esparza B, Sanchez H, Ruiz M, Barranquero M, Sabino E, Merino F. Neutrophil function in elderly persons assessed by flow cytometry. Immunol Invest 1996; 25:185–90.

23.  Rao KMK. Age-related decline in ligand-induced actin polymerisation in human leukocytes and platelets. J Gerontol 1986;41: 561–566.

24.  Emanuelli G, Lanzio M, Anfossi T, Romano S, Anfossi G, Calcamuggi G. Influence of age on polymorphonuclear leukocytes in vitro: phagocytic activity in healthy human subjects. Gerontology 1986; 32:308–16.

25.  Butcher SK, Wang Q, Lascelles D, Lord JM. Neutrophil ageing and immunosenescence. In "The Neuroendocrine Immune Network in Ageing", Neuroimmune Biology Volume 4. Straub RH, Mocchegiani E. Eds. Berczi I, Szentivanyi A. Series Eds. Elsevier, Amsterdam, 2004; pp. 41–55.

26.  Ferrari E, Casarotti D, Muzzoni B et al. Age-related changes of the adrenal secretory pattern: possible role in pathological brain aging. Brain Res Brain Res Rev 2001; 37: 294–300.

27.  Berczi I, Szentivanyi A. Immunodeficiency. In Neuroimmune Biology, Volume 3: The Immune-Neuroendocrine Circuitry. History and Progress. Berczi I, Szentivanyi A, Editors, Elsevier, Amsterdam, 2003b; pp.537–558.

28.  Tomlinson JW, Holden N, Hills RK et al. Association between premature mortality and hypopituitarism. West Midlands Prospective Hypopituitary Study Group. Lancet 2001; 357:425–31.

29.  Krammer PH. CD95's deadly mission in the immune system. Nature 2000;407 :789–795

30.  Stosic-Grujicic SS, Lukic ML. The production of TNF, IL-1 and IL-6 in cutaneous tissues during maturation and aging. Adv Exp Med Biol 1995;371:411–414.

31.  Screaton G, Xu X.-N. T cell life and death signaling via TNF-receptor family members. Cur. Opin Immunol 2000;12:316–3222.

32.  McLeod JD, Walker LSK, Patel YI., Boulougouris G, Sansom DM. Activation of human T cells with superantigen (SEB) and CD28 confers resistance to apoptosis via CD95. J Immunol 1998;160 :2072–2079.

33.  McLeod JD. Apoptotic capability in ageing T cells. Mech Age Dev 2000; 121:151–159

34.  Spaulding C, Guo W, Effros RB. Resistance to apoptosis in human CD8+ T cells that reach replicative senescence after multiple rounds of antigen-specific proliferation. Exp Gerontol 1999; 34:633–644.

35.  Fulop TJr, Foris G, Worum I, Leovey A. Age-dependent alterations of Fc receptor mediated effector functions of human polymorphonuclear leukocytes. Clin. Exp. Immunol. 1985;61: 425–432.

36.  Gardner ID, Lim ST, Lawton JWM. Monocyte function in ageing humans. Mech Ageing Dev 1981;16:233–239.

37.  Saurwein-Teissl M, Schonitzer D, Grubeck-Loebenstein B. Dendritic cell responsiveness to stimulation with influenza vaccine is unimpaired in old age. Exp Gerontol 1998;33: 625–631.

38.  Miyaji C, Watanabe H, Toma H, Akisaka M, Tomiyama K, Sato Y, Abo T. Functional alteration of granulocytes, NK cells, and natural killer T cells in centenarians. Hum

Immunol 2000;61:908–916.

39. Borrego F, Alonso MC, Galiani MD, Carracedo J, Ramirez R, Ostos B, Pena J, Solana R. NK phenotypic markers and IL2 response in NK cells from elderly people. Exp Gerontol 1999;34:253–265.

40. Facchini A, Mariani E, Mariani AR, Papa S, Vitale M, Manzoli FA. Increased number of circulating Leu 11+ (CD 16) large granular lymphocytes and decreased NK activity during human ageing. Clin Exp Immunol 1987;68:340–347.

41. Sansoni P, Cossarizza A, Brianti V, Fagnoni F, Snelli G, Monti D, Marcato A, Passeri G, Ortolani C, Forti E, et al. Lymphocyte subsets and natural killer cell activity in healthy old people and centenarians. Blood 1993;82:2767–2773.

42. Fagnoni FF, Vescovini R, Mazzola M, Bologna G, Nigro E, Lavagetto G, Franceschi C, Passeri M, Sansoni P. Expansion of cytotoxic CD8+ CD28- T cells in healthy ageing people, including centenarians. Immunology 1996;88:501–507.

43. Colonna-Romano G, Potestio M, Aquino A, Candore G, Lio D, Caruso C. Gamma/delta T lymphocytes are affected in the elderly. Exp Gerontol 2002;37:205–211.

44. Mariani E, Ravaglia G, Meneghetti A, Tarozzi A, Forti P, Maioli F, Boschi F, Facchini A. Natural immunity and bone and muscle remodelling hormones in the elderly. Mech Ageing Dev 1998;102:279–292.

45. Provinciali M, Donnini A, Re F. MHC-unrestricted cytotoxicity in ageing. In "The Neuroendocrine Immune Network in Ageing", Neuroimmune Biology Volume 4. Straub RH, Mocchegiani E. Eds. Berczi I, Szentivanyi A. Series Eds. Elsevier, Amsterdam, 2004; pp. 73–89.

46. Candore G, Caruso C, Romano GC, Lio D. Major histocompatibility complex polymorphisms and ageing. In "The Neuroendocrine Immune Network in Ageing", Neuroimmune Biology Volume 4. Straub RH, Mocchegiani E. Eds. Berczi I, Szentivanyi A. Series Eds. Elsevier, Amsterdam, 2004; pp. 91–101.

47. Hofman MA, Swaab DF. Neuroplascticity in the Human Hypothalamus During Aging. In "The Neuroendocrine Immune Network in Ageing", Neuroimmune Biology Volume 4. Straub RH, Mocchegiani E. Eds., Berczi I, Szentivasyi A. Series Eds. Elsevier, Amsterdam, 2004; pp. 105–121.

48. Berczi I, Szentivanyi A. The immune-Neuroendocrine circuitry. In Neuroimmmune Biology, Volume 3: The Immune-Neuroendocrine Circuitry. History and Progress. Berczi I, Szentivanyi A, Editors, Elsevier, Amsterdam, 2003c; pp. 561–592.

49. Bartke A, Heiman M, Turyn D, Dominici F. Kopchick JJ. The role of growth hormone signaling in the control of aging. In "The Neuroendocrine Immune Network in Ageing", Neuroimmune Biology Volume 4. Straub RH, Mocchegiani E. Eds., Berczi I, Szentivanyi A. Series Eds. Elsevier, Amsterdam, 2004; pp. 123–137.

50. Berczi I. The role of the growth and lactogenic hormone family in immune function. Neuroimmunomodulation 1994;1:201–216.

51. Berczi I, Szentivanyi A. Immunocompetence. In Neuroimmmune Biology, Volume 3: The Immune-Neuroendocrine Circuitry. History and Progress. Berczi I, Szentivanyi A, Editors, Elsevier, Amsterdam, 2003e; pp. 281–299.

52. Berczi I. Foreword: The Neuroimmune Biology of Growth and Lactogenic Hormones. In "Neuroimmune Biology Volume 2: Growth and Lactogenic Hormones". Matera L, Rapaport R, Editors, Elsevier, Amsterdam, 2002;pp. v–xiv.

53. Nagy E, Berczi I. Hypophysectomized rats depend on residual prolactin for survival. Endocrinology 1991;128:2776–2784.

54. Lamounier-Zepter V, Bornstein SR. Aging and the adrenal cortex. In "The Neuroendocrine Immune Network in Ageing", Neuroimmune Biology Volume 4. Straub RH, Mocchegiani E. Eds., Berczi I, Szentivanyi A. Series Eds. Elsevier, Amsterdam, 2004; pp. 139–152.

55. Nagy E, Baral E, Berczi I. Immune System.: In "Handbook of Experimental Pharmacology", Estrogens and anti-estrogens. Section IV. Physiology and Pathophysiology of Estrogens. M. Oettel, ed. in chief, Springer-Verlag 1999, 135/I:343–351 (A).

56. Plas E, Madersbacher S, Berger P. Hormonal changes in aging men. In "The Neuroendocrine Immune Network in Ageing", Neuroimmune Biology Volume 4. Straub RH, Mocchegiani E. Eds., Berczi I, Szentivanyi A. Series Eds. Elsevier, Amsterdam, 2004; pp. 153–163.

57. Vermeulen A, Deslypere JP, Kaufman JM. Influence of antiopioids on luteinizing hormone pulsatility in ageing men. J Clin Endocrinol Metab 1989;68: 68–72.

58. Bremner WJ, Vitiello V, Prinz PN. Loss of circadian rhythmicity in blood testosterone levels with ageing in normal men. J Clin Endocrinol Metab 1983;56:1278–1281.

59. Touitou Y, Haus E.: Aging and the endocrine circadian system. In "The Neuroendocrine Immune Network in Ageing", Neuroimmune Biology Volume 4. Straub RH, Mocchegiani E. Eds., Berczi I, Szentivanyi A. Series Eds. Elsevier, Amsterdam, 2004; pp. 165–193.

60. Riemersma RF, Mattheij CAM, Swaab DF, Van Someren EJW. Melatonin rhythms, melatonin supplementation and sleep in old age. In "The Neuroendocrine Immune Network in Ageing", Neuroimmune Biology Volume 4. Straub RH, Mocchegiani E. Eds., Berczi I, Szentivanyi A. Series Eds. Elsevier, Amsterdam, 2004; pp. 195–211.

61. Zeitzer JM, et al., Do plasma melatonin concentrations decline with age? Am J Med, 1999;107:432–6.

62. Agelink MW, Sanner D, Ziegler D. Age-related changes of the human autonomic nervous system. In "The Neuroendocrine Immune Network in Ageing", Neuroimmune Biology Volume 4. Straub RH, Mocchegiani E. Eds., Berczi I, Szentivanyi A. Series Eds. Elsevier, Amsterdam, 2004; pp. 215–231.

63. Bellinger DL, Madden KS, Lorton D. Age-Related Alterations in Autonomic Nervous Innervation. In "The Neuroendocrine Immune Network in Ageing", Neuroimmune Biology Volume 4. Straub RH, Mocchegiani E. Eds., Berczi I, Szentivanyi A. Series Eds. Elsevier, Amsterdam, 2004; pp. 233–255.

64. Felten SY, Bellinger DL, Collier TJ, Coleman PD, Felten, DL. Decreased sympathetic innervation of spleen in aged Fischer 344 rats. Neurobiol Aging 1987;8:159–165.

65. ThyagaRajan S, Felten SY, Felten DL. Restoration of sympathetic noradrenergic nerve fibers in the spleen by low doses of L-deprenyl treatment in young sympathectomized and old Fischer 344 rats. J Neuroimmunol 1998;81:144–157.

66. Caruso C, Candore G, Cigna D, DiLorenzo G, Sireci G, Dieli F, Salerno A. Cytokine production pathway in the elderly. Immunol Res 1996; 15:84–90.

67. Rink L, Cakman I, Kirchner H. Altered cytokine production in the elderly. Mech Ageing Dev 1998;15:199–209.

68. Hobbs MV, Weigle WO, Noonan DJ, Torbett BE, McEvilly RJ, Koch RJ, Cardenas GJ, Ernst DN. Patterns of cytokine gene expression by CD4+ T cells from young and old mice. J Immunol 1993;150:3602–3614.

69. Hobbs MV, Weigle WO, Ernst DN. Interleukin-10 production by splenic CD4+ cells and cell subsets from young and old mice. Cell Immunol 1994;154:264–272.

70. Utsuyama M, Hirokawa K, Kurashima C, Fukayama M, Inamatsu T, Suzuki K, Hashimoto W, Sato K. Differential age-change in the numbers of CD4+CD45RA+ and CD4+CD29+ T cell subsets in human peripheral blood. Mech Ageing Dev 1992;63:57–68.

71. Jackola DR., Ruger JK, Miller RA. Age-associated changes in human T cell phenotype and function. Aging 1994;6:25–34.

72. El-Salhy M. Aging and the neuroendocrine system of the gut. In "The Neuroendocrine Immune Network in Ageing", Neuroimmune Biology Volume 4. Straub RH, Mocchegiani E. Eds., Berczi I, Szentivanyi A. Series Eds. Elsevier, Amsterdam, 2004; pp. 257–271.

73. Kvetnoy I, Popuichiev V, Mikhina L, Anisimov V, Yuzhakov V, Konovalov S, Pogudina N, Franceschi C, Piantanelli L, Rossolini G, Zaia A, Kventania T, Hernandez-Yago J, Blesa JR. Gut neuroenocrine cells: relation to the proliferative activity and apoptosis of mucous epitheliocytes in aging. Neuroendocrinol Lett 2001;22:337–341.

74. El-Salhy M, Sandström O, Holmlund F. Age-induced changes in the enteric nervous system in mouse. Mech Ageing Dev 1999;107:93–103.

75. Bertoni-Freddari C, Meier-Ruge W, Ulrich J. Age-dependent effect of ethanol on the intranuclear ionic content in hippocampal CA1 pyramidal cells. Adv Biosci 71, 263–267, 1988.

76. Bertoni-Freddari C, Fattoretti P, Casoli T, Di Stefano G, Solazzi M, Giorgetti B, Balietti M. Modulating effects of nutrition on brain aging. In "The Neuroendocrine Immune Network in Ageing", Neuroimmune Biology Volume 4. Straub RH, Mocchegiani E. Eds., Berczi I, Szentivanyi A. Series Eds. Elsevier, Amsterdam; 2004; pp. 273–289.

77. Meier-Ruge. Senile dementia - a disease of exhaustion of functional brain reserve capacity. Excer. Med. Curr. Clin. Pract. Ser. 59, 371–375, 1990.

78. Meier-Ruge W, Iwangoff P, Bertoni-Freddari C. What is primary and what secondary for amyloid deposition in Alzheimer's disease. Ann NY Acad Sci 719, 230–237, 1994.

79. Mariotto S, Miscusi M, Persichini T, Colasanti M, Suzuki H. Ageing-related role of nitric oxide in the brain. In "The Neuroendocrine Immune Network in Ageing", Neuroimmune Biology Volume 4. Straub RH, Mocchegiani E. Eds., Berczi I, Szentivanyi A. Series Eds. Elsevier, Amsterdam, 2004, pp. 291–300.

80. Berczi I, Baragar FD, Chalmers IM, Keystone EC, Nagy E, Warrington RJ. Hormones in self tolerance and autoimmunity: a role in the pathogenesis of rheumatoid arthritis? Autoimmunity 1993;16:45–56.

81. Berczi I, Szentivanyi A. Nerve growth factor, leptin and neuropeptides. In "Neuroimmmune Biology, Volume 3: The Immune-Neuroendocrine Circuitry. History and Progress. Berczi I, Szentivanyi A, Editors, Elsevier, Amsterdam, 2003; pp. 181–189.

82. Berczi, I. Neurohormonal host defence in endotoxin shock. Ann NY Acad Sci 840: 787–802,1998.

83. Mocchegiani E, Giacconi R, Muti E, Muzzioli M, Cipriano C. Pasticity of neuroendocrine-thymus interactions during ontogeny and aging: Role of Zinc. In "The Neuroendocrine Immune Network in Ageing", Neuroimmune Biology Volume 4. Straub RH, Mocchegiani E. Eds., Berczi I, Szentivanyi A. Series Eds. Elsevier, Amsterdam, 2004; pp. 307–329.

84. Shiraishi J, Utsuyama M, Seki S, Akamatsu H, Sunamori M, Kasai M, Hirokawa K.Essential microenvironment for thymopoiesis is preserved in human adult and aged thymus. Clin Dev Immunol. 2003;10:53–9.

85. Berzins SP, Uldrich AP, Sutherland JS, Gill J, Miller JF, Godfrey DI, Boyd RL. Thymic regeneration: teaching an old immune system new tricks. Trends Mol Med. 2002;8:469–76.

86. Dinkel K, Sapolsky RM. Adverse glucocorticoid actions and their relevance to brain ageing. In "The Neuroendocrine Immune Network in Ageing", Neuroimmune Biology Volume 4. Straub RH, Mocchegiani E. Eds., Berczi I, Szentivanyi A. Series Eds. Elsevier,

Amsterdam, 2004; pp. 331–346.

87.  Gardner EM, Murasko DM. Age-related changes in Type 1 and Type 2 cytokine production in humans. Biogerontology 2002;3(5):271–90.

88.  Peterlik M. Neuroendocrine immune aspects of osteoporosis during the ageing process. In "The Neuroendocrine Immune Network in Ageing", Neuroimmune Biology Volume 4. Straub RH, Mocchegiani E. Eds., Berczi I, Szentivanyi A. Series Eds. Elsevier, Amsterdam, 2004; pp. 347–359.

89.  Cornish J, Callon KE, Mountjoy KG, Bava U, Lin JM, Myers DE, Naot D, Reid IR. {alpha}-Melanocyte stimulating hormone ({alpha}-MSH) is a novel regulator of bone. Am J Physiol Endocrinol Metab 2003;4:4.

90.  Li H, Matheny M, Nicolson M, Tumer N, Scarpace PJ. Leptin gene expression increases with age independent of increasing adiposity in rats. Diabetes 1997;46:2035–2039.

91.  Kudlacek S, Schneider B, Woloszczuk W., Pietschmann P, Willvonseder, R. Serum levels of osteoprotegerin increase with age in a healthy adult population. Bone 2003;32:681–686.

92.  Straub RH, Schölmerich J, Cutolo M. Neuroendocrine immune mechanisms of accelerated aging in patients with chronic inflammatory diseases. In "The Neuroendocrine Immune Network in Ageing", Neuroimmune Biology Volume 4. Straub RH, Mocchegiani E. Eds., Berczi I, Szentivanyi A. Series Eds. Elsevier, Amsterdam; 2004 pp. 363–374.

93.  Mariotti S, Pinna G, Pinchera A. Thyroid autoimmunity and ageing. In "The neuroendocrine Immune Network in Ageing", Neuroimmune Biology Volume 4. Straub RH, Mocchegiani E. Eds., Berczi I, Szentivanyi A. Series Eds. Elsevier, Amsterdam, 2004; pp. 375–382.

94.  Bruunsgaard H, Krabbe K. The clinical importance of proinflammatory cytokines in elderly populations. In "The Neuroendocrine Immune Network in Ageing", Neuroimmune Biology Volume 4. Straub RH, Mocchegiani E. Eds., Berczi I, Szentivanyi A. Series Eds. Elsevier, Amsterdam, 2004; pp. 383–395.

95.  Krabbe K, Bruunsgaard H, Hansen CM, Møller K, Fonsmark L, Quist J et al. Aging is associated with a prolonged fever response in human endotoxemia. Clin Diagn Lab Immunol 2001;8:333–338.

96.  Kroeger KM, Steer JH, Joyce DA, Abraham LJ. Effects of stimulus and cell type on the expression of the -308 tumour necrosis factor promoter polymorphism. Cytokine 2000; 12: 110–119.

97.  Jenny NS, Tracy RP, Ogg MS, Luong lA, Kuller LH, Arnold AM et al. In the elderly, interleukin-6 plasma levels and the -174G>C polymorphism are associated with the development of cardiovascular disease. Arterioscler Thromb Vasc Biol 2002; 22(12): 2066–2071.

98.  Pedersen M, Bruunsgaard H, Weis N, Hendel HW, Andreassen BU, Eldrup E et al. Circulating levels of TNF-alpha and IL-6 - Relation to truncal fat mass and muscle mass in healthy elderly individuals and patients with type 2 diabetes. Mech Ageing Dev 2003; 124: 495–502.

99.  Bruunsgaard H, Pedersen AN, Schroll M, Skinhoj P, Pedersen BK. TNF-α, Leptin, and lymphocyte function in human aging . Life Sci 2000; 67:2721–2731.

100. Straub RH, Miller LE, Schölmerich J, Zietz B. Cytokines and hormones as possible links between endocrinosenescence and immunosenescence. J Neuroimmunol 2000; 109: 10–15.

101. Bruunsgaard H, Østergaard L, Andersen-Ranberg K, Jeune B, Pedersen BK. Proinflammatory cytokines, antibodies to *Chlamydia pneumoniae* and age-associated

diseases in Danish centenarians – Is there a link? Scand J Infect Dis 2002;34:493–499.

102. Priø TK, Bruunsgaard H, Røge B, Skinhoj P, Pedersen BK. Asymtomatic bacteriuria in elderly humans is associated with increased levels of circulating TNF receptors. Experimental Gerontology 2002;37:693–699.

103. Mocchegiani E, Straub RA. Possible new anti aging strategies related to neuroendocrine immune interactions. In "The Neuroendocrine Immune Network in Ageing", Neuroimmune Biology Volume 4. Straub RH, Mocchegiani E. Eds., Berczi I, Szentivanyi A. Series Eds. Elsevier, Amsterdam, 2004; pp. 399–407.

# Index

## Symbols

## A

438